Natural Water Resources: Challenges and Concerns

Natural Water Resources: Challenges and Concerns

Edited by Herbert Lotus

SYRAWOOD
PUBLISHING HOUSE
New York

Published by Syrawood Publishing House,
750 Third Avenue, 9th Floor,
New York, NY 10017, USA
www.syrawoodpublishinghouse.com

Natural Water Resources: Challenges and Concerns
Edited by Herbert Lotus

International Standard Book Number: 978-1-68286-430-2 (Hardback)

Cataloging-in-publication Data

Natural water resources : challenges and concerns / edited by Herbert Lotus.
p. cm.
Includes bibliographical references and index.
ISBN 978-1-68286-430-2
1. Water. 2. Natural resources. 3. Water resources development. 4. Water supply. I. Lotus, Herbert.
GB665 .N38 2017
551.48--dc23

Printed in China.

TABLE OF CONTENTS

PREFACE

Natural water resources are the main source of the water on our planet. As we are facing a water crisis, it is important to utilize the current water resources judiciously and planning for the future as well. Natural water is rapidly depleting, and thus, new concepts and methods are required to compensate for the loss and to conserve the resources like rivers, lakes and swamps. This book will provide significant information on the challenges we face in achieving these goals and also the solutions required for the same. It will also provide interesting topics for research, which readers can take up. In the text, using case studies and examples, constant effort has been made to make the understanding of the difficult concepts of this subject, as easy and informative as possible, for the readers. A number of latest researches have been included to keep the readers up-to-date with the global concepts in this area of study.

The information contained in this book is the result of intensive hard work done by researchers in this field. All due efforts have been made to make this book serve as a complete guiding source for students and researchers. The topics in this book have been comprehensively explained to help readers understand the growing trends in the field.

I would like to thank the entire group of writers who made sincere efforts in this book and my family who supported me in my efforts of working on this book. I take this opportunity to thank all those who have been a guiding force throughout my life.

Editor

Recent History and Geography of Virtual Water Trade

Joel A. Carr[1]*, Paolo D'Odorico[1], Francesco Laio[2], Luca Ridolfi[2]

1 Department of Environmental Sciences, University of Virginia, Charlottesville, Virginia, United States of America, **2** Department of Environmental, Land and Infrastructure Engineering, Politecnico di Torino, Turin, Italy

Abstract

The global trade of goods is associated with a virtual transfer of the water required for their production. The way changes in trade affect the virtual redistribution of freshwater resources has been recently documented through the analysis of the virtual water network. It is, however, unclear how these changes are contributed by different types of products and regions of the world. Here we show how the global patterns of virtual water transport are contributed by the trade of different commodity types, including plant, animal, luxury (e.g., coffee, tea, and alcohol), and other products. Major contributors to the virtual water network exhibit different trade patterns with regard to these commodity types. The net importers rely on the supply of virtual water from a small percentage of the global population. However, discrepancies exist among the different commodity networks. While the total virtual water flux through the network has increased between 1986 and 2010, the proportions associated with the four commodity groups have remained relatively stable. However, some of the major players have shown significant changes in the virtual water imports and exports associated with those commodity groups. For instance, China has switched from being a net exporter of virtual water associated with other products (non-edible plant and animal products typically used for manufacturing) to being the largest importer, accounting for 31% of the total water virtually transported with these products. Conversely, in the case of The United states of America, the commodity proportions have remained overall unchanged throughout the study period: the virtual water exports from The United States of America are dominated by plant products, whereas the imports are comprised mainly of animal and luxury products.

Editor: Rodrigo Huerta-Quintanilla, Cinvestav-Merida, Mexico

Funding: The authors have no support or funding to report.

Competing Interests: The authors have declared that no competing interests exist.

* E-mail: jac6t@virginia.edu

Introduction

Food production requires adequate climate, soils, and water availability. Recently, there has been renewed focus on various issues surrounding water resources including: water for sanitation and drinking [1,2], water as a source of conflict [3,4], and water use in the growth of food [5]. In particular, with an ever increasing global population, the ability to maintain adequate food supplies with limited water resources has become a pressing concern [6,7]. When a society has no access to favorable conditions to produce all the food it needs, its demand can be met through importations from other regions. Food trade allows some populations to thrive in water deficit regions by avoiding or mitigating water stress conditions. It has been noted [8,9] that the trade of agricultural and industrial commodities is associated with a virtual transfer of the water resources used for the production of these goods [10]. Global trade virtually transfers large amounts of water resources from areas of production to far consumption regions, a phenomenon that has been named "the globalization of water" [11].

Virtual water trade is often considered as a solution to conditions of, overpopulation in regions with limited water availability, societal water stress, malnourishment and water wars [8]. Part of the global food security depends on virtual water trade [12], however, it has been noticed that virtual water trade has also negative implications, because, indirectly, it limits environmental stewardship [13], enhances inequality in the use of water resources [14,15], and erodes societal resilience to drought [6]. Moreover, the trade of virtually embedded water is complicated by competitive, absolute and comparative advantage in the production and trade of food commodities [16,17].

In the past two-to-three decades there has been an intensification of virtual water trade: the global volume of virtual water transfer and the number of links in the virtual water network (i.e., number of trade partnerships among countries) have almost doubled between 1986 and 2010, with an increasing dependence of the global virtual water trade on just a few exporting countries [18]. It is still unclear how the structural and functional properties of the virtual water network vary with the type of food commodity traded. Moreover, previous analyses [18,19] looked at general trends in the volumes of virtual water traded and in the number of connections in, and topological characteristics of, the network [18–20] without explicitly evaluating how different countries and regions of the world are contributing to important structural changes in the network [21]. Understanding the factors and countries contributing to changes in the redistribution of virtual water resources is crucial to the assessment of how the system might respond to conditions of stress resulting for example from droughts. This paper investigates the geography of virtual water flows though an explicit analysis of the global spatiotemporal patterns of virtual water trade associated with plant, animal, luxury and other products. We investigate how some changes in the network of virtual water trade are correlated with the gross domestic product (GDP) and with combined metrics of life

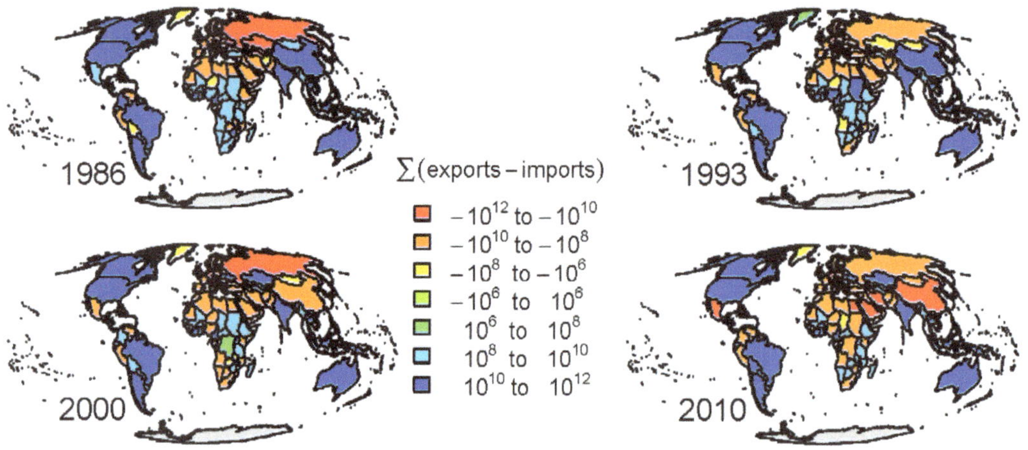

Figure 1. The virtual water import-export balance (m^3) for the years 1986, 1993, 2000 and 2010. The total flux along the network has more than doubled, and the number of nodes net importing has risen from 127 to 149 with a corresponding decrease in net exporters from 104 to 72. As such the ratio of net importers to exporters has changed from 1.2 to 2.0.

expectancy, education and income such as the Inequality adjusted Human Development Index (HDI).

Methods

Detailed international trade data (FAOSTAT) for the period 1986–2010 from the Food and Agricultural Organization of the United Nations [22] were used to reconstruct the global trade patterns of food products reported by the FAOSTAT data base, including crops, crop-derived food commodities, and animal products. For each product, m, and year t, a trade matrix, $\boldsymbol{T}_m(t)$ is generated, where the (i,j) element of the matrix is the export of that crop from country i, to country j. The reported trade data was rectified to political changes over the 25 year period to allow for clear trade node transitions (Appendix S1).

After the political rectification, by using for each food product the country-specific average blue and green virtual water content, WC_m, [23], each individual crop trade matrix, $\boldsymbol{T}_m(t)$, was converted to a crop specific virtual water trade matrix, $\boldsymbol{C}_m(t) = WC_m \bigcirc \boldsymbol{T}_m(t)$. The total virtual water trade matrix is simply $\boldsymbol{C}(t) = \Sigma_m \boldsymbol{C}_m(t)$. $\boldsymbol{C}(t)$ is non-symmetrical with two possible directed links between node i, and node j. The antisymmetrical undirected network matrix $\boldsymbol{C}_{sym}(t) = \boldsymbol{C}(t) - \boldsymbol{C}(t)^{\boldsymbol{T}}$ thus defines the virtual water balance for each connection (i,j),with the balance for each year t and node i, defined as $\Sigma_j \boldsymbol{C}(i, j, t)$. Virtual water contents for countries not reported in the estimates by [23] are based on the virtual water contents of the nearest neighbour ($\pm 10°$ of latitude and longitude). The global average values for the green and blue virtual water contents of that crop type were used for countries with no close neighbor.

Country-specific virtual water content estimates existed for 309 crops and animal products traded. Live animals were converted to live weight equivalents based on data from FAOSTAT [24]. The virtual water trade data were then further segregated into four primary groupings: edible crops, edible animal products, non-edible commodities (such as plant fibers, oils, oil cakes or animal hides) and luxury items (such as sugars, coffee and chocolate); the list of these commodities is detailed in Table S1. Populations of the nodes in the virtual water network over the period of study were obtained from Gapminder.org [25].

Results

There has been a strong increase in both the number of connections and total flux of virtual water over the 25 year period

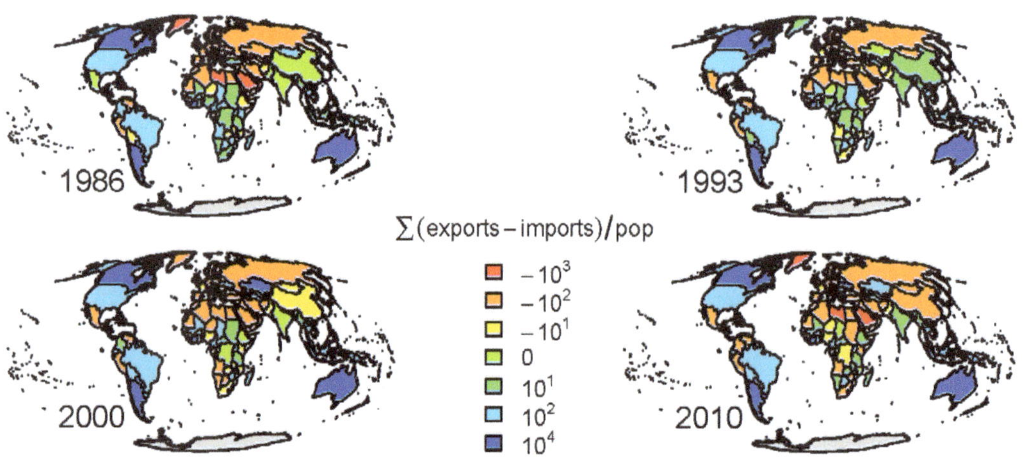

Figure 2. The virtual water import-export balance m^3 per capita for the years 1986, 1993, 2000 and 2010.

Table 1. The top 5 net importers (top) and exporters (bottom) and their virtual water export import balance (10^{11} m^3) for the years 1986, 1993, 2000 and 2010.

Rank	1986 Balance		1993 Balance		2000 Balance		2010 Balance	
1	Japan	−0.77	Japan	−0.85	Japan	−0.88	China, mainland	−2.43
2	USSR	−0.54	Germany	−0.48	Germany	−0.49	Japan	−0.91
3	Italy	−0.37	Italy	−0.36	Italy	−0.45	Germany	−0.64
4	Germany	−0.32	Republic of Korea	−0.34	Russian Federation	−0.44	Iran	−0.58
5	United Kingdom	−0.23	Russian Federation	−0.29	Republic of Korea	−0.38	Italy	−0.55
1	United States of America	0.85	United States of America	1.04	United States of America	1.15	Brazil	1.87
2	Australia	0.60	Argentina	0.53	Argentina	0.99	United States of America	1.1.55
3	Argentina	0.44	Australia	0.40	Australia	0.79	Argentina	1.48
4	Thailand	0.34	Brazil	0.39	Brazil	0.58	Indonesia	0.84
5	Brazil	0.33	India	0.33	Canada	0.44	Canada	0.72

[18]. The discrepancy between net exporting countries and net importing countries has also increased with the ratio between the number of net importers per net exporter shifting from 1.32 in 1986 to 2. 5 in 2010 (Figure 1). Interestingly, in general countries either net export or net import a substantial amount of virtual water, while only few countries have smaller virtual water fluxes (e.g., in the range -10^6 and 10^6 m^3). This gap persists even as countries shift in their virtual water import export balance from net importing to exporting or vice versa.

Recasting the import-export balance by population we see a slightly different picture (Figure 2). In 1986 many countries are close to 0 in terms of export-import balance per capita. This shifts by 2010 with the majority of the global population net importing virtual water. In 1986 there were 0.57 net importing people per net exporter. By 2010 this switches to 1.43 people net importing per net exporter.

Concurrent to the increase in total virtual water flux, the total number of connections that any country has (or degree), has increased over the 25 year period (Figure 3). This increase in connectivity and 'globalization' is less pronounced in the African continent, suggesting only limited participation to the global virtual water market, and more pronounced in the south Asian countries (China, India, Indonesia). The list of the top 5 net importing and exporting countries for the years 1986, 1993, 2000 and 2010 (Table 1) demonstrates the emergence of Brazil, Argentina, India, Indonesia and China as major players in the virtual water network. While most countries in the top 5 demonstrate an increase in strength (either net import or export), some countries within the top 5 such as Japan have demonstrated little (<5%) change in their net virtual water uptake over the 25 year period.

Separation of products into edible plant based products, edible animal based products, luxury products and other plant and animal derived products allows for examining the consumption change in virtual water as related to changes in commodity trade. The top 15 commodities by virtual water volume for the years 1986, 1993, 2000, and 2010, demonstrate the maintenance of cereal grains, oils and cotton lint being the predominant traded products (Table 2). However, luxury items such as coffee and chocolate remain consistently in the top 15 throughout the period and there is an increase in the virtual water trade associated with bovine meat. The virtual water volume associated with the top 15 products accounts for a relatively constant percentage (≈66%) of the total volume from 1986 to 2010. The remaining third of the

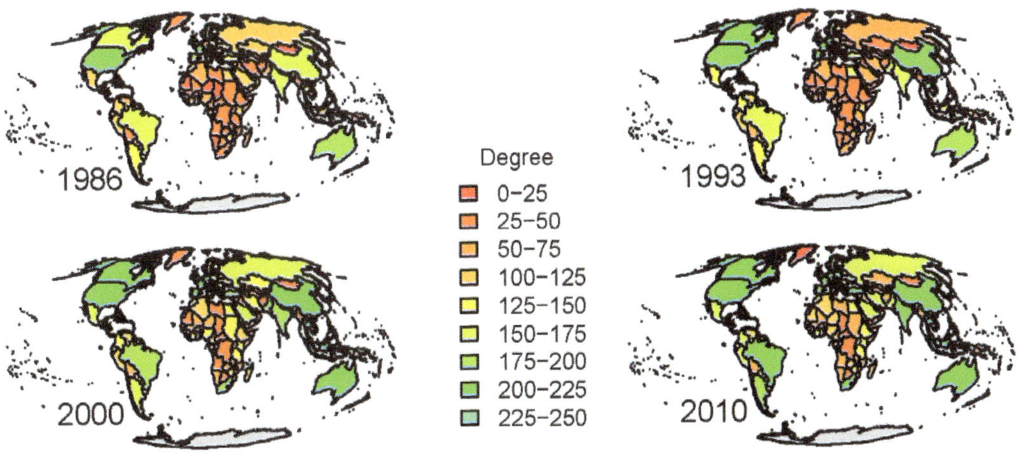

Figure 3. The degree (the sum of import and export connections) associated with each country for the years 1986, 1993, 2000, and 2010. While countries in general are more connected, this increase has not been as significant in most African countries.

Table 2. The top 20 products and corresponding virtual water flux (10^{11} m^3) for the years 1986, 1993, 2000 and 2010.

Rank	1986		1993		2000		2010	
1	Wheat	1.34	Wheat	1.45	Wheat	1.93	Wheat	2.30
2	Coffee, green	0.80	Coffee, green	0.89	Chocolate Prsnes	1.25	Soybeans	2.29
3	Cake of Soybeans	0.59	Cake of Soybeans	0.87	Soybeans	1.22	Chocolate Prsnes	2.02
4	Soybeans	0.56	Chocolate Prsnes	0.81	Cake of Soybeans	1.08	Cake of Soybeans	1.81
5	Maize	0.49	Soybeans	0.70	Coffee, green	0.99	Palm oil	1.56
6	Cotton lint	0.45	Cocoa beans	0.57	Cocoa beans	0.72	Maize	1.11
7	Cocoa beans	0.42	Maize	0.51	Maize	0.66	Coffee, green	1.04
8	Sugar Raw Centrifugal	0.36	Cotton lint	0.46	Palm oil	0.61	Cocoa beans	0.91
9	Chocolate Prsnes	0.34	Palm oil	0.37	Cotton lint	0.56	Cotton lint	0.83
10	Palm oil	0.26	Meat-Cattle Boneless (Beef & Veal)	0.35	Soybean oil	0.53	Meat-Cattle Boneless (Beef & Veal)	0.75
11	Sugar Refined	0.25	Sugar Refined	0.32	Meat-Cattle Boneless (Beef & Veal)	0.52	Rice Milled	0.69
12	Rice Milled	0.23	Rice Milled	0.32	Sugar Refined	0.50	Soybean Oil	0.68
13	Meat-Cattle Boneless (Beef & Veal)	0.23	Soybean oil	0.30	Rice Milled	0.45	Sugar Refined	0.6
14	Barley	0.21	Cocoa Butter	0.29	Cocoa Butter	0.41	Cocoa Butter	0.54
15	Soybean oil	0.20	Sugar Raw Centrifugal	0.22	Sugar Raw Centrifugal	0.38	Sugar Raw Centrifugal	0.52

flux is spread among 294 different commodities. While trade of virtual water associated with all commodities has increased (Figure 4), the increase in the virtual water trade associated with edible plant based products and luxury commodities is not matched by a comparable increases in the virtual water trade in edible animal products and other plant and animal derived

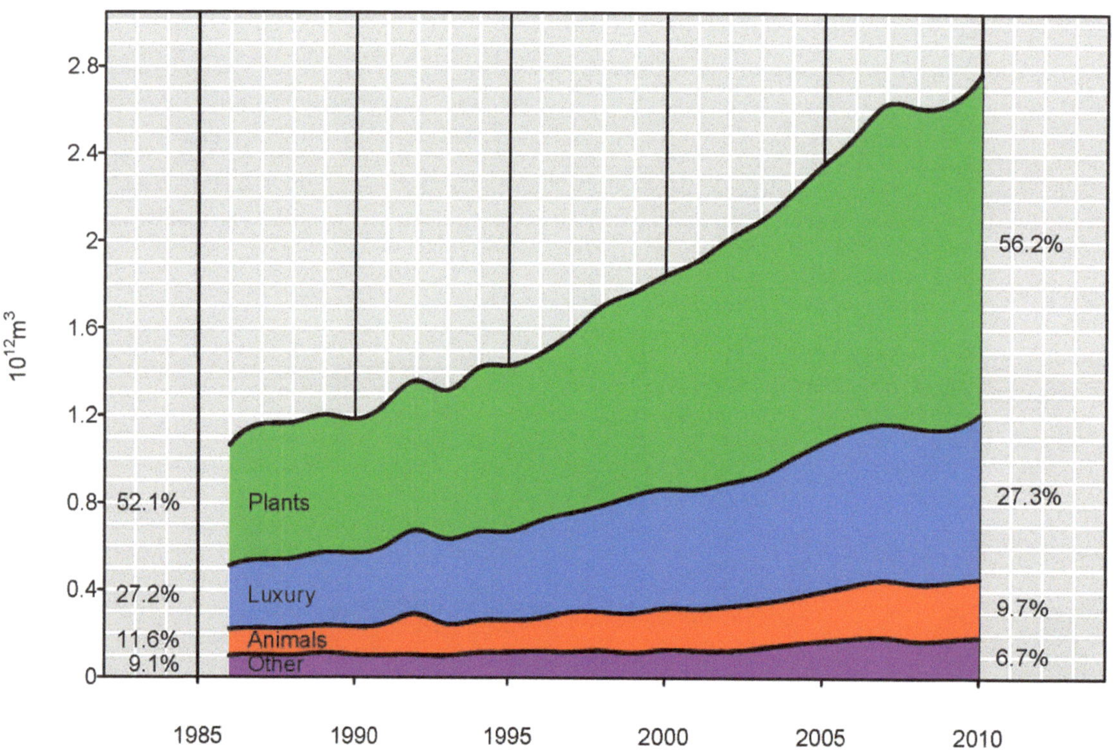

Figure 4. The increase in virtual water trade and the percentage of the total virtual water flux in the network corresponding to plant, animals, luxury, and other commodities. While there is an increase in the virtual water flux associated with all commodity categories, the fraction of the total flux associated with animals and other has decreased. There is a corresponding almost tripling of crop products and luxury products over the 25 year period with animal products and fibers increasing to a lesser extent.

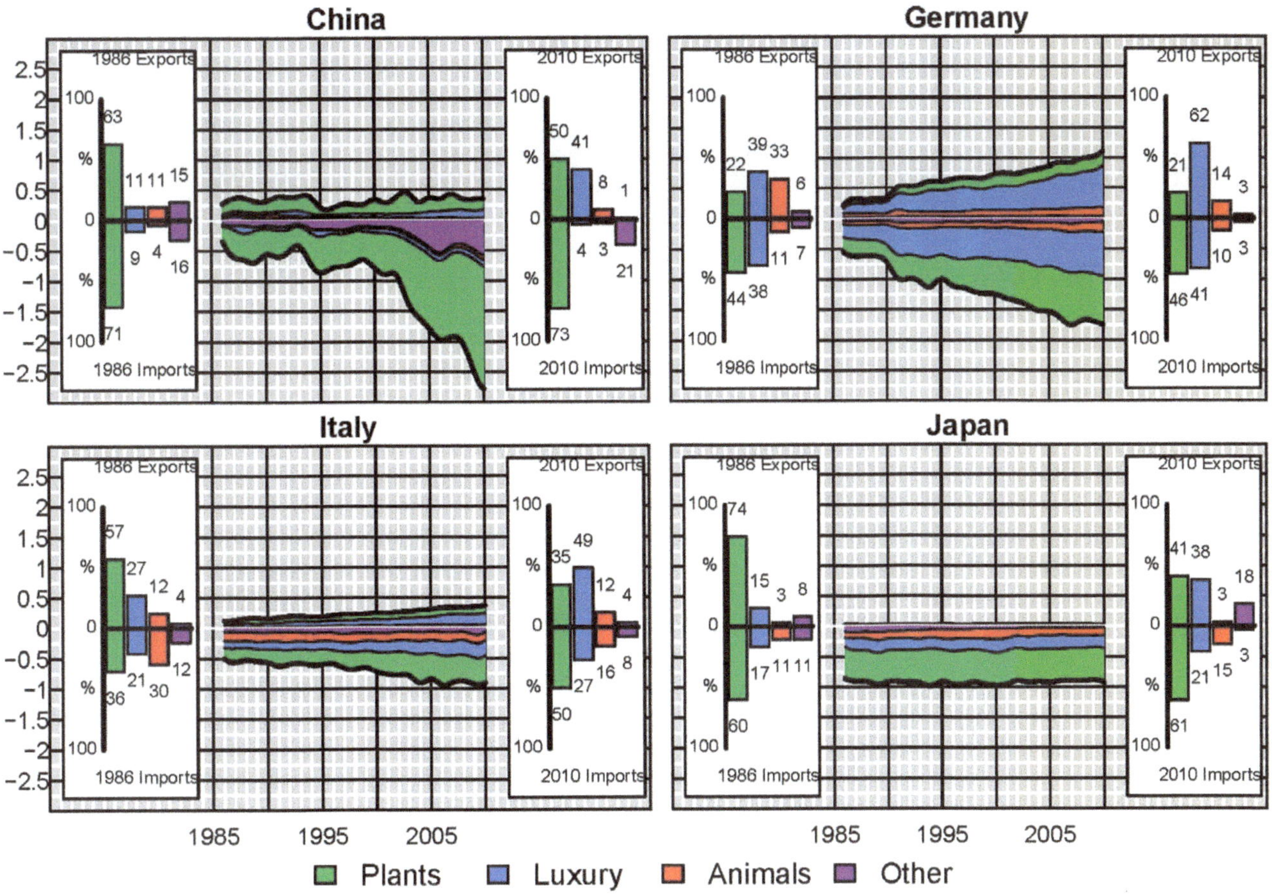

Figure 5. Detailed virtual water trade (10^{11} m^3) by commodity type for China (mainland), Germany, Italy and Japan. Positive values refer to exports and negative values to imports. The insets indicate the percentage of virtual water imports or exports contributed by plant, animal, luxury and other products.

products. The flux of virtual water associated with edible plant and luxury items has more than doubled over the 25 year period. The virtual water flux associated with edible animal products and other commodities has "only" doubled (Figure 3). Interestingly, the proportion of virtual water associated with the four commodity type groupings has remained relatively constant over the years with small increase associated with edible plant and luxury items and a corresponding decrease in animal and non-edible products.

Even though, globally, these proportions remain almost constant, at the country scale there have been important changes. Looking at a selection of some of the major player in the virtual water network (Table 1) we see distinct behavior differences (Figure 5 and 6). For example, in 1986 China was a net virtual water exporter with exports including other non-edible commodities and 9% of their virtual water imports associated with luxury items. By 2010, virtual water imports associated with non-edible plant and animal commodities comprise 214% of China's imports, and luxury items comprise 41% of the virtual water exports with the top 2 imported commodities in terms of virtual water being soybeans and cotton lint. China currently dominates not just the global virtual water imports, but specifically the virtual water associated with non-edible plant and animal commodities (31% of all global virtual water trade associated with non-edible plant and animal commodities). Turkey was the next largest importer of non-edible plant and animal products accounting for only 5% followed by Italy with ~3%. Similarly, China was importing 13% of all virtual water associated with edible plant products. This is contrasted by a country such as Japan (Figure 6), whose virtual water imports by commodity type have remained almost constant over the two and a half decades. These two countries are in turn contrasted by Germany and Italy. Like Japan, the proportional make up of Germany's virtual water imports has remained almost constant over the 25 year period; however, it has significantly decreased its meat exports while increasing its luxury product exports. Since 1991 and reunification, Germany has more than doubled its importation of virtual water, while not quite doubling its virtual water exports. Similarly Italy has more than doubled its importation of virtual water, however, it has shifted toward importing more virtual water associated with edible plant commodities and increased exportation of luxury items.

Examining a few of the major net exporters of virtual water (Figure 6), the United States of America continue to be a major importer of luxury products and the primary importer of virtual water associated with animal based products. By 2010 the USA accounts for ~8.5% of the virtual water trade in animal products, while the Russian federation accounts for 8%. Germany, Japan and Italy account each for ~ 5% of global virtual water imports in animal product. The United States of America and Germany are the largest importers of virtual water associated with luxury items each comprising 9% of the global luxury commodity virtual water imports. However, whereas virtual water exports from Germany do not exceed their heavy importation of virtual water, the import

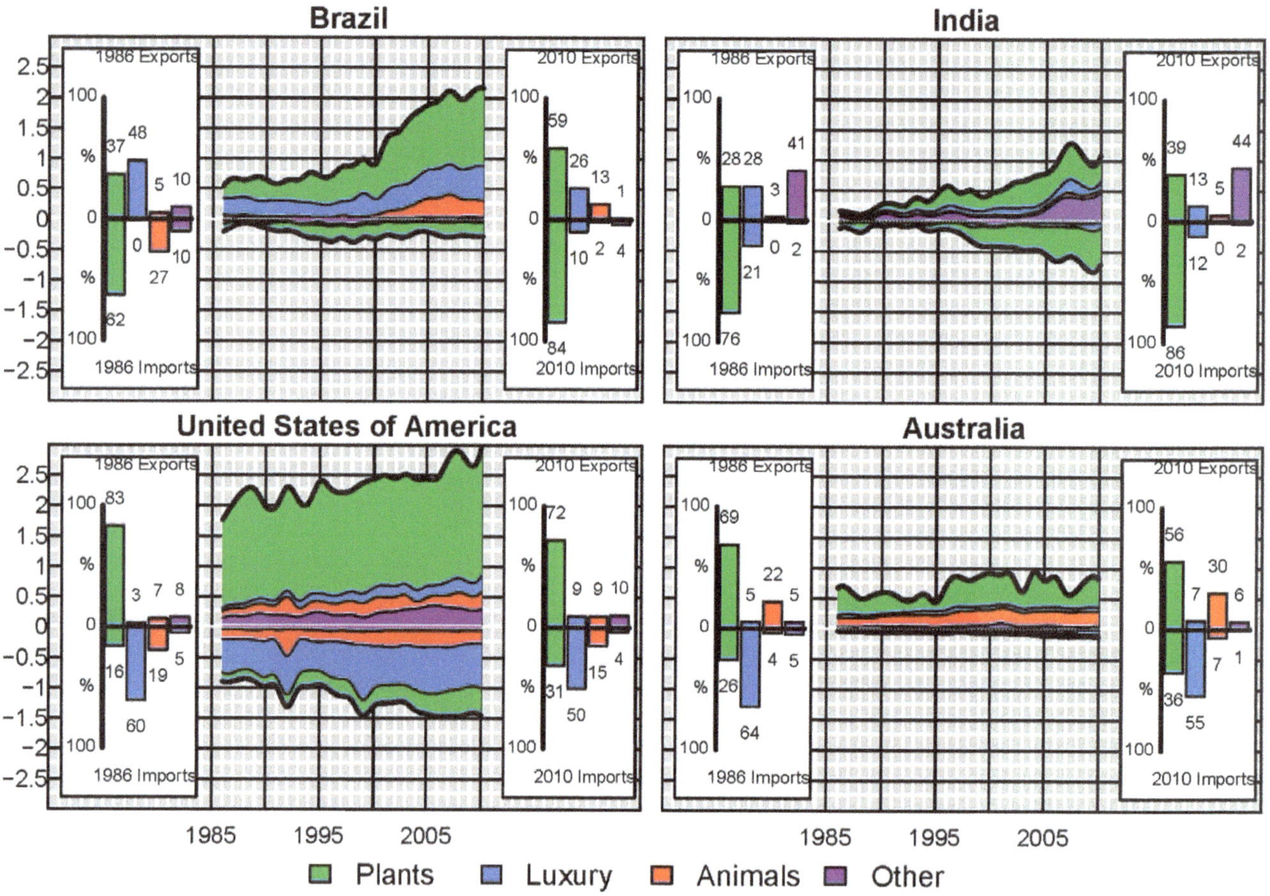

Figure 6. Detailed virtual water trade (10^{11} m^3) by commodity type for Brazil, India, The United States of America and Australia. Positive values refer to exports and negative values to imports. The insets indicate the percentage of virtual water imports or exports contributed by plant, animal, luxury and other products.

of virtual water associated with meat and luxury products in the United States of America is easily compensated for by the large export of plant based commodities, predominantly soybeans, wheat and maize. While the United States of America is not the largest net exporter of virtual water in 2010, they are the largest exporter of virtual water with a total exported volume of $\sim 10^{11}$ m^3. The large increases in the trade from the United States of America has not matched the changes that have occurred in the trade from Brazil and India. The virtual water exports from Brazil are three times their virtual water exports in 1986, whereas virtual water exports from India in 2010 are almost 5 times greater than those in 1986 (Figure 6). Australia in contrast has shown large variability in its exportation of virtual water associated with plant based commodities matching well with Australian precipitation records (Figure 6). These large discrepancies between commodity based imports and exports of virtual water for individual nodes indicate that, while globally the ratio of virtual water associated with the four commodity groups has changed only slightly over the period of study, virtual water trade in individual nodes is highly variable.

Figure 7 shows the top 50% of the virtual water fluxes (top 25% in heavy lines). It reveals a dynamic network, both in terms of connections as well as dominant flux along connections. Here each link is colored based on the commodity associated with the largest virtual water volume moving through that link. While there is a change, both in the links and major commodity along links, there is some consistency in the importation of virtual water associated with meat products into the United States and exportation of plant products from North and South America to Asia and Europe. Examining the net importers and exporters amongst the four commodity networks revels that major players in the global virtual water network are not necessarily a major contributor in all the commodity networks (Table 3). Moreover, the fluxes of virtual water moving into or out of Africa are consistently relatively small. However, the per capita export-import balance (Figure 2) of African nations is not as disparate in comparison to the rest of the world. In contrast, Australia and Argentina stand out as large per capita exporters with the Arabian Peninsula, and some of Europe being the largest net importers. Even though China is the largest net importer, its per capita imports are moderate.

Thus, while there has been a large increase in the virtual water flux over the 25 year period, the water burden (i.e., demand for virtual water exports) has been placed on a smaller and smaller portion of the world population. In 1986 68% of the world population was in countries that were net exporters of water. In the late 1980's China and India both were briefly net importers and the fraction of the world population net importing sharply increased. However, India has slowly become a larger net exporter of virtual water while China has continued to increase its virtual water imports. As a result, by the mid 1990's over half the world's population lived in countries that net imported virtual water. By 2010 roughly 60% of the world population lived in net importing

Table 3. The top 5 net importers (left) and net exporters (right) and their virtual water export import balance (10^{11} m^3) for the year 2010 in the four commodity type networks.

	Net importing	Net imports	Net Exporting	Net exports
Plant	China, mainland	−1.85	United States of America	1.71
	Germany	−0.56	Argentina	1.42
	Japan	−0.56	Brazil	1.04
	Egypt	−0.37	Canada	0.65
	Italy	−0.32	Indonesia	0.61
Animal	Russian Federation	−0.21	Brazil	0.27
	Japan	−0.14	Australia	0.24
	Italy	−0.10	New Zealand	0.11
	China, Hong Kong SAR	−0.07	Canada	0.09
	United Kingdom	−0.53	Denmark	0.08
Luxury	United States of America	−0.47	Brazil	0.54
	United Kingdom	−0.21	Indonesia	0.32
	Russian Federation	−0.20	Netherlands	0.27
	Japan	−0.19	Cote d'Ivoire	0.21
	Spain	−0.12	Ghana	0.18
Other	China, mainland	−0.58	India	0.37
	Turkey	−0.09	United States of America	0.20
	Italy	−0.06	Australia	0.04
	Republic of Korea	−0.05	Canada	0.04
	Indonesia	−0.05	Uzbekistan	0.03

countries. By weighting the population of each country i by what fraction of their virtual water exports is transferred to country j, the proportion of the population of country i that is supporting country j, $pop_{\text{sup},i,j} = \left(\frac{s_{i,j}}{\sum_j s_{i,j}}\right) pop_i$, can be calculated where $s_{i,j}$ is the strength of the link from country i to country j. Repeating for all countries, and scaling by the global population allows for examination of the fraction of the global population that is supplying virtual water to each given country. In general, any particular country obtains its imports from less than 9% of the global population. By only considering those countries that net import virtual water, we find that any of these individual countries are supplied by less than 4% of the global population. Sorting the net importers by their export-import balance and looking at the populations supplying those countries shows that 32% of the global population is providing 90% of the virtual water, while 13% is providing 50% of the virtual water imported by net importers. All commodities show a similar pattern, except for the virtual

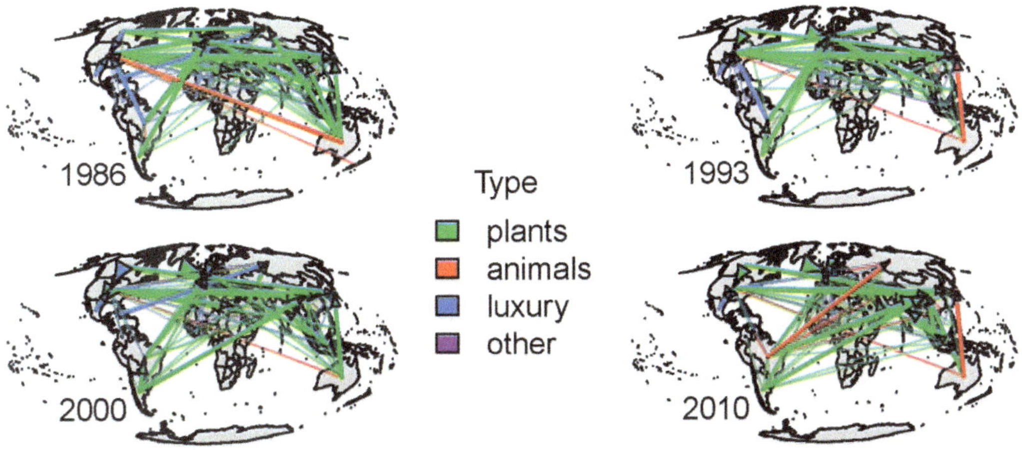

Figure 7. The Network associated with the top 50% of the virtual water flux for the years 1986, 1993, 2000 and 2010. Color in the links based on the most prevalent commodity along the link in terms of total flux through the link. The heavy lines indicate the links which comprise the top 25% of the virtual water flux.

Table 4. Parameters obtained by fitting a power law to import strength and IHDI values ($s_{i,imports} = a\ IHDI_i^{\,b}$).

IMPORT	b	a(m^3)	r^2
Full	10.31	2.20	0.29
Plant	9.91	1.76	0.16
Animal	9.32	3.07	0.39
Luxury	9.93	3.25	0.44
Other	8.90	3.46	0.26

water traded with animal and luxury products, which is supplied by a smaller fraction of the global population.

To investigate the drivers of virtual water trade, the import strength, $s_{i,imports} = \sum s_{ji}$, was related to the Inequality adjusted Human Development Index (IHDI) [26] through a power law relationship ($s_{i,imports} = a\ IHDI_i^{\,b}$). IHDI data is limited to 160 nodes in the network, with no preferential influence of nodal population. This relationship was found to be stronger for the luxury and animal product network (Table 4).

Discussion and Conclusions

The strong increase in virtual water trade over the past 25 year period is not as ubiquitous as the term "globalization" implies. Rather, there are distinct increases of virtual water flows associated with a few countries, while the majority of countries experienced only smaller fluctuations.

Overall, the results presented in this study indicate in multiple ways that there is a small global population controlling large amounts of the virtual water exports [18]. This indicates potential for limited resilience of the virtual network with respect to the possible decision of one of these suppliers to stop or reduce its exports. Most of the virtual water resources available for trade tend to remain concentrated in a small set of countries, controlled by a small fraction of the global populations similar to the Pareto principle. Thus, the importing countries are left with limited options for the diversification of their virtual water providers.

The controls on the strengths in both the total virtual water network, and the commodity networks remain poorly understood. There is a dependency of import strengths on the Inequality adjusted Human Development Index. This relationship is stronger in the case of luxury and animal products (Table 3), consistent with the so-called Bennett's Law, which postulates that diets become richer in fats, protein, and sugar as societies become wealthier [27]. In other words, the import of luxury and animal products is more intense in those countries who can afford them, which explains the stronger dependence of virtual water traded in animal and luxury products on the social development status. This indicates that the drivers of the full virtual water network are not only multi-factorial, but that they vary depending on the commodity classes. Other indices (e.g., Grass Domestic Product (GDP)) exhibited a weaker relation with import strength. Since

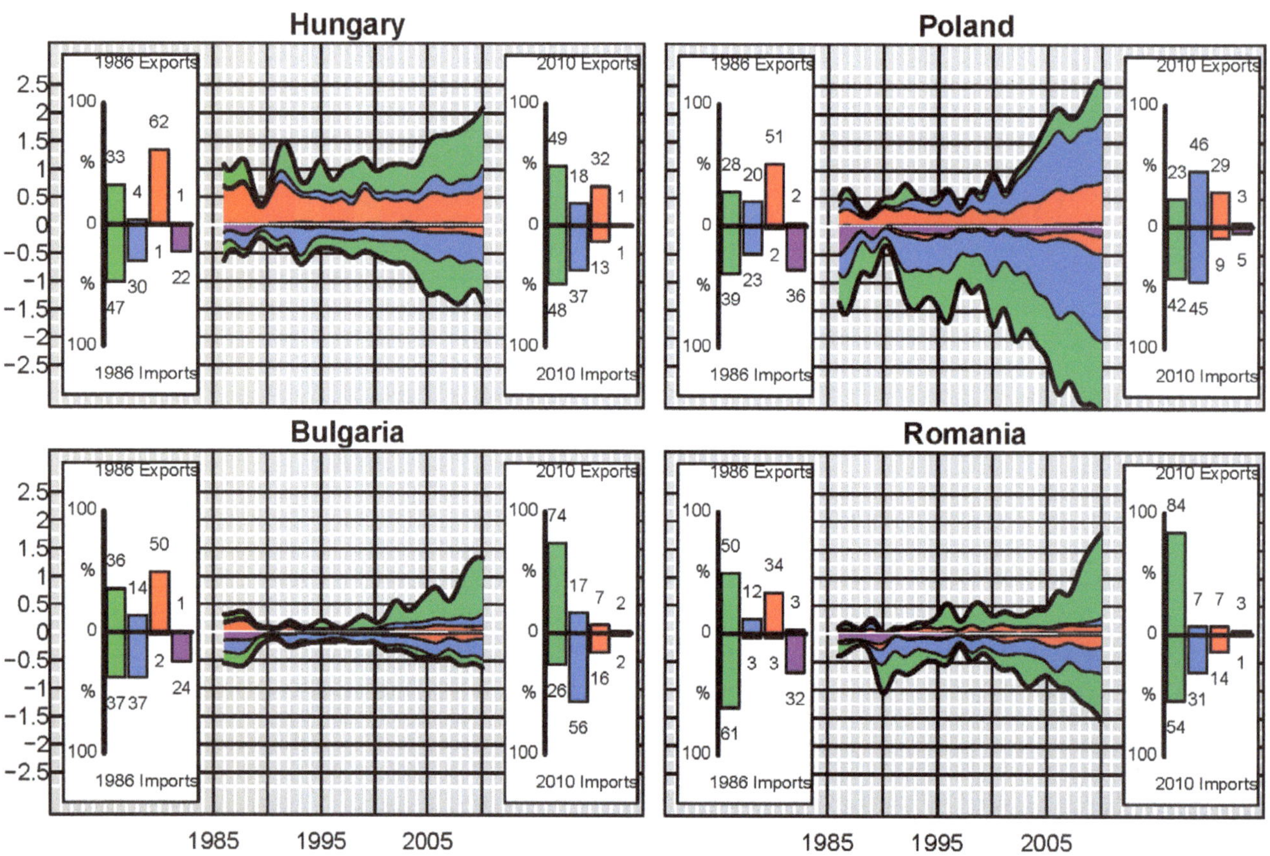

Figure 8. Detailed virtual water trade (10^{10} m^3) by commodity type for Hungary, Poland, Bulgaria and combined former USSR and Russian federation. The insets indicate the percentage of virtual water imports or exports contributed by plant, animal, luxury and other products.

IHDI incorporates many factors including gross national income, mortality rates, and education, it may allow for better cross comparison of countries. However, IHDI does not include any metric of water resource availability for the individual countries. These results are echoed in the findings of the community structure results of D'Odorico et al. [28], which demonstrated that the community structure of the virtual water network appears to not align well with any one factor (e.g., population, GDP, distance metrics or production).

Prior analysis has shown regional contributions to virtual water fluxes and their changes over time [19], however strong changes of individual countries rather than regions seem to be controlling the patterns and changes in the virtual water network. For example in 1992 there is globally a spike in the virtual water associated with the trade of animal products (Figure 4). This spike is also apparent in the temporal virtual water trade of the United States of America (Figure 6) and is due to a large increase in imports of cattle and pig meat from Mexico. Some of this results from the post cold war subsidy reforms which led to a significant decrease in livestock production [29] in the Russian Federation and other eastern European countries. This reform was combined with flooding in 1991 in China resulting in decreases in pig production [29]. Along with this potential increase in production, beef exports from the United States of America rapidly increased beginning in 1992. The corresponding spike in meat imports in the United States in 1992 was due to a net importation of pig meat in 1992. Alternately in 2003, the first reported case of mad cow disease in the United States of America resulted in significant decrease in meat exports (Figure 6). This change is not apparent in the global network (Figure 4). A sudden decline in importation and exportation of VW in the Eastern Bloc countries (Hungary, Poland, Romania and Bulgaria) (Figure 8) occurred around the 1990 fall of the Iron curtain, and resulted in a small dip in the global trade of VW (Figure 4), which was then followed by a substantial increase in VW importation, particularly in luxury items. Thus, while some of these node-specific changes relate to regional dynamics causing regional-scale changes, others are more global in scale. As such the resilience of the virtual water network to socioeconomic, political and environmental changes still needs to be explored.

While the proportional constituency in terms of commodity type of virtual transfers has remained relatively unchanged over two decades, the changes in virtual water proportions in any given node is variable. Moreover, the major contributors to the change in the virtual network do not seem to be responsible for the consistency in these global proportions and virtual water exports and imports do not seem to be driven by any single factor, but rather depend on multitude political, socio-economic and geographic factors.

Author Contributions

Analyzed the data: JAC. Wrote the paper: JAC PO FL LR.

References

1. Gleick PH (1999) The human right to water: Pacific Institute for Studies in Development, Environment, and Security.
2. Mason MB, Lyn KG, Quezon AU, Jamie B (2012) Implementing an evolving human right through water and sanitation policy.
3. Gleick PH (1993) Water and conflict: Fresh water resources and international security. International security: 79–112.
4. Wolf AT (1999) "Water Wars" and Water Reality: Conflict and Cooperation Along International Waterways. NATO SCIENCE SERIES 2 ENVIRONMENTAL SECURITY 65: 251–268.
5. Morison JIL, Baker NR, Mullineaux PM, Davies WJ (2008) Improving water use in crop production. Philosophical Transactions of the Royal Society B: Biological Sciences 363: 639–658.
6. D'Odorico P, Laio F, Ridolfi L (2010) Does globalization of water reduce societal resilience to drought? Geophysical Research Letters 37.
7. Schade C, Pimentel D (2010) Population crash: prospects for famine in the twenty-first century. Environment, Development and Sustainability 12: 245–262.
8. Allan JA (1998) Virtual water: A strategic resource global solutions to regional deficits. Ground Water 36: 545–546.
9. Allan T, editor (1993) Fortunately there are substitutes for water: otherwise our hydropolitical futures would be impossible, in Proceedings of the Conference on Priorities for Water Resources Allocation and Management,. Southampton, U. K: Overseas Dev. Agency. 13–26 p.
10. Hoekstra A (2002) Virtual water trade: Proceedings of the international expert meeting on virtual water trade. Delft, Netherlands: UNESCO-IHE.
11. Hoekstra A, Chapagain AK (2008) Globalization of water: Sharing the planet's freshwater resources. Oxford, UK: Blackwell Publishing.
12. Mekonnen MM, Hoekstra AY (2011) The green, blue and grey water footprint of crops and derived crop products. Hydrology and Earth System Sciences 15: 1577–1600.
13. Carr J, D'Odorico P, Laio F, Ridolfi L, Seekell D (2012) Inequalities in the networks of virtual water flow. *Eos Trans AGU* 93.
14. Seekell DA (2011) Does the Global Trade of Virtual Water Reduce Inequality in Freshwater Resource Allocation? Society & Natural Resources 24: 1205–1215.
15. Seekell DA, D'Odorico P, Pace ML (2011) Virtual water transfers unlikely to redress inequality in global water use. Environmental Research Letters 6.
16. Wichelns D (2004) The policy relevance of virtual water can be enhanced by considering comparative advantages. Agricultural Water Management 66: 49–63.
17. Dan W (2009) Optimization of water conservancy industry structure based on comparative advantage of water resources [J]. Journal of Economics of Water Resources 3: 019.
18. Carr JA, D'Odorico P, Laio F, Ridolfi L (2012) On the temporal variability of the virtual water network. Geophysical Research Letters 39.
19. Dalin C, Konar M, Hanasaki N, Rinaldo A, Rodriguez-Iturbe I (2012) Evolution of the global virtual water trade network. Proceedings of the National Academy of Sciences of the United States of America 109: 5989–5994.
20. Konar M, Dalin C, Suweis S, Hanasaki N, Rinaldo A (2011) Water for food: The global virtual water trade network. Water Resources Research 47.
21. Suweis S, Konar M, Dalin C, Hanasaki N, Rinaldo A (2011) Structure and controls of the global virtual water trade network. Geophysical Research Letters 38.
22. Food and Agriculture Organization of the United Nations Website. Available: http://faostat3.fao.org/home/index.html Accessed 2012 Oct 1.
23. Mekonnen MM, Hoekstra AY (2010) A global and high-resolution assessment of the green, blue and grey water footprint of wheat. Hydrology and Earth System Sciences 14: 1259–1276.
24. Food and Agriculture Organization of the United Nations Website. Available: http://www.fao.org/fileadmin/templates/ess/documents/methodology/tcf.pdf Accessed 2012 Feb 10.
25. Gapminder website. Available: http://www.gapminder.org. Accessed 2012 Aug 5.
26. Human Development Report Website. Available http://hdr.undp.org/en/statistics/ihdi/Accessed 2012 May 18.
27. (2011) Foresight. The Future of Food and Farming. London: The Government Office for Science.
28. D'Odorico P, Carr J, Laio F, Ridolfi L (2012) Spatial organization and drivers of the virtual water trade: A community-structure analysis. *Environmental Research Letters* 7.
29. United States department of Agriculture Website Available: http://www.ers.usda.gov/publications/aer813/aer813c.pdf Accessed 2012 May 20.

2

Shale Gas, Wind and Water: Assessing the Potential Cumulative Impacts of Energy Development on Ecosystem Services within the Marcellus Play

Jeffrey S. Evans[1,2]*, Joseph M. Kiesecker[3]

1 The Nature Conservancy, Fort Collins, Colorado, United States of America, 2 Department of Zoology and Physiology, University of Wyoming, Laramie, Wyoming, United States of America, 3 The Nature Conservancy, Fort Collins, Colorado, United States of America

Abstract

Global demand for energy has increased by more than 50 percent in the last half-century, and a similar increase is projected by 2030. This demand will increasingly be met with alternative and unconventional energy sources. Development of these resources causes disturbances that strongly impact terrestrial and freshwater ecosystems. The Marcellus Shale gas play covers more than 160,934 km^2 in an area that provides drinking water for over 22 million people in several of the largest metropolitan areas in the United States (e.g. New York City, Washington DC, Philadelphia & Pittsburgh). Here we created probability surfaces representing development potential of wind and shale gas for portions of six states in the Central Appalachians. We used these predictions and published projections to model future energy build-out scenarios to quantify future potential impacts on surface drinking water. Our analysis predicts up to 106,004 new wells and 10,798 new wind turbines resulting up to 535,023 ha of impervious surface (3% of the study area) and upwards of 447,134 ha of impacted forest (2% of the study area). In light of this new energy future, mitigating the impacts of energy development will be one of the major challenges in the coming decades.

Editor: Vanesa Magar, Plymouth University, United Kingdom

Funding: Funding was provided by The Robertson Foundation. The funders had no role in study design, data collection and analysis, decision to publish, or preparation of the manuscript.

Competing Interests: The authors have declared that no competing interests exist.

* E-mail: jeffrey_evans@tnc.org

Introduction

Global demand for energy has increased by more than 50 percent in the last half-century, and a similar increase is projected by 2030 [1]. Energy production to meet growing demand has resulted in impacts to human health and well-being as well as increased habitat fragmentation and stress on biological diversity worldwide [2–4]. In the United States, directives for renewable energy, energy security and technological advancements such as horizontal drilling in conjunction with hydraulic fracturing have spurred a rapid increase in alternative and unconventional energy production over the last decade [5–8]. Energy development is poised to continue its upward trajectory, with over 200,000 km^2 of new land projected to be developed in the U.S. alone by 2035 [1,4]. Development of "unconventional" gas dispersed in shale will be key, as the potential resource may be large [9]. In addition to shale gas, the U.S. Department of Energy's goals are to produce 20% of its electricity from terrestrial wind energy development (241 gigawatts of on-shore wind) by 2030 [8]. In light of this new energy future, understanding and mitigating the impacts of energy development will be one of the major challenges in the coming decades [10].

The Marcellus shale gas play covers approximately 160,934 km2 across eight states and contains both some of the largest technically recoverable shale gas resources is the U.S. and headwater watersheds that provide drinking water for over 22 million people in several of the largest U.S. metropolitan areas (e.g. New York City, Washington DC, Philadelphia & Pittsburgh) [11]. Land use change from shale gas and wind development is known to increase land clearing, impervious surface and increase deforestation[12–14]. Aquatic ecosystems are particularly vulnerable to land use change such as deforestation and change in impervious surface [11,15,16]. Deforestation and increases in impervious surface influence sediment, hydrologic, and nutrient regimes, which in turn influence aquatic biota and ecological processes in fresh waters [17]. Because sediment represents one of the most significant controlling variables on stream morphology and hydrology, deforestation and increases in impervious surface will be an important driver in stream health, integrity of headwater watersheds, and quality of drinking water. It is therefore essential to understand patterns of future deforestation and changes in impervious surface resulting from energy development.

Scenarios analysis has become a widespread approach in pursuit of sustainable development. However, it is used infrequently, at least in any formal way, in environmental impact assessment (EIA)[18]. This is surprising because EIA is a process designed specifically for exploring options for more-sustainable (i.e., less environmentally damaging) futures. Similarly predictive modeling techniques have been applied in recent years to project land cover changes and residential development [19–21] and to predict potential species habitat [22,23], but comparable techniques are rarely used to model anticipated energy development and proactively quantify environmental impacts. Although other studies have estimated future development scenarios for portion of the Marcellus Shale [12,24–26], this is the first to examine a comprehensive build out scenario for potential impacts associated with both shale gas and wind development across the entirety of the Marcellus Shale gas play.

Here we employ build-out scenarios for future energy development to quantify potential impacts on surface drinking water resources. First, we created prediction surfaces of wind and shale gas development potential for portions of six states in the Central Appalachians. Second, we modeled future build-out scenarios using these predictions and published projections from federal land management agencies [27]. Finally, we evaluated effects of build-out scenarios on ecosystem services by measuring impacts to watershed quality and identifying areas important for drinking water that are vulnerable to energy development. Studies have already documented potential point source impacts to surface water resources associated with shale gas fracking [13]. Here we focus on nonpoint source pollution across the entire landscape likely to change as a result of energy development [13,14,28]. Given expected development intensity, proactively addressing potential impacts can help develop strategies to mitigate issues related to biodiversity and surface water supply.

Methods

Study Area

We focused our analysis on the Interior Marcellus Shale Assessment Unit (IMSAU) and defined the project boundary using the entirety of the "subwatershed" (12-digit) watersheds ($n = 2003$) that intersected the IMSAU (Figure 1). These subwatersheds, on average, are 8498 ha in size. The area comprises a complex landscape mosaic of 17,134,045 ha, with 12,007,150 ha (70%) of the area forested. There are currently 4151 well pads in the IMSAU with up to six wells drilled per pad for a total of 10,419 current and permitted wells. We compiled spatial databases of wind and gas development along with covariates identified, thorough a multidisciplinary scoping process, as influential to development (table 1).

Data sources

We compiled gas and wind spatial databases from private, state and federal sources. The magnetic anomaly rasters (bouguerbougure, isograv and magnetic) were obtained from the USGS [29] as was the thermal maturity point data [30]. We obtained point-locations of well monitoring data from U.S. Department of Energy, National Energy Technology Laboratory (Kathy Bruner, personal communication), which were compiled from state-level databases. The thermal maturity and well monitoring data were used to build ordinary Kriging models [31], using the geoR library [32] to generate 1-km rasters of thermal maturity, shale depth and shale thickness.

Turbine locations used in the wind model were obtained through the U.S. Federal Aviation Administration obstruction evaluation database. All topographic variables were derived from the 30 m Digital Elevation Model available through the USGS National Elevation Dataset. All topographic variables were derived in ArcGIS 10.0 using the Geomorphometrics toolbox [33]. We utilized the U.S. Census Bureau TIGER road data and land cover, including forest and impervious surface, was derived from the 2006 USGS National Land Cover Dataset [28]. Wind production class data was acquired from the DOE-NETL [8] and rasterized to 30 m. Spatial data representing energy infrastructure was obtained from Ventyx [50]. Protected areas, where development is jurisdictionally precluded, were determined using PADUS v 1.3 [34] with gap code A-2 ("statutory or legally enforceable protection"). Watersheds and stream data were acquired from the USGS National Hydrography Dataset (NHD) (http://nhd.usgs.gov/index.html). Finally, data associated with water access and production related ecosystem services were obtained through the U.S. Forest Service, Forests to Faucets analysis [35] and associated with the NHD.

Statistical model

We implement a robust nonparametric model that provides stable spatial estimates of resource development probabilities. This is accomplished by independent model iteration with differing conditional random observations that act as the "absence" class in the model. The underlying model is an ensemble approach that allows us to combine model iterations into a final model after convergence is satisfied. Conditional random samples are generated by generating a Kernel Density Estimate (KDE) [36], using an isotropic Gaussian kernel intensity function in the R spatstat library [37] based on known resource development locations [38]. The inverse of this KDE is specified as the sample probability distribution for the NULL (i.e. absence observations) and conditional random samples are generated by using the probabilities as sample weights. The optimal KDE sigma parameter is selected using a cross validation.

The "positive" class is fixed and is represented by known observations. In a given iteration of the model, the number of random observations generated is the same size as the number known observations. A model is then built using Random Forests [39,40], using the R library randomForest [41], and the probabilities are predicted back to the training data. The estimated probabilities are then set aside, a new set of random points created and the process is repeated. At each iteration, where the modeling process is repeated, the new model is combined with the previous ones and the new probability estimated based on the prior ensemble. This is repeated until the resulting probability distribution is unchanged, when compared to prior combined ensembles, using a Kolmogorov-Smirnov [42] distributional equality test (significant at $p = 0.001$). Once the model converges, it is predicted to the final set of raster surface variables identified in the model yielding a raster surface of the estimated probabilities. Since Random Forests is a weak learner ensemble method, as long as the parameter space remains fixed, multiple model ensembles (independent models) can be combined [43]. This method ensures that statistical and spatial variability is captured across the random samples used to represent the negative case. The parameter space is fixed across iterations and our model selection follows methods presented in Murphy et al. 2010 [44]. R Code is provided in appendix A.

Development potential

The wind model was specified with a set of 17 candidate variables that are known to influence wind suitability and development (see Table 1) [8,45–47]. Generally, these variables were related to topographic, anthropogenic, transmission and wind production influences. The model selection procedure retained 7 of the 19 variables (Table 1) used in the final model. To remove areas where wind development is precluded, pixel-level probabilities associated with proximity to airports; [12], protected areas based on the PADUS [34] gap status code 1 ("An area having permanent protection from conversion of natural land cover"), streams and roads were recoded to zero. The gas model was specified with 13 candidate variables (Table 1) that are known to influence oil and gas potential [12,48]. Generally, these variables were related to geologic, topographic and current

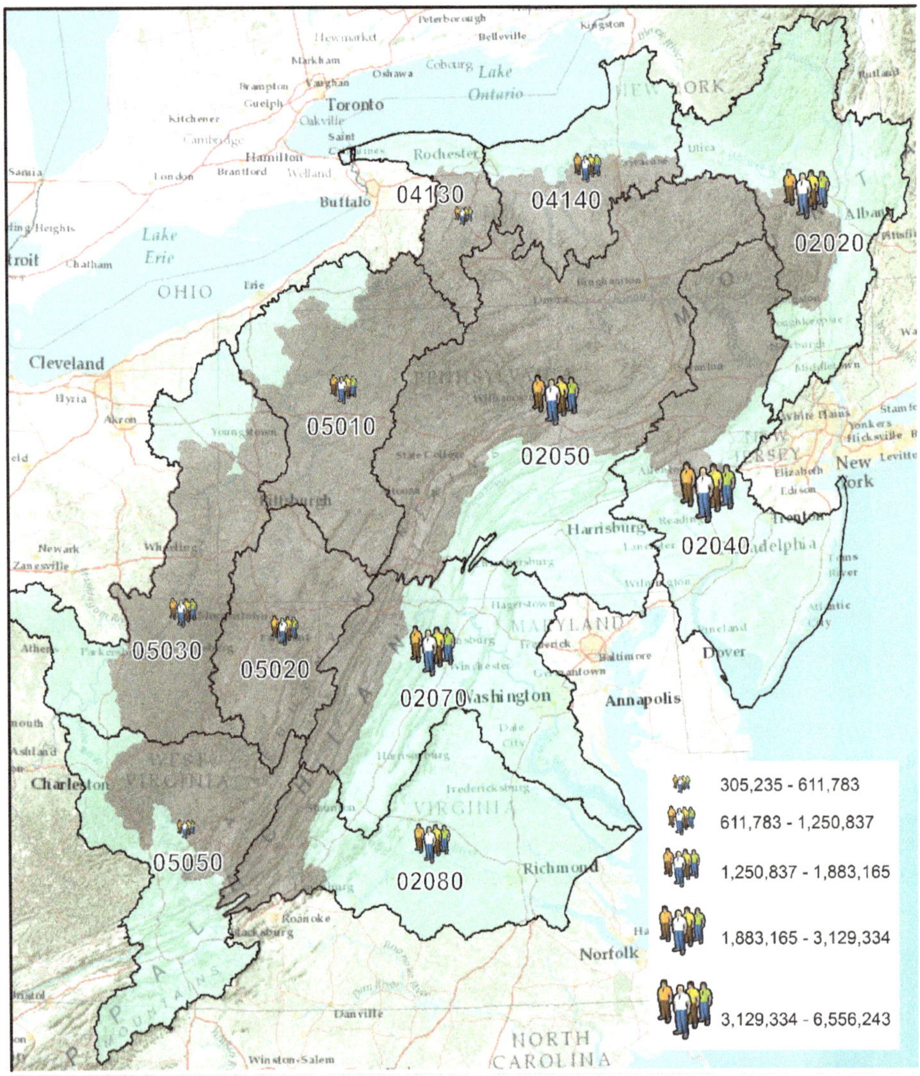

Watershed	Pop Served	2006 predevelopment	Current	EWITS+gas 8 wells/pad	EWITS+gas 4 wells/pad
02020	3,129,334	14,857	14,864	16,820	16,820
02040	6,556,243	98,321	98,515	101,252	101,471
02050	2,622,851	331,599	362,105	472,030	599,221
02070	2,636,379	36,489	37,206	49,220	49,343
02080	1,883,165	27,236	27,236	27,662	27,662
04130	305,235	23,907	24,532	31,390	34,689
04140	1,042,156	35,918	35,990	37,531	39,028
05010	978,965	190,404	198,551	228,146	257,449
05020	1,250,837	202,805	218,676	272,992	332,379
05030	841,748	249,387	260,559	297,795	348,720
05050	611,783	69,186	70,132	74,858	76,608
Total	21,858,696	1,280,111	1,348,366	1,609,696	1,883,389

Figure 1. Study area with basin-level watersheds and population served, labels correspond to the watershed number. The interior polygon shaded in brown represented the subwatershed-level watersheds that intersect the high pressurization area that defines our study area. The inset table represents population served for each watershed, impervious surface for 2006 baseline, current and two scenarios. The hectares are specific to the given impact and are not cumulative. For total impacted area under a scenario, add baseline, current and scenario.

Table 1. Variables used in the wind and gas models.

Variable	Description	Model
ASP	Linear transformation of slope direction	candidate wind
Bougure	Bougure gravity anomalies	included gas
CTI	Wetness index	candidate wind
Depth	Depth of shale	included gas
DPIPE	Distance in meters to nearest gas pipeline	candidate gas
DROADS	Distance in meters to nearest road	included wind, candidate gas
DTRANS	Distance in meters to nearest power transmission	included wind
ELEV	Elevation in meters	included wind
HLI	Heat Load Index	candidate wind
HSP	Slope position	candidate wind
Isograv	Isostatic gravity anomalies	included gas
Magnetic	Aeromagnetic gravity anomalies	included gas
PAG	Percent agriculture within 1 km radius	candidate wind
PDEV	Percent development within 1 km radius	candidate wind
PFOR	Percent forest within 1 km radius	candidate wind
SCOSA	Slope*COS(Aspect)	candidate wind
SLP	Slope intensity in degrees	included wind, candidate gas
SRR15	Surface Relief Ratio at 15×15 window	candidate wind, candidate gas
SRR27	Surface Relief Ratio at 27×27 window	included wind, considered gas
SRR3	Surface Relief Ratio at 3×3 window	candidate wind, candidate gas
TEX15	Variance of elevation at 15×15 window	candidate wind
TEX27	Variance of elevation at 27×27 window	candidate wind
TEX3	Variance of elevation at 3×3 window	included wind, candidate gas
Thickness	Shale thickness	included gas
Tmaturity	Geologic thermal maturity	included gas
WPC	Wind production classes	included wind

infrastructure characteristics. The final model retained six variables (Table 1) all associated with geological characteristics.

Development scenarios

To calculate number of wells needed to exploit the IMSAU, we used a projection of gas development from U.S. Energy Information Administration [27], which predicts 141 trillion cu ft of gas with an estimated ultimate recovery of 1.15 billion cu ft/well. We designated well pads as our experimental unit so we could account for expected development reflecting the expansion of current leases (i.e., additional wells drilled on an already established well pad). We created scenarios around two potential intensities of well pad development: 4 and 8 wells per pad, for a respective total of 26,501 and 11,175 new well pads, which accounts for current well pad development. These scenarios have the current well pads subtracted (i.e., [(4 wells/pad = 26,501 * 4) + (4151 current well pads * 4)] = 122,608 total wells) and are derived by solving for the maximum number of possible wells needed to exploit the resource, assuming that additional wells will first be drilled on established pads to reach the desired maximum density (4 or 8 wells per well pad). Local legal regulatory guidelines and lease availability were not addressed in our scenarios.

For the wind development scenario, we use projections from the Eastern Wind Integration and Transmission Study (EWITS) [49] and selected scenario 3 that emphasizes development of wind resources close to load centers in the northeastern US. This scenario represents an estimate on the high end of potential wind development for this region. We scaled the projection by the percentage of our study area intersecting the EWITS, using estimates of viable wind derived from wind production classes [8]. Taking currently installed capacity into account and, following the EWITS analysis, we estimate that our study area would need an additional 10,798 turbines to meet the EWITS projections. We assumed turbines would have a nameplate capacity of 2.5 MW and adjusted output (52%) which was consistent with the EWITS estimates and measures from current wind farm production compiled by Ventyx [50].

To create spatial data for a given wind or gas scenario, we generated random samples by iterating through the range of estimated sorted probabilities in the surface estimates, starting with the highest probability value(s), until the desired number of well pads or wind turbines is generated. We then take each simulated point location and buffer it to represent an expected footprint for each impact (Figure 2). Estimates of potential surface disturbance associated with gas wells and wind turbines were based on measurements of actual wind and shale gas development in Pennsylvania digitized from aerial photographs [12,26]. These simulated footprints were then use to modify impervious surface and forest rasters to illustrate potential impacts.

a)		
Wellpad footprint	1.3	11.6
Associated infrastructure	(2.3+8) 10.3	
Indirect impact	8.6	
Total	20.2	

b)		
Turbine footprint	0.57	0.77
Associated infrastructure	0.2	
Indirect impact	5.42	
Total	6.19	

Figure 2. Photographs of shale gas footprint and wind farm footprint. Inset table represents associated impacts in hectares used in the analysis. Estimates of potential surface disturbance associated with gas wells and wind turbines were based on measurements taken from aerial photographs from Johnson (2010) and Johnson et al. (2011). We also incorporated impacts associated with the pipelines needed to collect gas from well sites and transport it to storage areas. Measurements indicate that on average there are 2.66 km of pipeline with an average right of way of 30.48 m, where each mile of a 30.48 m right-of-way directly disturbs ~4.86 ha/well pad. The 8.019 ha of impact associated with offsite pipelines is included in the estimate of associated infrastructure for each well pad. Because we were unable to spatially configure the location of pipelines, we summed the surface disturbance associated with each simulated well in the watershed to estimate the amount of additional surface disturbance that pipelines would create at the watershed-level. Photographs by Mark Godfrey.

Surface impacts

Shale gas and wind development is known to increase land clearing, impervious surface and increase deforestation [12–14]. We evaluated effects of shale gas and wind energy development on ecosystem services by 1) measuring impacts to watershed quality based on imperviousness classifications [2,51] and identifying vulnerable areas based on an importance index for drinking water production [35]. Current research suggests that aquatic systems become very seriously impacted when watershed impervious cover exceeds 10% [52] and show significant declines in many stream taxa at even lower levels of imperviousness [53]. For example, significant declines in species have been documented between 0.5 and 2% imperviousness, with 40–45% declines in regional stream biodiversity (invertebrates, fish, amphibians) at imperviousness greater than 2–3% [51,53,54]. Costs of water treatment have also been shown to increase in line with increases in impervious surfaces [51]. To examine the impact of impervious surface in our study area, we summarized the amount of current impervious cover for each subwatershed in the study area using the National Land Cover Impervious Surface Dataset [28,55]. We grouped each subwatershed into one of four impact categories guided by the thresholds found in Schueler et al. [51]: Sensitive, 0<10% impervious; Impacted, 10–25% impervious; Nonsupporting, >=25–60% impervious; Urban drainage, >=60% impervious. We then used results from each development scenario to reclassify watersheds, and we summarized the percent change by impact class.

To quantify an area's value for surface drinking water, we utilized the Safe Drinking Water Information System, SDWIS [56] and mean annual drinking water supply [57] generated by the U.S. Forest Service's Forest to Faucets analysis [35]. The Forest to Faucets analysis produces an index of relative importance for water production, at the basin-level and subwatershed-level, which is based on the potential amount of water produced and the number of people who use that water. Since it is a measure of both production and population served, we use the metric as a measure of the ecosystem service. We highlighted forested areas important for drinking water production that might be susceptible to energy development impacts by identifying areas with both relative high water importance indexes (RIMP) scores and high percent forest cover (both at >=0.75).

Results

Model validation

All Kriging models (shale depth, shale thickness and thermal maturity), used to generate covariates, exhibited a Kriging variance smaller than 1 standard deviation of the predicted distribution, illustrating very good performance. Both the gas and wind models were very well supported. The AUC/ROC for the wind model was 0.98 with a sensitivity weighted Kappa [58] of 0.97. The ROC for the shale gas model is 0.93, indicating a very strong model. The sensitivity weighted Kappa is 0.93, demonstrating that a random chance and bias corrected percent correctly classified also supports the model. In the shale gas models, when probabilities were assigned back to the training and validation observations, we see 90% of the observations above $p = 0.65$ for the training data and 84% in the validation data.

Surface impacts

After our projections of energy development, we can expect 26,501 new well pads and 10,798 new wind turbines. In addition to the 68,255 ha attributed to current shale gas and wind development, new gas pads and wind turbines will results in an additional 329,585 (EWITS+8 wells/pad scenario) – 603,278 (EWITS+4 wells/pad scenario) ha of impervious surface (Figure 3d), and drive 268,503–495,357 ha of deforestation. These impacts will result in changes that will affect conditions at the subwatershed level. For example, at present 0.4% ($n = 9$) of the subwatersheds are already classified as urban drainages and 3%

($n = 67$) as nonsupporting subwatersheds. Predicted energy development will result in up to a 171% ($n = 130$) increase of subwatersheds in the urban drainage and nonsupporting category (Figure 3). Under the highest cumulative impact scenario, 10% ($n = 206$) of the study area subwatersheds will be classified as either a nonsupporting or urban drainage (Figure 3). Our analysis also predicts that energy development will be disproportional in high-value watersheds (Figure 3). Three of the basin-level watersheds (Figure 1, watersheds: "02050", "05020", "05030") will receive ~85% of the increased surface disturbance forecasted by our modeling (Figure 3). We see that 36% of the predicted energy expansion footprint in our study area is expected to occur in watersheds that are in the top quartile in terms of intactness and water importance.

Discussion

Our analysis indicates that in the Central Appalachians region, energy development will drive significant land use change, with shale gas development as the dominant driver of increased impervious surface and deforestation accounting for 94% of the projected footprint. This increase in surface disturbance and fragmentation will potentially impact the maintenance of biodiversity and the quality of surface water resources for ~22 million people. The increase in energy production forecasted by our analysis may be compatible with biodiversity if properly sited, but will still pose a challenge for surface water resources, both because of the strong link between surface water quality and surface disturbance and because of the high value for water production for watersheds in the study area. Watersheds in the study area scored among the highest relative "water importance" in the country because of both the high level of water produced and the number of people that consume the water (all >60/100) [35]. Here we utilize the Impervious Cover Model (ICM) [51,52,59] to relay potential impacts as a result of projected energy sprawl. The ICM has been extensively tested with more than 250 different reports reinforcing the basic model (see for review [51,52,59]). However, the ICM is not intended to predict the precise score of individual surface water quality metrics and given the scale of our analysis, we do not attempt to make specific site-level predictions. We use the ICM to highlight areas where the cumulative impact of increased impervious cover and deforestation could result in decreased water quality, changes in hydrology and geomorphology, increased storm runoff, and sedimentation.

We focused our analysis on the impacts of surface disturbance on water quality from nonpoint sources, however, hydraulic fracturing (fracking) used to extract shale gas poses additional risk for water quality. Fracking involves the injection of a mix of water, chemicals and sand underground at high pressure to fracture rock and release the oil or gas [60]. Risks to surface water supplies include depletion of fresh water supplies, spills of fracking chemicals, leaks of flow back fluids that can include fracking chemicals once the well is completed and discharge of treated flow back fluids from wastewater facilities [13,61]. For example, Olmsted et al. [13], showed that release of treated fracking wastewater increased chloride (Cl) by 10–11% resulting in higher risk to surface water. Threats to underground water supplies can also result when wells are poorly constructed and when old oil and gas wells that have been capped serve as a migration corridor for fracking fluids used in new wells nearby [61]. While we did not assess additional risk associated specifically with fracking, our analysis can serve to highlight areas where additional attention should be placed to monitor potential fracking related impacts.

In contrast to traditional gas development, shale gas is developed with multiple horizontal wells that can reach out 1524 m or more from one well pad [60]. The footprint associated with shale gas development is typically bigger than conventional shallow gas plays [12]. The well pads are typically more expansive (averaging just over 1.21 ha compared to a small fraction of 0.40 ha), the water used to fracture wells is much greater (~7–15 million liters vs. <500,000 liters), and the supporting infrastructure is much larger in scale (24″ diameter pipelines to gather gas from wells versus 2″ or 4″ pipelines in shallow fields). In addition, associated pipelines represent a significant source of fragmentation and, as our scenarios predict, can result in approximately half of the impervious surface associated with shale gas development. While the larger pad, greater water use, and more extensive infrastructure pose more challenges for conservation than shallow gas, the area "drained" by wells on each well pad is much larger than from shallow gas pads (200–400 ha versus 4–32 ha) since there are typically multiple lateral wells on a shale pad versus a single vertical well on a shallow gas pad [61]. The lateral reach of shale gas wells means there is more flexibility in where pads and infrastructure can be placed relative to shallow gas [12,61]. This increased flexibility in placing shale gas infrastructure can be used to avoid or minimize impacts to natural habitats in comparison to more densely spaced shallow gas fields. Given this flexibility, our results can be used to facilitate ecologically appropriate siting of development, ensuring that key ecological features that could be impacted are highlighted so they can be preserved. Impacts to water quality can also be mitigated by attention to siting. Siting well pads and wind turbines to reduce potential runoff will therefore be critical, and will involve appropriate setbacks from streams and wetlands as well as avoiding development in steep slopes and grades. In addition, slowing the rate of development to allow targeted regulatory oversight for watersheds projected to experience increased development will be critical to allow balance between development objectives and conservation goals.

In marked contrast wind development may have less flexibility than shale gas development to adjust siting in ways that will mitigate potential impacts within our study area. Developing wind energy on disturbed lands rather than placing new developments within large and intact habitats would reduce cumulative impacts to wildlife [46]. Kiesecker et al. [46] found that there are over 14 times the amount of wind energy potential on disturbed lands needed to meet the DOE goal of generating 20% of the United States electricity with wind [8]. Despite the extensive wind resources across the U.S., the states comprising the majority of our study area (WV, PA, VA,) are unable to meet DOE projections within areas already disturbed [46]. In these states wind energy potential is largely restricted to ridge tops, which often make up the heart of the last remaining intact natural ecosystems [62]. In addition impacts associated with wind development will be influenced by the placement of transmission lines [47]. While our analysis takes into account the local disturbance associated with individual turbines we do not account for large scale transmission necessary to facilitate wind development. To fully understand and estimate impacts associated with wind development it will be critical to incorporate the effect of new transmission lines.

Our scenarios used the most current information to estimate shale gas and wind development patterns in this region [8,27]. But the estimation of Marcellus Shale gas resources is highly uncertain, given the short production history. As additional data are released and the methodology for estimating the resource is refined, the ultimate estimated recovery for the Marcellus play may change. Also, as more wells are drilled over a broader area, and as

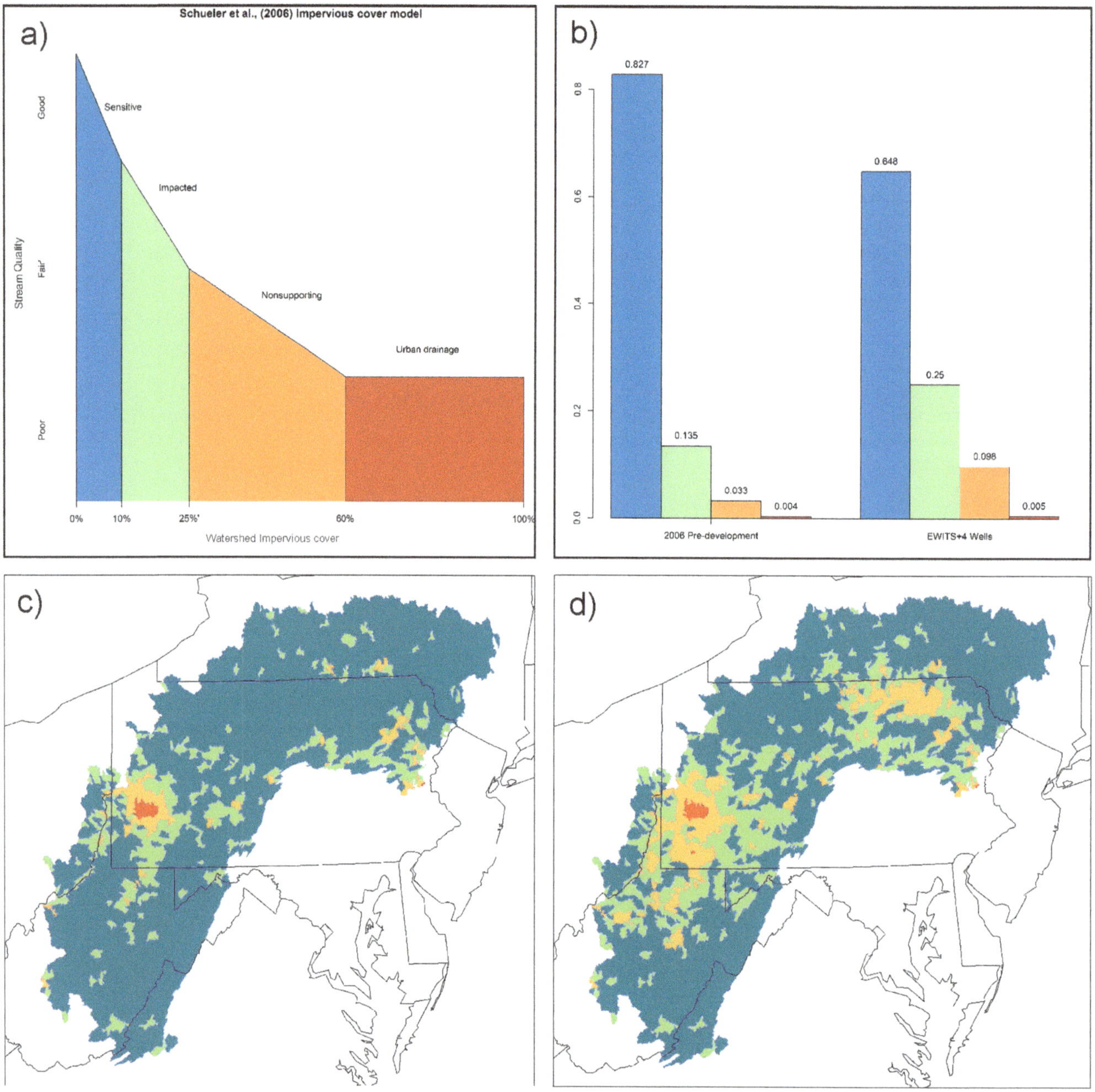

Figure 3. Impervious cover model at subwatershed-level; a) Graph of watershed impervious cover model (Schueler et al. 2009) representing classification of percent impervious surface. Colors in each impact class correspond to other panels in figure, b) bar graph showing percent subwatersheds in each class, c) 2006 "pre-development" subwatershed impacts, d) EWITS wind + 4 wells per pad subwatershed impacts.

operators optimize well spacing to account for evolving drilling practices, the assumption for average well spacing may also be revised. While several other studies have attempted to project future development scenarios within the Marcellus shale they have tended to focus on a subset of the area likely to be developed. Nonetheless these estimates are consistent with our estimate of the number of wells likely drilled. For example in their assessment of the economic impact of development in West Virginia the National Energy Technology Laboratory [25] project upwards of 40,000 new wells would be drilled by 2030. In a similar analysis focused on the economic impact of development in Pennsylvania, Considine et al. [24] project that upwards of 25,000 new wells would be drilled by 2020. The same can be said for our estimates of potential wind development. The estimation of wind development patterns is also highly uncertain, and will be influenced by a number of factors, i.e., investment in transmission, fuel costs and emission regulations [4,45]. We utilized an aggressive scenario that emphasizes development of wind resources close to load centers, which likely represents an estimate on the high end of potential wind development for this region [49]. Given the current scope of our modeling efforts, accounting for local and global market demand at specific time-scale is unrealistic. Because of this, we are not assuming a specific time scale for development, but rather the full potential build-out of each resource. This gives us a general idea of potential impacts of undirected development independent of market fluctuations.

In a 2003 report to the World Bank, Dudley and Stolton [63] concluded that to control erosion and sediment, maintain water quality, and in some cases capture and store water, protecting forest in water catchment areas is "no longer a luxury but rather a necessity." From an ecosystem services standpoint, the treatment cost of providing safe drinking water to urban areas increases dramatically with loss of forest [63]. In light of the new energy future in this landscape, as well as globally, understanding and mitigating the impacts of energy development will be one of the major challenges in the coming decade. Moving siting and mitigation decisions to a landscape scale rather than decisions made site by site, well by well will allow the regulators the ability to examine cumulative impacts like those discussed here [10].

Conclusions

The Marcellus Shale represents one of the fastest growing shale deposits in the world. With both wind and shale gas projected to expand dramatically in coming decades predicting patterns and impacts in the Marcellus could serve as a model for development that is likely to be replicated globally. Already, Argentina, Australia, China, and Colombia have identified large shale gas deposits that are in the planning stages of development. The impacts from individual gas wells/wind turbines or even those of a single wind farm or gas field are likely to be manageable and compatible with broader landscape level conservation goals. Our analysis reveals it will be the cumulative impacts that pose the greatest challenge for landscape level conservation goals. Unfortunately assessment of environmental impacts are made well by well or gas field by gas field with little or no attempt to assess cumulative impacts [64]. Scenarios and scenario analysis have become popular approaches in pursuit of sustainable development [18]. However, they are little used, at least in any formal way, in environmental impact assessment (EIA). Fostering the use of scenario modeling, like the approach outlined here, can allow regulators to examine the potential consequences of development objectives quickly and inexpensively. We conclude by encouraging EIA practitioners to learn about the promise of scenario-based analysis and implement scenario-based methods so that EIA can become more effective in fostering sustainable development.

Acknowledgments

Thanks to T. Minney and J. Dunscomb for logistical support; T.D. Gagnolet and N. Johnson, for their energy footprint analysis and helpful advice; S. Butler and F. Yates for guidance; C. Bruce for spatial data management and; S. Baruch-Mordo, D.L. Murphy, J. Fargione, P. Kareiva, M.A. Murphy, T. Moberg, and M. DePhilip for comments on the manuscript.

Author Contributions

Conceived and designed the experiments: JSE JMK. Performed the experiments: JSE. Analyzed the data: JSE. Wrote the paper: JSE JKM.

References

1. International Energy Agency (2009) World Energy Outlook. International Energy website. Available: http://www.worldenergyoutlook.org/media/weowebsite/2009/weo2009_es_english.pdf. Accessed 2013 May 15.
2. Chow J, Kopp RJ, Portney PR (2003) Energy resources and global development. Science 302: 1528–1531.
3. McDaniel C, Borton D (2002) Increased human energy use causes biological diversity loss and undermines prospects for sustainability. BioScience 52: 929–936.
4. McDonald R, Fargione J, Kiesecker J, Miller WM, Powell J (2009) Energy sprawl or energy efficiency: climate policy impacts on natural habitat for the United States of America. PLoS One 4: e680.
5. Erturk M (2011) Economic analysis of unconventional liquid fuel sources. Renewable & Sustainable Energy Reviews 12: 2766–2771.
6. Kerr RA (2010) Natural gas from shale bursts onto the scene. Science 328: 1624–1626.
7. Naugle D (2010)Energy Development and Wildlife Conservation in Western North America; Naugle D, editor Island Press: 344 p.
8. U.S. Department of Energy (2008) 20% Wind Energy by 2030 Increasing Wind Energy's Contribution to U.S. Electricity Supply. U.S. Department of Energy, National Renewable Energy Laboratory website. Available: http://www.nrel.gov/docs/fy08osti/41869.pdf. Accessed 2012 August 1.
9. Wood R, Gilbert P, Sharmina M, Anderson K, Footitt A, et al. (2011) Shale gas: a provisional assessment of climate change and environmental impacts. Tyndall Center for Climate Research website. Available: http://www.tyndall.ac.uk/sites/default/files/tyndall-coop_shale_gas_report_final.pdf. Accessed 2012 May 24.
10. Kiesecker JM, Copeland H, Pocewicz A, McKenney B (2010) Development by Design: Blending Landscape Level Planning with the Mitigation Hierarchy. Frontiers in Ecology and the Environment 8: 261–266.
11. Entrekin S, Evans-White M, Johnson B, Hagenbuch E (2011) Rapid expansion of natural gas development poses a threat to surface waters. Frontiers in Ecology and the Environment 9: 503–511.
12. Johnson N, Gagnolet T, Ralls R, Zimmerman E, Eichelberger B, et al. (2010) Pennsylvania Energy Impacts Assessment. Report 1: Marcellus Shale Natural Gas and Wind. The Nature Conservancy, Western Pennsylvania Conservancy and Audubon Pennsylvania. The Nature Conservancy website. Available: http://www.nature.org/ourinitiatives/regions/northamerica/unitedstates/pennsylvania/pa-energy-executive-summary.pdf. Accessed 2013 January 12.
13. Olmstead SM, Muehlenbachs LA, Shih J, Chu Z, Krupnick AJ (2013) Shale gas development impacts on surface water quality in Pennsylvania. Proceedings of the National Academy of Sciences of the United States of America 110: 4962–4967.
14. Vidic RD, Brantley SL, Vandenbossche JM, Yoxtheimer D, Abad JD (2013) Impact of Shale Gas Development on Regional Water Quality. Science 340.
15. Kaushal SS, Groffman PM, Likens GE, Belt KT, Stack WP, et al. (2005) Increased salinization of fresh water in the northeastern United States. Proceedings of the National Academy of Sciences of the United States of America 102: 13517–13520.
16. Rozell D, Reaven S (2012) Water pollution risk associated with natural gas extraction from the Marcellus Shale. Risk Analysis 32: 1382–1393.
17. Cuffney TF, Brightbill RB, May JT, Waite IR (2010) Responses of benthic macroinvertebrates to environmental changes associated with urbanization in nine metropolitan areas. Ecological Applications 20: 1384–1401.
18. Duinker PN, Greig LA (2007) Scenario analysis in environmental impact assessment: Improving explorations of the future. Environmental Impact Assessment Review 27: 206–219.
19. Pocewicz A, Nielsen-Pincus M, Goldberg CS, Johnson MH, Morgan P, et al. (2008) Predicting land use change: comparison of models based on landowner surveysand historical land cover trends. Landscape Ecology 23: 195–210.
20. Pontius G, Malanson J (2005) Comparison of the structure and accuracy of two land change models. International Journal of Geographical Information Science 19: 243–265.
21. Theobald DM, Hobbs T (1998) Forecasting rural land-use change: A comparison of regression- and spatial transition-based models. Geographical & Environmental Modelling 2: 65–82.
22. Guisan A, Zimmermann NE (2000) Predictive habitat distribution models in ecology. Ecological Modelling 135: 147–186.
23. Phillips SJ, Anderson RP, Schapire RE (2006) Maximum entropy modeling of species geographic distributions. Ecological Modelling 190: 231–259.
24. Considine TJ, Watson R, Blumsack S (2011) The Pennsylvania Marcellus Natural Gas Industry: Status, Economic Impacts and Future Potential. Marcellus Shale Coalition website. Available: http://marcelluscoalition.org/wp-content/uploads/2011/07/Final-2011-PA-Marcellus-Economic-Impacts.pdf. Accessed 2013 May 3.
25. U.S. Department of Energy (2010) Projecting the economic impact of Marcellus share gas development in West Virginia: A preliminary analysis using publicly avaliable data. DOE/NETL- 402033110. U.S. Department of Energy, National Renewable Energy Laboratory website. Available: http://www.netl.doe.gov/energy-analyses/pubs/WVMarcellusEconomics3.pdf
26. Johnson N, Gagnolet T, Ralls R, Stevens J (2011) Pennsylvania Energy Impacts Assessment. Report 2: Marcellus Shale Natural Gas and Wind, Natural Gas Pipelines. The Nature Conservancy, Western Pennsylvania Conservancy. The Nature Conservancy website. Available: http://www.nature.org/ourinitiatives/regions/northamerica/unitedstates/pennsylvania/ng-pipelines.pdf. Accessed 2013 January 12.

27. U.S. Energy Information Administration (2013) Annual Energy Outlook 2013. Washington, DC. U.S. Energy Information Administration website. Available: http://www.eia.gov/forecasts/ieo/. Accessed 2013 April 10.
28. Homer C, Dewitz J, Fry J, Coan M, Hossain N, et al. (2007) Completion of the 2001 National Land Cover Database for the Conterminous United States. Photogrammetric Engineering and Remote Sensing 73: 337–341.
29. Phillips JD, Duval JS, Ambroziak RA (1993) National geophysical data grids; gamma-ray, gravity, magnetic, and topographic data for the conterminous United States [dataset]. U.S. Geological Survey website. Available: http://mrdata.usgs.gov/catalog/cite-view.php?cite = 97.Accessed 2011 April 18.
30. Repetski JE, Ryder RT, Weary DJ, Harris AG, Trippi MH (2008) Thermal maturity patterns (CAI and %Ro) in Upper Ordovician and Devonian rocks of the Appalachian basin: A major revision of USGS Map I–917–E using new subsurface collections: U.S. Geological Survey Scientific Investigations Map 3006. U.S. Department of the Interior Geological Survey (USGS).
31. Cressie N (1991) Statistics for spatial data. New York, New York: Wiley-Interscience.
32. Ribeiro PJ, Diggle PJ (2001) geoR: A package for geostatistical analysis. R News 1.
33. Evans JS, Oakleaf J, Cushman SA, Theobald D (In prep) An ArcGIS toolbox for Geomorphometric and Gradient modeling. Ecography. Jeffrey S. Evans spatial ecology website. Available: http://evansmurphy.wix.com/evansspatial.Accessed 2013 February 20.
34. U.S. Geological Survey (2012) Protected Areas Database of the United States (PADUS) version 1.3. U.S. Geological Survey, GAP website. Available: http://gapanalysis.usgs.gov/padus/. Accessed 2012 December 3.
35. Barnes M, Todd A, Lilja R, Barten P (2009) Forests, Water and People: Drinking water supply and forest lands in the Northeast and Midwest United States. USDA Forest Service, Northeastern Area State and Private Forestry NA-FR-01-08. U.S. Forest Service website. Availiable: http://www.fs.fed.us/ecosystemservices/FS_Efforts/forests2faucets.shtml.Accessed 2012 July 25.
36. Diggle PJ (2003) Statistical analysis of spatial point patterns, Second edition. New York, New York: Arnold Oxford University Press.
37. Baddeley A, Turner R (2005) Spatstat: an R package for analyzing spatial point patterns. Journal of Statistical Software 12: 1–42.
38. Warton DI, Shepherd LC (2010) Poisson point process models solve the "pseudo-absence problem" for presence-only data in ecology. Annals of Applied Statistics 4: 1383–1402.
39. Breiman L (2001) Random forests. Machine Learning 45: 5–32.
40. Evans JS, Murphy MA, Holden ZA, Cushman SA (2011) Modeling species distribution and change using Random Forests. In: Drew CA, Wiersma YF, Huettmann F, editors. Predictive species and habitat modeling in landscape ecology: concepts and applications. New York, New York: Springer.
41. Liaw A, Wiener M (2002) Classification and Regression by randomForest. R News 2: 18–22.
42. Birnbaum ZW, Tingey FH (1951) One-sided confidence contours for probability distribution functions. The Annals of Mathematical Statistics 22: 592–596.
43. Evans JS, Cushman SA (2008) Gradient Modeling of Conifer Species Using Random Forest. Landscape Ecology 5: 673–683.
44. Murphy MA, Evans JS, Storfer A (2010) Quantifying Bufo boreas connectivity in Yellowstone National Park with landscape genetics. Ecology 91: 252–261.
45. Fargione J, Kiesecker J, Slaats MJ, Olimb S (2012) Wind and Wildlife in the Northern Great Plains: Identifying Low-Impact Areas for Wind Development. PLoS One 7(7):e41468. doi:10.1371/journal.pone.0041468.
46. Kiesecker JM, Evans JS, Fargione J, Doherty K, Foresman KR (2011) Win-Win for Wind and Wildlife: A Vision to Facilitate Sustainable Development. PLoS One 6: e1756
47. Obermeyer R, Manes R, Kiesecker J, Fargione J, Sochi K (2011) Development by Design: Mitigating Wind Development's Impacts on Wildlife in Kansas. PLoS One 6: e2669.
48. Copeland HE, Doherty KE, Naugle DE, Pocewicz A, Kiesecker JM (2009) Mapping oil and gas development potential in the U.S. Intermountain West and estimating impacts to species. PLoS One 4:e7400.http://dx.doi.org/10.1371/journal.pone.0007400.
49. U.S. Department of Energy (2011) Eastern Wind Integration and Transmission Study. U.S. Electricity Supply. U.S. Department of Energy, National Renewable Energy Laboratory website. Available: http://www.nrel.gov/docs/fy11osti/47078.pdf. Accessed 2013 March 10.
50. Ventyx Corporation (2013) Velocity Suite: EV Energy Map [dataset]. Database subscription service. Accessed 2013 February 8.
51. Schueler TR, Fraley-McNeal L, Cappiella K (2009) Is Impervious Cover Still Important? Review of Recent Research. Journal of Hydrologic Engineering 14: 309–315.
52. Center for Watershed Protection (2003) Impacts of impervious cover on aquatic systems. Ellicott City, MD: Center for Watershed Protection. 142 p. Center for Watershed Protection website. Available: http://clear.uconn.edu/projects/TMDL/library/papers/Schueler_2003.pdf.Accessed 2013 March 25.
53. King RS, Baker ME (2010) Considerations for analyzing ecological community thresholds in response to anthropogenic environmental gradients. Journal of the North American Benthological Society 29: 998–1008.
54. Baker ME, King RS (2010) A new method for detecting and interpreting biodiversity and ecological community thresholds. Methods in Ecology and Evolution 1: 25–37.
55. Yang L, Huang C, Homer C, Wylie B, Coan M (2003) An approach for mapping large-area impervious surfaces: Synergistic use of Landsat 7 ETM+ and high spatial resolution imagery. Canadian Journal of Remote Sensing 29: 230–240.
56. U.S. Environmental Protection Agency (EPA) (2009) Safe Drinking Water Information System (SDWIS) [dataset] Washington, DC: U.S. EPA. U.S. Environmental Protection Agency website. Available: http://www.epa.gov/enviro/facts/sdwis/search.html.
57. Brown T, Hobbins M, Ramirez J (2008) Spatial distribution of water supply in the coterminous United States. Journal of the American water Resources Association 44: 1474–1487.
58. Fleiss JL, Cohen J (1973) The equivalence of weighted kappa and the intraclass correlation coefficient as measures of reliability. Educational and Psychological Measurement 33: 613–619.
59. Schueler T (1994) The Importance of Imperviousness. Watershed Protection Techniques 2: 100–111.
60. Soeder DJ, Kappel WM (2009) Water Resources and Natural Gas Production from the Marcellus Shale. U.S. Geological Survey Fact Sheet 2009–3032,6 p.
61. Mooney C (2011) The Truth about Fracking. Scientific American 305: 80–85
62. Northrup JM, Wittemyer G (2013) Characterising the impacts of emerging energy development on wildlife, with an eye towards mitigation. Ecology Letters 6: 112–125.
63. Dudley N, Stolton S (2003) Running pure: the importance of forest protected areas to drinking water, a research report for the World Bank/WWF Alliance for Forest Conservation and Sustainable Use. Gland, Switzerland. 114 p. World Wildlife Foundation website. Available: http://wwf.panda.org/what_we_do/how_we_work/conservation/forests/publications/?uNewsID = 8443. Accessed 2011 June 19.
64. Canter L, Ross B (2010) State of practice of cumulative effects assessment and management: the good, the bad and the ugly. Impact Assessment and Project Appraisal 24: 261–268.

3

Influence of Arbuscular Mycorrhizae on Biomass Production and Nitrogen Fixation of Berseem Clover Plants Subjected to Water Stress

Sergio Saia, Gaetano Amato, Alfonso Salvatore Frenda, Dario Giambalvo, Paolo Ruisi*

Dipartimento di Scienze Agrarie e Forestali, Università degli Studi di Palermo, Palermo, Italy

Abstract

Several studies, performed mainly in pots, have shown that arbuscular mycorrhizal symbiosis can mitigate the negative effects of water stress on plant growth. No information is available about the effects of arbuscular mycorrhizal symbiosis on berseem clover growth and nitrogen (N) fixation under conditions of water shortage. A field experiment was conducted in a hilly area of inner Sicily, Italy, to determine whether symbiosis with AM fungi can mitigate the detrimental effects of drought stress (which in the Mediterranean often occurs during the late period of the growing season) on forage yield and symbiotic N_2 fixation of berseem clover. Soil was either left under water stress (i.e., rain-fed conditions) or the crop was well-watered. Mycorrhization treatments consisted of inoculation of berseem clover seeds with arbuscular mycorrhizal spores or suppression of arbuscular mycorrhizal symbiosis by means of fungicide treatments. Nitrogen biological fixation was assessed using the ^{15}N-isotope dilution technique. Arbuscular mycorrhizal symbiosis was able to mitigate the negative effect of water stress on berseem clover grown in a typical semiarid Mediterranean environment. In fact, under water stress conditions, arbuscular mycorrhizal symbiosis resulted in increases in total biomass, N content, and N fixation, whereas no effect of crop mycorrhization was observed in the well-watered treatment.

Editor: Lam-Son Phan Tran, RIKEN Center for Sustainable Resource Science, Japan

Funding: This study was funded by Regione Siciliana (Italy) – Project SI.CO.BIO.S. The funder had no role in study design, data collection and analysis, decision to publish, or preparation of the manuscript.

Competing Interests: The authors have declared that no competing interests exist.

* E-mail: paolo.ruisi@unipa.it

Introduction

Arbuscular mycorrhizal (AM) fungi are commonly occurring fungi that live in an obligate symbiotic status with the majority of land plants. AM symbiosis has a positive influence on plant growth, which is mainly attributable to the ability of AM fungi to take up nutrients from the soil — especially phosphorus [1,2] and to a lesser extent nitrogen (N) [3,4] — and deliver them to the roots of its host, and also to enhance the health of its host by protecting it from pathogens, pests, and parasitic plants [5]. In addition, in legume crops, AM symbiosis can improve the degree of dependence of the legume on atmospheric N_2 [6] by increasing the uptake of key nutrients to simultaneously feed plants and nodules or through direct effects on the functioning of the legume-*Rhizobium* symbiosis [7]. Moreover, AM symbiosis can mitigate the negative effects of water stress (WS) on plant growth [8], although the underlying mechanisms of this are still unclear [9]. Nevertheless, an extensive review by Augé [10] covering hundreds of studies (which were performed mainly in pots) highlights the fact that mycorrhizal effects on plant-water relationships are often subtle, transient, and probably circumstance- and symbiont-specific.

Berseem clover (*Trifolium alexandrinum* L.) is a drought-tolerant forage legume that is well adapted to semiarid conditions and widely cultivated under rain-fed conditions in the Mediterranean, central Asia, and, most recently, the United States [11]. In Mediterranean areas, it is usually sown in the autumn and grown until early summer as a multicut forage species, thanks to its ability to regrow after cutting [12]. However, during the spring, plant regrowth is often limited by water deficiency, which negatively affects photosynthesis and symbiotic N_2 fixation and leads to a reduction in accumulation and quality of biomass. In berseem clover, experiments performed under controlled conditions have shown that AM symbiosis can play an important role in protecting the host plant from both biotic and abiotic stresses [13,14]. Pellegrino et al. [15], in a study that included both pot and field trials, found that the use of AM fungal inocula may be very effective in improving berseem clover productivity and quality and that the choice of the most appropriate AM fungal inoculum is essential to the success of inoculation practices. Moreover, other studies have found that AM symbiosis plays a role in determining the N_2 fixation activity of berseem clover under low soil phosphorus concentration, at least during the early stages of the *Rhizobium*-legume interaction [16]. However, to our knowledge, no information is available about the effects of AM symbiosis on berseem growth and N_2 fixation under water shortage conditions. Thus, we performed a field experiment to determine whether symbiosis with AM fungi can mitigate the detrimental effects of drought stress (which in Mediterranean environments often occurs during the late period of the growing season) on forage yield, quality, and symbiotic N_2 fixation of berseem clover.

Materials and Methods

Ethics Statement

No specific permits were required for the described field study. The location is not protected in any way.

The experiment did not involve endangered or protected species.

Experimental site

The trial was conducted during the 2007–2008 growing season at the Pietranera farm, which is located about 30 km north of Agrigento, Sicily, Italy (37°30' N, 13°31' E; 178 m asl), on a deep, well-structured, clay soil classified as a Vertic Haploxerept. Soil characteristics are presented in table S1. The climate of the experimental site is semiarid Mediterranean, with a mean annual rainfall of 552 mm, mostly in the autumn and winter (74%) and in the spring (18%). The dry period lasts from May to September. The mean air temperature is 15.9°C in autumn, 9.8°C in winter, and 16.5°C in spring. The average minimum and maximum annual temperatures are 10.0°C and 23.3°C, respectively. The weather data were collected from a weather station located within 500 m of the experimental site.

During the 2007–2008 growing season (from September to May), total rainfall was 516 mm (slightly lower than the long-term average) and it was well distributed during fall and winter (Fig. 1). Total rainfall during the spring (91 mm) was markedly lower than normal (–30%). Average, minimum, and maximum temperatures were similar to long-term averages.

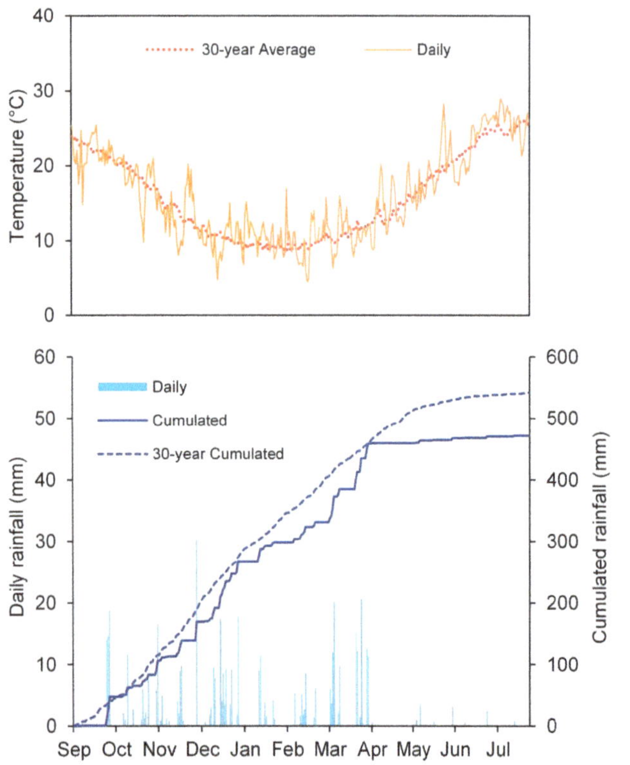

Figure 1. Cumulative daily rainfall and mean air temperatures at the experimental site during the 2007–2008 growing season. 30-year (1980–2010) average cumulative daily rainfall and temperatures are included.

Experimental design and crop management

A split-plot design with four replicates was adopted, with plot sizes of 36 m^2 each (12 rows 12 m long and 0.25 m apart). The main plot was soil water availability: crops were subjected to WS or well-watered (WW) conditions. The subplot was mycorrhization: inoculation of berseem clover seeds with AM spores (+AM) or suppression of AM symbiosis by means of fungicide treatments (–AM). In WS treatment, berseem clover was grown entirely under rain-fed conditions, whereas in WW three sprinkle irrigations, 30 mm each, were done by irrigating when soil water dropped below a value of about 0.7 of the available soil water capacity (116, 125, and 132 days after sowing, DAS). The amount of water applied at each irrigation point was sufficient to bring soil water content to field capacity. Soil moisture content was measured gravimetrically (in the 0–0.40-m layer) once per month from sowing to 110 DAS and every 3 days thereafter.

Natural AM spore population in the field was measured before sowing by the wet-sieving method [17] and consisted of a mycorrhizal consortium formed mainly by *Glomus*-group AM species and *Acaulospora*. Overall AM spore density was 5 spores per 10 g air-dried soil. AM hyphal mass in the soil was not detected. In –AM treatment, suppression of the natural AM symbiosis was achieved by spraying the plots/plants with systemic fungicides: Captan (20 mg a.i. m^{-2}), Carbendazim (20 mg a.i. m^{-2}), and Benomyl (20 mg a.i. m^{-2}) were used at sowing time to suppress AM spore germination; then, Benomyl, Fenpropimorph, and Carbendazim were applied once per month during the growth of the crop, at rates of 10, 5, and 5 mg a.i. m^{-2}, respectively, to reduce AM hyphal growth and symbiotic activity. These fungicides have no toxic effects on plant growth [18,19]. Inoculation with AM fungi involved the application of a commercial AM inoculum at a rate of 12 g kg^{-1} of seed. The inoculum consisted of a mixture of spores of *Rhizophagus irregularis* (formerly *Glomus intraradices*) and *Funneliformis mosseae* (formerly *G. mosseae*), each of which was present at a rate of 1000 spores g^{-1} of inoculum. Berseem clover seed was not inoculated with Rhizobium before planting because prolific inoculation occurs naturally at this site.

Soil, cropped in the previous growing season with durum wheat (*Triticum durum* Desf.), was plowed to a depth of 30 cm in the summer and then shallowly harrowed twice to control weeds and prepare suitable seedbed conditions. No herbicides were applied. Berseem clover (cv Lilibeo) was sown by hand at the end of December at a rate of 1200 viable seeds m^{-2}. Weeds were removed by hand during the growing season.

The entire plot was cut at a 5-cm stubble height at 76 DAS (C1) and again at 116 DAS (C2); at each cut, a subplot (2.25 m^2) was used to determine plant hypogeic (belowground) biomass. Starting at C2, water treatment was established, and crop regrowth was monitored by cutting different subplots (2.25 m^2) after 7 (R7), 14 (R14), 21 (R21), and 28 days (R28).

Symbiotic N_2 fixation was assessed at R14 and R28 using the ^{15}N-isotope dilution technique by applying a ^{15}N-labeled fertilizer ($[NH_4]_2SO_4$ with an isotopic composition of 10 atom% ^{15}N) at 116 DAS, soon after C2. ^{15}N-labeled fertilizer was applied as drench at a rate of 8 kg N ha^{-1} to the subplots assigned to R14 and R28. In all cases, the rest of the plot outside the ^{15}N-labeled areas received a topdressing of ammonium sulfate in an amount equivalent to that in the subplots. Ryegrass (*Lolium multiflorum* var. *Westerwoldicum* cv Elunaria) was chosen as a reference crop because its growth rate and regrowth ability are similar to those of berseem clover [20]; ryegrass received the same treatments as berseem clover.

Table 1. Field response of shoots and roots of berseem clover to AM inoculation at first cut (C1, at 76 days after sowing - DAS) and second cut (C2, at 116 DAS).

		C1 (76 DAS)			C2 (116 DAS)		
		–AM	+AM		–AM	+AM	
Total aboveground biomass	g m^{-2}	159	187	**	548	568	ns
Removed (Rem)	g m^{-2}	113	136	**	393	410	ns
Residual (Res)	g m^{-2}	46	51	*	155	158	ns
Leaves in Rem	% on Rem biomass	74.7	68.2	**	40.6	41.5	ns
Leaves in Res	% on Res biomass	32.2	32.8	ns	28.6	28.5	ns
Total LAI	m^2 m^{-2}	2.96	3.39	*	5.93	6.30	ns
LAI of Res	m^2 m^{-2}	0.44	0.51	*	1.31	1.33	ns
Root biomass	g m^{-2}	35	41	*	36	36	ns

+AM, inoculation with arbuscular mycorrhizal spores; –AM, suppression of arbuscular mycorrhizal symbiosis.
* and ** indicate significant differences at 0.05 and 0.01 probability level, respectively; ns indicates differences not significant.

Plant harvests and analyses

At C1 and C2, the removed biomass (above cutting height) was weighed, and two subsamples of 1 kg of fresh matter each were taken. The first was separated into leaves, stems, and heads, whereas the second was dried at 65°C until a constant weight was reached. Furthermore, on subplots assigned to destructive measures, both residual aboveground biomass (below cutting height) and hypogeic biomass (in the first 20 cm of soil) were measured. The residual aboveground biomass was separated into leaves and stems. All botanical fractions were weighed and dried at 65°C to a constant weight. The leaf area index (LAI) of the residual and removed leaves was immediately measured on a 10-g subsample. On subplots assigned to R7, R14, R21, and R28 cuts, total aboveground biomass (above cutting height + below cutting height) and hypogeic biomass were measured. Samples of total aboveground biomass were separated into botanical fractions, and LAI was measured using the same method described earlier. Biomass samples from R14 and R28 were analyzed for total N and ^{15}N enrichment by means of an elemental analyzer–isotope ratio mass spectrometer (Carlo Erba NA1500).

AM root colonization was measured according to Giovannetti and Mosse [21] on root samples (from five plants randomly chosen from each plot) collected at 55 DAS, at R14, and at R28; roots were stained with 0.05% trypan blue in lactic acid according to Phillips and Hayman [22].

Calculations and statistical analysis

Data on biomass ^{15}N enrichment were used to calculate the percentage of N_2 derived from symbiotic N_2 fixation (%Ndfa) according to Fried and Middelboe [23]:

$$\%\text{Ndfa} = \left(1 - \frac{\text{Atom}\%^{15}\text{N}_{\text{berseem}}}{\text{Atom}\%^{15}\text{N}_{\text{ryegrass}}}\right) \times 100,$$

Table 2. F values and significance levels for water availability (W) and crop mycorrhization (M) on measured traits of regrowth, conducted separately for each sampling date (R7, R14, R21, and R28).

Treatments	Biomass regrowth	Taproot biomass	LAI	Stems %	Heads %	Above ground N		Taproot N		Ndfa		Mycorrhizal infection
	g m^{-2}	g m^{-2}		on biomass regrowth		g kg^{-1}	g m^{-2}	g kg^{-1}	g m^{-2}	%	g m^{-2}	%
Water availabilty (W)												
R7	22.54*	0.14^{ns}	15.30*	0.13^{ns}	†	-	-	-	-	-	-	-
R14	20.03*	8.96^{ns}	262.25***	3.19^{ns}	33.14*	0.27^{ns}	25.67*	4.61^{ns}	50.03**	7.83^{ns}	57.77**	2.52^{ns}
R21	110.99**	2.63^{ns}	48.46**	38.79**	51.06**	-	-	-	-	-	-	-
R28	10.42*	0.03^{ns}	92.74**	233.98***	246.77***	3.81^{ns}	81.93**	0.06^{ns}	0.04^{ns}	16.14*	79.73**	19.87*
Crop mycorrhization (M)												
R7	0.08^{ns}	0.47^{ns}	3.42^{ns}	2.05^{ns}	†	-	-	-	-	-	-	-
R14	1.00^{ns}	2.82^{ns}	0.49^{ns}	0.13^{ns}	0.16^{ns}	0.03^{ns}	1.80^{ns}	0.47^{ns}	2.80^{ns}	0.00^{ns}	0.63^{ns}	1348.18***
R21	22.89**	1.03^{ns}	0.06^{ns}	0.14^{ns}	3.62^{ns}	-	-	-	-	-	-	-
R28	13.08*	0.01^{ns}	4.79^{ns}	1.15^{ns}	0.11^{ns}	0.82^{ns}	0.65^{ns}	0.17^{ns}	0.32^{ns}	14.38**	21.38**	887.47***
W × M												
R7	4.93^{ns}	3.14^{ns}	1.38^{ns}	0.23^{ns}	†	-	-	-	-	-	-	-
R14	0.86^{ns}	0.46^{ns}	0.06^{ns}	5.71^{ns}	0.16^{ns}	0.43^{ns}	2.20^{ns}	0.00^{ns}	0.38^{ns}	1.05^{ns}	4.74^{ns}	0.03^{ns}
R21	6.22*	0.01^{ns}	4.99^{ns}	0.73^{ns}	4.66^{ns}	-	-	-	-	-	-	-
R28	29.78**	1.49^{ns}	0.41^{ns}	0.01^{ns}	4.31^{ns}	0.04^{ns}	7.33*	0.00^{ns}	3.41^{ns}	6.68*	21.55**	14.88**

*, ** and *** indicate significant differences at 0.05, 0.01 and 0.001 probability level, respectively; ^{ns} indicates differences non significant.
†Heads were absent in all the treatments.
-, not detected.

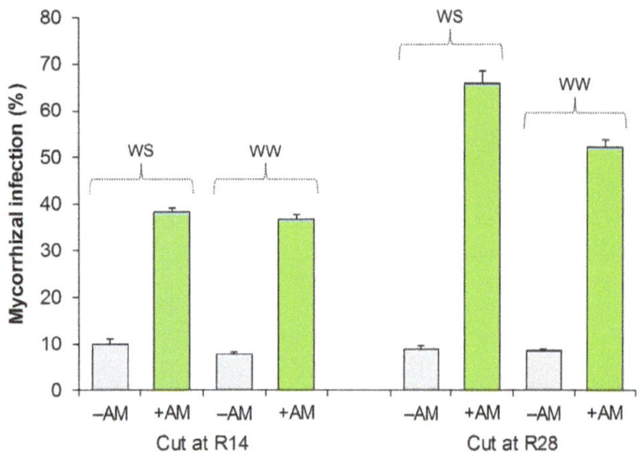

Figure 2. Mycorrhizal infection of berseem clover roots as affected by water availability and mycorrhization treatment. Data are means (n = 4) ± S.E. measured at R14 (14 days after C2) and R28 (28 days after C2). WW, well-watered; WS, water-stressed; +AM, inoculation with arbuscular mycorrhizal spores; -AM, suppression of arbuscular mycorrhizal symbiosis.

where Atom% $^{15}N_{berseem}$ represents the Atom% ^{15}N excess of berseem tissues and Atom% $^{15}N_{ryegrass}$ represents the Atom% ^{15}N excess of ryegrass tissue. The ^{15}N-natural abundance of the atmosphere (0.3663% ^{15}N) was used to calculate the Atom% ^{15}N excess of both crops. The amount of N fixed by berseem clover was estimated as:

$$\text{Ndfa} = \text{Total berseem N} \times \frac{\%\text{Ndfa}}{100}.$$

Analysis of variance [24] was performed separately for each cut according to the experimental design. All measured variables were assumed to be normally distributed. All variables corresponding to proportions were arcsine transformed before analysis to ensure a better fit with the Gaussian law distribution.

Results

Soil moisture did not drop below a value of about 0.7 of the available soil water capacity until C2. As expected, from C2 to R28, soil moisture content was always significantly higher in WW than WS conditions (25.4% vs. 19.4% across all sampling dates; P

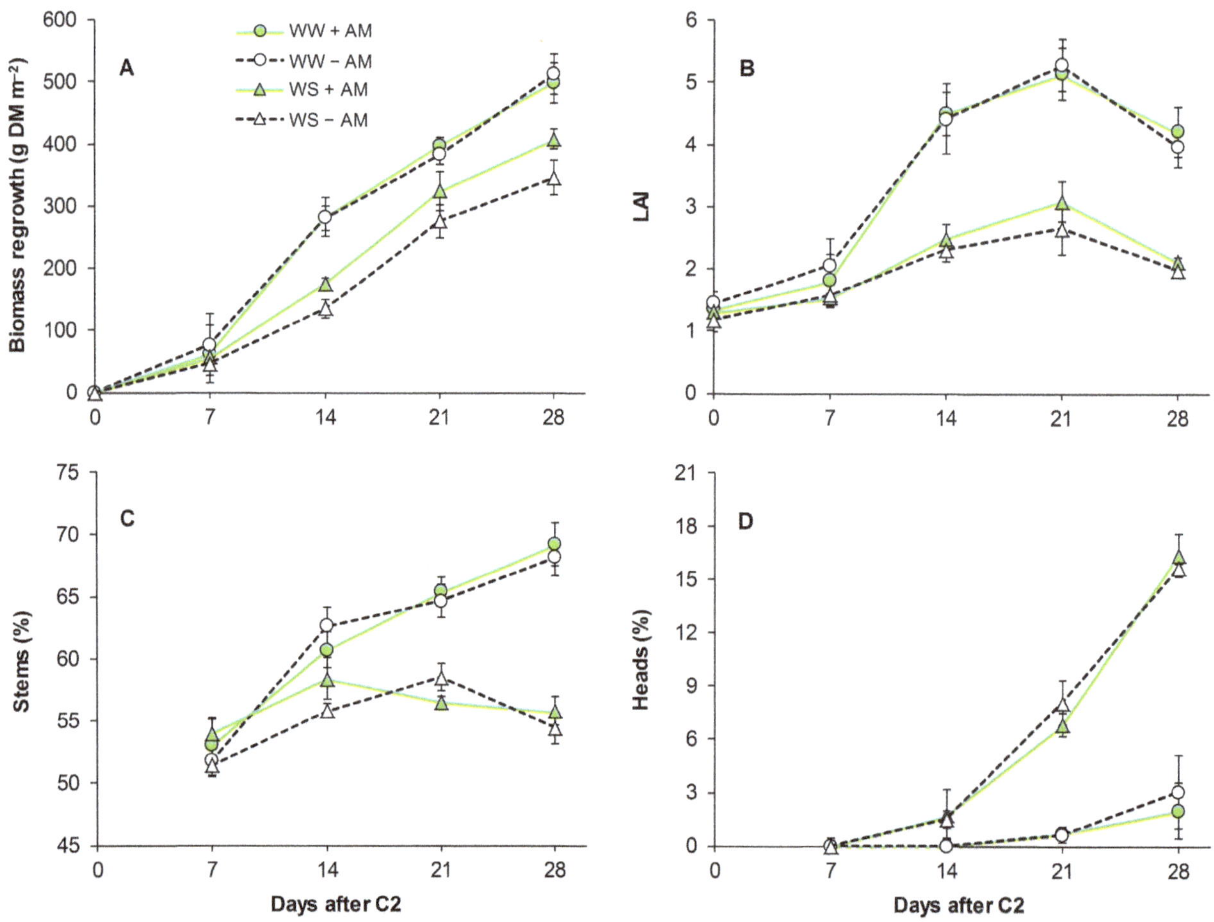

Figure 3. Biomass regrowth (A), LAI (B), and incidence of stems (C) and heads (D) as a percentage of berseem clover biomass as affected by water availability and mycorrhization treatment. Data are means (n = 4) ± S.E. measured after 7, 14, 21 and 28 days from second cut (C2). WW, well-watered; WS, water-stressed; +AM, inoculation with arbuscular mycorrhizal spores; -AM, suppression of arbuscular mycorrhizal symbiosis.

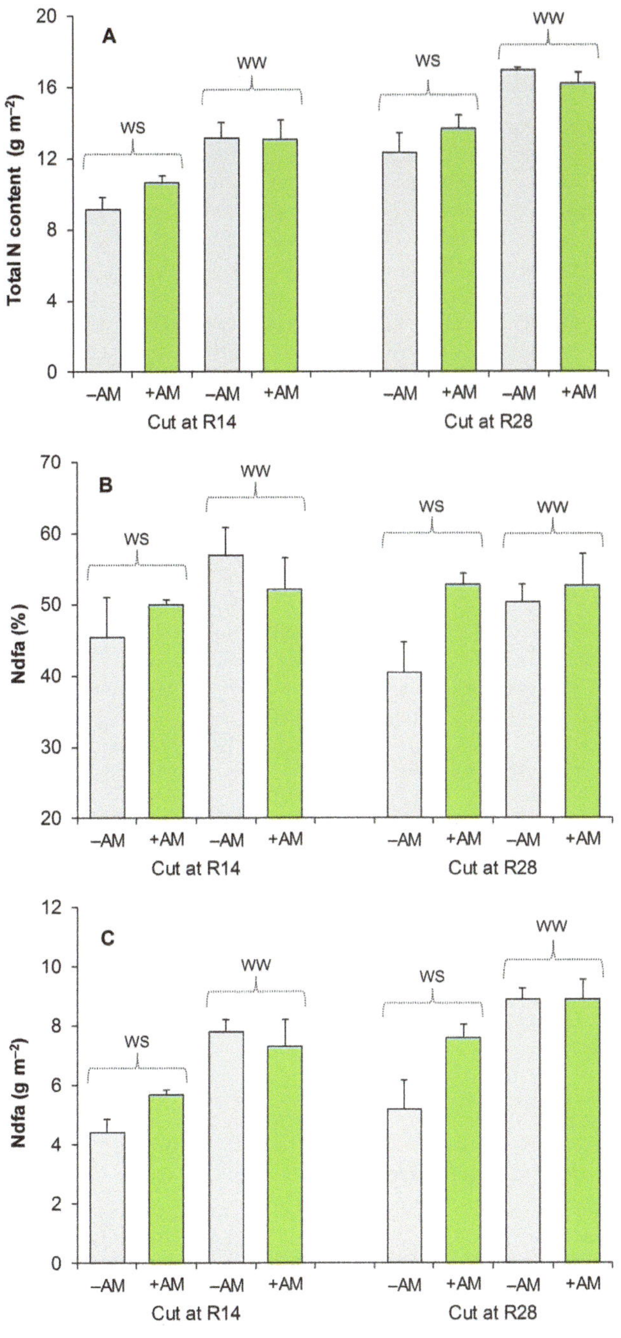

Figure 4. Total N content (A), Nitrogen derived from the atmosphere as a percentage of total N content (B) and total fixed N (C) of berseem clover as affected by water availability and mycorrhization treatment. Data are means (n=4) ± S.E. measured at R14 (14 days after C2) and R28 (28 days after C2). WW, well-watered; WS, water-stressed; +AM, inoculation with arbuscular mycorrhizal spores; -AM, suppression of arbuscular mycorrhizal symbiosis.

< 0.001), and the differences between the two treatments increased with time. Soil moisture content was not affected by crop mycorrhization, and the interaction with soil moisture regimen was never significant.

At 55 DAS, root colonization by AM fungi was significantly lower in the -AM crops than in the +AM crops (7.0% and 32.4%, respectively; $P < 0.001$). At C1, both total and residual biomass of berseem clover, and LAI of total biomass and of residual biomass, were significantly higher in the +AM than the -AM crops (Table 1). Also, root biomass (up to 20 cm depth) was significantly higher in +AM than in -AM. At C2, no significant effect of mycorrhization was observed on biomass accumulation, canopy traits (LAI and percentage of leaves), or root biomass.

Tests of significance for the effect of water treatment and crop mycorrhization on the traits measured during crop regrowth (at R7, R14, R21, and R28) are reported in Table 2. At R14 (14 days after the establishment of water treatment), AM infection was, on average, 8.8% in -AM and 37.7% in +AM (significantly different at $P < 0.001$); at this time, no effect of the soil moisture regime was observed on root colonization by AM fungi (Fig 2). At R28, root colonization by AM fungi was very low in the -AM treatment areas, irrespective of soil moisture regime (average of 8.7%), whereas in +AM treatment areas, root colonization was significantly higher in the WS areas than in the WW areas (66.0% and 52.4%, respectively).

The increase in water availability in WW compared to WS led to significant increases in biomass regrowth starting at R7 (Table 2 and Fig. 3A). The differences between WW and WS increased until R14 and then remained constant until R28. Crop mycorrhization increased biomass regrowth in WS but not in WW and the difference between +AM and -AM increased with time.

Irrigation led to an increase in LAI and stem percentage and a decrease in head percentage with differences between WW and WS increasing with time (Fig. 3B, 3C and 3D). Crop mycorrhization did not affect LAI or the incidence of stems and heads as a percentage of crop biomass in either WS or WW (Table 2 and Fig. 3B, 3C and 3D).

Both water treatment and crop mycorrhization almost never affected taproot biomass (38.4 g m^{-2} averaged across all sampling dates, with slight increases with time).

N concentration did not vary according to the treatment applied (on average 30.5 g kg^{-1} and 24.7 g kg^{-1} in aboveground biomass at R14 and R28, respectively, and 18.6 g kg^{-1} and 17.6 g kg^{-1} in taproot biomass at R14 and R28, respectively). Irrigation increased total N content both at R14 and R28 (Fig. 4A). Crop mycorrhization resulted in an increase in total N content in WS but not in WW.

Irrigation increased total N fixed both at R14 (on average 50% compared to WS) and at R28 (on average 39% compared to WS) (Fig. 4C). At R14, crop mycorrhization did not affect Ndfa (as both a percentage of total N content and the amount of N fixed on an area basis). On the contrary, at R28, Ndfa values were significantly higher in +AM than in -AM in the WS treatment, whereas no effects of crop mycorrhization were observed in the WW treatment.

Discussion

On the whole, when berseem clover did not experience WS (i.e., in WW treatments), aboveground biomass production was about 12 Mg dry matter ha^{-1}; this is similar to the biomass obtained in the same environment when rainfall during the growing season was 30% higher than the long-term average [12,20]. On average, no differences between -AM and +AM conditions were observed in terms of total aboveground biomass production (11.7 vs. 12.0 Mg dry matter ha^{-1}, calculated as a sum of C1+C2+R28 in WW treatments). This result is in contrast with the findings of Pellegrino et al. [15], who observed, in a field trial, an increase of berseem clover productivity (from 6.5 to 10.0 Mg dry matter ha^{-1}) following inoculation with AM fungi in comparison with a non-inoculated control. Because the main recognized plant benefit from AM symbiosis consists of improved access to limited soil resources (i.e., nutrients—especially phosphorus—and water), it is probable that, in our experiment, in contrast to that of Pellegrino et al. [15], the high level of soil fertility and the adequate water availability made mycorrhizal symbiosis superfluous. In such conditions, the only advantage of AM symbiosis that we observed occurred at the first cut (C1, at 76 DAS), at which time mycorrhiza inoculation resulted in increased root and shoot dry matter. According to Miller [25], during the early phase of the crop cycle, when roots are not yet well developed, AM symbiosis can make an important contribution to plant growth by enhancing nutrient acquisition.

In contrast to the WW conditions, under WS conditions, the AM symbiosis enhanced aboveground biomass production of berseem clover. This result is consistent with the findings of other authors who found, in pot-based studies, a beneficial effect of AM symbiosis under water restriction for both legumes [26,27] and non-legume plants [28,29]. The AM fungi contribution to plant drought tolerance is well documented (as reviewed by Augé [10]), but the underlying mechanisms are still unclear. These include hyphal water uptake [28], enhanced nutrient uptake —especially phosphorus— and enhanced plant osmotic adjustment [30,31], which favor plant water uptake from soil.

Numerous pot-based experiments have clearly indicated that AM symbiosis can greatly assist nodulation and N_2 fixation of legumes [32,33]. We observed significant increases in Ndfa (as both percentage of total N content and amount of N fixed on an area basis) only under WS conditions. This result may be attributable to the enhancement of nutrient uptake offered by AM fungi under WS conditions, which allowed the plants to satisfy both the nodule's and their own nutritional needs for growth. In this context, as is well known, an increase in phosphorus acquisition can play an important role in stimulating both nodulation and N_2 fixation [34]. Our research was performed in soil that is very rich in phosphorus; such conditions can strongly reduce the mycorrhizal benefit in terms of plant growth and phosphorus uptake [35]. However, in some cases, under WS conditions, AM plants are able to absorb more phosphorus than non-mycorrhizal plants, even when this element is highly available in the soil [9]; in fact, phosphorus effectively becomes less available with decreasing soil moisture [36,37], so that AM colonization and effective development of external mycelium can be extremely important for nutrient uptake in these situations. Moreover, other authors [27] highlighted that the advantages of the AM symbiosis in WS conditions are related to the defense of plant roots against the oxidative damage generated by drought, which can then ensure a certain level of protection against nodule senescence (with positive effects on the N_2 fixation process).

In WS treatments, the benefits of AM symbiosis were effective only at R28 (i.e., at 28 days after C2), whereas no differences were observed at R14 (i.e., 14 days after C2). This result can be explained by considering that, in legume forage crops subjected to cuts, regrowth ability is highly dependent on the plant's capacity to mobilize carbon and N reserves stored in roots and stubble and on photosynthetic activity of the residual leaf area [12,38,39]. Thus, during the early regrowth phase (from C2 to R14), carbon can represent a limiting factor, and in our experiment, it is likely that the high sink strength of the growing shoots limited the availability of resources for the AM fungus, counterbalancing the potential benefit offered by AM symbiosis; in the late regrowth phase (from R14 to R28), the photosynthetic apparatus reached an extent so great that plants were able to guarantee sufficient availability of the photosynthates to sustain both AM symbiosis and plant growth. The lower percentage of root colonization by AM fungi observed in R14 with respect to R28 seems to confirm the plant's difficulty in sustaining mycorrhizal symbionts during the early regrowth phase.

In conclusion, the results of our study confirm that, in field conditions, AM symbiosis can play an important role in the growth of plants under conditions of abiotic stresses such as WS. In fact, AM symbiosis in berseem clover subjected to WS mitigated the negative effects of drought on plant growth also through the stimulation of N_2 fixation. Moreover, AM symbiosis made an important contribution during the early growth phase, when nutrient uptake is limited by the relatively poorly developed plant root system.

Acknowledgments

The authors wish to thank Vincenzo Cannella and Francesco Cannella for their technical assistance in field management.

Author Contributions

Conceived and designed the experiments: SS GA DG PR. Performed the experiments: SS ASF PR. Analyzed the data: SS ASF DG PR. Contributed reagents/materials/analysis tools: GA ASF DG. Wrote the paper: SS GA DG PR.

References

1. Barea JM, Ferrol N, Azcón-Aguilar C, Azcón R (2008) Mycorrhizal symbioses. In: White PJ, Hammond JP, editors. The Ecophysiology of Plant-Phosphorus Interactions. Dordrecht: Springer Netherlands, Vol. 7. pp. 143–163.
2. Lambers H, Raven JA, Shaver GR, Smith SE (2008) Plant nutrient-acquisition strategies change with soil age. Trends Ecol Evol 23: 95–103.
3. Azcón R, Ruiz-Lozano JM, Rodríguez R (2001) Differential contribution of arbuscular mycorrhizal fungi to plant nitrate uptake (^{15}N) under increasing N supply to the soil. Can J Bot 79: 1175–1180.
4. Saia S, Benítez E, García-Garrido JM, Settanni L, Amato G, et al. (2014) The effect of arbuscular mycorrhizal fungi on total plant nitrogen uptake and nitrogen recovery from soil organic material. J Agric Sci In Press. doi: 10.1017/S002185961300004X. Published on line 07 February 2013
5. Jung SC, Martinez-Medina A, Lopez-Raez JA, Pozo MJ (2012) Mycorrhiza-induced resistance and priming of plant defenses. J Chem Ecol 38: 651–664.
6. Chalk P, Souza R, Urquiaga S, Alves B, Boddey R (2006) The role of arbuscular mycorrhiza in legume symbiotic performance. Soil Biol Biochem 38: 2944–2951.
7. Barea JM, Werner D, Azcón-Aguilar C, Azcón R (2005) Interactions of arbuscular mycorrhiza and nitrogen-fixing symbiosis in sustainable agriculture. In: Werner D, Newton WE, editors. Nitrogen Fixation in Agriculture, Forestry, Ecology, and the Environment. Springer Netherlands. pp. 199–222.
8. Jayne B, Quigley M (2013) Influence of arbuscular mycorrhiza on growth and reproductive response of plants under water deficit: a meta-analysis. Mycorrhiza In press doi: 10.1007/s00572-013-0515-x. Published on line 06 August 2013
9. Smith SE, Facelli E, Pope S, Andrew Smith F (2010) Plant performance in stressful environments: interpreting new and established knowledge of the roles of arbuscular mycorrhizas. Plant Soil 326: 3–20.
10. Augé RM (2001) Water relations, drought and vesicular-arbuscular mycorrhizal symbiosis. Mycorrhiza 11: 3–42.
11. Amato G, Giambalvo D, Ruisi P (2013) Cut and Post-Cut Herbage Management Affects Berseem Clover Seed Yield. Agron J 105: 1222.
12. Giambalvo D, Amato G, Stringi L (2011) Effects of Stubble Height and Cutting Frequency on Regrowth of Berseem Clover in a Mediterranean Semiarid Environment. Crop Sci 51: 1808.
13. Hasan N, Jain RK (1987) Parasitic nematodes and vesicular-arbuscular mycorrhizal (VAM) fungi associated with berseem (*Trifolium alexandrinum* L.) in Bundelkhand region. Indian J Nematol 17: 184–188.

14. Shokri S, Maadi B (2009) Effects of arbuscular mycorrhizal fungus on the mineral nutrition and yield of *Trifolium alexandrinum* plants under salinity stress. J Agron: 79–83.
15. Pellegrino E, Bedini S, Avio L, Bonari E, Giovannetti M (2011) Field inoculation effectiveness of native and exotic arbuscular mycorrhizal fungi in a Mediterranean agricultural soil. Soil Biol Biochem 43: 367–376.
16. Patterson NA, Chet I, Kapulnik Y (1990) Effect of mycorrhizal inoculation on nodule initiation, activity and contribution to legume productivity. Symbiosis 8: 9–20.
17. Gerdemann JW, Nicolson TH (1963) Spores of mycorrhizal Endogone species extracted from soil by wet sieving and decanting. Trans Br Mycol Soc 46: 235–244.
18. Udaiyan K, Manian S, Muthukumar T, Greep S (1995) Biostatic effect of fumigation and pesticide drenches on an endomycorrhizal-Rhizobium-legume tripartite association under field conditions. Biol Fertil soils 20: 275–283.
19. Schweiger P, Spliid N, Jakobsen I (2001) Fungicide application and phosphorus uptake by hyphae of arbuscular mycorrhizal fungi into field-grown peas. Soil Biol Biochem 33: 1231–1237.
20. Giambalvo D, Ruisi P, Di Miceli G, Frenda AS, Amato G (2011) Forage production, N uptake, N_2 fixation, and N recovery of berseem clover grown in pure stand and in mixture with annual ryegrass under different managements. Plant Soil 342: 379–391.
21. Giovannetti M, Mosse B (1980) An evaluation of techniques for measuring vesicular arbuscular mycorrhizal infection in roots. New Phytol 84: 489–500.
22. Phillips JM, Hayman DS (1970) Improved procedures for clearing roots and staining parasitic and vesicular-arbuscular mycorrhizal fungi for rapid assessment of infection. Trans Br Mycol Soc 55: 158–IN18.
23. Fried M, Middelboe V (1977) Measurement of amount of nitrogen fixed by a legume crop. Nitrogen Isot Tech 47: 713–715.
24. SAS Institute (2008) SAS/STAT® 9.2. User's Guide.
25. Miller MH (2000) Arbuscular mycorrhizae and the phosphorus nutrition of maize: A review of Guelph studies. Can J Plant Sci 80: 47–52.
26. Azcon R, El-Atrach F, Barea J (1988) Influence of mycorrhiza vs. soluble phosphate on growth, nodulation, and N_2 fixation (^{15}N) in alfalfa under different levels of water potential. Biol Fertil soils 7: 28–31.
27. Porcel R, Ruiz-Lozano JM (2004) Arbuscular mycorrhizal influence on leaf water potential, solute accumulation, and oxidative stress in soybean plants subjected to drought stress. J Exp Bot 55: 1743–1750.
28. Ruiz-Lozano JM, Azcon R (1995) Hyphal contribution to water uptake in mycorrhizal plants as affected by the fungal species and water status. Physiol Plant 95: 472–478.
29. Marulanda A, Azcon R, Ruiz-Lozano JM (2003) Contribution of six arbuscular mycorrhizal fungal isolates to water uptake by *Lactuca sativa* plants under drought stress. Physiol Plant 119: 526–533.
30. Goicoechea N, Szalai G, Antolín MC, Sánchez-Díaz M, Paldi E (1998) Influence of arbuscular mycorrhizae and *Rhizobium* on free polyamines and proline levels in water-stressed alfalfa. J Plant Physiol 153: 706–711.
31. Kubikova E, Moore JL, Ownley BH, Mullen MD, Auge RM (2001) Mycorrhizal impact on osmotic adjustment in Ocimum basilicum during a lethal drying episode. J Plant Physiol 158: 1227–1230.
32. Antunes PM, de Varennes A, Rajcan I, Goss MJ (2006) Accumulation of specific flavonoids in soybean (Glycine max (L.) Merr.) as a function of the early tripartite symbiosis with arbuscular mycorrhizal fungi and *Bradyrhizobium japonicum* (Kirchner) Jordan. Soil Biol Biochem 38: 1234–1242.
33. Olesniewicz KS, Thomas RB (1999) Effects of mycorrhizal colonization on biomass production and nitrogen fixation of black locust (*Robinia pseudoacacia*) seedlings grown under elevated atmospheric carbon dioxide. New Phytol 142: 133–140.
34. Ganry F, Diem HG, Dommergues YR (1982) Effect of inoculation with *Glomus mosseae* on nitrogen fixation by field grown soybeans. Plant Soil 68: 321–329.
35. Lekberg Y, Koide RT (2005) Is plant performance limited by abundance of arbuscular mycorrhizal fungi? A meta-analysis of studies published between 1988 and 2003. New Phytol 168: 189–204.
36. Viets FG (1972) Water deficits and nutrient availability. In: Kozlowski TT, editor. Water deficits and plant growth. Academic Press New York, NY, Vol. 3. pp. 217–239.
37. Gahoonia TS, Raza S, Nielsen NE (1994) Phosphorus depletion in the rhizosphere as influenced by soil moisture. Plant Soil 159: 213–218.
38. Avice J, Lemaire G, Ourry A, Boucaud J (1997) Effects of the previous shoot removal frequency on subsequent shoot regrowth in two *Medicago sativa* L. cultivars. Plant Soil 188: 189–198.
39. Meuriot F, Avice J-C, Simon J-C, Laine P, Decau M-L, et al. (2004) Influence of initial organic N reserves and residual leaf area on growth, N uptake, N partitioning and N storage in alfalfa (*Medicago sativa*) during post-cutting regrowth. Ann Bot 94: 311–321.

4

A Comparison of the Functional Traits of Common Reed (*Phragmites australis*) in Northern China: Aquatic vs. Terrestrial Ecotypes

Liping Li[1,2], Wenxuan Han[3], Niels Thevs[4], Xiuhong Jia[5], Chengjun Ji[6], Dongmei Jin[6], Ping He[7], Armin O. Schmitt[2], Giuseppe Tommaso Cirella[2], Stefan Zerbe[2]*

1 Key Laboratory of Bioactive Substances and Resources Utilization of Chinese Herbal Medicine (Peking Union Medical College), Ministry of Education, Institute of Medicinal Plant Development, Chinese Academy of Medical Sciences, Peking Union Medical College, Beijing, China, **2** Faculty of Science and Technology, Free University of Bozen-Bolzano, Bolzano, Italy, **3** Key Laboratory of Plant-Soil Interactions, Ministry of Education, College of Resources and Environmental Sciences, China Agricultural University, Beijing, China, **4** Institute of Botany and Landscape Ecology, University of Greifswald, Greifswald, Germany, **5** College of Horticulture and Forestry, Huazhong Agricultural University, Wuhan, China, **6** Department of Ecology, College of Urban and Environmental Sciences, Key Laboratory for Earth Surface Processes of the Ministry of Education, Peking University, Beijing, China, **7** State Environmental Protection Key Laboratory of Regional Eco-Process and Function Assessment, Chinese Research Academy of Environmental Sciences, Beijing, China

Abstract

Common reed (*Phragmites australis* (Cav.) Trin. ex Steud.) is distributed widely throughout the world with various ecotypes. This research compares the functional traits and biomass allocation patterns of two contrasting reed ecotypes. Twelve pairs of aquatic and terrestrial reed samples were collected in northern China. Significant differences in functional traits between the two reed ecotypes were observed, while biomass allocation patterns of reed organs did not differ significantly except for at the root. The dry matter content (DMC) in the whole of the reed plant, leaf, root, and rhizome was higher; while the specific leaf area (SLA) and specific root length (SRL) were lower in terrestrial versus aquatic reed. The biomass allocation in organs of the two forms of reed was isometric, only root in the terrestrial habitat increased faster with an increase in the whole plant biomass. It can be affirmed that aquatic and terrestrial reed that adapt to different environments generally has distinct leaf and root functional traits but isometric biomass allocation patterns. This suggests different resource acquisition strategies: (1) aquatic reed grows faster with high SLA and SRL and is more responsive to the environment, while (2) terrestrial reed with high DMC grows slower and is less responsive to the adverse environment (e.g. dry soil conditions).

Editor: Manuel Reigosa, University of Vigo, Spain

Funding: This study was funded by the Kurt Eberhard Bode Foundation, Germany (within the Stifterverband für die Deutsche Wissenschaft) and the National Natural Science Foundation of China (41173083). The funders had no role in study design, data collection and analysis, decision to publish, or preparation of the manuscript.

Competing Interests: The authors have declared that no competing interests exist.

* E-mail: Stefan.Zerbe@unibz.it

Introduction

It is of significant interest, within the field of plant ecology, the study of the relationships between plant functional traits [1]. As two of the most active metabolic organs, root and leaf are very important for plant functioning [2]. Their traits are associated closely with carbon and nutrient uptake and plant growth rates [3]. Compared to slow-growing species, plants with high growth rates usually have large specific root length (SRL), small root diameter, large specific leaf area (SLA), low dry matter content (DMC), low tissue longevity, low nutrient use efficiency and, generally, are better adapted to nutrient rich environments [4,5]. Leaf and root traits are functionally associated [6], as well as associated with other organs (e.g. stem and rhizome) [7–9]. Trait analyses that integrate leaf, stem and root across gradients of soil fertility and water availability are baseline for future research [8,10].

Biomass allocation patterns in different plant organs (e.g. flower, leaf, stem, root, and rhizome) can also reflect trade-offs between functional traits. Changes in plant biomass allocation patterns were considered phenotypic plasticity or allometric strategies [11]. Some studies found stable, i.e. isometric biomass allocation patterns [12–16], while others concluded that the allocation patterns were not stable and would change with plant growth and size (i.e. optimal partitioning) [17,18]. Meta-analysis showed that plants would alter the organ morphology rather than adjust the biomass allocations [9]. Large scale research conducted in the Tibetan grasslands of China found the isometric biomass allocation patterns [14,15]; moreover, the isometric patterns over time have been found globally throughout related forest ecosystems [16].

Plant functional traits were mainly studied in terrestrial plants (e.g. [19–21]); however, terrestrial and aquatic plants grow in completely differing environments and may face totally different stresses. For example, terrestrial plants are stressed more frequently by water shortage, while aquatic plants may often suffer from long-term oxygen deprivation [22]. The contrasting environmental stresses on aquatic and terrestrial plants supposedly can cause the differing in functional traits. Thus, the comparison of functional traits between plants in terrestrial and aquatic habitats may lead to a better understanding of the trade-offs between the two. Former studies found that there were big phenotypic differences between *Polygonum lapathifolium* when this plant was planted in dry and flooded soils, respectively [23].

Common reed (*Phragmites australis* (Cav.) Trin. ex Steud.) has wide distribution ranges throughout the world with regard to ground water level, nutrient supply, soil-water salinity, and land-use [24–26]. *P. australis* could also be a bio-indicator since its morphology changes with the growing environment [27]. It is an ideal plant for comparing functional traits in different environments. With twelve pairs of aquatic and terrestrial samples taken from the Wuliangsuhai Lake area and the wetlands in the Zhangye city in China, functional traits and biomass allocation patterns between the two ecotypes of *P. australis*, aquatic and terrestrial, were compared. The aim is to test whether there are differences in the functional traits (DMC of different organs, SLA, SRL, mean root diameter, and mean root area of unit mass) between aquatic and terrestrial reed. A further investigation is to find out if the organ biomass allocations of aquatic and terrestrial reed follow the optimal partitioning strategy or if it is isometric.

Materials and Methods

Study Sites

The study sites were located in northern China near the Wuliangsuhai Lake of Inner Mongolia and the wetlands in the Zhangye city, Gansu Province. Wuliangsuhai Lake is located around N41° and E109° and Zhangye N39° and E100°. The mean annual temperature in both areas is 7°C, while the mean annual precipitation is 216 mm in the Wuliangsuhai Lake area and 129 mm in the Zhangye city [28,29]. Wuliangsuhai Lake is an important wetland area in the Hetao Irrigation District watershed which flows into the Yellow River. Approximately 4.5~7.0 billion m^3 of annual water, within the forty year block of the 1960s to 1990s [30], from Yellow River goes into the Hetao Irrigation District watershed area; this traditionally has satisfied the water demand of crops and has kept the water level of Wuliansuhai Lake relatively stable. Wetlands in Zhangye are closely connected with the Hei River.

At both sites *P. australis* is widely spread as the dominant emergent species. Near the water habitat, *P. australis* grows with relatively lower stem density and stem height along extreme arid land. The locations of the two sites in China were mapped and the environmental conditions were recorded in reference [31].

Field Sampling and On-site Analysis

In August 2011, the reed plants were sampled in pairs, at the same spot, one in water habitat and the other on land. At this time of the year reed plants are at their peak biomass levels. Sampling sites were chosen in terms of representative reed communities with little competition from other species. We defined *P. australis* sampling within the two habitats as aquatic reed and terrestrial reed, respectively. Covering a wide environmental range, five pairs were sampled in the Wuliangsuhai Lake area and seven pairs in the wetlands around the Zhangye city (Table 1). The sampling work was supported by the local government that was Hetao Water Affairs Group, Inner Mongolia and Zhangye Environment Protection Bureau, Gansu. Mature and well-grown individuals were sampled. For water habitat, we chose *P. australis* individuals that had been established in a water body of more than 10 cm in depth, while for terrestrial habitat reed, *P. australis* individuals were within 10 m in distance from the water body source. The water habitat was saturated while the terrestrial one was not. Moreover, for terrestrial habitat the soil water content was very low due to a large number of individuals living along a bankside water source. In both sampled regions, potential evapotranspiration (PET) was very high [28,29]. Belowground parts were sampled 30 cm below the sediment or soil surface, with around a 10 cm radius from the main stem circle-center. Each individual was dissected into leaf, stem, flower, fine root (termed root hereafter), and rhizome. The fresh biomass of the five organs was weighed in the field immediately after sampling. Subsequently, a part of the five organs was brought to the laboratory for further analysis. In addition, sediment and water in the aquatic habitat and soil in the terrestrial habitat were collected and analyzed near the plant samples, in order to supply a background setting of the environmental conditions for the sampling sites. At least three samples were collected for each and mixed together for the further analysis. Both water electric conductivity (EC, ms cm^{-1}) and pH were measured on site with a Multi340i handheld meter (WTW, Weilheim, Germany).

Lab Analysis

A total ranging from 10–30 leaves and 0.5 g wet root (1–5 replicates) were scanned, with a Canon scanner (4400F), per sample. Then SLA, SRL, mean root diameter (R_{diam}) and mean root area of unit mass (R_{area}) were determined with WinFOLIA and WinRHIZO (Régent, Quebec, Canada). All samples were oven-dried at 60°C for 72 h to get the value of DMC. Sediment EC and pH were determined with a DDS-11A digital display conductivity meter (CSDIHO, Shanghai, China) and PH-3C (INESA, Shanghai, China). Total nitrogen (TN, mg g^{-1}) in the sediment and water was measured according to the Alkaline potassium persulfate digestion-UV spectro-photo-metric method [32]. Total phosphorous (TP, mg g^{-1}) in the sediment and water was measured following the molybdate stannous chloride method [32].

Data Analysis

We compared the functional traits of *P. australis* from both study sites and found no significant differences. Hence, we pooled together the data from the two study sites for further analyses. Since the sampling sites of the terrestrial and the aquatic habitat were in close vicinity (less than 10 m) we expected no other variations due to environmental conditions and therefore used the paired t-test to test for the differences in the traits of *P. australis* in the two habitats, as well as for characteristic differences between

Table 1. Characteristics of sediment, soil, and water in terrestrial and aquatic habitats.

	Aquatic habitat	Terrestrial habitat
Sediment/soil EC (ms cm^{-1})	$1.09^a \pm 0.63$	$2.22^b \pm 1.63$
Sediment/soil pH	$7.45^a \pm 0.18$	$7.49^a \pm 0.32$
Sediment/soil TN (mg g^{-1})	$1.13^a \pm 0.83$	$0.56^a \pm 0.36$
Sediment/soil TP (mg g^{-1})	$0.69^a \pm 0.25$	$0.52^a \pm 0.16$
Water EC (ms cm^{-1})	2.83 ± 4.54	–
Water pH	8.28 ± 0.75	–
Water TN (mg g^{-1})	4.39 ± 5.03	–
Water TP (mg g^{-1})	0.026 ± 0.068	–

Shown are the geometric means and the standard deviations (SD). The different superscripts indicate significant differences in the means (paired t-test, $P<0.05$).

the two habitats. One-sided paired t-tests were used for DMC, SLA, SRL, R_{area} and R_{diam}. DMC and R_{diam} were expected to be higher in terrestrial reed and SLA, SRL, and R_{area} to be higher in aquatic reed. Two-sided paired t-tests were used to test for any differences in biomass allocation in *P. australis* organs (i.e. flower, leaf, stem, root, and rhizome), and Root:Shoot ratio (RS). In this case, the proportion of root to the total biomass was also defined as Root Ratio (RR). The results of ANOVA for these characteristics of the two reed ecotypes were presented in Table S1.

Standardized major axis (SMA) analysis was used to detect relationships between biomass allocations in different organs. Slopes of SMA were compared between the two reed ecotypes [21]. We considered more the trait differences of *P. australis* due to the two types of habitats (mostly moisture differences) and ignored the differences of other environmental variables, for example, soil/sediment nutrient contents.

All analyses were carried out in R 2.12.2 [33]. The SMA analyses were performed with the 'smatr' package [34].

Results

Dry Matter Content (DMC) of Aquatic and Terrestrial Reed

The mean DMC (mean ± standard deviation, SD) ranged from 15.7%±6.8% in root to 42.4%±7.5% in stem of aquatic reed, and from 25.1%±6.5% in root to 45.8%±2.8% in leaf of terrestrial reed (Figure 1). The DMC was higher in terrestrial than in aquatic reed. This held true for both the whole plant and its parts (i.e. belowground part, leaf, root, and rhizome, $P<0.05$, Figure 1a,b,e,g,h). The DMC of the aboveground part was slightly higher for terrestrial than for aquatic reed ($P<0.1$, Figure 1c). There were no significant differences in DMC of flower and stem between the two forms ($P>0.1$, Figure 1d,f).

Leaf and Root Traits of Aquatic and Terrestrial Reed

The mean SLA (mean ± SD) were 14.9±2.0 and 11.7±1.8 m^2 kg^{-1} and the mean SRL for aquatic and terrestrial reed were 94.9±29.2 and 40.4±18.1 m g^{-1}, respectively. The SLA and SRL were both higher for aquatic than for terrestrial reed ($P<$ 0.05, Figure 2a,b). The mean root area of unit mass (R_{area}) was greater but the mean root diameter (R_{diam}) was smaller for aquatic than for terrestrial reed ($R_{area}=0.09\pm0.02$ and 0.05 ± 0.01 m^2 g^{-1}; and $R_{diam}=0.32\pm0.07$ and 0.42 ± 0.14 mm for aquatic and terrestrial reed, respectively, $P<0.05$, Figure 2c,d).

Biomass Allocations of Aquatic and Terrestrial Reed

The biomass proportions (dry weight) of the aboveground part, flower, leaf, stem, and rhizome did not differ significantly between the aquatic and terrestrial ecotypes (70% and 76% for aboveground biomass proportions, 3.9% and 8.9% for flower, 29.5% and 36.2% for leaf, 39.5% and 32.3% for stem, 21.4% and 21.9% for rhizome, for aquatic and terrestrial reed, respectively, all $P>$ 0.05, Table 2). The mean RS did not differ significantly between aquatic and terrestrial reed (0.45 and 0.32, $P>0.05$, Table 2 and Figure 2e); however, the biomass proportion of root was significantly higher for aquatic than for terrestrial reed (5.5% and 0.7%, $P<0.05$, Table 2 and Figure 2f).

There was positive correlation between biomass of aboveground and belowground for both terrestrial and aquatic reed ($P<0.05$). The slopes for the two regressions were not significantly different and neither were significantly different from 1.0 (SMA regression, $P>0.05$, Table 3 and Figure 3a). The positive correlation also existed between leaf/root biomass and the whole plant biomass in which the SMA slopes for the two ecotypes were not significantly different (Table 3 and Figure 3b,c). The slope of the root against whole plant biomass of the aquatic form did not differ significantly

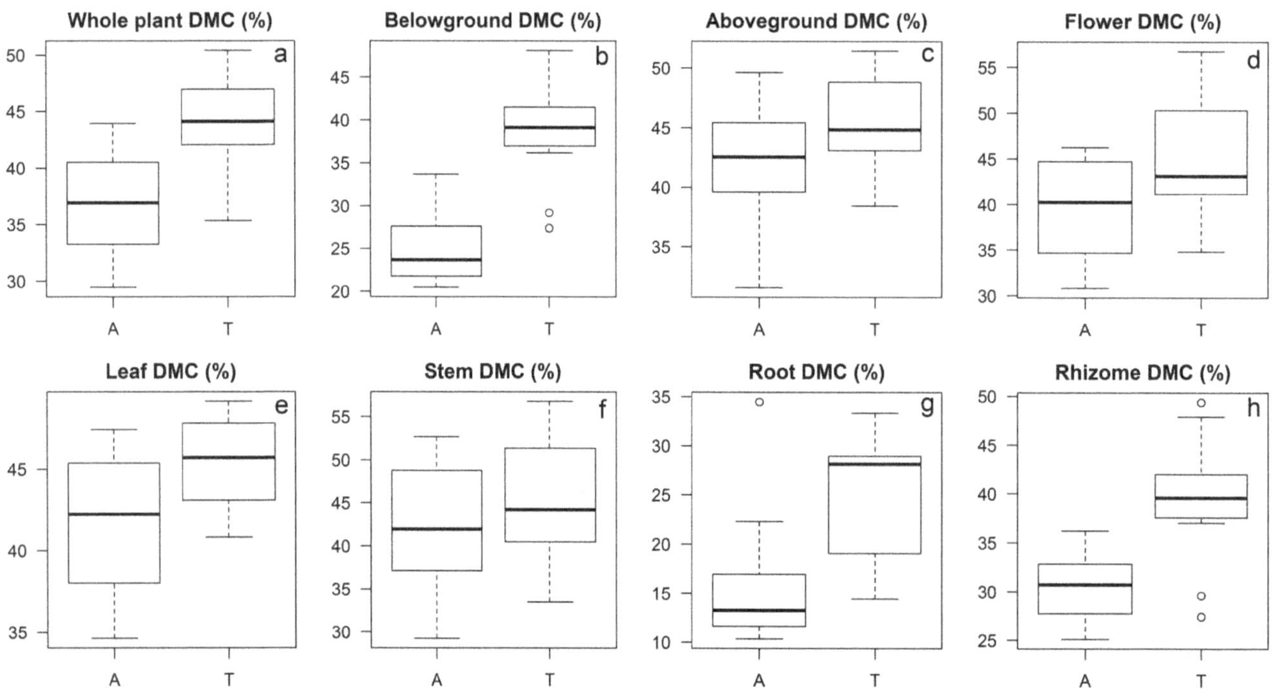

Figure 1. The dry matter content (DMC) of aquatic (A) and terrestrial (T) reed for the (a) whole plant, (b) belowground part, (c) aboveground part and (d–h) separate organs (the solid lines in the boxes indicate median values, the upper and lower ranges of the boxes show the third and first quartiles whereas upper and lower lines out of the boxes indicate the maximum and minimum values, extreme values are shown as dots, analogously for Figure 2).

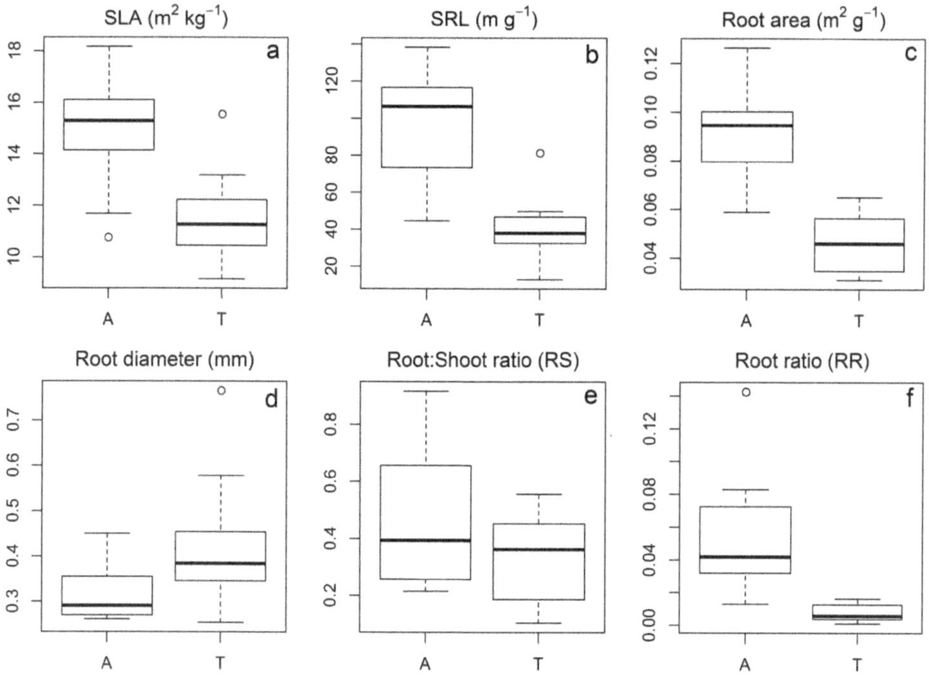

Figure 2. The functional traits and biomass allocations of aquatic (A) and terrestrial (T) reed (a, specific leaf area (SLA); b, specific root length (SRL); c, mean root area of unit mass (R_{area}); d, mean root diameter (R_{diam}); e, Root:Shoot ratio (RS); f, root ratio (RR)).

from 1.0 while the corresponding slope of the terrestrial form was significantly greater than 1.0 (Table 3 and Figure 3b). The slope of the leaf against whole plant biomass of both forms did not differ significantly from 1.0 (Table 3 and Figure 3c).

The ratio of root vs. rhizome biomass was greater for aquatic than for terrestrial reed (0.22 and 0.02, $P<0.05$, Table 2), illustrating that one unit of mass of rhizome supported more root

Table 2. Biomass allocations of aquatic and terrestrial reed ecotypes.

	(%)	Flower	Leaf	Stem	Root*	Rhizome
Aquatic ecotype	Mean	3.9[a]	29.5[a]	39.5[a]	5.5[a]	21.4[a]
	SD	2.9	9.5	9.6	3.5	7.0
Terrestrial ecotype	Mean	8.9[a]	36.2[a]	32.3[a]	0.7[b]	21.9[a]
	SD	5.9	8.1	8.5	0.6	8.0
		RS	Leaf: Stem	Root: Rhizome	Root: Leaf	Stem: Rhizome
Aquatic ecotype	Mean	0.45[a]	0.83[a]	0.26[a]	0.22[a]	2.09[a]
	SD	0.24	0.47	0.13	0.18	0.95
Terrestrial ecotype	Mean	0.33[a]	1.28[a]	0.04[b]	0.02[b]	1.84[a]
	SD	0.15	0.79	0.02	0.02	1.22

Leaf:Stem indicates the biomass ratio of leaf and stem, analogously for the Root:Rhizome, Root:Leaf, and Stem:Rhizome. The different superscripts indicate significant differences in the means (paired t-test, $P<0.05$). SD, standard deviation.
*The proportion of root biomass is also defined as root ratio (RR).

in aquatic than in terrestrial reed. The ratio of leaf vs. stem biomass was slightly higher for terrestrial than for aquatic reed ($P<0.1$, Table 2) indicating that one unit of mass of stem appeared to support slightly greater leaf biomass in terrestrial than in aquatic reed. However, the slopes of the SMA regressions were not significantly different for both leaf vs. stem and root vs. rhizome from the two identified forms (Table 3 and Figure 3d,e). In addition, the slopes of the two regressions did not differ from 1.0, possibly suggesting the isometric scaling of leaf vs. stem biomass and root vs. rhizome biomass in the two reed ecotypes.

Discussion

Dry Matter Content

When dealing with plant traits, DMC is one of the most important indicators for resource use [35]. Leaf DMC is often negatively correlated with leaf growth [36]. An example of this relates to annual plants with higher growth rates having lower leaf DMC than perennial herb plants [37]. The organ DMC of terrestrial reed is significantly higher than aquatic reed in leaf, root, and rhizome, suggesting that aquatic reed with low DMC has a higher growth rate than terrestrial reed. In contrast, terrestrial reed, developed on land, grows slower and most likely resists better environmental stresses (e.g. frequent water shortage) due to the effect of high DMC.

Leaf and Root Traits

Plants with high growth rates have high SLA, for example, annual plants have higher SLA than perennial herb plants [37]. Relatively low SLA under water shortage can be beneficial because this means that water use efficiency is enhanced [38]. It was found that plant SLA is lower in desert steppe (with lower water availability) than in meadow steppe and typical steppe [5]. Terrestrial reed has low SLA, i.e. it is adapted to adverse environments; for example it grows slower to cope with low water

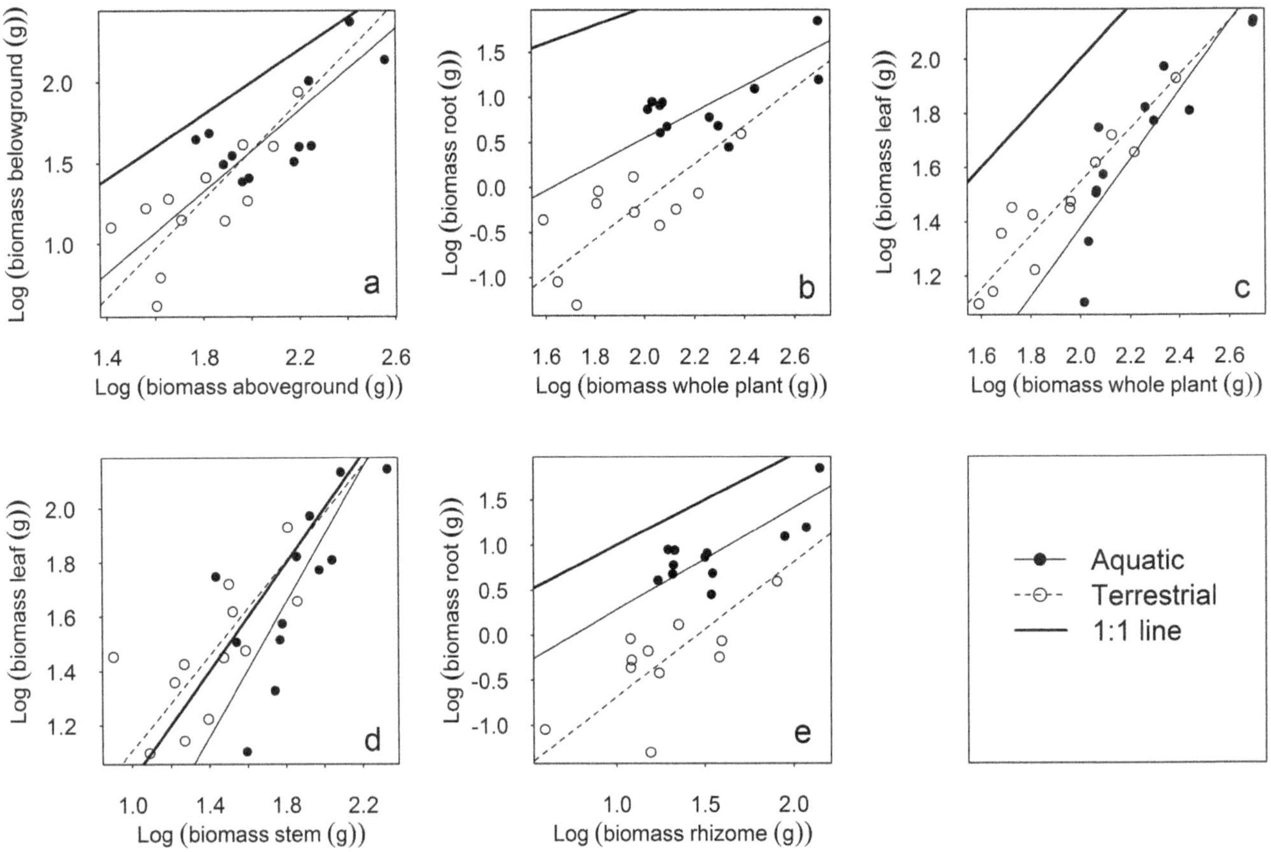

Figure 3. The biomass allocation patterns between organs and the belowground/whole plant of *P. australis* (a, aboveground and belowground biomass; b, root and whole plant biomass; c, leaf and whole plant biomass; d, leaf and stem biomass; e, root and rhizome biomass; with data log_{10}-transformed).

Table 3. The relationships of biomass allocation in different *P. australis* organs in the aquatic and terrestrial ecotypes (data were log_{10}-transformed).

	Ecotype	Slope	Lower	Upper	r^2	*P*
Biomass below- and aboveground	Aquatic	1.26	0.78	2.04	0.49	0.01
	Terrestrial	1.52	0.97	2.37	0.58	<0.01
Biomass root and whole plant	Aquatic	1.45	0.86	2.46	0.39	0.03
	Terrestrial	2.10	**1.24**	**3.55**	0.46	0.02
Biomass leaf and whole plant	Aquatic	1.26	0.91	1.77	0.77	<0.01
	Terrestrial	0.99	0.76	1.28	0.86	<0.01
Biomass leaf and stem	Aquatic	1.24	0.78	1.99	0.52	<0.01
	Terrestrial	0.88	0.54	1.43	0.47	0.01
Biomass root and rhizome	Aquatic	1.13	0.72	1.76	0.57	<0.01
	Terrestrial	1.50	0.89	2.53	0.47	0.02

Lower and upper indicate the 95% confidence interval (CI) of the SMA regression slope.

availability and soil salinization problems (terrestrial habitats have higher soil EC than that of sediment in aquatic habitat, Table 1). In contrast, aquatic reed that has high SLA which grows and expands faster. Low light availability caused by high stem density in aquatic habitat could also lead to high SLA of aquatic reed. Finally, aquatic reed that has high SRL, high R_{area} and low R_{diam}, suggests a higher growth rate and nutrient use efficiency.

Species with high maximal relative growth rates occupy fertile habitat and are more sensitive to environmental nutrient changes [3]. It was found that competition decreased biomass of *Myriophyllum spicatum* in the fertile sediment more than in the infertile sediment [39]. But the potentially slow-growing species exhibited a higher phenotypic plasticity than the potentially fast-growing species when they were planted in soils with different nitrate contents [40], i.e. potentially slow-growing species changed the morphology more than fast-growing species to adapt to the change in soil nutrient availability. In this study, the aquatic reed is supposed to be more sensitive to environmental changes than terrestrial reed considering that the variations of both SLA and SRL are higher for aquatic than for terrestrial reed (Figure 2). This could be due to the sediment TN and TP (in which varying water pollution levels exist) of aquatic reed was highly variable, compared to terrestrial reed in which samples grew primarily on extremely arid and salinized soils where no other primary species grew.

Biomass Allocation

Biomass allocation has profound implications for plant growth [17]. According to the hypothesis of optimal partitioning, decreased nutrient supply can increase biomass allocation to roots while decreased irradiance can increase biomass allocation to leaves [17]. The hypothesis suggests that plant biomass allocation (e.g. RS) is determined by factors that influence environmental nutrient, water and light availability and the allocation patterns possibly can change with varying environmental factors. Plants with low RS ratio are usually expected to grow in environments that are rich in nutrient and water availability [11] (i.e. RS is higher in nutrient- or water-stressed than in the more favorable environments for terrestrial plants). The hypothesis of isometric allocation suggests stable biomass allocation patterns that are insensitive to environmental conditions [13]. Isometric scaling was found at different scales and in various vegetation types, such as between aboveground and belowground biomass in China's grasslands [14,15], and over time in temporal forest ecosystems [16]. With a meta-analysis, it was found that woody and herbaceous taxa had different nutrient contents, while the statistical scaling did not differ both within and across organs [1]. In 2000, Müller et al. investigated the biomass allocations in 27 clonal species via different soil nutrient availabilities and found a similar scaling law in 21 species, albeit the scaling exponents were not equal to one [11]. Although there were some significant differences, (i.e. saturated and extreme arid of the two habitats) (Table 1), we did not find significant differences in RS between aquatic and terrestrial reed. The biomass allocation in different reed organs, except for root, did not differ significantly between the two forms of reed, indicating a generally similar biomass allocation pattern in the two reed ecotypes.

Conclusion

With twelve pairs of samples, this study compared the functional traits and biomass allocation patterns of *P. australis* in aquatic and terrestrial habitats. We conclude that aquatic and terrestrial reed, established in contrasting environments, have generally similar biomass allocation patterns but distinct leaf and root functional traits which suggests different resource acquisition strategies, i.e. aquatic reed grows faster due to high SLA and SRL and is more responsive to the environment, while terrestrial reed grows slower and could resist more adverse environment due to the high DMC and may be less responsive to the environment at large.

The findings of this study could have implications for better understanding trade-offs of plant functional traits in changing climatic conditions. Climate change will become increasingly pronounced [41], especially for high latitude areas, like in northern China, where there has been a strong warming in recent years [42]. Water plants are more sensitive to the impacts of climatic change than terrestrial plants; recent research suggested a vast loss of species and community restructuring of wetland regions in the last four decades [43]. To better understand this impact, differences in the responses of functional traits of water and terrestrial plants to environmental change are important for predictive and scenario-based learning. *P. australis* dominated wetlands may have a considerable effect on and be affected by climatic change as it is species that is extensively found throughout the planet [44]. *P. australis* could be a model plant to exemplify wetland ecosystem changes in a dynamic global climate. Further studies, including genetics, are also needed to clearly understand different syndromes that correspond to the relating ecotypes.

Supporting Information

Table S1 The ANOVA results of characteristics of environmental variables, dry matter content, functional traits, biomass allocations, and allometry of *P. australis* in aquatic and terrestrial habitat.

Acknowledgments

We thank Zhong Zhao, Yafei Yang, and Hongwei Wang from Hetao Water Affairs Group, Inner Mongolia for field work support in the Wuliangsuhai Lake area; and Bingtang Guo as well as Zongqi Che from the Zhangye Environment Protection Bureau for field work support in the wetlands of Zhangye city. Extended appreciation also goes out to Chengyong Yu, Junliang Meng, and Jing Liu for driving during our field work expeditions and special thanks to Fei Li from Water and Soil Environment Monitoring Center, Bayannaoer Hydro Research Institute, Inner Mongolia for the measurement of water, sediment and soil samples.

We thank Yining Liu, Elizabeth George, Xiaofeng Wang, Christian Ceccon, and Jie Xu for the help of sampling, data analyses and comments on an early version of the manuscript. We also would like to thank Manuel Reigosa and two anonymous reviewers for the valuable comments, and give special thanks to Yongke Li and Jacques Haury for their guidance.

Author Contributions

Conceived and designed the experiments: LL SZ WH NT. Performed the experiments: LL WH XJ CJ DJ PH. Analyzed the data: LL AOS. Contributed reagents/materials/analysis tools: SZ NT WH XJ CJ DJ. Wrote the paper: LL WH SZ NT AOS XJ CJ DJ PH GTC.

References

1. Kerkhoff AJ, Fagan WF, Elser JJ, Enquist BJ (2006) Phylogenetic and growth form variation in the scaling of nitrogen and phosphorus in the seed plants. The American Naturalist 168: E103–E122.
2. Eissenstat DM, Yanai RD (1997) The ecology of root lifespan. Advances in Ecological Research 27: 1–60.
3. Grime JP (1977) Evidence for the existence of three primary strategies in plants and its relevance to ecological and evolutionary theory. The American Naturalist 111: 1169–1194.
4. Baruch Z (2011) Leaf trait variation of a dominant neotropical savanna tree across rainfall and fertility gradients. Acta Oecologica 37: 455–461.
5. Zhou P, Geng Y, Ma WH, He JS (2010) Linkages of functional traits among plant organs in the dominant species of the Inner Mongolia grassland, China. Chinese Journal of Plant Ecology 34: 7–16.
6. Birouste M, Kazakou E, Blanchard A, Roumet C (2012) Plant traits and decomposition: are the relationships for roots comparable to those for leaves? Annals of Botany 109: 463–472.
7. Fortunel C, Fine PVA, Baraloto C (2012) Leaf, stem and root tissue strategies across 758 Neotropical tree species. Functional Ecology 26: 1153–1161.
8. Freschet GT, Cornelissen JHC, van Logtestijn RSP, Aerts R (2010) Evidence of the 'plant economics spectrum' in a subarctic flora. Journal of Ecology 98: 362–373.
9. Poorter H, Niklas KJ, Reich PB, Oleksyn J, Poot P, et al. (2011) Biomass allocation to leaves, stems and roots: meta-analyses of interspecific variation and environmental control. New Phytologist 193: 30–50.
10. Baraloto C, Timothy Paine CE, Poorter L, Beauchene J, Bonal D, et al. (2010) Decoupled leaf and stem economics in rain forest trees. Ecology Letters 13: 1338–1347.
11. Müller I, Schmid B, Weiner J (2000) The effect of nutrient availability on biomass allocation patterns in 27 species of herbaceous plants. Perspectives in Plant Ecology, Evolution and Systematics 3: 115–127.
12. Cheng DL, Wang GX, Li T, Tang QL, Gong CM (2007) Relationships among the stem, aboveground and total biomass across Chinese forests. Journal of Integrative Plant Biology 49: 1573–1579.
13. Enquist BJ, Niklas KJ (2002) Global allocation rules for patterns of biomass partitioning in seed plants. Science 295: 1517–1520.
14. Yang YH, Fang JY, Ji CJ, Han WX (2009) Above- and belowground biomass allocation in Tibetan grasslands. Journal of Vegetation Science 20: 177–184.

15. Yang YH, Fang JY, Ma WH, Guo DL, Mohammat A (2010) Large-scale pattern of biomass partitioning across China's grasslands. Global Ecology and Biogeography 19: 268–277.
16. Yang YH, Luo YQ (2011) Isometric biomass partitioning pattern in forest ecosystems: evidence from temporal observations during stand development. Journal of Ecology 99: 431–437.
17. Shipley B, Meziane D (2002) The balanced-growth hypothesis and the allometry of leaf and root biomass allocation. Functional Ecology 16: 326–331.
18. Weiner J (2004) Allocation, plasticity and allometry in plants. Perspectives in Plant Ecology, Evolution and Systematics 6: 207–215.
19. Han WX, Fang JY, Reich PB, Woodward FI, Wang Z (2011) Biogeography and variability of eleven mineral elements in plant leaves across gradients of climate, soil and plant functional type in China. Ecology Letters 14: 788–796.
20. Reich PB, Oleksyn J (2004) Global patterns of plant leaf N and P in relation to temperature and latitude. Proceedings of the National Academy of Sciences, USA 101: 11001–11006.
21. Wright IJ, Reich PB, Westoby M, Ackerly DD, Baruch Z, et al. (2004) The worldwide leaf economics spectrum. Nature 428: 821–827.
22. Willby NJ, Abernethy VJ, Demars BOL (2000) Attribute-based classification of European hydrophytes and its relationship to habitat utilization. Freshwater Biology 43: 43–74.
23. Sultan SE (2000) Phenotypic plasticity for plant development, function and life history. Trends in Plant Science 5: 537–542.
24. Patuzzi F, Mimmo T, Cesco S, Gasparella A, Baratieri M (2013) Common reeds (*Phragmites australis*) as sustainable energy source: experimental and modelling analysis of torrefaction and pyrolysis processes. GCB Bioenergy 5: 367–374.
25. Thevs N, Zerbe S, Gahlert F, Mijit M, Succow M (2007) Productivity of reed (*Phragmites australis* Trin. ex Steud.) in continental-arid NW China in relation to soil, groundwater, and land-use. Journal of Applied Botany and Food Quality 81: 62–68.
26. Zerbe S, Thevs N (2011) Restoring Central Asian floodplain ecosystems as natural capital and cultural heritage in a continental desert environment. In: Hong S-K, Wu J, Kim J-E, Nakagoshi N, editors. Landscape Ecology in Asian Cultures, Ecological Research Monographs: Springer.
27. Engloner AI (2004) Annual growth dynamics and morphological differences of reed (*Phragmites australis* [Cav.] Trin. ex Steudel) in relation to water supply. Flora - Morphology, Distribution, Functional Ecology of Plants 199: 256–262.
28. Ding HW, Yao JL, He JH (2009) Environmental isotope characteristics and groundwater recharge in groundwater level rise area in Zhangye City. Arid Land Geography 32: 1–8.
29. Duan XN, Wang XK, Feng ZZ, Ouyang ZY (2004) Study on the photosynthetic and transpiration properties of wild *Phragmites australis* in the Wuliangsuhai Lake, Inner Mongolia. Arid Land Geography 27: 637–641.
30. Jiang XH, Qian YP, Jiang XH, He W, Zhou JB (2001) Research on real-time monitoring of water diverting quantity in Inner Mongolia loop area of Yellow River irrigation district. Northwest Water Resources & Water Engineering 12: 24–27.
31. Li LP, Zerbe S, Han WX, Thevs N, Li WP, et al. (2014) Nitrogen and phosphorus stoichiometry of common reed (*Phragmites australis*) and its relationship to nutrient availability in northern China. Aquatic Botany 112: 84–90.
32. He JS, Fang JY, Wang Z, Guo DL, Flynn DFB, et al. (2006) Stoichiometry and large-scale patterns of leaf carbon and nitrogen in the grassland biomes of China. Oecologia 149: 115–122.
33. R Development Core Team (2011) R: a language and environment for statistical computing. Vienna, Austria: R Foundation for Statistical Computing.
34. Warton DI, Wright IJ, Falster DS, Westoby M (2006) Bivariate line-fitting methods for allometry. Biological Reviews 81: 259–291.
35. Garnier E, Laurent G, Bellmann A, Debain S, Berthelier P, et al. (2001) Consistency of species ranking based on functional leaf traits. New Phytologist 152: 69–83.
36. Arredondo JT, Schnyder H (2003) Components of leaf elongation rate and their relationship to specific leaf area in contrasting grasses. New Phytologist 158: 305–314.
37. Zhao HY, Li YL, Wang XY, Mao W, Zhao XY, et al. (2010) Variations in leaf traits of 52 plants in Horqin sand land. Journal of Desert Research 30: 1292–1298.
38. Wright IJ, Reich PB, Westoby M (2001) Strategy shifts in leaf physiology, structure and nutrient content between species of high- and low-rainfall and high- and low-nutrient habitats. Functional Ecology 15: 423–434.
39. Wang J, Yu D, Wang Q (2008) Growth, biomass allocation, and autofragmentation responses to root and shoot competition in *Myriophyllum spicatum* as a function of sediment nutrient supply. Aquatic Botany 89: 357–364.
40. Van de Vijver CADM, Boot RGA, Poorter H, Lambers H (1993) Phenotypic plasticity in response to nitrate supply of an inherently fast-growing species from a fertile habitat and an inherently slow-growing species from an infertile habitat. Oecologia 96: 548–554.
41. McCarthy JJ, Canziani OF, Leary NA, Dokken DJ, White KS (2001) Climate Change 2001: Impacts, Adaptation, and Vulnerability. Cambridge University Press.
42. Piao SL, Ciais P, Huang Y, Shen ZH, Peng SS, et al. (2010) The impacts of climate change on water resources and agriculture in China. Nature 467: 43–51.
43. Fang JY, Wang Z, Zhao SQ, Li YK, Tang ZY, et al. (2006) Biodiversity changes in the lakes of the Central Yangtze. Frontiers in Ecology and the Environment 4: 369–377.
44. Engloner AI (2009) Structure, growth dynamics and biomass of reed (*Phragmites australis*) – A review. Flora - Morphology, Distribution, Functional Ecology of Plants 204: 331–346.

Identification of Differentially Expressed Genes in Leaf of *Reaumuria soongorica* under PEG-Induced Drought Stress by Digital Gene Expression Profiling

Yubing Liu[1,2]*⁹, Meiling Liu[1,2,3]⁹, Xinrong Li[1,2], Bo Cao[1,2,3], Xiaofei Ma[2]

1 Shapotou Desert Research & Experiment Station, Cold and Arid Regions Environmental and Engineering Research Institute, Chinese Academy of Sciences, Lanzhou, P. R. China, **2** Key Laboratory of Stress Physiology and Ecology in Cold and Arid Regions of Gansu Province, Cold and Arid Regions Environmental and Engineering Research Institute, Chinese Academy of Sciences, Lanzhou, P. R. China, **3** University of Chinese Academy of Sciences, Beijing, China

Abstract

Reaumuria soongorica (Pall.) Maxim., a resurrection semi-shrub, is a typical constructive and dominant species in desert ecosystems in northwestern China. However, the gene expression characteristics of *R. soongorica* under drought stress have not been elucidated. Digital gene expression analysis was performed using Illumina technique to investigate differentially expressed genes (DEGs) between control and PEG-treated samples of *R. soongorica*. A total of 212,338 and 211,052 distinct tags were detected in the control and PEG-treated libraries, respectively. A total of 1,325 genes were identified as DEGs, 379 (28.6%) of which were up-regulated and 946 (71.4%) were down-regulated in response to drought stress. Functional annotation analysis identified numerous drought-inducible genes with various functions in response to drought stress. A number of regulatory proteins, functional proteins, and proteins induced by other stress factors in *R. soongorica* were identified. Alteration in the regulatory proteins (transcription factors and protein kinase) may be involved in signal transduction. Functional proteins, including flavonoid biosynthetic proteins, late embryogenesis abundant (LEA) proteins, small heat shock proteins (sHSP), and aquaporin and proline transporter may play protective roles in response to drought stress. Flavonoids, LEA proteins and sHSP function as reactive oxygen species scavenger or molecular chaperone. Aquaporin and proline transporters regulate the distribution of water and proline throughout the whole plant. The tolerance ability of *R. soongorica* may be gained through effective signal transduction and enhanced protection of functional proteins to reestablish cellular homeostasis. DEGs obtained in this study may provide useful insights to help further understand the drought-tolerant mechanism of *R. soongorica*.

Editor: Cynthia Gibas, University of North Carolina at Charlotte, United States of America

Funding: This work was financially supported by the State Key Development Program for Basic Research of China (973 Program, Grant No. 2013CB4229904) and the National Science Foundation of China (Grant Nos. 31070358, 30800122). The funders had no role in study design, data collection and analysis, decision to publish, or preparation of the manuscript.

Competing Interests: The authors have declared that no competing interests exist.

* E-mail: ybliu13@163.com

⁹ These authors contributed equally to this work.

Introduction

Water deficit is one of the most significant abiotic stresses that influence germination, growth, development, and productivity of plants [1]. Drought stress causes stomatal closure, limited gas exchange, and reduced photosynthesis in plants [2]. Over-reduction of photosynthetic electron transport chain induces the generation of reactive oxygen species (ROS), such as singlet oxygen ($^1O_2$), superoxide anion (O_2^-), hydrogen peroxide (H_2O_2), and hydroxyl radical (•OH), which damage cellular structures and macromolecules [3]. Plants have enzymatic and non-enzymatic systems to eliminate ROS. Moreover, plants have developed drought-resistance strategies such as succulent leaves, formation of osmophilic globules, stomatal movement, and reduction of leaf water potential [4,5]. The understanding of plant responses to water deficit have improved because of the application of molecular techniques. The expression of numerous genes in response to drought stress was described in previous studies [6,7]. Seki et al. [8] classified the genes expressed during stress into two groups: (i) genes encoding proteins involved in signal transduction (protein kinases and transcription factors) and (ii) genes with products, such as late embryogenesis abundant (LEA) proteins, chaperone, osmoprotectants, and detoxification enzymes, that directly protect cells against stress. Genomic and transcriptomic analyses revealed that various transcriptional regulatory systems are involved in stress-responsive gene induction. Several different sets of *cis*- and *trans*-acting factors are known to be induced by drought stress at the molecular level [9]. A large number of metabolites and proteins have also been reported to be up regulated in response to drought stress [10].

Reaumuria soongorica (Pall.) Maxim. is an extreme xerophytic semi-shrub and a typical constructive and dominant species in the desert vegetation in China. *R. soongorica* forms the zonal landscape and is widely distributed in northwest China [11]. Water supply is one of the main limiting factors in the habitat of this species. The unique adaptive strategies in the morphology and physiology of *R. soongorica*, such as thick cuticle, hollow stomata, and accumulation of some low-molecular-weight metabolites, are attributed to its special distribution area [4]. *R. soongorica* is a vascular flowering plant with desiccation-tolerance. *R. soongorica* leaves wither and enter a state of dormancy during dehydration but is revived when water becomes available. Thus, *R. soongorica* is utilized as a valuable non-model species in exploring drought-tolerance mechanisms. Liu et al. [4] found sucrose, malate, and proline, which play important roles in osmoregulation in *R. soongorica* during water loss. Studies on the protection mechanism for photosynthetic properties, activity of antioxidant enzyme, and metabolite changes under drought stress have also been conducted [12,13]. However, the molecular mechanism of drought tolerance in *R. soongorica* is still poorly understood. A preliminary genome survey, which included genome size, chromosome number, and karyotype, was conducted with *R. soongorica* [14]. In our previous study, the mitogen-activated protein kinase gene (*RsMPK2*) was isolated and its participation as possible mediator under different stresses was investigated [15]. We also identified and characterized the response of flavonone 3-hydrolase gene (*RsF3H*), a key enzyme-encoding gene involved in flavonoid biosynthesis pathway in drought treatment [16]. Recently, the transcriptome of *R. soongorica* was sequenced and analyzed to obtain valuable genetic information for the investigation on the molecular mechanism of drought tolerance [17]. Genetic information from the transcriptome of *R. soongorica* was also used to identify differentially expressed genes (DEGs) under drought stress to understand the mechanism of drought tolerance.

In recent years, with the increasing availability of sequence data, expression profiling has been used to identify DEGs in different tissues, organs, developmental stages or mutants and determine the expression patterns induced by stress in a large number of model and non-model organisms. Microarray analysis and suppression subtractive hybridization are powerful tools for isolating differentially expressed cDNAs that mediate complex biological processes in higher eukaryotes [18]. These technologies have been used in a number of animals and plants, such as *Aedesaegypti* [19], soybean [20], *Populuscanadensis* [21], *Scylla paramamosain* [22], cucumber [23], and navel orange [24]. Next-generation sequencing technologies offer new approaches for global measurements of gene expression with the advantage of high efficiency and low cost. Digital gene expression (DGE) tag profiling is based on ultra-high-throughput sequencing of cDNA fragments that uniquely tag the corresponding gene, thereby allowing direct quantification of transcript abundance [25]. DGE technology allows identification of millions of DEGs without the need for prior annotations and permits the analysis of organisms that lack genomic information. Thus, DGE has been widely utilized to monitor the differences in transcriptional responses among different tissues and organs in response to stress in silkworm [26], *Sagittariatrifolia* [27], *Dugesia japonica* [28], *Populus* [29], cotton [30], spruce [31], *Brassica napus* [32], and moss [33].

In this study, DEGs in leaves of *R. soongorica* in response to PEG-induced drought stress were examined using DGE tag profiling technology. Analysis of gene expression related to stress response provides further insight into the molecular mechanisms of stress tolerance in *R. soongorica.* Based on the putative functions of the identified genes, some important genes may be cloned. Moreover, the cloning of stress tolerance genes and determination of their expression patterns under drought stress may reveal attractive candidate genes and valuable information to improve drought stress tolerance of plants through genetic engineering.

Materials and Methods

Ethics Statement

R. soongorica is widely distributed in northwest China and is not listed as endangered or protected species. No specific permissions are required for sample collection in the Northern Mountain of Lanzhou, and that the field studies did not involve any endangered or protected species.

Plant Materials and Drought Treatment

Seeds of *R. soongorica* were collected from the Northern Mountain of Lanzhou, Gansu, China (36°17′N, 103°48′E, 1700–1900 m elevation). Seeds were planted in the breeding base at the foothill. Three-year-old plants were transferred to a greenhouse under the following conditions: 150 μmol $m^{-2}s^{-1}$ (fluorescent tubes), photoperiod 16 h light/8 h darkness, day/night temperature of 25°C/20°C. The plants were washed with deionized water to remove surface contaminants. Soil attached to roots was also rinsed off with deionized water. The plants were then cultivated in 0.125×Murashige and Skoog medium solution (without vitamins and organics) with insufflating air for two weeks to accommodate the liquid environment during treatments. The samples were classified into untreated and PEG-treated groups. The untreated samples were maintained in cultivating conditions as previously described. For drought treatment, 15% PEG6000 was added to the culture solution of *R. soongorica.* All plants were treated for 12 h, from 6:00 to 18:00 local time. Leaves from each treatment were collected and immediately stored in liquid nitrogen for later use.

DGE Library Preparation and Sequencing

Total RNA samples were isolated according to modified CTAB-method [34]. Quality and quantity analysis of total RNA, library construction, and sequencing were performed at the Beijing Genome Institute at Shenzhen, China. An extract of 8 μg of total RNA was obtained and treated with oligo (dT) magnetic bead adsorption to purify mRNA. Oligo (dT) was then used as primer in the synthesis of first- and second-strand cDNA. The 5′-ends of tags can be generated by two types of endonucleases, namely, NlaIII or DpnII. The bead-bound cDNA was subsequently digested with restriction enzyme NlaIII, which recognizes and severs CATG sites. Fragments other than 3′-cDNA fragments connected to oligo (dT) beads were washed away, and the Illumina adaptor 1 was ligated to the sticky 5′-end of the digested bead-bound cDNA fragments. The junction of the Illumina adaptor 1 and the CATG site constitutes the recognition site of MmeI, which is a type of endonuclease with separate recognition and digestion sites. MmeI breaks 17 bp downstream of the CATG site, producing tags with adaptor 1. After removing 3′ fragments with magnetic bead precipitation, the Illumina adaptor 2 was ligated to the 3′-ends of tags, thereby acquiring tags with different adaptors of both ends to form a tag library. After linear PCR amplification, fragments were purified by PAGE gel electrophoresis. During the quality control steps, Agilent 2100 Bioanalyzer and ABI Step One Plus Real-Time PCR System were used in the quantification and qualification of the sample library. Finally, the library was sequenced using Illumina HiSeq 2000 sequencer. The available raw data was then deposited in the NCBI Gene Expression Omnibus (GEO) database (http://www.ncbi.nlm.nih.gov/geo/query/acc.cgi?acc =GSE55412).

Analysis and Mapping of DGE Tags

Sequencing-received raw image data were transformed by base calling into sequence data. Prior to mapping the readings from the reference database, all sequences were filtered to remove 3′ adaptor sequence, low-quality sequences (tags with unknown sequences 'N'), tags with a copy number of one (probably sequencing error), and tags that were long or too short, leaving clean tags of 21 nt. For annotation, all clean tags were mapped in reference sequences, and a mismatch of only 1 bp was considered. We obtained the combined transcriptome of *R. soongorica* with (suffix is "_3″) and (suffix is "_A") (http://www.ebi.ac.uk/arrayexpress/experiments/E-MTAB-1543/) [17], which was used as reference database. Clean tags were mapped in reference sequences. The clean reads mapped in reference database from multiple genes were filtered. The remaining clean tags were used as unambiguous clean tags. The number of unambiguous clean tags for each gene was calculated and normalized to TPM (transcripts per million clean tags).

Identification and Functional Annotation of DEGs

We used a rigorous algorithm method to identify DEGs between two samples. False discovery rate (FDR) was applied to determine the threshold of P-value in multiple tests and analysis. The DEGs were obtained through FDR $\leq$0.001 and $|\log_2\text{Ratio}| \geq 1$. Gene ontology (GO) and pathway classification was used to determine the possible functions of all DEGs by mapping the GO (http://www.geneontology.org/) and KEGG (http://www.genome.jp/kegg/) databases.

Results

Analysis of DGE Libraries

The Illumina platform was used to perform high throughput tag-seq analysis on *R. soongorica* to investigate transcriptome responses to PEG-induced drought stress. Total RNA isolated from untreated and PEG-treated groups were named as libraries C and P, respectively. The major characteristics of the two libraries are summarized in Table 1. An average of 7,231,417 total sequence tags per library was obtained with 466,957 distinct tag sequences from the two libraries. Prior to mapping tag sequences to the reference sequences, adaptor tags were filtered (low-quality tags and tags with one copy) and produced approximately 6,969,782 total clean sequence tags per library with 211,695 distinct clean tag sequences. The distributions of the total and distinct clean tag copy numbers showed highly similar tendencies in the two libraries (Figure 1). Among the distinct clean tags, more than 60% of the distinct clean tags had 2 to 5 copies, 34% of the distinct clean tags had 5 to 100 copies, and 4.5% had copy numbers higher than 100.

Analysis of Tag Mapping

A reference gene database that included 94,878 sequences of the *R. soongorica* unigene was preprocessed for tag mapping. Among the sequences, genes with a CATG site accounted for 77.89%. To obtain the reference tags, all CATG+17 tags were used as gene reference tags. Finally, 218,791 total reference tag sequences with 182,184 unambiguous reference tags were obtained. Approximately 40.44% and 38.75% of the distinct clean tags were mapped unambiguously in the unigene database, and 47.94% and 50.2% of the distinct clean tags were not mapped to the unigene virtual tag database in libraries C and P, respectively (Table 1).

The sequencing saturation was analyzed in the two libraries to estimate whether the sequencing depth was sufficient for the transcriptome coverage. The genes that were mapped by all clean tags and unambiguous clean tags increased with the total number of tags. However, when the sequencing counts reached three million tags or higher, the number of detected genes was saturated (Figure S1). Given that Illumina sequencing can distinguish transcripts originating from both DNA strands and the strand-specific nature of the sequencing tags obtained, we found approximately 27% and 26% of distinct tags, which were mapped insense transcripts of libraries C and P, respectively (Figure S2). Approximately 25% and 24% of distinct tags perfectly matched antisense transcripts, which suggested that antisense genes play important roles in the transcriptional regulation of drought response in *R. soongorica*.

Drought-caused Changes in the Global Gene Expression in *R. Soongorica*

To obtain the transcriptional changes in PEG-treated *R. soongorica*, a rigorous algorithm method was applied to identify DEGs from normalized DGE data using pairwise comparison between the two groups (C and P). Results showed that 1,325 genes had FDRs $\leq$0.001, and $|\log_2\text{Ratio}| \geq 1$, which were declared to be the DEGs between the control and PEG-treated plants. Among these genes, 379 (28.6%) genes were up-regulated and 946 (71.4%) were down-regulated in response to drought stress. The expression of most genes was suppressed after drought treatment.

Functional annotation was performed to assign the genes of *R. soongorica* with GO terms. The main GO terms included "cellular component", "molecular function", and "biological process". A total of 535 DEGs were annotated as "cellular component". The categories of "cell part" and "cell" had the same proportion (88.0%), followed by "intracellular" and "intracellular part", which accounted for 78.5% and 77.9% of assignments, respectively (Table S1). In addition, a total of 570 DEGs were annotated to "molecular function" terms. The categories "catalytic activity", "organic cyclic compound binding", and "heterocyclic compound binding" were prominent with 60.5%, 25.3% and 25.3% of assignments, respectively (Table S1). Furthermore, a total of 521 DEGs were annotated to "biological process" terms, in which 71.4% and 63.6% of the assignments belonged to "metabolic process" and "cellular process", respectively, followed by "cellular metabolic process" and "primary metabolic process" (Figure 2).

To characterize the functional consequences associated with drought response, a pathway analysis of the DEGs based on the KEGG database was performed. The results indicated that genes participating in the biosynthesis of secondary metabolites, plant hormone signal transduction, plant-pathogen interaction, amino sugar and nucleotide sugar metabolism, and ubiquitin-mediated proteolysis were differentially expressed in PEG-treated group (Table S2).

Genes Associated with the Major Functional Group were Affected by PEG-induced Drought Stress

The global gene expression results showed that a great number of genes were significantly affected by PEG-induced drought stress. We classified the DEGs into three groups.

Group A was composed of some regulatory proteins, such as transcription factors, protein kinase, and other signaling molecules (Table 2). Twenty genes encoding protein kinase, including CBL-interacting protein kinase (CIPK), threonine protein kinase, receptor-like protein kinase (RPK), mitogen-activated protein kinase (MAPK) cascade, phosphoenolpyruvate carboxylase kinase, and other kinases, were remarkably changed under drought stress. The results showed that six genes were up-regulated, and 14 genes were down-regulated. Meanwhile, a total of 14 drought-activated or repressed transcription factor genes were identified from several different families including WRKY, NAC, MYC, TCP, and bZIP.

Table 1. Categorization and abundance of tags.

Summary		C	P
Raw tags	Total	7266032	7196802
	Distinct tags	445588	488325
Clean tags	Total number	7026443	6913120
	Distinct tag numbers	212338	211052
All tag mapping to gene	Total number	5869765	5356098
	Total % of clean tags	83.54%	77.48%
	Distinct tag numbers	110536	105096
	Distinct tag % of clean tags	52.06%	49.8%
Unambiguous tag mapping to gene	Total number	4261924	3722859
	Total % of clean tags	60.66%	53.85%
	Distinct tag numbers	85870	81776
	Distinct tag % of clean tags	40.44%	38.75%
All tag-mapped genes	Number	39354	38413
	% of ref genes	41.48%	40.49%
Unambigous tag-mapped genes	Number	27207	26381
	% of ref genes	28.68%	27.81%
Unknown tags	Total number	1156678	1557022
	Total % of clean tags	16.46%	22.52%
	Distinct tag numbers	101802	105956
	Distinct tag % of clean tags	47.94%	50.2%

Clean tags are tags that remained after filtering out dirty tags (low quality tags) from raw data. Distinct tags are different kinds of tags. Unambiguous tags are clean tags remaining after the removal of tags mapped to reference sequences from multiple genes. C represents the control group; P represents the PEG-treated group.

In our results, five transcription factors were identified as up-regulated, whereas nine transcription factors were repressed. The expression profiles exhibited that the majority of regulatory proteins were repressed after 12 h of drought treatment.

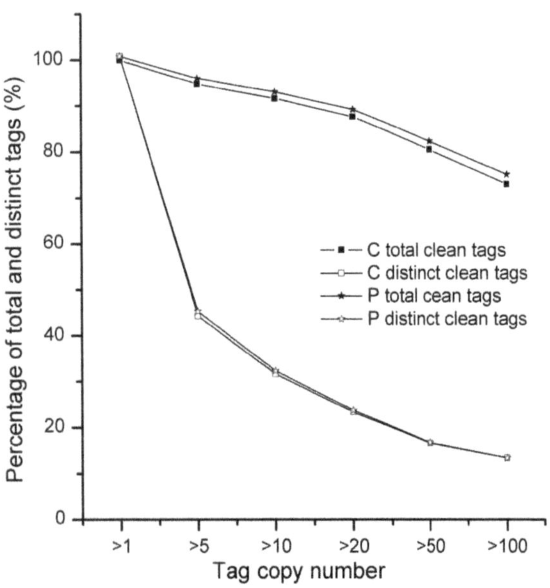

Figure 1. Distribution of total clean tag and distinct clean tag copy numbers from the libraries C and P. C represents the control group; P represents the PEG-treated group.

Group B was composed of 13 functional protein genes, which included genes involved in flavonoid biosynthesis (6 DEGs), LEA protein (2 DEGs), small heat shock protein (sHSP) (2 DEGs), aquaporin (2 DEGs), and proline transporters (1 DEG) (Table 2). Contrary to the expression of regulatory protein genes, most of the functional protein genes were up-regulated after 12 h of drought treatment.

Group C was composed of 14 proteins that were annotated as low-temperature-induced protein (1 DEG), dehydration-induced protein (1 DEG), defensing precursor (1 DEG), resistance protein (4 DEG), universal stress protein (1 DEG), and proteins involved in chitinase biosynthesis (4 DEGs) (Table 2). Similar to the alteration of functional protein genes, the expression of these 14 genes also exhibited a general upward trend.

Discussion

Plants, which are sessile organisms, try to respond and adapt to stress by altering their gene expression patterns when they are subjected to adverse environments [7]. Therefore, the genes that showed different expressions are probably involved in stress response. Shinozaki and Yamaguchi-Shinozaki [10] reported that the drought-inducible genes identified can be classified into regulatory and functional proteins. The regulatory proteins include various transcription factors, protein phosphatases, protein kinases, and other signaling molecules that participate in further regulation of signal transduction and downstream gene expression. The functional proteins are composed of molecules, such as chaperones, water channel proteins, LEA proteins, mRNA-binding proteins, sugar and proline transporters, detoxification enzymes, and various proteases. In the present study, the DEGs between the leaves of drought-stressed and unstressed *R. soongorica*

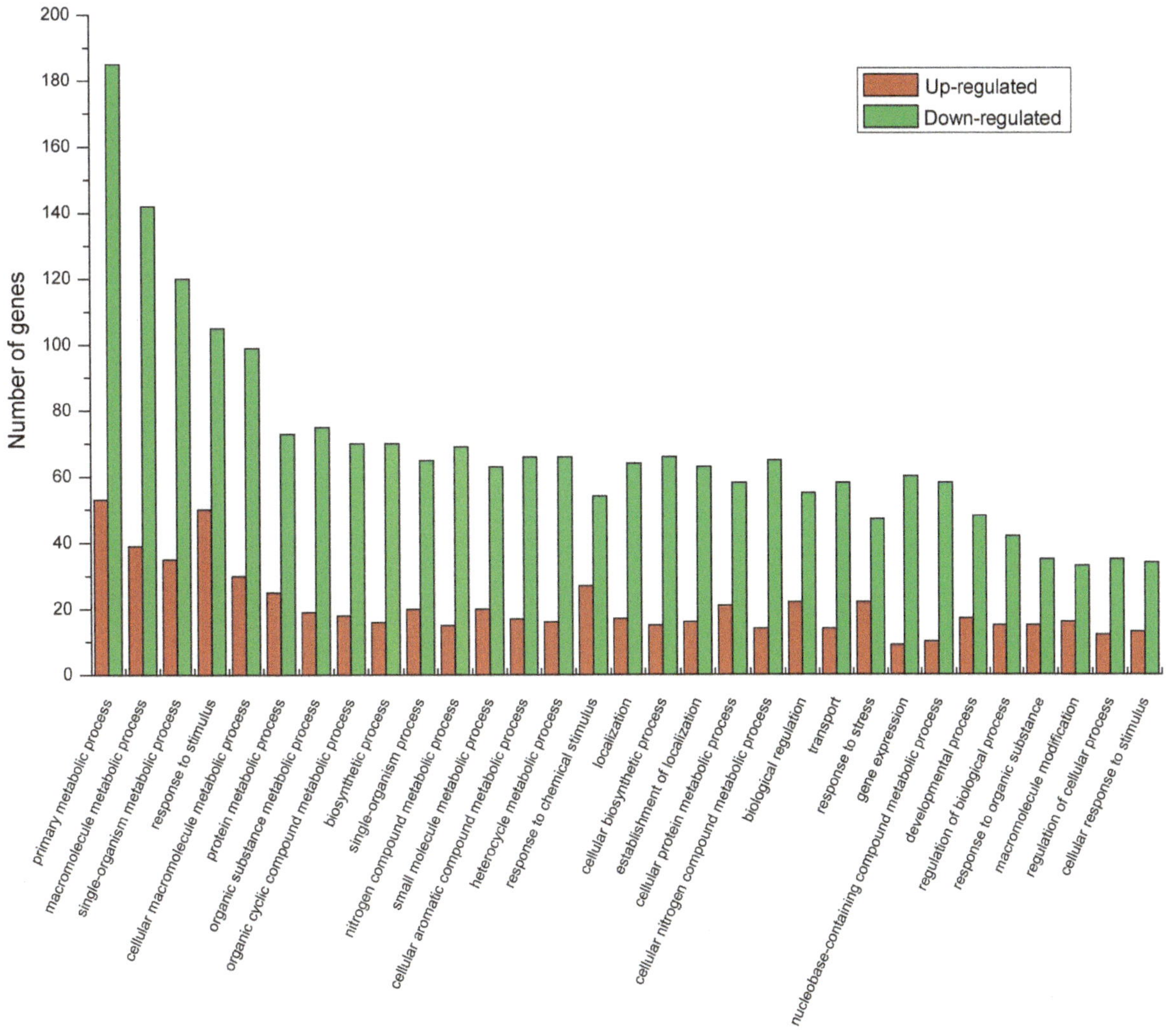

Figure 2. Gene classification based on gene ontology (GO) for DEGs in C and P libraries of *R. soongorica*. Only the biological processes were used for GO analysis. C represents the control group; P represents the PEG-treated group.

were investigated to conduct a global analysis of the transcriptomic response, which can facilitate our understanding of drought tolerance mechanisms. In our results, 1325 DEGs with 379 up-regulated and 946 down-regulated genes were detected. Our results are similar to that of Fan et al. [6], who also reported that down-regulated genes were more abundant than up-regulated in soybean leaves after drought exposure. Functional annotation analysis results indicated that many drought-inducible genes with various functions were identified in *R. soongorica*, including a number of regulatory proteins, functional proteins, and other inducible proteins in response to drought stress.

Regulatory proteins are involved in signal transduction pathways and in controlling the expression of functional genes in stress responses. In higher plants, many regulatory genes, such as those encoding protein kinases and transcription factors, are induced by environmental signals or stresses [35]. In this study, 20 DEGs encoding protein kinases were detected after drought treatment. Various protein kinases including CIPK protein, threonine protein kinase, MAPK, calcium-dependent protein kinases, and RPK are important regulatory proteins in the regulation of some stress-inducible genes [36]. Several experiments have demonstrated that protein kinase signaling conferred plant resistance to stress by regulating downstream functional genes [37,38]. In addition, protein kinase also functions in regulating transcription factors [39]. A total of 14 genes encoding transcription factors, including WRKY, NAC, MYC, TCP, and basic region leucine zipper (bZIP), were also significantly differentially expressed. Transcription factors are important in regulating plant responses to biotic and abiotic stresses [40]. Altering the expression of certain transcription factors can greatly change the expression of a number of stress-related target genes that influence plant stress tolerance in transgenic plants [9,10]. Thus, the alteration of protein kinases and transcription factors may function in the signal transduction process by regulating the expression of various functional genes in response to drought stress.

Aside from signaling transduction, plants adapt to adverse environment through the activation or up-regulation of genes that directly function in stress resistance [7]. Flavonoids are a class of secondary metabolites with diverse functions in growth, development, reproduction, and are also involved in diverse stress responses in plants. Flavonoids accumulate rapidly and play a protective role when plants are exposed to abiotic stress [41]. In recent years, several genes encoding important enzymes, such as phenylalanine ammonia-lyase, F3H, dihydroflavonol 4-reductase

Table 2. Selected DEGs between the C and P libraries.

Funtional group	Unigene ID	Gene annotation	TPM-C	TPM-P	Log_2Ratio (C/P)
Protein kinases					
	Unigene 9485_A	CBL-interacting protein kinase	2.99	12.87	2.11
	CL2201. Contig1_3	CBL-interacting protein kinase	25.9	59.45	1.20
	CL4525. Contig1_A	threonine protein kinase	4.98	15.62	1.65
	Unigene 2338_A	threonine protein kinase	2.85	7.96	1.48
	Unigene 14518_A	serine/threonine-protein kinase	78.13	35.73	-1.13
	Unigene 32052_3	receptor-like protein kinase	2.56	0.01	-8
	Unigene 39297_A	receptor-like protein kinase	2.13	0.01	-7.73
	Unigene 52174_A	receptor-like protein kinase	6.26	1.3	-2.27
	CL2777. Contig2_3	receptor-like protein kinase	36.43	14.9	-1.29
	CL6483. Contig1_3	receptor-like protein kinase	16.94	7.67	-1.14
	Unigene 43240_A	receptor-like protein kinase	76.71	38.19	-1.01
	Unigene 21624_3	mitogen-activated protein kinase kinase kinase	7.26	1.59	-2.19
	CL1583.Contig2_A	mitogen-activated protein kinase kinase kinase	7.26	1.59	-2.19
	Unigene 9581_3	mitogen-activated protein kinase 2	25.62	10.7	-1.26
	Unigene 633_3	AMP-activated protein kinase	24.76	10.7	-1.21
	Unigene 21618_3	calcineurin B-like protein-interacting protein kinase	25.33	12.01	-1.08
	Unigene 10286_A	phosphoenolpyruvate carboxylase kinase	0.57	4.63	3.02
	Unigene 27156_3	pyruvate, phosphate dikinase	35.44	85.2	1.27
	CL4464. Contig2_3	pyruvate kinase	85.82	29.36	-1.55
	Unigene 50230_A	histidine kinase	40.28	17.94	-1.67
Transcription factors					
	Unigene 30342_3	WRKY transcription factor 7	4.7	14.75	1.65
	Unigene 10063_3	WRKY transcription factor 6–1	1.99	0.01	-7.64
	Unigene 27699_3	MYC protein	14.23	47.74	1.75
	CL7944. Contig1_3	NAC domain-containing protein	14.23	4.48	-1.67
	Unigene 31521_3	NAC domain-containing protein	12.52	4.19	-1.58
	Unigene 25674_3	transcription factor TCP20-like	13.66	33.7	1.30
	CL7341. Contig1_3	transcription factor TCP20-like	45.26	16.92	-1.42
	Unigene 22142_3	transcription factor TCP7-like	31.88	15.04	-1.08
	Unigene 21994_3	DNA binding protein	6.55	14.32	1.13
	Unigene 23673_3	transcription factor bZIP124	51.24	20.83	-1.29
	Unigene 1336_3	transcription factor bZIP12	4.7	11.57	1.29
	Unigene 28584_3	transcription factor UNE12	37.86	12.01	-1.66
	CL8097. Contig1_A	transcription factor UNE12	8.11	3.18	-1.35
	CL5168. Contig1_3	transcription factor UNE12	12.95	5.35	-1.28
Flavonoids biosynthesis					
	CL93. Contig1_A	UDP-glucose: glucosyltransferase	22.06	50.19	1.19
	CL1676. Contig1_A	Flavonol sulfotransferase-like protein	3.56	9.4	1.40
	CL2370. Contig2_3	flavonoid 3′-hydroxylase	10.39	25.6	1.30
	Unigene 1473_3	flavonoid 3′-hydroxylase	84.96	34.72	-1.29
	Unigene 19573_A	dihydroflavonol 4-reductase	1.14	17.65	3.95
LEA proteins					
	CL4558. Contig1_3	late embryogenesis abundant protein	5.84	12.73	1.12
	Unigene 442_3	late embryogenesis abundant protein group 9	7.54	22.57	1.58
Heat shock proteins					
	Unigene 18824_A	mitochondrial small heat shock protein	3.84	10.99	1.51
	CL12936. Contig1_A	heat shock protein 17.5 cytosolic class II	1.85	8.25	2.15
aquaporin					
	Unigene 39258_A	aquaporin	1.71	9.69	2.50

Table 2. Cont.

Funtional group	Unigene ID	Gene annotation	TPM-C	TPM-P	Log_2Ratio (C/P)
	Unigene 19242_A	aquaporin	426.96	176.77	-1.27
Proline transporter					
	CL1109. Contig2_A	proline transporter	7.54	27.34	1.86
Cold shock protein					
	Unigene 233_3	cold shock protein	11.53	58.44	2.34
low temperature-induced protein					
	Unigene 24454_A	low temperature-induced protein	0.01	2.03	7.67
Dehydration-induced proteins					
	Unigene 922_A	dehydration-induced proteins	10.67	21.99	1.04
universal stress protein					
	Unigene 25385_3	universal stress protein	93.22	263.56	1.50
Resistance protein					
	CL8753. Contig4_3	disease resistance protein	1.85	7.23	1.97
	CL6875. Contig2_3	disease resistance protein	0.01	2.03	7.67
	CL5118. Contig1_3	disease resistance protein	4.7	1.01	-2.22
	Unigene 26781_3	natural resistance-associated macrophage protein	1.28	16.49	3.69
	Unigene 28389_3	nematode resistance protein	36.86	77.68	1.08
defensin precursor					
	Unigene 26068_3	defensin precursor	20.92	87.08	2.06
chitinase					
	CL630. Contig2_3	class IV chitinase	3.7	67.84	4.20
	CL5695. Contig1_A	chitinase	6.4	41.8	2.71
	Unigene 30551_3	class IV chitinase	13.52	6.08	-1.15
	CL630. Contig3_3	class IV chitinase	60.91	27.63	-1.14

C represents the control group; P represents the PEG-treated group. The DEGs was selected through FDR $\leq$0.001 and $|log_2Ratio| \geq 1$.

(DFR), and anthocyanin synthase, which are involved in the flavonoid biosynthetic pathway, were found to be up-regulated in response to different kinds of stresses, such as salinity, UV-B irradiation, water deficit, and nitrogen stress [41–44]. Increased expression of genes involved in the flavonoid pathway was detected in our previous study on *R. soongorica* [16]. In the present study, most DEGs encoding key enzymes involved in flavonoid biosynthesis were also up-regulated, especially DEGs that encode UFGT, DFR, and F3′H. These findings are consistent with the results of other studies conducted on birch, potato, and rice after stress exposure [45–47]. The up-regulated genes may result in accumulation of flavonoids. Flavonoids may serve as signaling molecules in signaling cascades. Flavonoids also affect cellular function and modulate gene expression in response to adverse conditions [48]. Agati et al. [49] found that flavonoids directly scavenged the ROS brought about by stress. Therefore, up-regulation of flavonoid pathway genes may enhance the protective role of *R. soongorica*, thus leading to better drought tolerance.

Secondary metabolites and stress-responsive proteins, such as LEA proteins, accumulate naturally in the seeds of some desiccation-tolerant plants, and are also induced in vegetative tissues when subjected to water-limited conditions [50,51]. LEA proteins, largely attributed to their extreme hydrophilic nature, have also been implicated in detoxification, sequestration of ions, maintenance of protein or membrane structure, binding water, and operation as molecular chaperones during dehydration [52]. In this study, two DEGs were identified as up-regulated LEA proteins. Our results are in agreement with the findings of Gechev et al. [53], who found high expression of LEA genes during drought and desiccation in leaves of *Haberlea rhodopensis*. We concluded that dehydration-induced expression of LEA transcripts is a common response in resurrection and desiccation sensitive plants [54]. Improved resistance to drought and salinity was also observed in transgenic rice when LEA was overexpressed [55]. Expression of LEA proteins in yeast confers improved resistance to high salinity and freezing [56]. These studies demonstrated that LEA proteins functions in binding water, maintenance of protein or membrane structure, or operation as molecular chaperones to enhance tolerance to dehydration, whereas few additional clues to the mechanism of protein function are provided.

Plant HSPs facilitate protein folding or assembly under diverse developmental and adverse environmental conditions. In this study, two small HSPs (sHSPs) encoding DEGs showed up-regulation under drought stress. One sHSP was annotated to mitochondrial sHSP (MT-sHSP), and the other was annotated as cytosolic sHSP17.5. Although the precise functional mechanism of sHSPs is still unclear, in vivo studies have shown that sHSPs function as molecular chaperones in stressful conditions and in normal development [57]. Overproduction of sHSP increased drought tolerance in transgenic rice seedlings [58]. MT-sHSP protects the electron transport complex I and prevents damage by ROS [59]. Lee et al. [60] found that transgenic plants with over-expression of MT-sHSP efficiently maintained membrane stability by scavenging ROS, and were conferred with enhanced tolerance to salinity. Ectopic expression of cytosolic sHSP 17.1 in *Arabidopsis thaliana* exhibited higher resistance to drought [61]. In a study conducted by Sato et al. [62], sHSP 17.5 showed molecular chaperone activity in vitro and represented an example of an HSP capable of protecting cells against thermal extremes. Therefore, these results indicated that the sHSPs may play roles in scavenging ROS or functioning as molecular chaperone to respond to drought stress.

The DEGs encoding aquaporins and proline transporter were also identified. Aquaporins are channel proteins of intracellular and plasma membranes that function in transporting water, small solutes (urea, boric acid, and silicic acid), and gases (ammonia and carbon dioxide). A specific role for aquaporins in embolism refilling and recovery of stem axial conductance after drought was proposed in grapevine [63]. In our results, two aquaporin genes were identified. One aquaporin gene was up-regulated, whereas the other gene was down-regulated. An overall tendency of down regulation for aquaporin gene was observed. However, up regulation of certain aquaporin transcripts was also detected in rice and Arabidopsis leaves [64,65]. Therefore, transcriptional control of aquaporins in drought-stressed leaves appears to be more complex [66]. Interestingly, transport of proline may also be affected by water deficit. A gene encoding proline transporter was up-regulated under drought stress in the present study, which is in agreement with the result of Rentsch et al. [67], who also observed the induction of a specific proline transporter in Arabidopsis under water deficit. The DEGs encoding aquaporins and proline transporter in *R. soongorica* may serve as an adaptive strategy by regulating water and proline distribution throughout the whole plant.

Moreover, in addition to two dehydration-induced 19 homologs, proteins that can be induced by other stresses were also detected in the study and annotated to low-temperature-induced (LT) protein, cold shock protein (CSP), defensin precursor, chitinase, disease resistance protein, natural resistance-associated macrophage protein, nematode resistance protein, and universal stress protein. LT proteins and CSPs are commonly induced by cold stress. Various genes are induced by both drought and cold stress, suggesting the existence of crosstalk between drought and cold-stress signaling pathways [10]. Defensins, chitinase, disease resistance proteins, natural resistance-associated macrophage protein, and nematode resistance protein were all found to be involved in the defense against phytopathogens [68–72]. Therefore, the identification of the genes in this study suggested that a crosstalk between drought and pathogen attack. The universal stress proteins can confer the ability to respond and adapt to environmental changes [73]. The identification of all these genes suggests crosstalk between responsive genes or pathways and multiple abiotic or even biotic stresses [23]. Moreover, these identified genes may provide gene resources for understanding the drought tolerance mechanism of *R. soongorica*.

Among the identified DEGs, most regulatory proteins were down-regulated, whereas the majority of functional proteins were up-regulated. The different expression patterns shown by the two groups are interesting and worthy to be explored. Therefore, we tried to discuss the probable interaction of regulatory proteins and functional proteins. Most gene regulators exhibit either antecedent or simultaneous change in the expression level when compared with their targets [74]. One reason is that it takes time for the regulator gene to express its protein product and affect (directly or indirectly) the transcription of its target gene [75]. Other reasons may have been caused by the different mRNA half-lives of the regulator and target genes, and the different profiles between regulator genes and functional genes may also indicate a feedback loop, where a target gene can regulate its regulator [76]. Given that signaling pathways are complex dynamic events that occur over time, single time point expression profiles in our study are insufficient to elucidate temporal events. The different expression profiles of regulatory and functional proteins observed in this study are interesting. Expression profiling experiment with a series of time points should be performed to elucidate this problem in further studies.

Conclusions

Based on the putative function analysis of annotated DEGs, DEGs are classified into three groups. The first group, which includes protein kinases (CIPK and threonine protein kinase) and transcription factors (WRKY, MYC and TCP), regulates signal transduction. The second group includes genes involved in the biosynthesis of flavonoids in scavenging ROS, sHSP, LEA proteins, aquaporins, and proline transporter that represents functional genes that directly enhance drought-stress tolerance. The sHSP maintains protein or membrane structure and act as molecular chaperones that protect cells from injury under drought stress. Aquaporins and proline transporter regulates water or proline distribution throughout the whole plant. The third group contains proteins induced by other stress factors, which suggests that a crosstalk between different stresses. In conclusion, the tolerant ability of *R. soongorica* may be gained through effective signal transduction and enhanced protection of functional proteins to reestablish cellular homeostasis.

Supporting Information

Figure S1 Sequencing saturation analysis of the two libraries. C represents the control group; P represents the PEG-treated group.

Figure S2 Mapping of distinct clean tags in the two DGE libraries. C represents the control group; P represents the PEG-treated group.

Table S1 Gene classification based on gene ontology (GO) for DEGs in C and P libraries of *R. soongorica*. C represents the control group; P represents the PEG-treated group.

Table S2 KEGG analysis for DEGs in C and P libraries of *R. soongorica*. C represents the control group; P represents the PEG-treated group.

Acknowledgments

We thank the technical support for Illumina sequencing and initial data analysis from Beijing Genome Institute at Shenzhen, P. R. China, and appreciate the constructive and helpful suggestions from the two anonymous reviewers.

Author Contributions

Conceived and designed the experiments: XRL YBL. Performed the experiments: MLL BC. Analyzed the data: MLL. Contributed reagents/materials/analysis tools: XRL YBL XFM. Wrote the paper: MLL YBL.

References

1. Okcu G, Kaya MD, Atak M (2005) Effects of salt and drought stresses on germination and seedling growth of pea (*Pisumsativum* L.). Turk J Agric For 29: 237–242.
2. Lawson T, Oxborough K, Morison JIL, Baker NR (2003) The responses of guard and mesophyll cell photosynthesis to CO_2, O_2, light, and water stress in a range of species are similar. J Exp Bot 54: 1743–52.
3. Dat JF (1998) Parallel changes in H_2O_2 and catalase during thermotolerance induced by salicylic acid or heat acclimation in mustard seedlings. Plant Physiol 116: 1351–1357.
4. Liu YB, Wang G, Liu JY, Zhao X, Tan HJ, et al. (2007) Anatomical, morphological and metabolic acclimation in the resurrection plant *Reaumuria soongorica* during dehydration and rehydration. J Arid Environ 70: 183–194.
5. McDowell N, Pockman WT, Allen CD, Breshears DD, Cobb N, et al. (2008) Mechanisms of plant survival and mortality during drought: why do some plants survive while others succumb to drought? New Phytol 178: 719–739.
6. Fan XD, Wang JQ, Yang N, Dong YY, Liu L, et al. (2013) Gene expression profiling of soybean leaves and roots under salt, saline–alkali and drought stress by high-throughput Illumina sequencing. Gene 512: 392–402.
7. Hazen SP, Pathan MS, Sanchez A, Baxter I, Dunn M, et al. (2005) Expression profiling of rice segregating for drought tolerance QTLs using a rice genome array. Funct Integr Genomic 5: 104–116.
8. Seki M, Kameiy A, Yamaguchi-Shinozakiz K, Shinozaki K (2003) Molecular responses to drought, salinity and frost: common and different paths for plant protection. Curr Opin Biotech 14: 194–199.
9. Yamaguchi-Shinozaki K, Shinozaki K (2005) Organization of *cis*-acting regulatory elements in osmotic- and cold-stress-responsive promoters. Trends Plant Sci 10: 88–94.
10. Shinozaki K, Yamaguchi-Shinozaki K, Seki M (2003) Regulatory network of gene expression in the drought and cold stress responses. Curr Opin Plant Biol 6: 410–417.
11. Ma JY, Chen T, Qiang WY, Wang G (2005) Correlations between foliar stable carbon isotope composition and environmental factors in desert plant *Reaumuria soongorica* (Pall.) Maxim. J Intergr Plant Biol 47: 1065–1073.
12. Liu YB, Zhang TG, Wang J (2008) Photosynthesis and metabolite leaves levels in dehydrating leaves of *Reaumuria soongorica*. Acta Biol Cracov Bot 50: 19–26.
13. Bai J, Gong CM, Chen K, Kang HM, Wang G (2009) Examination of antioxidative systems responses in the different phases of drought stress and during recovery in desert plant *Reaumuria soongorica* (Pall.) Maxim. J Plant Biol 52: 417–425.
14. Wang XH, Zhang T, Wen ZN, Xiao HL, Yang ZJ, et al. (2011) The chromosome number, karyotype and genome size of the desert plant diploid *Reaumuria soongorica* (Pall.) Maxim. Plant Cell Rep 30: 955–964.
15. Liu YB, Li XR, Tan HJ, Liu ML, Zhao X, et al. (2010) Molecular characterization of RsMPK2, a C1 subgroup mitogen-activated protein kinase in the desert plant *Reaumuria soongorica*. Plant Physiol Biochem 48: 836–844.
16. Liu ML, Li XR, Liu YB, Cao B (2013) Regulation of flavanone 3-hydroxylase gene involved in the flavonoid biosynthesis pathway in response to UV-B radiation and drought stress in the desert plant, *Reaumuria soongorica*. Plant Physiol Biochem 73: 161–167.
17. Shi Y, Yan X, Zhao PS, Yin HX, Zhao X, et al. (2013) Transcriptomic analysis of a tertiary relict plant, extreme xerophyte *Reaumuria soongorica* to identify genes related to drought adaptation. PLoS One 8: e63993.
18. Diatchenko L, Laue YF, Campbell AP, Chenchik A, Moqadam F, et al. (1996) Suppression subtractive hybridization: a method for generating differentially regulated or tissue-specific cDNA probes and libraries. P Natl Acad Sci USA 93: 6025–6030.
19. Lertkiatmongkol P, Pethuan S, Jirakanjanakit N, Rongnoparut P (2010) Transcription analysis of differentially expressed genes in insecticide-resistant *Aedes aegypti* mosquitoes after deltamethrin exposure. J Vector Ecol 35: 197–203.
20. Asakura T, Tamura T, Terauchi K, Narikawa T, Yagasaki K, et al. (2012) Global gene expression profiles in developing soybean seeds. Plant Physiol Biochem 52: 147–153.
21. Caruso A, Chefdor F, Carpin S, Depierreux C, Delmotte FM, et al. (2008) Physiological characterization and identification of genes differentially expressed in response to drought induced by PEG 6000 in *Populus Canadensis* leaves. J Plant Physiol 165: 932–941.
22. Chen FY, Liu HP, Bo J, Ren HL, Wang KJ (2010) Identification of genes differentially expressed in hemocytes of *Scylla* paramamosain in response to lipopolysaccharide. Fish Shellfish Immun 28: 167–177.
23. Zhou XG, Wu FZ (2009) Differentially expressed transcripts from cucumber (*Cucumis sativus* L.) root upon inoculation with *Fusarium oxysporum* f. sp. *cucumerinum* Owen. Physiol Mol Plant P 74: 142–150.
24. Da XL, Yu KQ, Shen SH, Zhang YJ, Wu JX, et al. (2012) Identification of differentially expressed genes in a spontaneous altered leaf shape mutant of the navel orange [*Citrus sinensis* (L.) Osbeck]. Plant Physiol Biochem 56: 97–103.
25. Hong LZ, Li J, Schmidt-Küntzel A, Warren WC, Barsh GS (2011) Digital gene expression for non-model organisms. Genome Res 21: 1905–1915.
26. Gao P, Chen AL, Zhao QL, Shen XJ, Qiu ZY, et al. (2013) Differentially expressed genes in the ovary of the sixth day of pupal "Ming" lethal egg mutant of silkworm, *Bombyxmori*. Gene 527: 161–166.
27. Cheng LB, Li SY, Xu XY, Hussain J, Yin JJ, et al. (2013) Identification of differentially expressed genes relevant to corm formation in *Sagittaria triflia*. PLoS One 8: e54573.
28. Qin YF, Fang HM, Tian QN, Bao ZX, Lu P, et al. (2011) Transcriptome profiling and digital gene expression by deep-sequencing in normal/regenerative tissues of planarian *Dugesia japonica*. Genomics 97: 364–371.
29. Chen S, Jiang J, Li HY, Liu GF (2012) The salt-responsive transcriptome of *Populussimonii* × *Populusnigra* via DGE. Gene 504: 203–212.
30. Wei MM, Song MZ, Fan SL, Yu SX (2013) Transcriptomic analysis of differentially expressed genes during anther development in genetic male sterile and wild type cotton by digital gene-expression profiling. BMC genomics 14: 97.
31. Albouyeh R, Farzaneh N, Bohlmann J, Ritland K (2010) Multivariate analysis of digital gene expression profiles identifies a xylem signature of the vascular tissue of white spruce (*Piceaglauca*). Tree Genet Genomes 6: 601–611.
32. Jiang JJ, Shao YL, Du K, Ran LP, Fang XP, et al. (2013) Use of digital gene expression to discriminate gene expression differences in early generations of resynthesized *Brassica napus* and its diploid progenitors. BMC Genomics 14: 72.
33. Nishiyama T, Miyawaki K, Ohshima M, Thompson K, Nagashima A, et al. (2012) Digital gene expression profiling by 5-end Sequencing of cDNAs during reprogramming in the moss physcomitrella patens. PLoS One 7: e36471.
34. Wang XH, Xiao HL, Chen GX, Zhao X, Huang CH, et al. (2011) Isolation of high-quality RNA from *Reaumuria soongorica*, a desert plant rich in secondary metabolites. Mol Biotechnol 48: 165–172.
35. Shinozaki K, Yamaguchi-Shinozaki K (1996) Molecular responses to drought and cold stress. Curr Opin Biotech 7: 161–167.
36. Zhu JK (2002) Salt and drought stress signal transduction in plants. Annu Rev Plant Biol 53: 247–273.
37. Boller T, Felix G (2009) A renaissance of elicitors: perception of microbe-associated molecular patterns and danger signals by pattern-recognition receptors. Annu Rev Plant Biol 60: 379–406.
38. Lacombe S, Rougon-Cardoso A, Sherwood E, Peeters N, Dahlbeck D, et al. (2010) Interfamily transfer of a plant pattern-recognition receptor confers broad-spectrum bacterial resistance. Nat Biotechnol 28: 365–369.
39. Tena G, Boudsocq M, Sheen J (2011) Protein kinase signaling networks in plant innate immunity. Curr Opin Plant Biol 14: 519–529.
40. Chen WQ, Provart NJ, Glazebrook J, Katagiri F, Chang HS, et al. (2002) Expression profile matrix of Arabidopsis transcription factor genes suggests their putative functions in response to environmental stresses. Plant Cell 14: 559–574.
41. Castellarin SD, Matthews MA, Gaspero GD, Gambetta GA (2007) Water deficits accelerate ripening and induce changes in gene expression regulating flavonoid biosynthesis in grape berries. Planta 227: 101–112.
42. Park JS, Choung MG, Kim JB, Hahn BS, Kim JB, et al. (2007) Genes up-regulated during red coloration in UV-B irradiated lettuce leaves. Plant Cell Rep 26: 507–516.
43. Walia H, Wilson C, Condamine P, Liu X, Ismail AM, et al. (2005) Comparative transcriptional profiling of two contrasting rice genotypes under salinity stress during the vegetative growth stage. Plant Physiol 139: 822–835.
44. Han YP, Vimolmangkang S, Soria-Guerra RE, Rosales-Mendoza S, Zheng D, et al. (2010) Ectopic expression of apple F3′H genes contributes to anthocyanin accumulation in the Arabidopsis tt7 mutant grown under nitrogen stress. Plant Physiol 153: 806–820.
45. Morales LO, Tegelberg R, Brosché M, Keinänen M, Lindfors A, et al. (2010) Effects of solar UV-A and UV-B radiation on gene expression and phenolic accumulation in *Betulapendula* leaves. Tree Physiol 30: 923–934.
46. André CM, Schafleitner R, Legay S, Lefèvre I, Aliaga CAA, et al. (2009) Gene expression changes related to the production of phenolic compounds in potato tubers grown under drought stress. Phytochemistry 70: 1107–1116.
47. Lenka SK, Katiyar A, Chinnusamy V, Bansal KC (2011) Comparative analysis of drought-responsive transcriptome in Indica rice genotypes with contrasting drought tolerance. Plant Biotechnol J 9: 315–327.
48. Williams RJ, Spencer JPE, Rice-Evans C (2004) Flavonoids: antioxidants or signaling molecules? Free Radical Biol Med 26: 838–849.
49. Agati G, Biricolti S, Guidi L, Ferrni F, Fini A, et al. (2011) The biosynthesis of flavonoids is enhanced similarly by UV radiation and root zone salinity in L. *vulgare* leaves. J Plant Physiol 168: 204–212.
50. Battaglia M, Olvera-Carrillo Y, Garciarrubio A, Campos F, Covarrubias AA (2008) The enigmatic LEA proteins and other hydrophilins. Plant Physiol 148: 6–24.
51. Tunnacliffe A, Wise MJ (2007) The continuing conundrum of LEA proteins. Naturwissenschaften 94: 791–812.
52. Bray EA (1996) Plant responses to water deficit. Trends Plant Sci 96: 1360–1385.
53. Gechev TS, Benina M, Obata T, Tohge T, Sujeeth N, et al. (2013) Molecular mechanisms of desiccation tolerance in the resurrection glacial relic *Haberlea rhodopensis*. Cell Mol Life Sci 70: 689–709.
54. Challabathula D, Bartels D (2013) Desiccation tolerance in resurrection plants: new insights from transcriptome, proteome and metabolome analysis. Front Plant Sci 4: 482.
55. Xu D, Duan X, Wang B, Hong B, Ho THD, et al. (1996) Expression of a late embryogenesis abundant protein gene, HVA1, from barley confers tolerance to water deficit and salt stress in transgenic rice. Plant Physiol 110: 249–257.

56. Imai R, Chang L, Ohta A, Bray EA, Takagi M (1996) A lea-class gene of tomato confers salt and freezing tolerance when overexpressed in *Saccharomyces* cerevisiae. Gene 170: 243–248.
57. Sun WN, Montagu MV, Verbruggen N (2002) Small heat shock proteins and stress tolerance in plants. Biochim Biophys Acta 1577: 1–7.
58. Sato Y, Yokoya S (2008) Enhanced tolerance to drought stress in transgenic rice plants overexpressing a small heat-shock protein, sHSP17.7. Plant Cell Rep 27: 329–334.
59. Banzet N, Richaud C, Deveaux Y, Kazmaier M, Gagnon J, et al. (1998) Accumulation of small heat shock proteins, including mitochondrial HSP22, induced by oxidative stress and adaptive response in tomato cells. Plant J 13: 519–527.
60. Lee KW, Cha JY, Kim KH, Kim YG, Lee BH, et al. (2012) Overexpression of alfalfa mitochondrial HSP23 in prokaryotic and eukaryotic model systems confers enhanced tolerance to salinity and arsenic stress. Biotechnol Lett 34: 167–174.
61. Zhang L, Gao YK, Pan HT, Hu WJ, Zhang QX (2013) cloning and characterization of a *Primula* heat shock protein gene, PfHSP17.1, which confers heat, salt and drought tolerance in transgenic *Arabidopsis thaliana*. Acta Physiol Plant 35: 3191–3200.
62. Sato A, Allona I, Collada C, Guevara MA, Casado R, et al. (1999) Heterologous expression of a plant small heat-shock protein enhances Escherichia coli viability under heat and cold stress. Plant Physiol 120: 521–528.
63. Lovisolo C, Schubert A (2006) Mercury hinders recovery of shoot hydraulic conductivity during grapevine rehydration: evidence from a whole-plant approach. New Phytol 172: 469–478.
64. Alexandersson E, Fraysse L, Sjövall-Larsen S, Gustavsson S, Fellert M, et al. (2005) Whole gene family expression and drought stress regulation of aquaporins. Plant Mol Biol 59: 469–484.
65. Guo L,Wang ZY, Lin H, Cui WE, Chen J, et al. (2006) Expression and functional analysis of the rice plasma-membrane intrinsic protein gene family. Cell Res 16: 277–286.
66. Maurel C, Verdoucq L, Luu DT, Santoni V (2008) Plant aquaporins: membrane channels with multiple integrated functions. Annu Rev Plant Biol 59: 595–624.
67. Rentsch D, Hirner B, Schmelzer E, Frommer WB (1996) Salt stress-induced proline transporters and salt stress-repressed broad specificity amino acid permeases identified by suppression of a yeast amino acid permease-targeting mutant. Plant Cell 8: 1437–1446.
68. Thomma BP, Cammue BP, Thevissen K (2002) Plant defensins. Planta 216: 193–202.
69. Hamid R, Khan MA, Ahmad M, Ahmad MM, Abdin MZ, et al. (2013) Chitinases: an update. J Pharm Bioallied Sci 5: 21–29.
70. McDowell MJ, Woffenden BJ (2003) Plant disease resistance genes: recent insights and potential applications. Trends Biotechnol 21: 178–183.
71. Thomine S, Wang R, Ward JM, Crawford NM, Schroeder JI (2000) Cadmium and iron transport by members of a plant metal transporter family in Arabidopsis with homology to Nramp genes. Proc Natl Acad Sci USA 97: 4991–4996.
72. Williamson VM (1999) Plant nematode resistance genes. Curr Opin Plant Biol 2: 327–331.
73. Isokpehi RD, Simmons SS, Cohly HHP, Ekunwe SIN, Begonia GB, et al. (2011) Identification of drought-responsive universal stress proteins in viridiplantae. Bioinform Biol Insights 5: 41–58.
74. Yu H, Luscombe NM, Qian J, Gerstein M (2003) Genomic analysis of gene expression relationships in transcriptional regulatory networks. Trends Genet 19: 422–427.
75. Ferrell JE Jr (2002) Self-perpetuating states in signal transduction: positive feedback, double-negative feedback and bistability. Curr Opin Chem Biol 6: 140–148.
76. Zou M, Conzen SD (2005) A new dynamic Bayesian network (DBN) approach for identifying gene regulatory networks from time course microarray data. Bioinformatics 21: 71–79.

6

Video Surveillance Captures Student Hand Hygiene Behavior, Reactivity to Observation, and Peer Influence in Kenyan Primary Schools

Amy J. Pickering[1,2*], Annalise G. Blum[2], Robert F. Breiman[3,5], Pavani K. Ram[4], Jennifer Davis[1,2]

1 Woods Institute for the Environment, Stanford University, Stanford, California, United States of America, 2 Environment and Water Studies, Civil and Environmental Engineering, Stanford University, Stanford, California, United States of America, 3 Global Disease Detection Program, Kenya Office of the Centers for Disease Control and Prevention, Nairobi, Kenya, 4 Social and Preventative Medicine, State University of New York at Buffalo, Buffalo, New York, United States of America, 5 Global Disease Detection Branch, Division of Global Health Protection, Center for Global Health, Centers for Disease Control and Prevention, Atlanta, Georgia, United States of America

Abstract

Background: In-person structured observation is considered the best approach for measuring hand hygiene behavior, yet is expensive, time consuming, and may alter behavior. Video surveillance could be a useful tool for objectively monitoring hand hygiene behavior if validated against current methods.

Methods: Student hand cleaning behavior was monitored with video surveillance and in-person structured observation, both simultaneously and separately, at four primary schools in urban Kenya over a study period of 8 weeks.

Findings: Video surveillance and in-person observation captured similar rates of hand cleaning (absolute difference <5%, $p = 0.74$). Video surveillance documented higher hand cleaning rates (71%) when at least one other person was present at the hand cleaning station, compared to when a student was alone (48%; rate ratio = 1.14 [95% CI 1.01–1.28]). Students increased hand cleaning rates during simultaneous video and in-person monitoring as compared to single-method monitoring, suggesting reactivity to each method of monitoring. This trend was documented at schools receiving a handwashing with soap intervention, but not at schools receiving a sanitizer intervention.

Conclusion: Video surveillance of hand hygiene behavior yields results comparable to in-person observation among schools in a resource-constrained setting. Video surveillance also has certain advantages over in-person observation, including rapid data processing and the capability to capture new behavioral insights. Peer influence can significantly improve student hand cleaning behavior and, when possible, should be exploited in the design and implementation of school hand hygiene programs.

Editor: Vineet Gupta, University of Pittsburgh Medical Center, United States of America

Funding: This study was funded by CDC-Kenya/KEMRI (http://www.cdc.gov/globalhealth/countries/kenya/), and the Freeman Spogli Institute for International Studies at Stanford University (http://fsi.stanford.edu/). The findings and conclusions in this report are those of the authors and do not necessarily represent the official position of the Centers for Disease Control and Prevention. The funders had no role in study design, data collection and analysis, decision to publish, or preparation of the manuscript.

Competing Interests: The authors have declared that no competing interests exist.

* E-mail: amyjanel@stanford.edu

Introduction

Hand hygiene promotion programs are increasingly common around the world. Measurement of hand hygiene behavior is important for evaluating program effectiveness, as well as for understanding the relationship between hygiene and health. Few methods exist, however, to measure hand hygiene behavior reliably, accurately, and efficiently [1]. One of the most commonly used and least costly methods is to obtain self-reported data through in-person interviews. Such data are often biased by social desirability effects, particularly when respondents are participating in a hand hygiene intervention [2,3]. Rapid observation (i.e., spot checks) of the presence and location of hand cleaning supplies (*e.g.* soap, water) has been shown to correlate with handwashing behavior [4]. Motion sensors placed inside bars of soap have also been used to measure use[5]. Rapid observations and sensors provide no information, however, about which household member is using the soap, whether it is being used for handwashing exclusively, or whether the soap is used at critical times, such as after using the toilet or before feeding a child. In-person structured observation, in which a human observer spends several hours watching and documenting a subject's behavior, is often considered the gold standard for measuring hand hygiene behavior. Observation is time consuming and expensive relative to other data collection strategies, however, and has been shown to alter subjects' behavior[5–9].

Some studies have used video surveillance to capture hand hygiene behavior in high-income countries. Video cameras were used to capture hand hygiene behavior in hospitals in the United States, Japan, and China as early as the 1990s [10–12]. A more recent study of health care workers in a United States intensive care unit found video-recorded rates of hand cleaning to be less than 10%, as compared to rates of 60% as measured during in-person observation conducted in the same location[13]. Video monitoring of hand hygiene has also been used outside hospitals to measure hand cleaning at a petting zoo in Canada[14] and to monitor food hygiene in Australian kitchens[15]. However, video monitoring of hand hygiene has yet to be validated against in-person structured observation, or employed to monitor hand hygiene behavior in low-income settings.

The main objective of this analysis was to evaluate the viability of using video cameras to measure hand-cleaning behavior accurately within a low-income school setting. At primary schools in Kenya, video cameras recorded footage of hand cleaning stations located next to latrines. We compare data on hand cleaning rates after toileting events captured by video camera with data collected through in-person structured observation to assess concordance between the two methods. We also explore student reactivity to video surveillance and in-person observation by assessing rates of hand cleaning during simultaneous monitoring *versus* rates captured by each method independently. In addition, by comparing rates of hand cleaning when students are alone *versus* in the presence of others, we assess peer influences on hand hygiene behavior. Finally, the benefits and drawbacks of using cameras to measure hand hygiene behavior are discussed.

Methods

This study was conducted in four primary schools located within the informal settlement of Kibera, located in Nairobi, Kenya. Data collection occurred between September and November 2010. The four schools were participating in a school-based hand hygiene evaluation of alcohol-based hand sanitizer compared to traditional soap; schools were selected based on similar student populations, water supply, and sanitation characteristics. The full hand hygiene study design is described elsewhere[16]. Eligible schools had at least 100 enrolled students and latrines located on the school premises. Two of the schools were randomly assigned to receive a hand sanitizer intervention, and the other two schools assigned to receive a hand washing with soap and water intervention. A wall dispenser filled with hand sanitizer or liquid hand soap (depending on intervention assignment) was installed outside the latrines at each school. Schools assigned to the soap and water intervention were also provided with water tanks, but were responsible for obtaining water to fill the tanks.

Video cameras were positioned such that the camera surveillance frames included the doors to the latrines as well as the hand cleaning station (Figure 1). The cameras had motion sensors to initiate recording when any movement was detected within the frame, such as when students entered or exited latrines and/or used the hand cleaning stations. Over a period of 8 weeks, cameras were placed at each school 2–4 days per week to record between the hours of 8:30am and 4:00pm. At the end of each recording day, the cameras were removed to download footage files and recharge the batteries. The cameras were enclosed in plastic electrical boxes (Supercircuits, Inc. Austin, TX), mounted on wooden boards, and padlocked in place. The electrical box was designed to conceal the presence of the enclosed camera (Figure 2); however, students were informed about the presence of the cameras by school staff.

A rotating field schedule was employed to capture toileting events simultaneously and separately by video surveillance and in-person observation (Figure 3). During a subset of video recording days, in-person observation was conducted concurrently by trained Kenyan enumerators between the hours of 10:30am and 1:30pm. In-person observation was also conducted on days that video cameras were not recording, to enable comparison of hygiene behavior that was captured by in-person observation alone ($N_{P,I}$, Figure 3), video surveillance alone ($N_{V,I}$), and simultaneous in-person observation and video surveillance ($N_{P,S}$ and $N_{V,S}$). In addition, on days with simultaneous monitoring, video cameras recorded from 9:30am–10:30am, the hour preceding the start of in-person observation.

To assess teacher awareness of student reactivity to observation, in-depth interviews were conducted with 3 teachers per school (N = 12). These semi-structured interviews included open-ended questions about whether and how the CCTVs and in-person observation affected hand hygiene behavior at the school. The interviews were voice recorded, transcribed, and then translated into English.

In-person observation and video surveillance coding were generated with the use of personal digital assistants (PDAs) loaded with a survey instrument programmed in The Survey System (Creative Research Systems, Petaluma, CA). Trained observers sat within view of the latrines during in-person observation and coded toileting events in real time using PDAs. Observers did not initiate interaction with students or teachers during data collection. Video footage was coded by 4 research assistants at Stanford University. Surveillance footage without students in the camera frame was viewed at increased speed using the fast-forward feature in media viewing software. When a student subject appeared in the frame, the footage was typically observed in real time, paused if necessary, and rewound to capture details if needed. Qualitative notes were also recorded to document relevant details and behaviors not captured by the survey instrument on the PDA.

For both in-person observation and video surveillance, each student exit from a latrine was considered a "toileting event." When multiple students cleaned their hands simultaneously, both in-person and video coders were instructed to select one student arbitrarily to follow. Observers recorded if a student cleaned his or her hands with water, soap, and/or sanitizer following each toileting event. Observers also documented the student's gender (by the school uniform), the duration of hand cleaning in seconds, and the method of drying hands. Video coders also collected data on the number of other students visible within the camera frame.

Data analysis

In-person observation and video footage data for each school were compared in aggregate across all observations. The difference in rates captured by the two methods (rate ratios) of any type of hand cleaning and hand cleaning with product were estimated using Poisson regression, including binary variables to identify the individual school at which the data were collected (school fixed effects). The modeling results were also stratified by intervention type. In-person observation and video footage data were then matched by concurrent 3-hour time blocks. A date and time stamp of each toileting event was recorded automatically by the PDA during in-person observations, and by the camera itself for the video recordings. To compare in-person structured observation with video observation on the same day, video data were coded as occurring on a day without in-person observation or on a day with in-person observation. The number of toileting and hand cleaning events was summed within each of these periods, then merged to obtain a complete data set for each school day. For these simultaneous observation periods, independent sample t-tests were used to assess significant differences in overall hand cleaning rates, hand cleaning rates with soap, and the number of toileting events captured by each method.

Figure 1. Still frame from a camera positioned above a handwashing with soap station next to a latrine (door is open).

To explore reactivity to video surveillance, hand cleaning rates captured by in-person observation were compared between days with and without concurrent video camera surveillance. A similar analysis was conducted with video surveillance data to estimate reactivity to structured observation. Rate ratios (between independent and simultaneous monitoring) were calculated using Poisson regression, controlling for the individual school the data were collected from. Modeling results were also stratified by intervention type to examine any differences in reactivity between sanitizer and soap intervention schools. In addition, on days with simultaneous monitoring, hand cleaning rates captured by video surveillance in the hour preceding in-person observation were compared to rates captured once in-person observation had begun.

Ethics Statement

Administrators from the study schools gave written consent for video cameras to be placed in public locations on the school

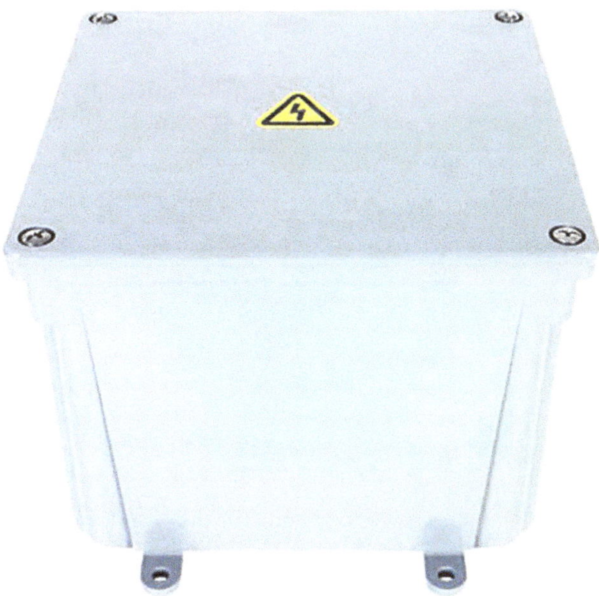

Figure 2. Electrical box camera case; camera lens is concealed by yellow electrical sticker in center (image courtesy of Supercircuits, Inc.).

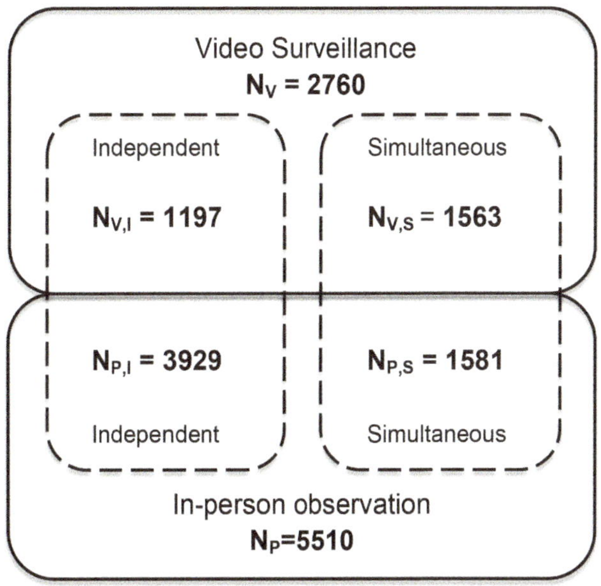

Figure 3. Schematic of study design. Number of toileting events captured on video surveillance (N_V) and through in-person observation (N_P), during independent (single-method) *versus* simultaneous monitoring periods.

premises. Written consent to observation was also obtained from teachers and the parents of enrolled students. Video surveillance was not specifically mentioned in parental consents for the following reasons: a) identities of individuals could not be recognized in the video surveillance due to limited resolution of the footage; b) the cameras were placed in public locations; and c) there was concern that explicitly publicizing video surveillance to parents could amplify student reactivity to the surveillance. It should be noted that both video surveillance and in-person observation captured non-target behaviors at times, such as students fighting or urinating/defecating outside the latrines. Teachers and school administrators were informed that the cameras could not be used to monitor illicit behavior since students were not identifiable in the footage. Ethical approval for the project and all consent procedures described above (including not informing parents of video surveillance) was obtained from the Stanford University Institutional Review Board (PR#: 19143) and the Kenya Medical Research Institute (KEMRI) Scientific and Ethical Research Committees (No. 1840).

Results

A total of 5,510 student toileting events were captured by in-person observation during 46 unique school-days at sanitizer intervention schools and 51 unique school-days at soap and water intervention schools. Video surveillance captured 2760 toileting events over 21 school-days at sanitizer schools and 20 school-days at soap and water schools. Overall, hand cleaning rates (with water only, soap, and/or sanitizer) captured by video observation were higher (64%) than hand cleaning rates documented during in-person observation (56%); however, the rates were not significantly different when controlling for the individual school at which the data were collected (Table 1). Trends were similar for hand cleaning rates with a product (soap or sanitizer) (Table 1). When stratified by intervention, video surveillance and in-person observation captured similar hand cleaning rates at sanitizer schools (82–84%, RR = 0.99, 95% CI [0.92–1.06]) (Table 1). At soap intervention schools, video surveillance captured significantly higher rates of hand cleaning (42%) as compared to in-person observation data (37%) (RR = 1.14, 95% CI [1.03–1.26]) (Table 1).

Both methods found that female students cleaned their hands slightly more often than boys; the hand cleaning rate after toileting was 4% higher among girls according to video surveillance ($p = 0.02$) and 3% higher according to in-person observation ($p = 0.01$). Average hand cleaning time recorded during video observation was significantly higher (49 seconds, SD 44, N = 515 hand cleaning events) than that recorded during in-person observation (30 seconds, SD 24, N = 2622 hand cleaning events; $p<0.01$). Both video observation and in-person observation demonstrated longer hand cleaning times for handwashing with soap than rubbing with sanitizer (Table 1).

In-person observation and video surveillance data were directly matched and compared for 22 school-days, each with 3-hours of observation overlap (13 school-days at soap intervention schools, 9 school-days at sanitizer intervention schools). The absolute difference in mean hand cleaning rates captured by in-person observation *versus* video observation was not significant (difference = 0.04, $p = 0.74$, N = 44 school-days). Similar results were obtained for comparisons of hand cleaning rates with soap (difference = 0.03, $p = 0.81$, N = 44 school-days) and with sanitizer (difference = 0.00, $p = 0.99$, N = 44 school-days). However, the mean number of toileting events captured by in-person observation over a 3-hour period was significantly higher than that captured by video observation (mean difference = 25, $p = 0.02$, N = 44).

Reactivity to video surveillance

Overall, in-person observation data collected with and without concurrent video surveillance show significantly higher hand cleaning rates during video surveillance periods (RR = 1.11, 95% CI 1.03–1.20, Table 2). Students at soap intervention schools were 1.3-fold more likely to wash their hands during concurrent video surveillance when compared with periods of in-person observation alone (RR = 1.28, 95% CI 1.14–1.43). By contrast, students at sanitizer intervention schools were not significantly more likely to clean their hands during simultaneous video and in-person observation (RR = 0.99, 95% CI 0.89–1.11, Table 2). Video

Table 1. Student hand cleaning rates (% of toileting events) and duration captured by in-person structured observation *versus* video observation.

	Video surveillance	In-person observation	
	All hand cleaning (with or without product) [%, N]		RR (95% CI)[φ]
All schools	64%, N = 2760	56%, N = 5510	1.04 (0.98–1.12)
Sanitizer schools	82%, N = 1474	84%, N = 2130	0.99 (0.92–1.06)
Soap schools	44%, N = 1286	38%, N = 3380	1.16 (1.05–1.28)*
	Cleaned hands with soap or sanitizer [%, N]		RR (95% CI)[φ]
All schools	63%, N = 2760	55%, N = 5510	1.04 (0.98–1.10)
Sanitizer schools	82%, N = 1474	84%, N = 2130	0.99 (0.92–1.06)
Soap schools	42%, N = 1286	37%, N = 3380	1.14 (1.03–1.26)*
	Mean duration of hand cleaning in seconds [mean (SD), N]		p-value (t-test)
All schools	49 (44), N = 515	30 (24), N = 2622	P<0.001
Sanitizer schools	34 (26), N = 310	21 (9), N = 1676	P<0.001
Soap schools	71 (55), N = 205	47 (33), N = 946	P<0.001

*p<0.05.
[φ]Rate-ratios (RR) and p-values reported for Poisson regression analysis of rate differences between video and in-person data, while controlling for the individual school at which the data were collected.

Table 2. Hand cleaning rates (% of toileting events) captured by video surveillance data (left), and by in-person observation (right).

	Video surveillance data			In-person observation data		
	Simultaneous monitoring [%, $N_{V,S}$]	Video surveillance only [%, $N_{V,I}$]	RR (95% CI)[φ]	Simultaneous monitoring [%, $N_{P,S}$]	In-person observation only [%, $N_{P,I}$]	RR (95% CI)[φ]
All schools	65%, N = 1563	64%, N = 1197	1.08 (0.98–1.19)	60%, N = 1581	54%, N = 3929	1.11 (1.03–1.20)*
Sanitizer schools	83%, N = 738	81%, N = 736	1.02 (0.91–1.14)	84%, N = 593	83%, N = 1537	0.99 (0.89–1.11)
Soap schools	47%, N = 825	39%, N = 461	1.22 (1.02–1.46)*	45%, N = 988	36%, N = 2392	1.28 (1.14–1.43)*

Simultaneous monitoring refers periods when in-person observation and video surveillance were conducted concurrently.
*$p<0.05$.
[φ]Rate-ratios (RR) and p-values reported for Poisson regression analysis of rate differences between simultaneous and independent monitoring, while controlling for the individual school at which the data were collected.

footage analysts noted instances of students staring directly at cameras while hand cleaning, and documented students pointing and gesturing towards the cameras.

During in-depth interviews, several teachers mentioned that video surveillance may have served as a reminder for students to practice hand cleaning, and as a source of motivation for teachers to ensure compliance. One teacher at a sanitizer intervention school stated "once a camera is put in place, it acts as a spy so even if you tend to forget [you] will see it and wash your hands." A teacher at a soap intervention school said, "you know when there is supervision you cannot be lazy. When the camera was there we made sure there was water in the tank." A few teachers thought the camera did not affect behavior long-term; one teacher stated: "even when [the camera] was not there it had become a routine for people to wash their hands."

Reactivity to in-person observation

Overall, hand cleaning rates were higher during concurrent video and in-person observation compared to during video surveillance alone, although the difference was not statistically significant (RR = 1.08, 95% CI 0.98–1.19, Table 2). Students at soap intervention schools were significantly more likely to wash their hands during concurrent video and in-person observation (RR = 1.22, 95% CI 1.02–1.46); for students at sanitizer intervention schools no significant difference was found (RR = 1.02, 95% CI 0.91–1.14, Table 2). On those days with concurrent in-person observation and video surveillance, hand cleaning rates were lowest prior to the arrival of the field observer (55%), after which they increased by about 10 percentage points during in-person observation (Table 3).

The majority of teachers interviewed mentioned that the presence of study field staff at hand cleaning stations may have improved hand hygiene behavior by students. One teacher noted, "by you sitting there and observing made it easier for me because you helped me watch [the students]." It was noted that the influence of in-person observation was greater at the beginning of the study. One teacher said "with time we got used to you and also to the sanitizer," while another noted, "the good thing is even after you leave we will still clean our hands as now it is a habit."

Peer influence

Only one student was visible within the camera view frame in 28% of toilet use events captured by video surveillance (N = 2811). Overall, when students were alone at a hand cleaning station, hand cleaning rates averaged 48%, compared to 71% when at least one other student was present (Table 4). At sanitizer schools, students were significantly more likely to clean their hands when one or more other students were viewed near the hand cleaning station (RR = 1.18, 95% CI: 1.02–1.37). This trend was not statistically significant at soap intervention schools (Table 4). Hand cleaning rates showed an overall trend of increasing as the number of other people present at hand cleaning stations increased, with the exception of a slight decrease in hand cleaning when greater than 10 subjects were observed (Figure 4).

Video coders noted the following scenarios when more than one person was present at the hand cleaning station: 1) Teachers instructing groups of students on proper hand cleaning technique; 2) Older students assisting younger students (*e.g.* lifting young students up to reach wall dispenser, placing product on hands); 3) Students reminding other students to clean hands and/or demonstrating proper technique; 4) Groups of students engaged in conflict over access to hand cleaning materials (*e.g.* pushing, hitting); 5) Students being called away from hand cleaning stations by teachers to attend class.

Discussion

Analysis of video surveillance footage captured by cameras placed at shared school latrines yielded similar rates of hand cleaning by students post-toileting to those found by in-person structured observation during the same time period. This work suggests that video surveillance of hand hygiene behavior yields

Table 3. Average rate of hand cleaning (% of toileting events) captured by video surveillance preceding and during in-person observation (days with concurrent video/in-person observation only).

Hand cleaning, as observed by video surveillance	Preceding in-person observation, N = 368	During in-person observation, N = 1022
All hand cleaning (with or without product)	55%	66%
Hand cleaning with soap or sanitizer	54%	65%

Table 4. Hand cleaning rates (% of toileting events) when the subject was observed to be alone in the video frame *versus* when other students were present in the frame, as captured by video surveillance.

	Subject alone in video frame, N = 768	Two or more individuals in video frame, N = 1991	RR (95% CI)
All data	48%	71%	1.14 (1.01–1.28)*
Sanitizer schools	71%	85%	1.18 (1.02–1.37)*
Soap schools	32%	51%	1.08 (0.89–1.30)

valid results that are comparable to in-person observation in a resource-constrained setting.

We found evidence of student reactivity to both video surveillance and in-person observation, in the form of higher hand cleaning rates at handwashing with soap intervention schools. Reactivity was not detected at sanitizer intervention schools; this may be explained by the fact that hand cleaning rates with sanitizer were over 80%, leaving less room for compliance rates to shift upwards. Reactivity to simultaneous video and in-person observation increased handwashing rates by similar magnitudes (28% and 22%, respectively). These data should be interpreted as the marginal impact of each observation method on behavior when the other method is also in effect, as reactivity to each method could be higher when compared to a period during which no observation is occurring. Reactivity measured in this study was less than the 35% increase in hand cleaning frequency due to structured observation found by Ram and colleagues (2010) among female caregivers of young children [5]. Previous studies documenting reactivity to in-person observation have been principally conducted with adult subjects [17–19]. It is possible that students may be less likely than adults to exhibit reactivity to in-person observation, as they are accustomed to being watched by teachers. Examining changes in reactivity over time (*e.g.* decreased reactivity as the novelty of observation wore off) for each method was outside the scope of this study, and would be a valuable area for future research.

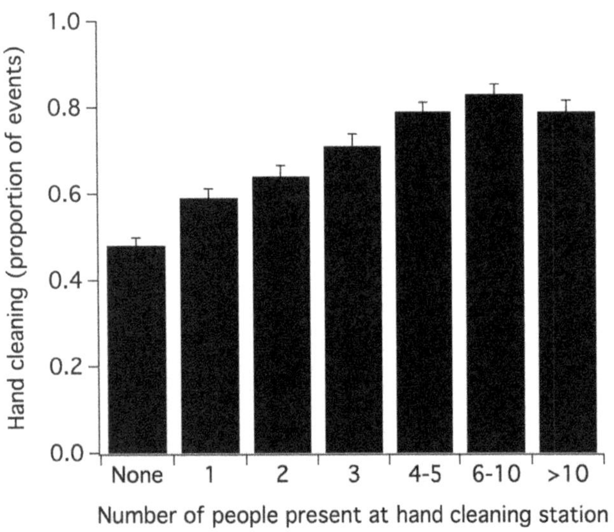

Figure 4. Proportion of student toileting events followed by hand cleaning, shown by number of other people present at the hand cleaning station (visible within the camera view frame). Error bars show standard error of the mean.

The presence of at least one other person at the hand cleaning station increased student hand cleaning rates by 14% among all schools, indicating that peer influence can significantly impact student hand cleaning compliance. Reaction to peer presence has previously been documented to increase hand hygiene compliance among adults. A previous study found that female students cleaned their hands 91% of the time after using the bathroom when someone else was present, compared to 55% of the time when they were alone in the sink area [20]. We also found that hand cleaning rates increased proportionally with the number of additional people observed in the immediate vicinity of the hand cleaning station, up to a threshold of 10 people (Figure 4). Higher rates of hand cleaning in groups indicate that the intervention established hand cleaning as a social norm among the students, an important motivation for hand cleaning behavior [21]. It is also possible that the presence of other individuals at the hand cleaning station served as a reminder or cue, whether conscious or unconscious, for a student to stop and hand clean. Video analysts recorded a number of phenomena that could explain the drop in hand cleaning rates during crowding events (>10 people), including students being unwilling to wait in queues to use the hand cleaning facilities, physical conflicts between students, and teachers instructing students to go to class.

The positive influence of peer presence on student hand hygiene behavior observed in this study has key implications for the design and implementation of future hand hygiene programs in schools. Placement of hand cleaning materials in public locations, along with the scheduling of specific times for bathroom breaks between classes, could significantly improve student hand hygiene compliance rates. Strategies to instill hand hygiene as a social norm may also be effective, such as designating specific students to be hand hygiene "champions," or the formation of student clubs to demonstrate and promote hand hygiene to their classmates [22].

Some disadvantages to using video surveillance were evident. Cameras had to be padlocked during the day and removed nightly to avoid theft and vandalism. Minimal available space and flimsy school walls made installation and stable positioning of the camera difficult. On some occasions, cameras were knocked or fell into positions that did not capture the full view of the latrine entrance and hand cleaning station. Notably, it was infeasible to utilize cameras to record hand cleaning at control schools in this study, because one school did not keep its hand cleaning materials (i.e. portable basin and soap) in a specific place. Video surveillance is thus best suited for monitoring compliance at fixed hand cleaning stations.

We found that video surveillance has several potential advantages as an observational method. First, video surveillance significantly reduced the amount of local staff time necessary for data collection. The cameras had motion sensors to trigger recording only when movement was detected at hand cleaning stations, thus eliminating staff time spent observing the hand cleaning station when latrines were not in use. Moreover, video footage could be analyzed at increased speeds. Video surveillance also allowed off-site research staff to view intervention compliance directly, to replay footage to capture additional information, and to witness unexpected reasons for non-compliance. Video surveillance is a useful method for monitoring hand hygiene behavior that enables rapid synthesis of rich behavioral data.

Acknowledgments

We would like to thank Jenna Scalmanini, Beryl Oyier, Harun Njuguna, Lillyan Githinji, and Mark Buyu Luta for excellent field assistance. We also appreciate the cooperation of the study school administrators and study participants.

Author Contributions

Conceived and designed the experiments: AJP AGB RFB PKR JD. Performed the experiments: AJP AGB. Analyzed the data: AJP AGB. Contributed reagents/materials/analysis tools: PKR RFB. Wrote the paper: AJP AGB RFB PKR JD.

References

1. Ram P (2013) Practical Guidance for Measuring Handwashing Behavior: 2013 Update: 1–36.
2. Manun'Ebo M, Cousens S, Haggerty P, Kalengaie M, Ashworth A, et al. (1997) Measuring hygiene practices: a comparison of questionnaires with direct observations in rural Zaire. Tropical Medicine & International Health 2: 1015–1021. doi:10.1046/j.1365-3156.1997.d01-180.x.
3. Stanton BF, Clemens JD, Aziz KMA, Rahman M (1987) Twenty-four-hour recall, Knowledge-attitude-practice questionnaire, and direct observations of sanitary practices: a comparative study. Bulletin of World Health Organization 65: 217–222.
4. Luby SP, Halder AK, Tronchet C, Akhter S, Bhuiya A, et al. (2009) Household Characteristics Associated with Handwashing with Soap in Rural Bangladesh. Am J Trop Med Hyg 81: 882–887. doi:10.4269/ajtmh.2009.09-0031.
5. Ram PK, Halder AK, Granger SP, Jones T, Hall P, et al. (2010) Is Structured Observation a Valid Technique to Measure Handwashing Behavior? Use of Acceleration Sensors Embedded in Soap to Assess Reactivity to Structured Observation. Am J Trop Med Hyg 83: 1070–1076. doi:10.4269/ajtmh.2010.09-0763.
6. Cousens S, Kanki B, Toure S, Diallo I, Curtis V (1996) Reactivity and repeatability of hygiene behaviour: Structured observations from Burkina Faso. Soc Sci Med 43: 1299–1308.
7. Gittelsohn J, Shankar AV, West KP, Ram RM, Gnywali T (1997) Estimating reactivity in direct observation studies of health behaviors. Human organization 56: 182–189.
8. McCarney R, Warner J, Iliffe S, van Haselen R, Griffin M, et al. (2007) The Hawthorne Effect: a randomised, controlled trial. BMC Med Res Methodol 7: 30. doi:10.1186/1471-2288-7-30.
9. Pan S-C, Tien K-L, Hung I-C, Lin Y-J, Sheng W-H, et al. (2013) Compliance of Health Care Workers with Hand Hygiene Practices: Independent Advantages of Overt and Covert Observers. PLoS ONE 8: e53746. doi:10.1371/journal.pone.0053746.s002.
10. Nishimura S, Kagehira M, Kono F, Nishimura M, Taenaka N (1999) Handwashing before entering the intensive care unit: what we learned from continuous video-camera surveillance. Am J Infect Control 27: 367–369.
11. Brown J, Froese-Fretz A, Luckey D, Todd JK (1996) High rate of hand contamination and low rate of hand washing before infant contact in a neonatal intensive care unit. The Pediatric Infectious disease journal 15: 908–910.
12. Zhang L, Chen J, Chang L, Xu M (2009) Hand Hygiene Management by Video Monitoring in Neonatal Intensive Care Unit. Chinese Journal of Nosocomiology.
13. Armellino D, Hussain E, Schilling ME, Senicola W, Eichorn A, et al. (2011) Using High-Technology to Enforce Low-Technology Safety Measures: The Use of Third-party Remote Video Auditing and Real-time Feedback in Healthcare. Clinical Infectious Diseases 54: 1–7. doi:10.1093/cid/cir773.
14. Anderson M, Weese JS (2012) Video observation of hand hygiene practices at a petting zoo and the impact of hand hygiene interventions. Epidemiol Infect 140: 182.
15. Jay LS, Comar D, Govenlock LD (1999) A video study of Australian domestic food-handling practices. J Food Prot 62: 1285–1296.
16. Pickering AJ, Davis J, Blum AG, Scalmanini J, Oyier B, et al. (2013) Access to Waterless Hand Sanitizer Improves Student Hand Hygiene Behavior in Primary Schools in Nairobi, Kenya. American Journal of Tropical Medicine and Hygiene. doi:10.4269/ajtmh.13-0008.
17. Russell A, Russell G, Midwinter D (1992) Observer influences on mothers and fathers: Self-reported influence during a home observation. Merrill-Palmer Quarterly (1982-): 263–283.
18. Pedersen DM, Keithly S, Brady K (1986) Effect of an Observer on Conformity to Handwasing Norm. Perceptual and Motor Skills 62: 169–170.
19. Maury E, Moussa N, Lakermi C, Barbut F, Offenstadt G (2006) Compliance of health care workers to hand hygiene: awareness of being observed is important. Intensive Care Med 32: 2088–2089. doi:10.1007/s00134-006-0398-9.
20. Drankiewicz D, Dundes L (2003) Handwashing among female college students. Am J Infect Control 31: 67–71. doi:10.1067/mic.2003.6.
21. Curtis VA, Danquah LO, Aunger RV (2009) Planned, motivated and habitual hygiene behaviour: an eleven country review. Health Educ Res: cyp002. doi:10.1093/her/cyp002.
22. O'Reilly CE, Freeman MC, Ravani M, Migele J, Mwaki A, et al. (2007) The impact of a school-based safe water and hygiene programme on knowledge and practices of students and their parents: Nyanza Province, western Kenya. Epidemiol Infect.

7

A Methodology for Adaptable and Robust Ecosystem Services Assessment

Ferdinando Villa[1]*, Kenneth J. Bagstad[2], Brian Voigt[3], Gary W. Johnson[4], Rosimeiry Portela[5], Miroslav Honzák[5], David Batker[6]

1 Basque Centre for Climate Change (BC3), IKERBASQUE, Basque Foundation for Science, Bilbao, Bizkaia, Spain, **2** Geosciences & Environmental Change Science Center, U.S. Geological Survey, Denver, Colorado, United States of America, **3** Gund Institute for Ecological Economics, Rubenstein School of Environment and Natural Resources, University of Vermont, Burlington, Vermont, United States of America, **4** Department of Computer Science, University of Vermont, Burlington, Vermont, United States of America, **5** Conservation International, Arlington, Virginia, United States of America, **6** Earth Economics, Tacoma, Washington, United States of America

Abstract

Ecosystem Services (ES) are an established conceptual framework for attributing value to the benefits that nature provides to humans. As the promise of robust ES-driven management is put to the test, shortcomings in our ability to accurately measure, map, and value ES have surfaced. On the research side, mainstream methods for ES assessment still fall short of addressing the complex, multi-scale biophysical and socioeconomic dynamics inherent in ES provision, flow, and use. On the practitioner side, application of methods remains onerous due to data and model parameterization requirements. Further, it is increasingly clear that the dominant "one model fits all" paradigm is often ill-suited to address the diversity of real-world management situations that exist across the broad spectrum of coupled human-natural systems. This article introduces an integrated ES modeling methodology, named ARIES (ARtificial Intelligence for Ecosystem Services), which aims to introduce improvements on these fronts. To improve conceptual detail and representation of ES dynamics, it adopts a uniform conceptualization of ES that gives equal emphasis to their production, flow and use by society, while keeping model complexity low enough to enable rapid and inexpensive assessment in many contexts and for multiple services. To improve fit to diverse application contexts, the methodology is assisted by model integration technologies that allow assembly of customized models from a growing model base. By using computer learning and reasoning, model structure may be specialized for each application context without requiring costly expertise. In this article we discuss the founding principles of ARIES - both its innovative aspects for ES science and as an example of a new strategy to support more accurate decision making in diverse application contexts.

Editor: Antoni Margalida, University of Lleida, Spain

Funding: The development of ARIES was originally funded by the US National Science Foundation (grant 9982938), and receives support from the ASSETS project funded by ESPA/NERC (grant NE-J002267-1). UNEP-WCMC and Conservation International also provided support for the development of specific components. Mendenhall Postdoctoral Research Program funded Bagstad's work at the U.S. Geological Survey. The World Bank WAVES partnership supported the case study in Madagascar. Funding for Brian Voigt was provided by the US Agency for International Development (EEM-A-00-10-00001). The funders had no role in study design, data collection and analysis, decision to publish, or preparation of the manuscript.

Competing Interests: The authors have declared that no competing interests exist.

* E-mail: ferdinando.villa@bc3research.org

Introduction

The advantages of an ecosystem services (ES) view of coupled human-natural systems have been widely recognized in science, management and governance [1]. Focusing on both the biophysical mechanisms of ES provision and the socioeconomic implications of their use can allow decision makers to directly link natural capital to the societies and economies that depend on it. An ES approach can also facilitate understanding and communication of the projected consequences of resource competition in the face of scarcity as well as global and local change. Once discussions on ES became mainstream - thanks largely to the seminal Millennium Ecosystem Assessment (MEA) [2] - and lessons from many individual case studies were learned [3], a first generation of integrated, multi-ES assessment methodologies and tools has been striving to meet the needs of an audience that cuts across the academic, governmental, NGO, and corporate sectors [4,5]. Rapid assessment methods have come to command wide interest from all these communities [5]. Yet it is generally recognized that systematic use of ES in decision- and policy-making requires a degree of accuracy that is rarely met in practice [6,7]. Most early assessment studies [8,9] and some recent methods [1,10] infer ES values through production functions whose driving input is land cover type, alone or complemented by limited other structural information (e.g., vegetation type). Other methods [3] have proposed models of a more functional nature to more accurately represent the mechanistic underpinnings of ES dynamics [11–14].

Much less well explored remain the issues of non-linearity, incongruent scales of provision and use, thresholds, feedbacks and tipping points [15,16] in and between the ecological and social systems that define ES dynamics [17–19]. In practice, both ecological and social sciences have struggled to understand and predict changes in large and complex dynamic systems [18–20]. Even with adequate methodologies, the specificity and cost of assessments increase quickly with model sophistication [10], due to the need for both domain expertise and accurate time-series data for model calibration and parameterization. This makes detailed ES quantification impractical in most institutional contexts, particularly since decision making requires timely analysis [21]. Applications are further complicated by the need for true interdisciplinarity to link the underlying science and policy sides of ES and to communicate results effectively across institutional and societal boundaries. Indeed, effective translation to policy is often as challenging as the search for suitable quantification methods.

Despite these difficulties, better consideration of the dynamic aspects of ES provision is needed to understand the consequences of policy decisions impacting ES [7,11,12,22]. Overlooking temporal dynamics and spatial modes of service delivery to users makes it difficult to understand and communicate *actual* values of ES accrued by societies; instead, *potential* values are commonly estimated (e.g., the amount of floodwater potentially retained by green infrastructure instead of the amount of water actually prevented from impacting flood-prone people and property). This issue may be at the source of common criticisms of overstatement of value [23] and of a subsequent lack of confidence in ES-informed policy.

Reconciling calls for simplicity and intuitiveness with the need for accuracy, specificity and dynamic resolution is challenging and risky. If on one hand decision makers (or even the scientific community) are skeptical of methods they see as complex and opaque, on the other hand the oversimplification of complex and highly diverse processes and trade-offs may yield ineffective assessments. True methodological innovation could result from incorporating enough flexibility to adapt models to diverse social, economic, and policy contexts without overly complicating their application. A previously published set of evaluative criteria for ES methods (Table 1, modified from [24]) enumerates characteristics that we believe crucial to wider consideration of ES in public- and private-sector decision making. ES assessment studies in the recent literature have begun to address these criteria, for example accounting for spatial aspects of ES dynamics [7,12] and attempting to quantify uncertainty, e.g., using Bayesian techniques [25,26]. Only a few of these advances have made their way into user-ready "tools" for decision-making [14,21,27]. The methodology proposed in this article aims to consolidate these principles by providing them with a systematic scientific foundation, so that technologies can be designed that improve their accessibility to decision makers.

The effort described in the next sections was structured around a set of core goals. We strived first of all to improve the ES narrative so that services are consistently quantified from the viewpoint of their beneficiaries. This approach emphasizes the spatial dynamics of ES flow and use by beneficiary groups, thereby distinguishing between potential ES values and actually accrued ones. We also aimed to explicitly quantify uncertainty by modeling ES supply and demand probabilistically when appropriate. To reduce the burden of data gathering and pre-processing for the end-user, we prioritized automatic retrieval of input data from web-accessible datasets using open standards wherever possible. Finally, we devised ecoinformatics advances that support more flexible, computer-aided model assembly, including the option of transparent integration with pre-existing process models (e.g., for hydrology or nutrient dynamics) when needed input data are available and in contexts where these models have proven their effectiveness. The rest of this article discusses the first results of the 6-year, multi-investigator effort that grew from these initial goals, by addressing in each following section:

- The founding principles and technological innovations that underlie the design of ARIES.
- Results of pilot applications that exemplify both the conventional and the novel ES assessments that can be conducted with it.
- Innovations, limitations and ways forward with reference to the fast-changing ES assessment landscape.
- The positioning of ARIES in context with the criteria of Table 1, in terms of achievements and outstanding goals.

The ARIES Methodology

The ARIES (ARtificial Intelligence for Ecosystem Services) methodology has been in development since 2007 and in use, through case studies using prototypes of evolving sophistication, since 2010. ARIES aims to quantify ES in a manner that acknowledges dynamic complexity and its consequences, but keeps its models sufficiently simple to remain tractable, general and scalable to varying levels of detail and data availability. The method is founded on two main innovations:

1. An extension of ES science intended to enrich the dominant MEA narrative with a renewed focus on beneficiaries, probabilistic analysis, and spatio-temporal dynamics of flows and scale. The result can heighten awareness of important distinctions such as that between potential and actual benefits.
2. The capability to automatically assemble the most appropriate ES models based on a library of modular components, driven by context-specific data and machine-processed ES knowledge. A model structure fitting the goals, the context and the available data can thus be used for each situation, avoiding the pitfalls of the common "one model fits all" paradigm.

The advancements made in these two research directions are interrelated, as the theoretical extensions (1) are designed to fit a modular modeling approach (2). The subsections that follow provide methodological detail on each point.

Improving the ES Narrative: from Static Service Values to Dynamic Benefits

The MEA defined ES using a typology of "supporting," "regulating," "provisioning," and "cultural" services [8,28–30]. As this original classification became dominant in framing the discourse on ES, social components of ES were emphasized less strongly than ecological ones. As a result, the diversity of social values and uptake modes for the same ES in different social and geographical contexts remained relatively understudied [31]. Problems with such ecosystem-centric classifications (e.g., the potential for "double counting" of ES values [32–34]) prompted suggestions to shift focus to the beneficiary side [25,35] and to better characterize the spatial locations of ES provision, beneficiaries, and spatial flows [7,12,22].

Spatially explicit benefit flows. A redefinition of the key terms of the MEA language, now deeply ingrained in scientific and policy dialogue, would be impractical and undesirable. In order to extend the mainstream MEA conceptual model to support a systematic emphasis on beneficiaries while remaining compatible with its underlying framework, an ARIES assessment begins with the mapping of concrete and spatially explicit beneficiary groups, each uniquely characterized by their demand type and conceptualization of value (this and other key terms are defined in a glossary available as Glossary S1). We define an ES *benefit* as the outcome of the *set of processes that join a beneficiary group with specified source ecosystem(s) through a clearly identified spatio-temporal flow*. A *service*

Table 1. Evaluative criteria to improve uptake and utility of ES quantification methods in decision-making.

Criterion	Justification
Quantitative	Quantitative results are needed to compare trade-offs. Quantitative character includes providing spatially explicit results accompanied by uncertainty measures.
Time/resource requirements	A less time-intensive method can be more practically applied on a widespread scale.
Open source or proprietary	Methodologies delivered through open-source software and services are more transparent and can be independently applied, tested and improved.
Development and documentation	Methods that are well developed and documented have greater transparency and credibility, improving trust with decision makers and the public.
Scalability	Methods that can be applied across multiple spatial and temporal scales are more versatile and able to address trade-offs whose significance varies with scale.
Adaptability vs. Generality	Methods that can be applied in diverse ecological and socioeconomic contexts can be more consistently and inexpensively applied than place-specific approaches. A versatile methodology should operate with measurable accuracy across the continuum between general (low-cost, rapid assessment) and custom-tailored to specific needs and situations.
Amenability to multiple valuation systems	Strictly monetarily-based valuation methods are inadequate to account for all value types.

in MEA parlance corresponds conceptually to a collection of such benefits. For example, the water supply service would include separate benefits for each type of water use in an area, such as irrigation, domestic, or industrial use. Aggregate service values can then be obtained by combining the values assigned to each benefit, which may be modeled at independent spatial and temporal scales and translated into human well-being in different ways (Figure 1). Emphasizing and spatially locating beneficiaries also helps to systematically identify appropriate spatial boundaries for ES analysis. The region of interest for each benefit can be identified in space by determining the supply area capable of providing a flow of benefits that intercepts the beneficiary groups under study.

Emphasis on beneficiaries allows us to improve the detail of ES models and to clarify their scale, dynamics, and eventual valuation. Once beneficiaries have been identified spatially, quantifying and mapping ES flows that reach them becomes key to distinguishing between the *potential* for benefit provision by ecosystems and the benefits *actually accrued* by society. This approach can substantially improve the accuracy of ES valuation [15,22] and expand the value of ES assessments to decision makers [7]. In resource management scenarios, beneficiary-based maps of ES provision can be crucial for influencing management decisions that appropriately address distributional equities among "winners" and "losers" [22,36]. Yet a mechanistic understanding of the dynamics of ES provision, use, and flow for such diverse and complex services as sediment regulation, pollination, or recreation is very challenging, and model outputs may be difficult for decision makers to understand. Like other mainstream ES models (see the review in [3]), ARIES currently uses production functions to model its main elements (source, use and sink regions: see Figure 2 and description below). Such outputs can be independently compared to those provided by other methods. However, unlike in other methods, the supply component is intended to quantify *potential* benefit provision, as it does not account for society's use of ES.

ES benefit transport and delivery in time and space are handled in ARIES through dynamic *flow models*, whose algorithms use the production function output along with quantification of demand as inputs. This multi-stage approach is illustrated in Figure 2, where amounts of a *service carrier* produced in *source* (supply) regions flow to beneficiaries situated in *use* (demand) regions. Flows reach beneficiaries along physical or informational *flow paths*, which result from spatially explicit and dynamic physical processes. Demand may be rival (each user reduces the flow available for others) or non-rival (use does not appreciably reduce availability to others). The benefit connected with these flows may be *provisioning* (supplying a valuable good or service to users, such as scenic views, food, or drinking water) or *preventive* (where the contact with a biophysical flow is detrimental to human well-being, and the actual benefit is supplied by an ecosystem's mitigation of that damaging effect, as in the mitigation of flood water, sediment, nutrients, disease, or wildfire). Along flow paths, *sink* regions may absorb or deplete the service-carrying medium – a beneficial process in the case of preventive benefits but a detrimental process for provisioning benefits. It should be noted that although the MEA ES classification uses the similar term *provisioning services*, we are not seeking to classify services when we distinguish between provisioning and preventive benefits, but instead to classify flow behaviors to enable a systematic description of how ecosystems provide benefits to people. As inputs, most ARIES flow models use the spatial distributions of the sources of service-carrying medium, beneficiary demand, and potential sinks, specified as maps covering the area under study.

The conceptual model shown in Figure 2 does not necessarily depict a spatial reality directly: regions may overlap or be remote, and, depending on the type of benefit, flows may take place in diverse ways, for example through hydrologic, informational, or transportation networks. The amount of carrier that actually reaches the beneficiaries (for provisioning benefits) or is absorbed by ecosystems on its way to beneficiaries (for preventive benefits) is the basis for assessing accrued value. Areas where the flow trajectories for one or more benefit concentrate ("area of critical ES flow" in Figure 2) can be critical to the delivery of the service even if they are not included in either source or use regions [13,22,27,37].

Values accrued by society are, as introduced above, the result of the flow of a beneficial or detrimental carrier. The carrier may be physical (e.g., water in the case of water supply or flood regulation, CO_2 in carbon storage and sequestration) or informational (e.g., visual information in aesthetic services). The mode of transmission and resulting spatial patterns of ES flow are determined by the nature of the carrier, the type of benefit

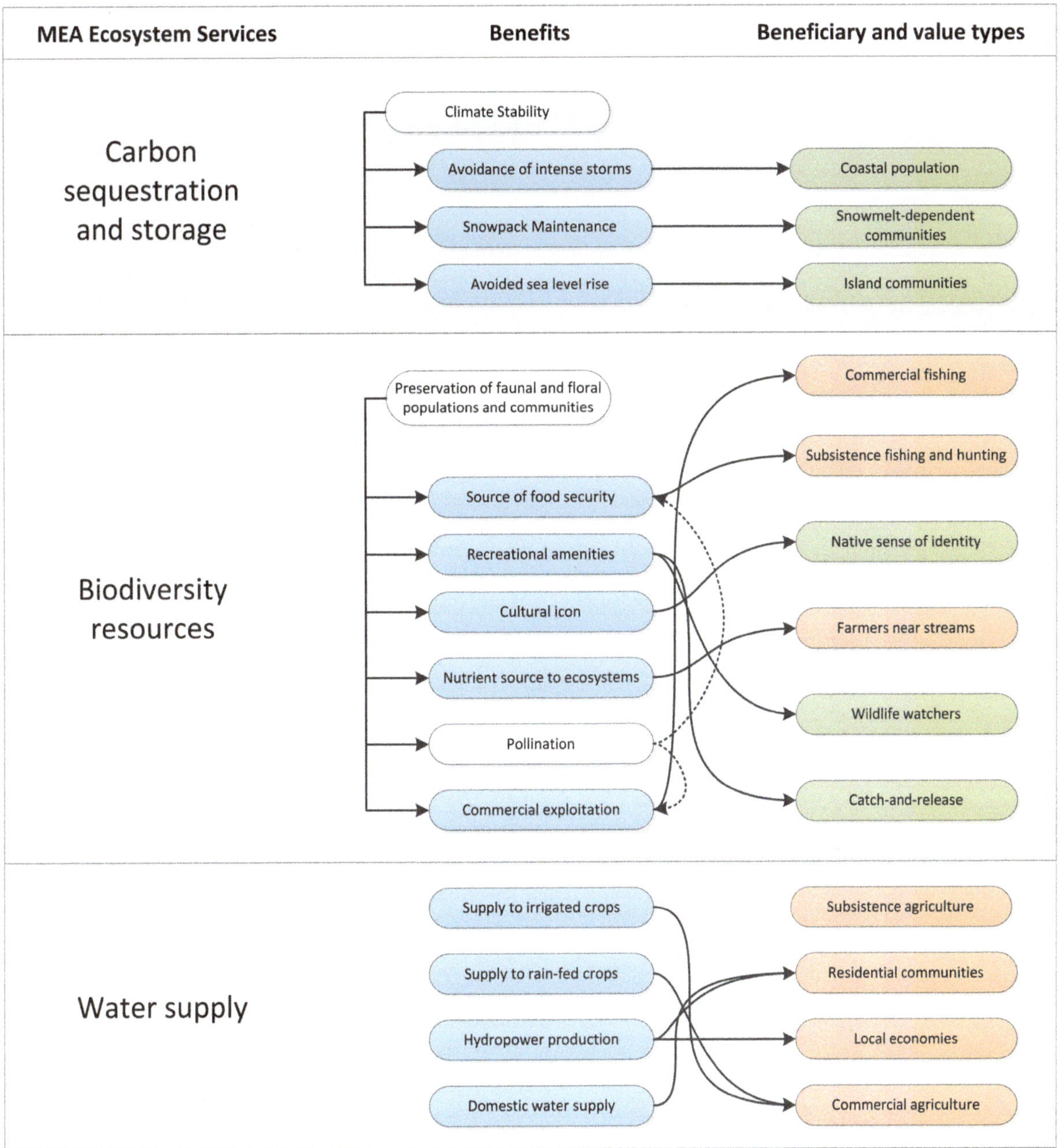

Figure 1. A simplified image of a small part of the ARIES knowledge base. The MEA ES categories on the left are broken down into the benefits in the middle, only some of which (in blue) are directly connected to beneficiaries. Dashed lines exemplify indirect relationships that, when taken as the description of legitimate ecosystem services, have the potential of causing "double counting" by identifying benefits that are "intermediate" and not "final", i.e., not directly linked to beneficiaries. Beneficiaries are depicted on the right, with non-rival benefits in green and rival benefits in orange.

(provisioning or preventive), the physical characteristics of the landscape, and the presence of human or natural features that act as sinks. ARIES quantifies such flows using a family of network flow propagation models [38] termed Service Path Attribution Networks (SPANs: [7,12,13,26]) that simulate carrier movement, absorption and delivery for different classes of carriers (Table 2). The appropriate SPAN models are chosen and linked during model assembly depending on the benefit type.

The SPAN framework, covered in detail in Johnson et al. [12,13], reduces the great diversity of ES benefits to a small and general taxonomy of flow types [7], based on a uniform model of connectivity and accessibility across spatial networks. This approach offers less mechanistic accuracy than could be provided by a detailed physical model, such as hydrological or sediment transport, but offers the advantages of representing all ES in a unified way and, more importantly, of being able to run in most situations with probabilistic initial conditions and manageable data requirements. The interpretation of time in SPAN models is explicit but is intentionally not temporally referenced: an initial condition progresses toward completion of an ES flow across space

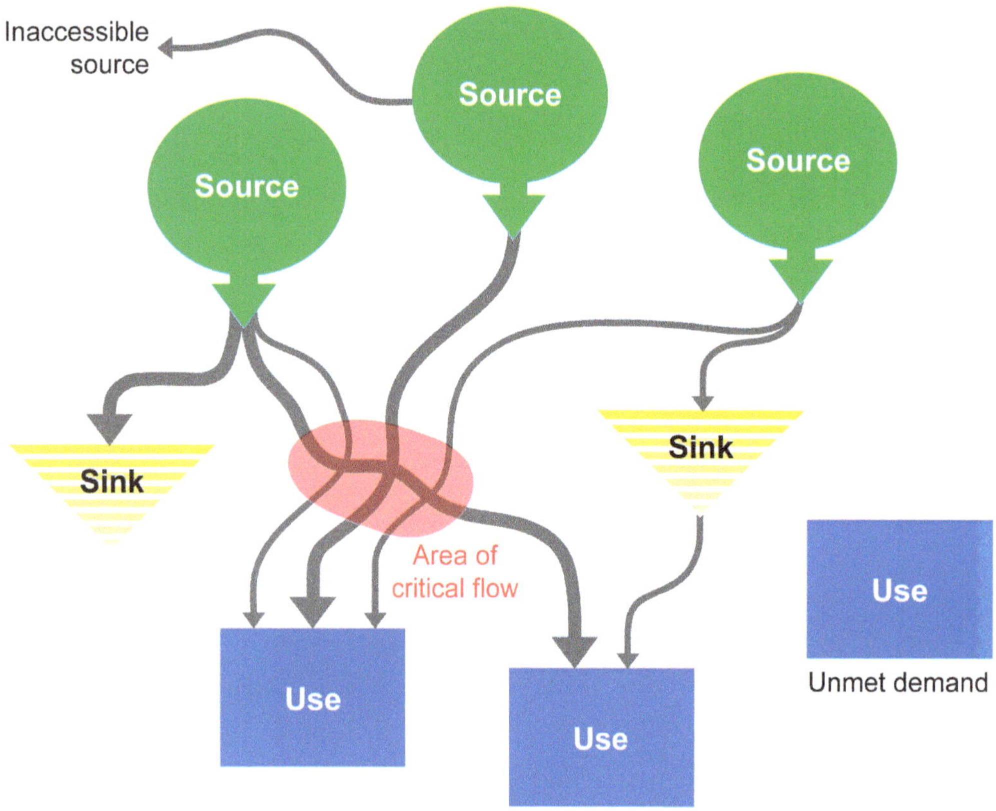

Figure 2. The ARIES conceptual model of ecosystem service flow dynamics.

and time without attempting to reference the specific time when a carrier completes its traversal of a flow path. The model stops when the entire study area has been populated with flow trajectories. This approach, while more abstract and limited in predictive capabilities than a full process-based approach, facilitates spatial analysis of benefit distribution with the data available to most management situations. When data allow, the ARIES infrastructure can integrate more sophisticated models as explained in the subsection that follows.

Outputs of an ARIES assessment. The flow trajectories simulated are processed into different groups of mapped results. For provisioning benefits, a *flow density map*, displaying the amount of ecosystem benefit that has traversed each location during the course of the simulation, highlights high-value areas that are most critical to maximizing the transmission of a benefit to beneficiary groups (exemplified later on in Figures 3 and 4). For preventive benefits, flow density highlights areas where the damaging medium concentrates and can help spot areas where intervention is needed. Such maps can greatly aid planning, as in most cases it is difficult to relate flow information to either source or use areas. Because each trajectory is modeled individually, the specific amount of benefit flowing from a particular source region or to a particular beneficiary group can be determined; this can greatly aid targeted policy making. For example, targeted flow modeling could inform polluter- or beneficiary-pays policies based on service degradation or use [39].

Other decision-relevant maps produced by an analysis of benefit flows are summarized in Table 3. *Theoretical* source and sink maps show the amount of value that could be produced in ideal situations for provisioning and preventive benefits, respectively, assuming that there is a demand for all of the service produced and that the benefit flow can reach all people. *Possible* source maps show the amount that has flow paths to reach beneficiaries, but may not due to the action of sink regions. *Actual* source maps show the source of the value that actually reaches users (provisioning benefit) or of the damaging medium that actually impacts them (preventive benefit). Comparison of these maps can aid in understanding the efficiency of service delivery: for example if the possible values of a provisioning benefit are much higher than the actual, there may be room for policy-driven improvements. The objective function for scenario analysis can be the actual value accrued or another metric, for example the distributional equity of a delivered benefit.

Other maps link supply and demand in ways that may help spot problem areas in need of policy attention. The *blocked* source map shows the value from provisioning benefits that is produced by ecosystems but cannot get to people, because of issues such as pollution or stream diversion, or flows of preventive benefits, such as the amount of threat reduction provided to source regions by intervening landscapes (e.g., disease control, wildfire mitigation). *Inaccessible* source maps show the value that is produced by the ecosystem but cannot be accessed by people because of a lack of flow connections on the landscape. Blocked ES maps can be used, for provisioning benefits, to spot areas where intervention may be called for to restore service delivery. Inaccessible ES maps highlight areas where service production may be "underutilized".

Result maps are produced in pairs, describing both the ecosystem sources and the human beneficiaries of each benefit. Depending on policy priorities, one or the other may be more relevant. For example, the blocked use map for water supply shows the spatial distribution of unmet demand by water users which could be met by ecosystems if benefits were not diverted to natural

Table 2. Flow characteristics for selected ecosystem services. Types are P (provisioning) or R (preventive). Rivalness is R (rival) or N (non-rival).

Service	Type	Rivalness	Benefit carrier	Extent	Mode of transmission	Beneficiary types in ARIES
Carbon sequestration & storage	R	R	CO_2	Global	Atmospheric mixing	Greenhouse gas emitters
Riverine flood regulation	R	N	Runoff	Watershed	Hydrologic flow	Resident lives Built infrastructure Agriculture Industrial assets
Coastal flood regulation	R	N	Storm surge	Coastal zone	Wave run-up	Resident lives Coastal infrastructure
Nutrient regulation	R	N	Nutrients in water	Watershed	Hydrologic flow	Commercial fishing, recreational fishing, other water-based recreation, waterfront property owners
Sediment regulation	P,R	R	Sediment	Watershed	Hydrologic flow	Farmers (P or R) Reservoirs (R)
Water supply	P	R	Water	Watershed	Hydrologic flow	Residents Industry Agriculture
Fisheries	P	R	Fish biomass	Accessible fisheries	Travel simulation	Subsistence fishermen
Pollination	P	R	Pollen	Pollinator range	Pollinator movement	Farmers
Aesthetic	P	N	Scenic	Viewshed	Line of sight	Property owners
value			Quality (relative ranking)			Recreational users
Open space proximity	P	N	Open-space quality (relative ranking)	Accessible amenities	Human-powered access	Recreational users
Recreation	P	N	Recreational enjoyment (relative ranking)	Recreation travel	Travel simulation	Tourism Resident users

processes such as evapotranspiration, re-routed by infrastructure, or polluted beyond the point of usability. Conversely, the blocked source map identifies areas that produce water that is unusable for the reasons listed above. The inaccessible use map will show those areas in need that cannot be served without major structural intervention on the landscape to alter flow dynamics. For example, large water diversion projects were built over the last century to reroute previously inaccessible water to users in arid and semiarid environments. With training, a decision maker could learn to use these outputs to gain a deeper understanding of the actual service values, available policy opportunities, and location and extent of demand, both met and unmet, plus value provided for each beneficiary group [7,26].

Treatment of uncertainty. To achieve the goal of making uncertainties explicit to the user, the initial conditions that enter a flow model (source, use and sink distribution) are often modeled in ARIES using a Bayesian Network (BN) approach [25,40,41]. This accounts for and quantifies part of the uncertainty inherent in the data and model structures that generate them. In the common occurrence of data scarcity, this also makes it possible to use informed prior probabilities gathered from local experts or prior statistical analysis [42]. The model specialization algorithms in ARIES ensure that BN models used in each context reflect available local knowledge. One practical advantage of BN models is their intuitive visual presentation, which helps decision-makers relate and contribute to the conceptual phase of model development in focus groups and participatory sessions [25,41]. The process of *training* [42] lends BN models their data-driven character. A BN is trained to replicate correlations in trusted datasets via an iterative process that adjusts the probabilities of each outcome to best reflect the evidence seen in data. The trained BN has "learned" the correlations in the data and is more accurate in predicting the probability distributions of each outcome given the input data submitted to the model. In cases where data are missing, these distributions will exhibit greater uncertainties, maintaining in most cases some value for decision support. When source, use, or sink distributions are modeled with BNs, ARIES aims to make their interpretation intuitive by visualizing each model result as a pair of maps. Viewed side by side, they show the most likely outcome per location along with its associated level of uncertainty, computed as the coefficient of variation (for numeric predictions) or as the Shannon index of diversity (for categorical ones). Because explicit uncertainty is valuable for decision-making, the uncertainties computed in spatial BN models are carried through the flow models in the SPAN algorithms, using methods such as Monte Carlo simulation and variance propagation [26] so that flow models can also produce uncertainty information, at the expense of longer run times.

While probabilistic models do allow communication and quantification of *some* uncertainty, their linearity prevents them

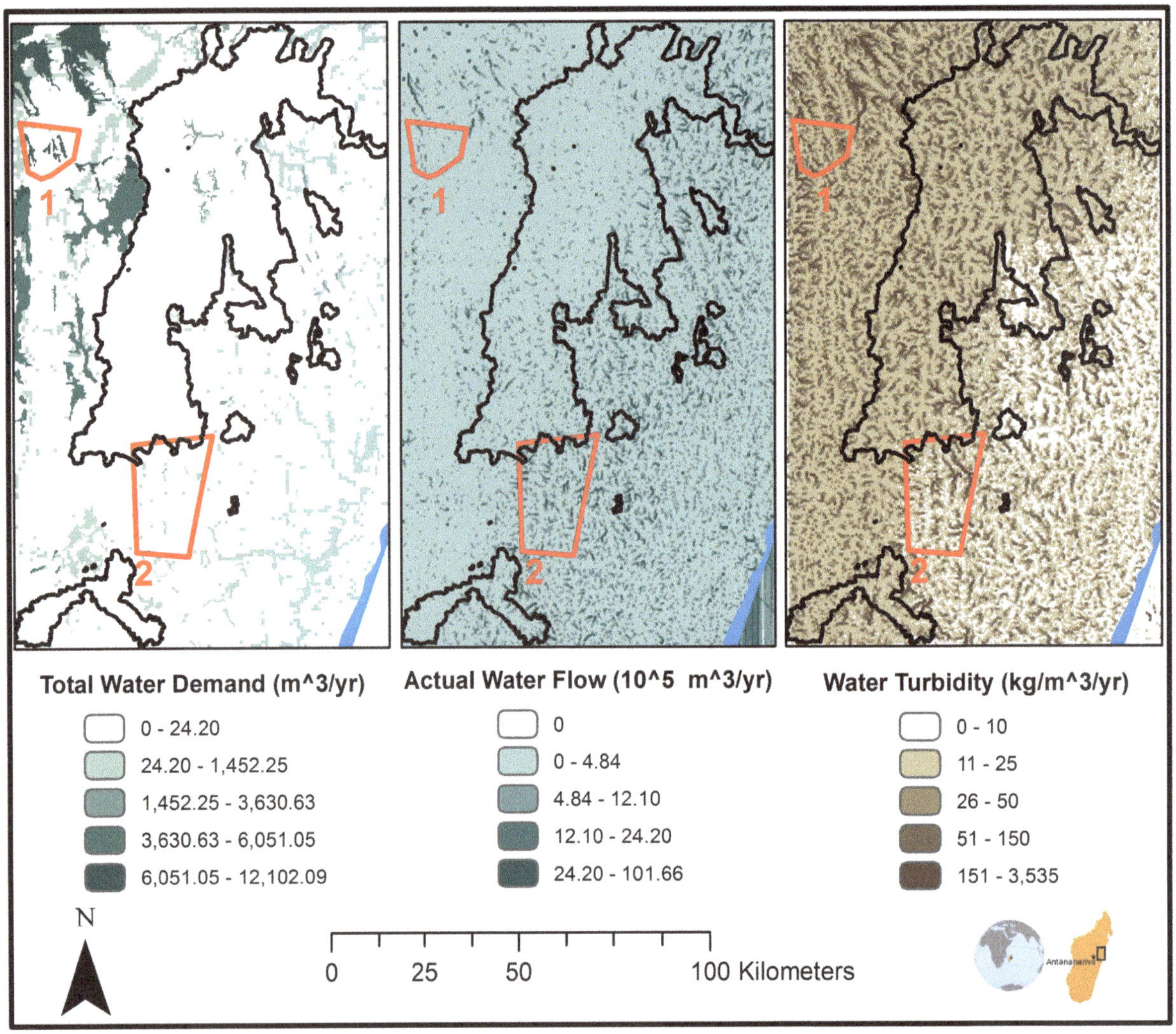

Figure 3. Water supply and quality in the CAZ area of Madagascar. From the left: total water demand across sectors, surface-water flow that is used by beneficiaries, and amount of sediment that is transported by hydrologic flows. Regions 1 and 2 (outlined in red) show the areas selected for comparison; the CAZ boundary is shown in black.

from incorporating and expressing feedback processes that affect natural and social systems. To partially alleviate this problem, ARIES only uses BN models to quantify initial conditions used as inputs for non-Bayesian dynamic flow models. This approach does address some of the complexity of ES flows, but of course does not completely honor the intrinsic complexity of coupled human-natural systems. In addition, the effectiveness of BN models depends on the availability of training datasets. Proper training of data-driven models can be problematic when training data are available at different resolutions and levels of reliability.

Integrated Intelligent Modeling

A "one model fits all" approach, which relies on parameterization of a fixed model structure to accommodate differences in social and ecological contexts between case studies, is common in contemporary ES methodologies. In contexts as complex and diverse as those that characterize ES studies, the trivialization caused by such a model's structural rigidity can compromise its utility in informing decision needs [43] and in addressing highly context-specific values and associated trade-offs [44,45]. On the other hand, customizing model assumptions, variables and equations to match complex decision contexts usually requires great amounts of knowledge, time and expertise, an investment that is often impractical [21]. In an attempt to alleviate this near-universal limitation of modeling applications, ARIES incorporates advances in ecoinformatics that allow model structures to vary "intelligently" based on the contexts in which they are run. This is accomplished through semantic meta-modeling [46], a technique that automatically selects model components from an extensible repository reflecting data availability and the specific features of ES in each application context. Although this method is not tied to any specific conceptualization of ES (or even to ES problem space in general), the view of ES as linked, independently described source, sink and use initial conditions joined through a flow process fits an automatic model assembly method optimally by virtue of its inherent modularity.

Modularity, structural variability and structural validation of models have been hard-sought goals in modeling for decades [46–49]. Despite some success in the areas of model integration and synchronization [50,51] no previously established methods were available that could directly address the needs of ARIES. The

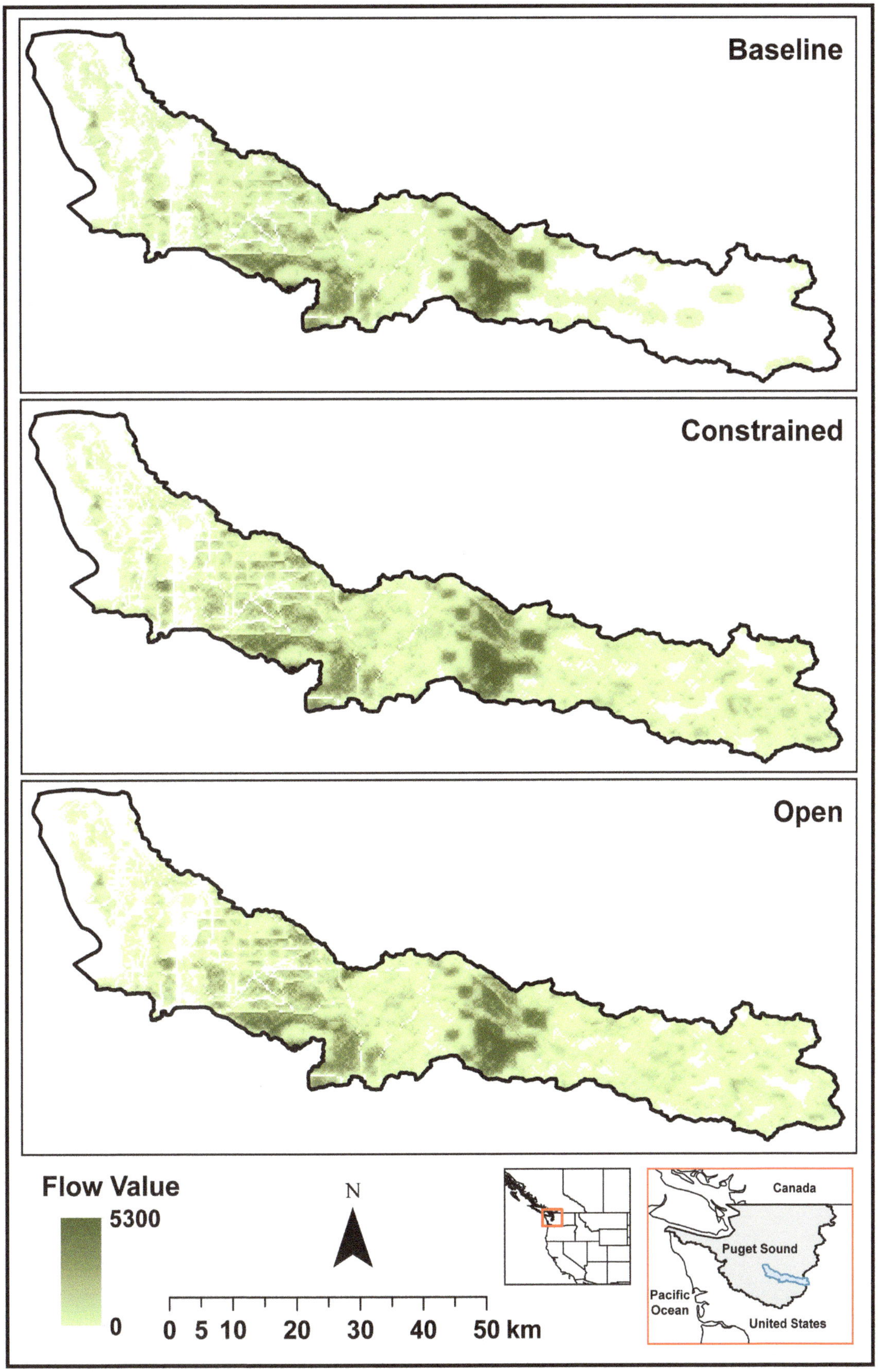

Figure 4. Open space proximity flows in the Green-Duwamish watershed under baseline conditions and constrained and open urban-growth scenarios. Theoretical values are in relative rankings, ranging from 0 to 100 for each cell. When multiple users have access to one

source of proximity value, the value for this non-rival service is multiplied by the number of users, so total flow values can exceed 100.

scope of the ecoinformatics innovations pioneered in ARIES, collectively termed *semantic meta-modeling* [46,49], is three-fold:

1. **Adaptive modeling**. Model structure is not defined *a priori*, but is built for each simulation to represent the best and most problem-specific knowledge available for the context of interest. Knowledge in this sense refers to both models and data, both of which are chosen at run time by ranking the available "building blocks" for degree of fit to the context and assembling them through a process driven by artificial intelligence (termed a resolution algorithm or, for brevity, a *resolver*). Context-specific expert opinion collected and organized by ARIES modelers influences the choice of variables, algorithms, scale, and input data that determines the final model structure in each assessment. This approach relies on extensible, distributed data and model repositories made available as online services by the ARIES team and other independent research groups. Once a model has been assembled by the resolver, complete provenance information [52] is recorded, allowing a user to audit all data sources and model choices made by the system.
2. **End-user simplification**: Adoption of any technology is dependent on the simplicity of the user-side workflow. Simplifying usage without compromising accuracy and detail has been a priority in developing ARIES, in an effort to sidestep the unavoidable tension between "keeping it simple" and producing the most effective decision aid. The availability of an independently extensible model and data repository network (distributed over multiple servers) coupled with ARIES' ability to automatically adapt data to models makes it possible to run many models without the user having to input additional data. Users need only provide data when otherwise unavailable, or to create *ex-ante* scenarios based on locally predicted changes. At the same time, the semantic validation provided by the ARIES resolver (see Model Resolution and Assembly) ensures that important factors and necessary variables are not overlooked.
3. **Independent extensibility**: The modularity of the semantic meta-modeling approach implies that multiple models can be developed for very general concepts (such as ES benefits, individually or in bundles) or more specific variables (such as soil texture or land cover). Each repository can provide data and models for any concept, and all of the available knowledge is ranked prior to selection at the assembly phase. This paradigm facilitates extension of the model base through a community process where no top-level coordination is required beyond agreement on common ontological concepts (see Model Resolution and Assembly). The knowledge base can therefore grow independently with use. Provided that knowledge repositories adhere to an agreed semantics, interoperability and consistency of the assembled models will be ensured by the model resolution process.

The main instrument to achieve these innovations is the ARIES *knowledge base*, which includes abstract concepts, models and data. In ARIES, two levels of description exist: *abstract* and *model* knowledge.

Abstract knowledge is composed of individual ontologies [47,53] that organize concepts and relationships relevant to ES and biophysical modeling. Ontologies are a standardized way of representing conceptualizations as interdependent definitions of concepts and relationships [53] that can be easily extended and merged. They consist of computer-readable files that are used by widely available machine reasoning algorithms [47,54] and remain a very active research field in artificial intelligence. ARIES' core ontologies are based on efforts led by NASA [55] and on foundational ontologies of recognized generality [56]. In the ES-specific ontologies developed by the authors, the MEA categories are broken down into an extensible classification of ES benefits (Figure 1), which is further broken into model-relevant ecological, social and economic concepts. Each ES benefit is interpreted according to the conceptualization described in the Methods section, using source, use, sink and flow concepts for organization. An extensive analysis of case studies across multiple ES and

Table 3. ARIES flow model outputs generated by the SPAN algorithm.

a	Definition	Estimation methods	Applications
Theoretical source, sink, use maps	*In situ* provision, depletion, or use of a service	Values calculated without the SPAN model, not considering service flows	Understand maximum ES supply and demand independent of ES flow paths
Possible source, use, flow maps	Service dynamics when accounting for flows but not sinks	Values calculated by the SPAN model considering flows but not sinks	Understand ES flows in the absence of sinks
Actual source, sink, use, flow maps	Service dynamics when accounting for sinks and flows	Values calculated by the SPAN model considering sinks and flows	Understand actual ES delivery (provisioning benefits) or damage (preventive benefits) and values
Inaccessible source, sink, use maps	Service flows not delivered due to a lack of flow connections	Calculated by subtracting actual from theoretical sink values and possible from theoretical source and use values	Understand unused ES supply or demand based on inaccessibility
Blocked source, use, flow maps	Service flows blocked by sinks	Calculated by subtracting actual from possible values	Understand ES scarcity due to sinks in provisioning benefits, or provision of preventive benefits, where sinks are beneficial

contexts was the starting point for the assembly of these ontologies, which assist in the identification of particular beneficiaries and benefits within a target area and in model selection and assembly to best simulate their behavior.

Model knowledge pairs *models* to individual concepts, subject to rules that guide model selection based on specific characteristics of the application context (e.g., geographic location or the values of observations like precipitation, biome, or per capita income). In semantic meta-modeling, a model can be defined as a *strategy to observe its associated concept*, and can simply consist of a semantically annotated dataset (the preferred alternative when available) or of an algorithm of varying complexity, which may in turn require observation of other concepts in order to be computed. Models are specified in a dedicated modeling language (documented so far only in early release forms [46]) also capable of "wrapping" external models by providing semantics for their inputs and outputs, bridging to the concepts in the abstract knowledge base.

Model resolution and assembly. After users choose their geographical context and the concept corresponding to their ES of interest (e.g., "water supply to domestic users"), the ARIES resolver uses the abstract description of the concept to assemble the most detailed model allowed by the model knowledge (including available data). Model resolution proceeds top-down, identifying models according to the semantics of the context (e.g. a mountain watershed and a wetland region will trigger different ways of observing surface water flow), the scale of the observation (e.g. features that are only meaningful at large scales, like mountains in assessing aesthetic value, will not be included unless the region of interest is wide enough to allow their observation) and the available data (models that need input that cannot be observed will not be chosen). The resolution process builds a decision tree that resolves the principal concept to the most suitable model and, in turn, any other concepts required by the models chosen at each step, until all concepts are resolved into a computable algorithm. More than one model may be chosen for the same concept when the area of interest contains features that are different enough to require distinct model formulations in different spatial or temporal segments of the same run (e.g., if both water bodies and land areas are present). Because the model base is multi-purpose and distributed, it is common for one concept to be linked to more than one possible model; therefore it is critical to rank models by their suitability for the context. ARIES adopts a sophisticated, multiple criteria ranking algorithm that can mix objective criteria (such as spatio-temporal resolution or currency) with user-provided rankings of reliability and quality. Table 4 provides an overview of the criteria currently used for ranking. The relative weighting of these criteria is important to the outcome of model resolution: while modelers can provide customized weights on a model-by-model base, current research in ARIES is directed to devising adaptive weighting schemata that can use both objective and subjective metrics of quality of the models produced, as described briefly in the Discussion section. All other ranking criteria being equal, the algorithm prioritizes specific, detailed models that have been tagged as appropriate for the region of interest over more general, coarser alternatives, as long as data to support them exist. This integrated modeling approach supports the mixing of data-driven and hypothesis-driven models to produce the overall model structure most suited to the application context [4]. A data-driven approach such as BNs is prioritized by ARIES to compute static components, like ES production functions, wherever accepted dynamic models or the data to populate them are unavailable. A hypothesis-driven approach (such as flow models or trusted external models that have gained decision-maker confidence through repeated application and refinement) is preferred where the dynamic complexity of the phenomena (e.g., sediment or water transport) is well understood and adequate data are available for parameterization. The modeling language (named *Thinklab*) and the infrastructure implementing the semantic meta-modeling approach are open source software [57], developed concurrently with ARIES and available under the terms of the GNU General Public License [58].

Results from Application Examples

Models addressing eight ecosystem services – carbon sequestration and storage, riverine flood regulation, coastal flood regulation, aesthetic views and open space proximity, water supply, sediment regulation, subsistence fisheries, and recreation - have been developed so far using ARIES. Model components were developed and parameterized based on literature reviews and expert elicitation [27]. Efforts are underway to streamline and standardize expert elicitation procedures via surveys that can be used with future ARIES case studies. Case studies to date have focused on several locations in the USA [24,59], Latin America, and Africa [60,61]. Initial case study locations were selected to represent a range of ecological and socioeconomic conditions, data availability, and scientific expertise, in order to best serve as a foundation for the development of globally available models (see the Discussion section). To provide examples of early ARIES outputs, in this section we summarize key results from two contrasting case studies: water supply and quality in Eastern Madagascar and open space values in Washington State, USA. Such studies also provide examples of potential *storylines* (see the Discussion section) that users may replicate in a single action in different contexts, supported by ARIES' automatic model building algorithms.

Example 1: Water Quality and Quantity in Eastern Madagascar

In an integrated ecological and economic study for the World Bank WAVES (Wealth Accounting and the Valuation of Ecosystem Services) program [60], the values of ecosystem services were evaluated near a key conservation area termed the Ankeniheny-Zahamena Corridor (CAZ). ARIES flow analysis enabled comparison of water quantity and quality values for four classes of beneficiaries located within and outside a protected area. The CAZ, a largely forested, newly established protected area on Madagascar's eastern escarpment, includes a population of nearly 350,000 people in rural communities and also supplies water to the national capital of Antananarivo. To understand the value of a protected area in providing benefits to downstream water users, we compared water budgets and erosion for an area near the CAZ but hydrologically unconnected to the protected area and with intensive agriculture (area 1) versus another adjacent and hydrologically connected to the CAZ protected area (area 2, Figure 3). Using the models from the 2011 ARIES release [27], we modeled spatially explicit water demand, simulating water-delivery dynamics when accounting for precipitation, evapotranspiration, infiltration, runoff, and rival use [60]. We then computed a preliminary water budget for the region and aggregated the spatial results to provide total figures for beneficiaries located within the two areas. Water demand for irrigation, livestock, residential consumption and tourism was estimated separately, using best practice manuals and heuristic criteria to obviate the lack of primary data for most sectors [60]. Total water demand estimates by sector are shown in Table 5.

Erosion and sedimentation are significant environmental processes in Madagascar and strongly affect water quality [62–

Table 4. Current criteria for ranking model components and data selected during model assembly.

Scoring criterion	Explanation
Semantic specificity	Prioritizes data and models that are specifically defined as applying to the semantics of the context of interest; e.g. "carbon content in top soil layer" over more generically described "carbon soil content" when the requesting model is defined to apply to the top layer.
Scale specificity	Prioritizes data and models that are more specific for the selected spatial and/or temporal context, by comparing the relative proportion of coverage for the data or models with the context chosen for simulation.
Detail and resolution	All else being equal, data and models of higher temporal and spatial resolution will be given priority.
Semantic distance	Data and models whose definition is closer to that of the model they are being applied to, for example by belonging to the same project or coming from related ontologies.
Currency	If no specific time period is specified for the simulation, the most current data and models are chosen preferentially.
User-attributed quality rankings	Users may attribute numeric ranks (0 to 100) to perceived data and model reliability. The value 50 is used if no value is specified. Other user-defined rankings can be used at the discretion of the modeler, for example for prioritizing public data over non-disclosable ones if the model needs to be audited externally.

64]. To better understand the CAZ's role in regulating sediment, a model was developed to quantify and map areas of erosion and the hydrologic flow paths of waterborne sediment. Water supply and sediment sources and sinks were quantified, which enabled comparative analysis and future valuation when economic data become available. Soil erosion was computed using a hybrid approach, automatically applied by ARIES model specialization algorithms: the Revised Universal Soil Loss Equation (RUSLE [65]) was used on gentle slopes and a probabilistic model considering soil, precipitation, and vegetation factors [27] took over when the slope was higher than 20%, too high for the RUSLE to be defensible [66]. The surface water supply and sediment transport flow models employed [27] simulate (i) the movement of surface water throughout the basins and (ii) transport and deposition of eroded sediment via hydrologic flows. Our models generated maps for which summary values of demand, supply and flow can be calculated for both sample areas and the CAZ overall (Figure 3).

To estimate the sustainability of water supply in the region, the model was run repeatedly with increasing demand levels; this allowed us to estimate approximate critical thresholds of water supply in both sample areas. Results of this analysis for the largest water use (rice agriculture) are summarized in Table 6, which shows that while current levels of demand are essentially met in both sample areas, only sample area 2 has the potential to sustain much greater demand in a future when deforestation and climate change are likely to further strain water supplies.

Results of the soil erosion and deposition analysis allowed us to estimate levels of sediment contamination in water, which are approximately 6 times higher outside the CAZ (area 1, 11.3 kg/m^3.year) than adjacent to it (area 2, 1.9 kg/m^3.year). As before, due to the great approximation and lack of primary data in all model components, these results must be considered only as comparative indicators. Still, the analysis supports the hypothesis that CAZ's natural features are important in protecting water quality for its productive use by downstream beneficiaries, results that were previously highlighted in more in-depth studies for agricultural [67] and economic [68] productivity in other parts of Madagascar. While such results must be interpreted with caution, flow analysis enables a rapid semi-quantitative assessment of supply threats and mitigation effects that could not be obtained using other mainstream ES methodologies [3].

Example 2: Aesthetics and Open Space Values in the Puget Sound

A case study in the Puget Sound, Washington State, USA stands in stark contrast to the previous one, in terms of both differing ecological and socioeconomic contexts and data quality and availability. The Puget Sound, the second largest estuary in the USA, is a defining social, cultural and economic feature of Washington State and home to a human population of 4.4 million, including 15 American Indian tribes and the major port cities of Seattle and Tacoma. In a recent study, we differentiated between the theoretical provision of ES (i.e., ecosystems' capacity to supply services) and their actual delivery when accounting for the location

Table 5. Total estimated water budget (m^3/year) for sample areas outside (1) and adjacent to (2) CAZ.

	Total in CAZ	Sample area 1	Sample area 2
Rice agriculture	512,187,528	15,943,889	5,958,885
Non-rice agriculture	31,718,842	444,689	6,512,517
Livestock water use	684,499	206,041	54,484
Residential use	17,173,088	3,206,662	4,426,315
Annual precipitation	16,619,520,610	1,074,244,347	7,476,712,388

Table 6. Water supply sustainability (m^3/year) for rice agriculture in the two areas considered.

	Sample area 1	Sample area 2
Current water need	15,943,889	5,958,885
Maximum potential	15,443,129	304,155,269
Ratio potential/need	97%	5104%

of beneficiaries and flow paths [59]. Here we map the value provided by ecosystems to property owners via open-space proximity within the Green-Duwamish watershed, which rises from Seattle to the slopes of the Cascade Mountains (Figure 4). Although hedonic analysis can be used to monetarily value open-space proximity for property owners, spatially explicit flow models enable a comparison of theoretical and actual service delivery values under landscape change scenarios (i.e., as a result of conversion of open space to developed land use and movement of new beneficiaries into the region). To address this, we compared year 2000 (baseline) conditions to managed (constrained) and unmanaged (open) development scenarios for the year 2060 [69]. We expected widespread development to reduce theoretical open space sources that can provide value to residents, but at the same time increase the number of users and the total flow of proximity value to property owners.

Open space proximity values were mapped by identifying: 1) ecosystems providing high-quality open space (sources), 2) features that impede access to open space (sinks), and 3) housing (use) locations [27]. The values of source and sink features were ranked using a relative scale (0–100) where higher values represent the most valuable open space based on hedonic valuation studies [70,71]. Sources, sinks and users were connected by a flow model simulating physical access to desirable spaces [27].

While the total number of new users was greater in the open development scenario, there was less high-quality open space available to provide value to property owners than in the constrained development scenario (Table 7, Figure 4). As a result, urban growth increased open space flow values, but by a greater percentage in the constrained development scenario than the open development scenario (25.7% and 23.3% increases, respectively; Table 7, Figure 4). Theoretical source values increased slightly in the development scenarios because both increase the value of open space through the designation of new parkland.

These results illustrate how placing more beneficiaries across the landscape may have the effect of increasing ecosystem service flows and, by consequence, actual values, but may simultaneously degrade the ecosystem's underlying ability to provide the same services (i.e., theoretical values). In this case, an expansion of the urban footprint yields an increase in beneficiaries in locations where ecosystem service flows were previously inaccessible. However, land-cover change associated with new development often reduces an ecosystem's capacity to provide services, i.e., their theoretical source values [7]. Thus a more measured approach to urban growth may actually lead to greater ecosystem service values and protection of a region's underlying natural capital than an unconstrained development pattern.

The differences between development scenarios become far sharper when a richer set of ES and management alternatives is considered, including for example when endangered species restoration (Chinook salmon and others), flood risk reduction, retention of local farmlands, port expansion, transportation development, carbon emissions reduction, wildlife conservation, and high-tech and industrial development are all pursued in the Puget Sound Basin. Ongoing ARIES applications in the Puget Sound can be used to better quantify geographical and temporal tradeoffs in ES flows for diverse beneficiary groups.

Discussion

In this section, we discuss the benefits and drawbacks of the different areas of innovation in ARIES and how they relate to the state of the art in spatially explicit ES modeling and valuation. We conclude by comparing our efforts to date against the evaluative criteria for ES assessment tools listed in the introduction.

Improving Model Detail, Coverage and Transparency

The ES assessment framework proposed in this article calls for a more detailed and systematic view of ES problems than that used in many of today's ES applications (Table 1). Yet the practical value of this new approach remains undemonstrated outside of the case studies where it has been applied so far. The value of this or other approaches to quantitative ES assessment should be seen as comparative rather than absolute; the difficulty in validation and

Table 7. Relative values for open space proximity source, use, and flows source, under alternative urban growth scenarios Green-Duwamish watershed, WA, USA.

	Constrained development	Open development
Theoretical source	+12.8%	+6.2%
Theoretical use	+16.3%	+19.4%
Actual source	+24.5%	+21.6%
Actual sink	+51.2%	+39.6%
Actual use	+24.6%	+21.8%
Actual flow	+25.7%	+23.3%

the necessarily simple nature of the models involved make them more suitable for comparing scenario outcomes than for obtaining reliable physical estimates. Widespread testing of ES tools against discipline-specific models is still needed to understand the limitations of ES-based approaches for accurate, consistent quantification of the biophysical processes underlying ES provision [24]. Yet incorporation of dynamics and uncertainty, and the possibility of composing contextually specific models, allows for improved precision and may make the methods described suitable for more sophisticated and rigorous ES-informed decision support.

ARIES' development has followed a bottom-up approach, starting with case studies of considerable detail conducted with partner institutions, then generalizing that knowledge to yield "global" models offering a bird's-eye characterization of many ES in most locations, limiting data input requirements from users. The potential for large-scale adoption will clearly depend on completion of these generalized models, whose development is ongoing. These simpler models will provide a "bottom line" to which the artificial intelligence in ARIES can default, allowing it to produce results of variable detail in almost any geographic region using global data, but automatically switching to more detailed models when the knowledge base and data allow. When this stage of development is completed, ARIES will meet the needs of a larger share of potential users, growing in utility and sophistication over time. Although the integration of large model libraries poses many challenges, such libraries exist within academic institutions and agencies (such as Environmental Protection agencies or the US Army Corps of Engineers) that can be connected to ARIES. The potential of ARIES as a large-scale meta-modeling framework will be revealed as case studies and multiple ES analysis are further tested.

ARIES runs on an extensive (3.5TB of data alone at the time of this writing) and fast-growing knowledge base that includes data from both public sources and local institutions. Where data are privately owned or protected by privacy laws, the ARIES team has obtained clearance from the respective sources to release the ES assessments as derived products, as long as the original data cannot be directly downloaded or displayed. This allows public use of information that would otherwise need to be obtained by users on a case-by-case basis, adding to the up-front user investment needed to conduct an assessment. On the model side, a variety of well-known, open source physical process models are being integrated into the ARIES model base. Among these, the CAESAR-LISFLOOD flooding and erosion model [72], the PRMS water balance model [73] and a general ecosystem model broadly inspired by LPJ-GUESS [74] are being included to improve the quality of descriptions of flood, sediment, nutrient, carbon and primary production dynamics. Adding these tools will not affect the usability and complexity of the system on the user and decision-maker side, although it may produce more outputs. As a corollary of the system's built-in flexibility, ARIES is designed to be amenable to transparent updates of the knowledge base, making it possible to perform earlier assessments repeatedly while automatically benefitting from any improved data and modeling knowledge. This also provides a mechanism for testing relative gains obtained from using complex vs. more simple models.

A precondition for effectively using ES in decision-making is acknowledging, quantifying and communicating the uncertainties that are inherent to any modeling endeavor [16]. ARIES is designed to use probabilistic initial conditions for most of its models, through the adoption of BNs, and to carry the uncertainty through the dynamic parts of its models, using methods including Monte Carlo simulation and variance propagation, so that uncertainties can be communicated to the user [26]. Importantly, only the components of overall uncertainty that relate to missing data or *known* data quality issues can be dealt with effectively in such a probabilistic model. No accounting is possible for the uncertainty that relates to the *structure* of the causal dependencies that define the Bayesian model, although this can be alleviated to some extent by adopting context-specific model assemblage rules.

Understanding strengths and limitations of these uncertainty estimates is necessary to avoid engendering a false sense of security in the decision-making process, leading users to ignore other realms of uncertainty that may have a crucial influence on outcomes in those parts of the model where the explicit uncertainty is relatively low. Like any quantitative modeling approach, BNs can lead to subjectivity and error. Guidelines and best practices for BN development have been developed [45] and are being adhered to during the ongoing development of ARIES models.

Simplifying User Workflows

The multiple levels of detail in the ARIES model base are intended not only to produce more suitable models, but also to streamline workflows for end users, who will be able to query the system in simple ways and obtain results that automatically reflect the best available knowledge for their context. In doing so, ARIES hopes to overcome a common hurdle in ES applications today: complex usage of simple models, where models of relatively simple methodological sophistication still place a heavy burden on users for data pre-processing and parameterization. ARIES is designed to automate data choice, algorithm choice, data pre-processing and optimal attribution of levels of resolution for all ES, and to allow users to query "bundles" of ES with exactly the same workflow as single ES assessments, requesting user inputs only for knowledge that is not available. The artificial intelligence-assisted process pioneered in ARIES emphasizes user simplification without trivializing the application, a paradigm that could also be valuable for broader application in modern environmental and economic decision-making. Provenance information complements the outputs of each model run, describing data sources and model structure in the interest of transparency and traceability of problems.

To further simplify user workflow, ARIES adopts the metaphor of *storylines* to organize and present observable concepts (e.g., carbon sequestration) for the user and guide visualization after their resolution to models. Storylines connect observable concepts from the knowledge base with metadata that include explanatory descriptions of each concept, references, links, and descriptions of related case studies. Storylines are assembled automatically as models are built, documenting each model component and the rationale for its choice, and provide a blueprint for generating user-friendly documentation of results that can be viewed immediately or downloaded as digital media. ARIES is currently developed and used through a software interface geared towards modelers, which is a primary focus of current development. The intended end-user interface, however, is a web 2.0 application that allows users to search for storylines that describe concepts of interest in standard "web search" fashion, based on concept names as well as descriptions, locations, references and other metadata. Within a graphical user interface, the act of dragging and dropping one of the storyline links resulting from a search on an interactive world map specifies the geographical context and initiates model assembly and execution. The existing ARIES web interface prototype available at www.ariesonline.org [75] is, however, only demonstrational and based on an older release of the software; the described user workflow will become available in an updated application once the development of global models has been completed.

A user workflow that hides complexities under familiar metaphors and can transparently produce sophisticated models can carry subtle but important disadvantages. Early pilot tests with users have highlighted that while even limited user-level complexity is poorly tolerated by users, a lack of it can be perceived as lack of sophistication in the approach and lead to incorrect assumptions that influence decisions. To obviate this, more feedback from the modeling system can be communicated to document choices and computations as they are made, at the cost of "more output to be understood." Further, choices made by an automated system are typically questioned by users, even if they reflect the best available expert knowledge. Incorporation of user feedback on model results may provide additional ranking criteria for model selection (Table 4), and allow a degree of "crowdsourcing" to be incorporated in the model assembly process. This can have important implications in a field where objective validation is often not an option; it is of course paramount that all subjective criteria are used in a controlled and transparent way.

ARIES and ES Valuation

ES valuation has typically been interpreted monetarily. The realization that value is a highly multi-dimensional issue with deep ethical implications has become central in recent ES literature [76,77]. The definition of value can vary widely in different ES applications, for example in food security [61,78] vs. payments for ecosystem services [79], or the values reflected by indigenous communities vs. centralized governments [80]. Much as the biophysical science underlying ES provision demands careful contextualization while modeling, ES valuation cannot be trivialized with a simple definition and deserves case by case consideration within the comprehensive framework of equity [81]. The development of ARIES has so far concentrated on the biophysical modeling of ES, leaving the translation to economic value and its implications to the end user. In some circumstances, biophysical outputs (particularly the possible and actual estimates, Table 3) can be seen as representing value directly, or provide a base for inferring value beyond mere ES supply quantification [60]. Yet, the many facets of equity [81] and value [82] make the problem of case-study specific value attribution difficult to resolve in general. As a consequence, a priority in ARIES is to remain as agnostic and pluralistic as possible in its approach to valuation.

Independent of the definition, value in ARIES will ultimately be the result of a model, and the ideal representation of value for a system adopting an "intelligent" modeling approach would consist of a context-specific and customizable model of societal well-being that reflects locally established demand, customs and beliefs. If the application requires monetization, that can be accommodated via current valuation methodologies. ARIES is capable of incorporating monetization models, but no value model of this kind is currently included in the knowledge base. For generality, value is best represented as a multiple-criteria problem whose definition greatly depends on the application context [82] and target stakeholder community. Still, the need for a "common currency" to compare the effects of decisions across multiple ES and social sectors means that demand for economic estimates of ES value remains very high, no matter how controversial the surrounding issues. An often relied upon technique for estimating economic value of ES is value transfer [83,84], based on the adaptation of primary ES valuation studies done elsewhere to a new context of interest using a set of transfer criteria [85]. Although land-cover type has often been used as the sole transfer criterion [9], function transfer – the use of mathematical functions, often derived from meta-analyses – is considered a much more robust approach [86,87]. While ARIES does not offer benefit transfer algorithms, research on valuation is an important part of the project's background [88–90] and the ARIES team is working with partners to provide bridges to valuation databases that can help generate more rigorous transfer functions [91]. For such approaches, beneficiary-based biophysical estimates of ES flows, like the actual or possible estimates computed through flow modeling [7], may assist in the development of transfer functions. This would improve on a state of the art that most commonly uses theoretical values or static proxies for these purposes.

Economic value changes in a highly nonlinear fashion in the vicinity of ecological thresholds as ES supply declines [92,93]. Assessment methods must therefore be able to provide sufficient quantitative accuracy in assessing supply and demand, so that thresholds can be anticipated before they are encountered. Without such consideration, economic estimates should only be used when the relationship between service supply and marginal prices is known to be predictable and stable. The information on threats to supply that biophysical modeling can provide is key to the assessment of whether the linkage to economic values or prices is appropriate. For example, in the Madagascar pilot study illustrated above, supply threats have been discussed as a way to improve on the quality of the decision-making information that results from economic assessments.

Dynamic Complexity and Trade-offs

Non-linear dynamics determine catastrophic behaviors that are of utmost interest when investigating the consequences of policies on ES delivery. They include, for example, the sudden loss of homeostatic behavior ("tipping points") that may be encountered as particular amounts of development-induced change are reached. Despite extensive research in both ecological and social sciences on the dynamic behavior of complex systems [18,19], our understanding of their general properties remains limited. As a result, we are unlikely to meet the various challenges related to accounting for complex human-natural system behaviors, like those listed by Carpenter et al. [16], without substantial further research. The question of how completely such non-linear dynamics can be realistically represented by ES assessment methods remains paramount. The adequacy of any model in predicting non-linear dynamics is very difficult to assess when detailed historical data are lacking, and few modeling studies exist that incorporate sufficient detail (e.g., employing a detailed and accurately calibrated hydrological assessment) to serve as a basis for qualitative cross-calibration. ARIES does account for flows using dynamic models, and the processes that underlie such flows are one of many sources of dynamic complexity. Analyzing spatially explicit and temporally referenced linkages between ecosystems and societies can provide a degree of dynamic description and the opportunity to address feedbacks, all of which can be useful for improved resource management. Yet, many other sources of ecological and social complexity remain unaccounted for when production functions are used for quantification. Agent-based models [38] that incorporate feedback on ecological systems from the societal side have begun to appear [94] and may prove useful to account for more dynamic complexity than the current state of the art in ES. The Ecosystem Services for Poverty Alleviation-Attaining Sustainable Services from Ecosystems through Trade-off Scenarios (ESPA-ASSETS) project [78] on food security and poverty alleviation through access to ES is one area where the ARIES methodology is being extended to consider such methods.

Trade-offs are ubiquitous in ES assessments and are of central importance in decision-making. Decisions can result in trade-offs between users of the same service, between different ES for the same users, or any combinations thereof, and have different meaning and relevance when considered over different space and time horizons [95]. There is currently no systematic methodology for addressing ES trade-offs, although guidelines meant for application with specific methods are appearing [96]. Systematic trade-off analysis is obviously not practical without a fully quantitative account of beneficiaries and accrued benefits. But modeling the multiple beneficiaries of a single ES can be difficult due to the collinearities resulting from rival services and societal feedbacks. An integrated approach where all such effects are modeled explicitly and simultaneously can help address the dual problem of access to and distribution of limited goods and services ("winners and losers"). For example, relatively simple outputs that hint at trade-offs can be obtained by intersecting multiple flow path outputs for a "bundle" of different ES, identifying landscape locations that are responsible for the transmission of a disproportionate amount of one or more benefit within the area of interest. Such results can, however, only be obtained if multiple ES are modeled simultaneously – i.e., subjected uniformly to the influence of each scenario and their mutual effects on each other. This is difficult with most methodologies in use today, which are typically applied separately for each service. A common limitation of current ES practice is a narrow focus on single ES, or their "bundling" without explicit regard for their inextricable interactions. While the problem is widely recognized, even comprehensive and well-funded programs (such as REDD+ [97]) remain largely focused on a single ES and only hint at the integrated perspectives.

By virtue of its computer-assisted modeling infrastructure, ARIES can produce integrated ES models with the same effort as those for single services. Land cover and other policy-controlled variables entering the models as inputs typically affect more than one service; the ARIES infrastructure ensures that a single dependency chain exists across the integrated model, so that a simulated policy intervention input affects all ES outputs. The granularity provided by ARIES in accounting for each class of beneficiaries also allows trade-offs between different stakeholders to be represented unambiguously, as each benefit and beneficiary is counted as an individual sub-model in the overall simulation.

Even with improved methodologies, important limitations remain in the face of real-life, multi-stakeholder problems. For example, the different spatial and temporal scales that accompany each conflict or policy window require careful consideration of the assumptions made in an integrated model when planning scenarios and analyzing results. The ability to quantify flow paths and address individual beneficiaries does not solve all the difficulties inherent in modeling ES trade-offs. However, such results could be used within techniques such as multiple criteria analysis [98] to assess the concordance or discordance of a set of simulated outcomes with spatially explicit or aggregated social priorities [99]. Such techniques can help minimize conflict and rank the likelihood of successful outcomes when competing policy choices must be considered. The approach can assist when defining and analyzing complex scenarios, which from the user's point of view simply become collections of modified data or models reflecting *ex-ante* predictions, such as modified land-use data or IPCC climate-change model outputs [100]. When a particular scenario in ARIES is selected for comparison with the baseline, the data and models contained in the alternative scenario override baseline conditions that would otherwise be used to model the same concepts. This way, scenario specifications will affect all levels of the model chain, and comparing results against baseline values will highlight all trade-offs when a full ES portfolio is evaluated.

Increasing Participation and Valuing Community Knowledge

Modularity and extendibility in modeling have two main goals: enabling a more flexible model assembly process and making it possible for independent communities to contribute knowledge that is reusable and linkable by design. ARIES was conceived from the start as a community process, where an initial model base contributed by the core team can be extended by modelers from diverse disciplinary backgrounds and application contexts. Because the knowledge and model base are theoretically unlimited in size, the system can constantly "learn" to produce better models as new data and information are accumulated. Independent case studies can reuse model components or contribute new ones to expand the model base that, given its ability to rank and switch model components, can allow the system to grow and improve with application.

Model validation through community adoption of complex paradigms is difficult and only possible over the medium- to long-term. Hopes of "viral" adoption for scientific projects in their beginnings are usually overenthusiastic at best. Yet interest in ES modeling is widespread and growing, and initiatives are in place to test this approach in practice. An intensive 2-week modeling school [101] is held annually by ARIES investigators; the models and ontologies resulting from associated workshops help address local problems brought in by participants while extending the existing model base for the benefit of future users. The first edition of the school, held in Spring 2013 [101], saw a large number of applications and the participation of thirty modelers and decision-makers from six continents with encouraging results. Preparations for the 2014 edition are currently underway. To support the growth of this community, ARIES development has prioritized the building of tools for model designers rather than model end users; the toolset used to develop and test ARIES models has been the primary focus of development, while the web 2.0 application designed for end users remains in an earlier stage.

If the goal of building a model base with extensible coverage through community participation is achieved, the role of the model-ranking algorithms that select the most context-suitable model components will become increasingly important. Just as the success of web search engines depends on the page ranking algorithms they adopt [102], mechanisms that choose the "best" model strategy can deeply affect model outcomes. This occurs despite the fact that models are merely assembled from well-tested and self-contained model components, that model choices are documented, and that ranking criteria can be inspected and modified by the user. In such cases, inclusion of crowd-sourced ratings of outputs from end users may prove useful. The current approach to ARIES capacity building, based on intensive application-driven case studies and courses focusing on ES problems of diverse nature, can also assist in developing and testing the effectiveness of different model-ranking strategies. Approaches based on careful community screening have proved very successful with relatively unstructured content (e.g., Wikipedia) but have not seen application in eliciting scientific data and models, where issues of validation are more nuanced and important. The consequences of choices in unsupervised, rule-driven model building may extend far and wide, and their full advantages and disadvantages will become clear only with continued use and development. In all cases, structural validation (both of semantic constraints through machine reasoning and by solicited community feedback on the resulting models) is expected to play a large role in the future development, use, and delivery of ARIES as a service for decision makers.

Conclusions

Recent years have seen enormous demand for, and a growing supply of, decision maker-ready "tools" capable of quantifying ES for scenario analysis and improved decision-making [21,103]. ARIES would most accurately be classified as a long-term research project based on a specific scientific interpretation of ES and a new modeling paradigm. Yet the method also addresses decision-makers as end users, which legitimizes the view of ARIES as another "tool on the market." Moving a scientific target forward when potential users are so hungry for usable products is difficult, and efforts often had to be made to prevent community pressure for rapid developments from interfering with rigorous scientific thinking. The combination of a complex problem area where solutions are urgently needed but unifying theories, and even common definitions, are lacking is a serious hindrance to most applied sciences; this is particularly true for ES. In such cases, incompatible or weak assumptions may often be treated as equally legitimate and common validation criteria are difficult to identify and agree upon.

Evaluation of the current ARIES effort against the criteria we laid out in the introduction shows both met and outstanding goals. While the goal of providing quantitative, spatially explicit results that account for uncertainty is addressed extensively, there are dimensions of "quantitative" – namely those relating to predicting non-linear dynamics and its policy implications – that can only be partially met at this stage. Also, while uncertainty information can be of great value in decision-making, our pilot experiences suggest that many end users are not yet equipped to consider it in the decision-making process. Identifying the best ways to use such information and to ensure its correct interpretation at the user end remains an open problem.

In terms of resource requirements, ARIES proposes a different paradigm where user investments can be reduced drastically compared to methods that require extensive data gathering, pre-processing and supervision during model runs; yet, the achievement of this paradigm depends on complex technology that will require time to reach a level of stability that satisfies user expectations, and abundant field testing will be needed before goals of parsimony and simplicity can be definitively achieved. The issue of open source and verifiability has similar corollaries: ARIES' modeling platform and model base are both distributed under an open source license and every model is fully documented, but the novelty of the approach, use of BNs, and the considerably more expansive vision of ES emerging from ARIES outputs may yield initial impressions of opacity for some users. Areas where we believe ARIES can bring important innovations to the ES field are those of scalability and generalizability, due to the structural variability built into the modeling approach. This is of course contingent on satisfactorily addressing the novel challenges posed by the growth of a dynamic, modular model base. With respect to valuation, ARIES is currently not addressing the definition of "currencies" for ES results, either economic or non-economic, and relies on end users to interpret biophysical flow results as they see most appropriate. As integration proceeds with models that can more directly express value, such as those based on multiple-criteria analysis and value transfer, it will be easier to directly assess the added value of ARIES in meeting demand for valuation outputs.

We believe that proper accounting for ES is one of the key scientific challenges for the new millennium, and that finding space for return-free intellectual development in this field can be as important as delivering methods to decision makers in the short term. The end goal of ARIES is both to seed new discussion in the science behind ES and to pioneer innovations in technology that can allow this science to become more readily usable to inform sustainable development. Important hurdles remain in achieving a vision as ambitious as the one described here. Yet, the field of ES has so far seen very simplified approaches aimed towards rapid assessment and quick policy advice. No current ES modeling effort has attempted to fully account for a coupled human-natural system dynamics that can only be investigated with in-depth and long-term scientific study. Indeed, the complex and multiple-scale modeling required for such assessments is likely to remain impossible or impractical, at least on a routine basis, for some time. Yet, with increased availability of remotely sensed data and low-cost computing power, more refined instruments to assist decision-making even in data- and resource-limited policy contexts become increasingly practical, and it is our hope that this work contributes to the scientific and policy discourse surrounding ES and their values.

Acknowledgments

The work documented in this article has been possible thanks to the advice and collaboration of many contributors. A necessarily partial list needs to acknowledge at least the insight, collaboration and support of Sergey Krivov, Marta Ceroni, Ioannis Athanasiadis, Joshua Farley, Sissel Waage, Stefano Balbi, Elena Perez-Miñana, Jennifer Harrison-Cox, Luke Scott, Maya Kocian, Zach Ancona, Iris Goodman, Paulo A.L.D. Nunes, Andy Rosenberg, Celia Harvey, Nalini Rao, Glenn-Marie Lange, Jon Erickson and Darius Semmens. The 30 participants to the first International Spring University on Ecosystem Service Modeling (Bilbao, Spain, March 2013) provided crucial feedback on both the existing model base and the functionality and usability of the software infrastructure. Dave Raffaelli and an anonymous reviewer greatly helped improve the clarity and structure of the manuscript. We thank Jeremy Havens for assistance in preparing Figure 2. Finally, this project would not have been possible without the enormous investment by developers worldwide in creating excellent open source software; among the many open source projects on which ARIES depends, we must mention at least Eclipse, OWL-API, Geotools, GeoServer, the Clojure language and the Linux operating system.

Author Contributions

Conceived and designed the experiments: FV KB BV GJ RP. Performed the experiments: FV KB BV GJ. Analyzed the data: KB FV BV RP. Wrote the paper: FV KB BV GJ DB RP MH. Design and implementation of the modeling language and software: FV. Design and implementation of flow models: GJ FV. Managed Madagascar case study: RP MH. Managed Puget case study: DB KB.

References

1. Daily GC, Polasky S, Goldstein J, Kareiva PM, Mooney HA, et al. (2009) Ecosystem services in decision making: time to deliver. Frontiers in Ecology and the Environment 7: 21–28.
2. Millennium Ecosystem Assessment (2005) Millennium Ecosystem Assessment: living beyond our means - Natural assets and human well-being. 28 p.

3. Martinez-Harms MJ, Balvanera P (2012) Methods for mapping ecosystem service supply: a review. International Journal of Biodiversity Science, Ecosystem Services and Management 8: 17–25.
4. Vigerstol KL, Aukema JE (2011) A comparison of tools for modeling freshwater ecosystem services. Journal of Environmental Management 92: 2403–2409.
5. Waage S, Kester C, Armstrong K (2013) Global Public Trends in Ecosystem Services, 2009–2012. San Francisco: BSR. 25 p.
6. Eigenbrod F, Armsworth PR, Anderson BJ, Heinemeyer A, Gillings S, et al. (2010) The impact of proxy-based methods on mapping the distribution of ecosystem services. Journal of Applied Ecology 47: 377–385.
7. Bagstad KJ, Johnson GW, Voigt B, Villa F (2013) Spatial dynamics of ecosystem service flows: A comprehensive approach to quantifying actual services. Ecosystem Services 4: 117–125.
8. Costanza R, dArge R, deGroot R, Farber S, Grasso M, et al. (1997) The value of the world's ecosystem services and natural capital. Nature 387: 253–260.
9. Troy A, Wilson MA (2006) Mapping ecosystem services: Practical challenges and opportunities in linking GIS and value transfer. Ecological Economics 60: 435–449.
10. Tallis H, Polasky S (2009) Mapping and Valuing Ecosystem Services as an Approach for Conservation and Natural-Resource Management. Annals of the NY Academy of Sciences 1162: 265–283.
11. Fisher B, Turner RK, Burgess ND, Swetnam RD, Green J, et al. (2011) Measuring, modeling and mapping ecosystem services in the Eastern Arc Mountains of Tanzania. Progress in Physical Geography 35: 595–611.
12. Johnson GW, Bagstad K, Snapp R, Villa F (2012) Service Path Attribution Networks (SPANs): A Network Flow Approach to Ecosystem Service Assessment. International Journal of Agricultural and Environmental Information Systems 3: 54–71.
13. Johnson GW, Bagstad KJ, Snapp R, Villa F (2010) Service Path Attribute Networks (SPANs): Spatially quantifying the flow of ecosystem services from landscapes to people. Lecture Notes in Computer Science 6016: 238–253.
14. Kareiva PM, Tallis H, Ricketts T, Daily GC, Polasky S (2011) Natural capital : theory and practice of mapping ecosystem services. New York: Oxford University Press. 365 p.
15. Burkhard B, Kroll F, Nedkov S, Muller F (2012) Mapping ecosystem service supply, demand and budgets. Ecological Indicators 21: 17–29.
16. Carpenter SR, Mooney HA, Agard J, Capistrano D, Defries RS, et al. (2009) Science for managing ecosystem services: Beyond the Millennium Ecosystem Assessment. Proceedings of the National Academy of Sciences of the United States of America 106: 1305–1312.
17. Scheffer M, Bascompte J, Brock WA, Brovkin V, Carpenter SR, et al. (2009) Early-warning signals for critical transitions. Nature 461: 53–59.
18. Scheffer M, Carpenter S, Foley JA, Folke C, Walker B (2001) Catastrophic shifts in ecosystems. Nature 413: 591–596.
19. Holling CS (2001) Understanding the complexity of economic, ecological, and social systems. Ecosystems 4: 390–405.
20. Daniel TC, Muhar A, Arnberger A, Aznar O, Boyd JW, et al. (2012) Contributions of cultural services to the ecosystem services agenda. Proceedings of the National Academy of Sciences 109: 8812–8819.
21. Bagstad KJ, Semmens DJ, Waage S, Wintrop R (2013) A comparative assessment of decision-support tools for ecosystem services quantification and valuation. Ecosystem Services 5: 27–39.
22. Syrbe RU, Walz U (2012) Spatial indicators for the assessment of ecosystem services: Providing, benefiting and connecting areas and landscape metrics. Ecological Indicators 21: 80–88.
23. TEEB (2010) The Economics of Ecosystems and Biodiversity Ecological and Economic Foundations.; Kumar P, editor. London and Washington: Earthscan.
24. Bagstad K, Semmens D, Winthrop R, Jaworski D, Larson J (2012) Ecosystem Services Valuation to Support Decision Making on Public Lands - A Case Study of the San Pedro River Watershed, Arizona. U.S. Geological Survey. 93 p.
25. Haines-Young RH, Potschin M (2011) Common International Classification of Ecosystem Services (CICES): 2011 Update.: European Environmental Agency. EEA/BSS/07/007 EEA/BSS/07/007.
26. Johnson GW, Snapp R, Villa F, Bagstad K (2012) Modelling Ecosystem Services Flows under Uncertainty with Stochastic SPAN; Leipzig, Germany.
27. Bagstad K, Villa F, Johnson GW, Voigt B (2011) ARIES - ARtificial Intelligence for Ecosystem Services: a guide to models and data. 122 p.
28. Daily GC (1997) Nature's services : societal dependence on natural ecosystems. Washington, DC: Island Press. 392 p.
29. DeGroot RS, Wilson MA, Boumans RMJ (2002) A typology for the classification, description and valuation of ecosystem functions, goods and services. Ecological Economics 41: 393–408.
30. Carpenter SR, Castilla JC, deGroot RS, Mooney H, Naeem S, et al. (2003) Ecosystems and their Services. Millennium Ecosystem Assessment: Ecosystems and Human well-being: a Framework for Assessment 245 pp. NY: Island Press. 49–70.
31. Daniel TC, Muhar A, Arnberger A, Aznar O, Boyd JW, et al. (2012) Contributions of cultural services to the ecosystem services agenda. Proceedings of the National Academy of Sciences of the United States of America 109: 8812–8819.
32. Boyd J, Banzhaf S (2007) What are ecosystem services? The need for standardized environmental accounting units. Ecological Economics 63: 616–626.
33. Fisher B, Turner K, Zylstra M, Brouwer R, de Groot R, et al. (2008) Ecosystem Services and Economic Theory: Integration for Policy-Relevant Research. Ecological Applications 18: 2050–2067.
34. Wallace KJ (2007) Classification of ecosystem services: Problems and solutions. Biological Conservation 139: 235–246.
35. Nahlik AM, Kentula ME, Fennessy MS, Landers DH (2012) Where is the consensus? A proposed foundation for moving ecosystem service concepts into practice. Ecological Economics 77: 27–35.
36. Palomo I, Martín-López B, Potschin M, Haines-Young R, Montes C (2013) National Parks, buffer zones and surrounding lands: Mapping ecosystem service flows. Ecosystem Services 4: 104–116.
37. Villa F, Bagstad K, Johnson GW, Voigt B (2011) Scientific instruments for climate change adaptation: estimating and optimizing the efficiency of ecosystem services provision. Economia Agraria y Recursos Naturales 11: 83–98.
38. Bonabeau E (2002) Agent-based modeling: Methods and techniques for simulating human systems. Proceedings of the National Academy of Sciences of the United States of America 99: 7280–7287.
39. Salzman J (2005) Creating markets for ecosystem services: notes from the field. New York University Law Review 80: 870–961.
40. McCann RK, Marcot BG, Ellis R (2006) Bayesian belief networks: applications in ecology and natural resource management. Canadian Journal of Forest Research 36: 3053–3062.
41. Marcot BG, Thompson MP, Runge MC, Thompson FR, McNulty S, et al. (2012) Recent advances in applying decision science to managing national forests. Forest Ecology and Management 285: 123–132.
42. Cowell RG, Dawid AP, Lauritzen SL, Spiegelhalter DJ (1999) Probabilistic Networks and Expert Systems. New York: Springer.
43. van Nes EH, Scheffer M (2005) Implications of Spatial Heterogeneity for Catastrophic Regime Shifts in Ecosystems. Ecology 86: 1797–1807.
44. Martín-López B, Montes C, Benayas J (2008) Economic Valuation of Biodiversity Conservation: the Meaning of Numbers. Conservation Biology 22: 624.
45. Sagoff M (2011) The quantification and valuation of ecosystem services. Ecological Economics 70: 497.
46. Villa F (2009) Semantically-driven meta-modelling: automating model construction in an environmental decision support system for the assessment of ecosystem services flow. In: Athanasiadis IN, Mitkas PA, Rizzoli AE, Marx Gomez J, editors. Information Technology in Environmental Engineering. New York, NY: Springer. 23–36.
47. Villa F, Athanasiadis IN, Rizzoli AE (2009) Modelling with knowledge: a review of emerging semantic approaches to environmental modelling. Environmental Modelling & Software 24: 577–587.
48. Argent RM (2004) An overview of model integration for environmental application - components, frameworks and semantics. Environmental Modelling & Software 19: 219–234.
49. Villa F (2007) A semantic framework and software design to enable the transparent integration, reorganization and discovery of natural systems knowledge. Journal of Intelligent Information Systems 29: 79–96.
50. Moore RV, Tindall CI (2005) An overview of the open modelling interface and environment (the OpenMI). Environmental Science & Policy 8: 279–286.
51. David O, Ascough Ii JC, Lloyd W, Green TR, Rojas KW, et al. (2013) A software engineering perspective on environmental modeling framework design: The Object Modeling System. Environmental Modelling & Software 39: 201–213.
52. Simmhan YL, Plale B, Gannon D (2005) A survey of data provenance in e-science. SIGMOD Rec 34: 31–36.
53. Gruber TR (1995) Toward principles for the design of ontologies used for knowledge sharing. International Journal of Human-Computer Studies 43: 907–928.
54. Madin JS, Bowers S, Schildhauer MP, Jones MB (2008) Advancing ecological research with ontologies. Trends in Ecology & Evolution 23: 159–168.
55. Raskin RG, Pan MJ (2005) Knowledge representation in the semantic web for Earth and environmental terminology (SWEET). Computers & Geosciences 31: 1119–1125.
56. Gangemi A, Guarino N, Masolo C, Oltramari A, Schneider L (2002) Sweetening ontologies with DOLCE. Knowledge Engineering and Knowledge Management, Proceedings 2473: 166–181.
57. Villa F, Scott JL, Athanasiadis IN (2013) Thinklab software repository. Online: bitbucket.org/ariesteam/thinklab.
58. German D, González-Barahona J (2009) An Empirical Study of the Reuse of Software Licensed under the GNU General Public License. In: Boldyreff C, Crowston K, Lundell B, Wasserman A, editors. Open Source Ecosystems: Diverse Communities Interacting: Springer Berlin Heidelberg. 185–198.
59. Bagstad KJ, Villa F, Batker D, Harrison-Cox J, Voigt B, et al. (In review) From theoretical to actual ecosystem services: Mapping beneficiaries and spatial flows in ecosystem service assessments. Ecology and Society.
60. Portela R, Nunes PALD, Onofri L, Villa F, Shepard A, et al. (2012) Assessing and Valuing Ecosytem Services in the Ankeniheny-Zahamena Corridor, Madagascar: A Demonstration Case Study for the Wealth Accounting and the

Valuation of Ecosystem Services (WAVES) Global Partnership. New York: World Bank.
61. Villa F, Voigt B, Erickson J (2013) New perspectives in ecosystem services science as instruments to understand environmental securities. Philosophical Transactions of the Royal Society of London Series B-Biological Sciences Forthcoming.
62. Wendland KJ, Honzák M, Portela R, Vitale B, Rubinoff S, et al. (2010) Targeting and implementing payments for ecosystem services: Opportunities for bundling biodiversity conservation with carbon and water services in Madagascar. Ecological Economics 69: 2093–2107.
63. Kull CA (2000) Deforestation, erosion, and fire: degradation myths in the environmental history of Madagascar. Environment and History: 423–450.
64. Randrianarijaona P (1983) The erosion of Madagascar. Ambio: 308–311.
65. Renard KG, Foster GR, Weesies GA, Porter JP (1991) RUSLE: Revised universal soil loss equation. Journal of soil and Water Conservation 46: 30–33.
66. Roose E (1996) Land husbandry: components and strategy: FAO Rome.
67. Solonitompoarinony J (2000) Dommage hors site de l'érosion: les effets de l'ensablement sur la production rizicole. Etude de cas dans la Commune d'Ambohitrarivo. Antananarivo, Madagascar: MINEV, ONE, USAID, CFSIGE, PAGE, FAC DEGS.
68. Brand J, Minten B, Randrianarisoa J (2002) Etude d'impact de la Déforestation sur la Riziculture Irriguée: Cas de Petits Bassins Versants de Maroantsetra, Nord-Est Madagascar. Antananarivo, Madagascar: FOFIFA.
69. Bolte JP, Vache K (2010) Envisioning Puget Sound Alternative Futures. Oregon State University. 50 p.
70. Bourassa SC, Hoesli M, Sun J (2006) A simple alternative house price index method. Journal of Housing Economics 15: 80–97.
71. McConnell V, Walls MA (2005) The value of open space: Evidence from studies of nonmarket benefits: Resources for the Future Washington, DC, USA.
72. Coulthard TJ (2013) CAESAR-Lisflood. Available: www.coulthard.org.uk/CAESARLisflood.html.
73. Leavesley G, Stannard L, Singh V (1995) The precipitation-runoff modeling system-PRMS. Computer models of watershed hydrology: 281–310.
74. Smith B, Samuelsson P, Wramneby A, Rummukainen M (2011) A model of the coupled dynamics of climate, vegetation and terrestrial ecosystem biogeochemistry for regional applications. Tellus A 63: 87–106.
75. ARIES Consortium (2012) ARtificial Intelligence for Ecosystem Services (ARIES). Available: www.ariesonline.org.
76. Jax K, Barton DN, Chan KMA, de Groot R, Doyle U, et al. (2013) Ecosystem services and ethics. Ecological Economics 93: 260–268.
77. Chan KMA, Satterfield T, Goldstein J (2012) Rethinking ecosystem services to better address and navigate cultural values. Ecological Economics 74: 8–18.
78. Poppy GM, Chiotha S, Eigenbrod F, Harvey C, Honzak M, et al. (2013) Understanding food security in a perfect storm: an ecosystem services approach. Philosophical Transactions of the Royal Society of London Series B-Biological Sciences Forthcoming.
79. Corbera E, Pascual U (2012) Ecosystem services: heed social goals. Science 335: 655–656.
80. Farley J (2012) Ecosystem services: The economics debate. Ecosystem Services 1: 40–49.
81. McDermott M, Mahanty S, Schreckenberg K (2013) Examining equity: A multidimensional framework for assessing equity in payments for ecosystem services. Environmental Science and Policy 33: 416–427.
82. Wegner G, Pascual U (2011) Cost-benefit analysis in the context of ecosystem services for human well-being: A multidisciplinary critique. Global Environmental Change 21: 492–504.
83. Wilson MA, Hoehn JP (2006) Valuing environmental goods and services using benefit transfer: The state-of-the art and science. Ecological Economics 60: 335–342.
84. Liu S, Portela R, Ghermandi A, Rao N, Wang X (2011) Environmental Benefit Transfers of Ecosystem Service Valuation. In: Wolanski E, McLusky DS, editors. Treatise on Estuarine and Coastal Science. New York: Elsevier.
85. Bateman I, Mace G, Fezzi C, Atkinson G, Turner K (2011) Economic Analysis for Ecosystem Service Assessments. Environmental and Resource Economics 48: 177–218.
86. Ghermandi A, Nunes PALD (2013) A global map of coastal recreation values: Results from a spatially explicit meta-analysis. Ecological Economics 86: 1–15.
87. Loomis JB (1992) The evolution of a more rigorous approach to benefit transfer: Benefit function transfer. Water Resources Research 28: 701–705.
88. Boumans R, Costanza R, Farley J, Wilson MA, Portela R, et al. (2002) Modeling the dynamics of the integrated earth system and the value of global ecosystem services using the GUMBO model. Ecological Economics 41: 529–560.
89. Villa F, Ceroni M, Krivov S (2007) Intelligent databases assist transparent and sound economic valuation of ecosystem services. Environmental Management 39: 887–899.
90. Villa F, Wilson M, de Groot R, Farber S, Costanza R, et al. (2002) Designing an integrated knowledge base to support ecosystem services valuation. ECOLOGICAL ECONOMICS 41: 445–456.
91. Earth Economics (2013) Ecosystem Valuation Toolkit. Online: esvaluation.org.
92. Farley J (2008) The Role of Prices in Conserving Critical Natural Capital. Conservation Biology 22: 1399–1408.
93. Farley J, Schmitt F, Alvez J, de Freitas NR Jr (2012) How valuing nature can transform agriculture. Solutions 2: 64–73.
94. Bolte JP, Hulse DW, Gregory SV, Smith C (2007) Modeling biocomplexity – actors, landscapes and alternative futures. Environmental Modelling & Software 22: 570–579.
95. Rodríguez JP, Beard TD Jr, Bennett EM, Cumming GS, Cork S, et al. (2006) Trade-offs across space, time, and ecosystem services. Ecology and Society 11: 28.
96. Tallis HT, Ricketts T, Guerry AD, Wood SA, Sharp R, et al. (2011) InVEST 2.2.0 User's Guide. Stanford University.
97. Miles L, Kapos V (2008) Reducing Greenhouse Gas Emissions from Deforestation and Forest Degradation: Global Land-Use Implications. Science 320: 1454–1455.
98. Figueira J, Greco S, Ehrgott M (2005) Multiple criteria decision analysis: State of the art. New York: Springer.
99. Villa F, Tunesi L, Agardy T (2002) Zoning marine protected areas through spatial multiple-criteria analysis: the case of the Asinara Island National Marine Reserve of Italy. Conservation Biology 16: 515–526.
100. Conservation International (2013) Future Climates: climate projections for assessing the conservation impacts of climate change. Online: futureclimates.-conservation.org.
101. Basque Centre for Climate Change (BC3) (2013) International Spring University on Ecosystem Services Modelling. Available: www.bc3research.org/springuniversity.
102. Page L, Brin S, Motwani R, Winograd T (1999) The PageRank Citation Ranking: Bringing Order to the Web. Stanford InfoLab.
103. Schägner JP, Brander L, Maes J, Hartje V (2013) Mapping ecosystem services' values: Current practice and future prospects. Ecosystem Services 4: 33–46.

8

Solar Disinfection of *Pseudomonas aeruginosa* in Harvested Rainwater: A Step towards Potability of Rainwater

Muhammad T. Amin[1]*, Mohsin Nawaz[1], Muhammad N. Amin[2], Mooyoung Han[3]

1 Alamoudi Water Research Chair, King Saud University, Riyadh, Kingdom of Saudi Arabia, **2** Department of Civil and Environmental Engineering, College of Engineering, King Faisal University, Al-Hofuf Al-Ahsa, Kingdom of Saudi Arabia, **3** Civil and Environmental Engineering Department, Seoul National University, Shinrimdong, KwanakGu, Seoul, South Korea

Abstract

Efficiency of solar based disinfection of *Pseudomonas aeruginosa (P. aeruginosa)* in rooftop harvested rainwater was evaluated aiming the potability of rainwater. The rainwater samples were exposed to direct sunlight for about 8–9 hours and the effects of water temperature (°C), sunlight irradiance (W/m^2), different rear surfaces of polyethylene terephthalate bottles, variable microbial concentrations, pH and turbidity were observed on *P. aeruginosa* inactivation at different weathers. In simple solar disinfection (SODIS), the complete inactivation of *P. aeruginosa* was obtained only under sunny weather conditions (>50°C and >700 W/m^2) with absorptive rear surface. Solar collector disinfection (SOCODIS) system, used to improve the efficiency of simple SODIS under mild and weak weather, completely inactivated the *P. aeruginosa* by enhancing the disinfection efficiency of about 20% only at mild weather. Both SODIS and SOCODIS systems, however, were found inefficient at weak weather. Different initial concentrations of *P. aeruginosa* and/or *Escherichia coli* had little effects on the disinfection efficiency except for the SODIS with highest initial concentrations. The inactivation of *P. aeruginosa* increased by about 10–15% by lowering the initial pH values from 10 to 3. A high initial turbidity, adjusted by adding kaolin, adversely affected the efficiency of both systems and a decrease, about 15–25%; in inactivation of *P. aeruginosa* was observed. The kinetics of this study was investigated by Geeraerd Model for highlighting the best disinfection system based on reaction rate constant. The unique detailed investigation of *P. aeruginosa* disinfection with sunlight based disinfection systems under different weather conditions and variable parameters will help researchers to understand and further improve the newly invented SOCODIS system.

Editor: Anil Kumar, University of Missouri-Kansas City, United States of America

Funding: Authors would like to thank and extend their sincere appreciation to the National Plan of Science and Technology (NPST) at King Saud University for the financial support through Project # 11-WAT1875-02. The funders had no role in study design, data collection and analysis, decision to publish, or preparation of the manuscript.

Competing Interests: The authors have declared that no competing interests exist.

* E-mail: mtamin@ksu.edu.sa

Introduction

Water shortage and adequate cost-effective water treatment technologies are the major concerns of this era around the globe. According to the latest available information by the joint study of UNICEF and WHO, still 783 million people do not have access to safe water resources [1]. The use of polluted water for ingestion, especially in many areas of developing countries, is the major cause of diarrhea and various other water-borne diseases. Over 1.2 million children, most being less than five years old, die due to diarrheal diseases each year [2]. Currently the alternate water sources and efficient water treatment technologies are needed. Rainwater harvesting (RWH) as safe water source and cost-effective solar disinfection (SODIS) system has significant importance in this regard [3–5]. For efficient utilization of harvested rainwater, the RWH system should consist of its catchment area, a storage tank, a treatment facility, a supply facility and pipes [6]. Roof being an impermeable catchment area is a good source to collect rainwater. However the microbial quality of harvested rainwater can be deteriorated due to the possible contamination upon contact with catchment which is one of the major constraints of using rainwater for potable purposes [7,8]. *P. aeruginosa* is a part of microbial contamination which is considered responsible for nosocomial infection in immune-compromised people [9]. The presence of this microorganism is unacceptable as it has been implicated in waterborne and food borne diseases and is now considered primary infectious agent [10]. *P. aeruginosa* is detected in stored rainwater by many researchers which is of serious health concerns for water users [11–15].

SODIS is being used as point-of use treatment method in many parts of the world. It is believed that solar energy can be used in improving water quality in those parts of the world that experience hot, sunny climates [16,17]. The portion of solar radiation which plays a major role in generating the antimicrobial activity is believed to be UV-A, 315–400nm, and to a lesser extent visible violet and blue light in the range of 400–490nm [18,19]. In addition, the increase in water temperature through solar collector is known to work synergistically with solar radiation to further enhance the efficiency of SODIS process [20,21]. The inactivation of high populations of microorganisms in highly turbid water

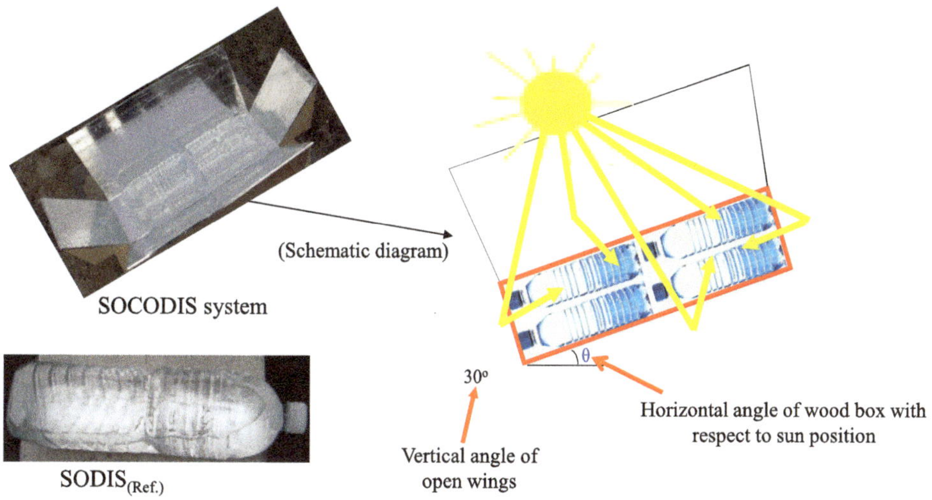

Figure 1. Demonstration of SODIS$_{(Ref.)}$ and SOCODIS systems.

(approximately 200 nephelometric turbidity units, NTU) can be achieved within 7h if the water temperature reached at least 55°C [22]. Previous studies found that children, between the ages of 5–16 years, who stored their drinking water in 1.5 liters polyethylene terephthalate (PET) bottles in direct sunlight were less prone to diarrhea and other water borne diseases compared to the children who kept their water bottles out of the sun [23,24]. The efficiency of SODIS and solar collector disinfection (SOCODIS) systems was tested against total and fecal coliform, *Escherichia coli* (*E. coli*) and HPC previously [4] while the efficiency of SODIS and SOCODIS systems was checked against the inactivation of *P. aeruginosa* in this study.

The current study focuses on harvested rainwater as a potential potable water source by using cost-effective solar disinfection of *P. aeruginosa*. The objectives include the enhancement of *P. aeruginosa* inactivation using different rear surface of PET bottles and to investigate the efficiency of solar collector under mild and weak weather conditions with different concentrations of the tested organism. The effects of different initial pH and turbidity values (adjusted by adding kaolin) on *P. aeruginosa* inactivation are also the important part of this study in addition to investigating the effects of the presence of *E. coli* in different concentrations.

Materials and Methods

Sampling sites and disinfection setup

The rainwater storage tank was installed in one of the educational buildings (engineering building #39) at the Campus of Seoul National University, South Korea. The harvested rainwater from the rooftop of the building was used mainly for toilet flushing inside building. Rainwater samples were collected in 8 sterile PET bottles of 2L capacity from the supply point which is situated at an height of 1.35 m from the base of the underground rainwater storage tank. The PET bottles were filled ¾ full and were shaken for about 20 seconds to oxygenate the water and an air space of about 15% of the bottle volume was still kept for aeration [16].

Table 1. Reference values of stored rainwater samples.

Physicochemical parameters					Microbial parameter	
Turbidity	Temperature	pH	EC	DO	*P. aeruginosa*	*E. coli*
NTU	*°C*		*μs/cm*	*mg/l*	*CFU/100 ml*	*CFU/100 ml*
2-5	22-26	7-8	150-450	6-9	350-900	200-450

In simple SODIS, 3 PET bottles with transparent, absorptive (rear surface painted black) and reflective (rear surface covered with aluminium foil) rear surfaces were exposed to natural sunlight while 4 transparent PET bottles kept inside the box were exposed to sunlight at the same time in SOCODIS system. The detailed design of the SOCODIS system has already been described in a previous study [4]. The actual and schematic diagram of SOCODIS system and reflective SODIS (SODIS$_{(Ref.)}$) in this study is shown as Fig. 1 [25] while the transparent and absorptive SODIS (SODIS$_{(Abs.)}$) cases are not presented. Both SODIS and SOCODIS system was exposed to direct sunlight on the rooftop of the same building where the rooftop RWH system was installed and it was located 37°27'18.31"N, 126°57'3.25"E. There was no need of any special permission to carry out the exposure experiments at the rooftop of the building and the activities did not involve endangered or protected species.

The duration of natural sunlight exposure in both systems, SODIS and SOCODIS was about 9 hours. Experiments were performed at all weathers in case of simple SODIS system but for SOCODIS system, the analysis was made only under mild and weak weather conditions. Sunlight radiations, at all occasions within a year, were monitored on-site with a SP-110 Pyranometer (Apogee Instruments Inc., Logan, USA) connected to a data logger (DT80 Series 2) recording 1 minute averages in Watt/m^2 (W/m^2). The spectral range of the Pyranometer used was 0–1750 W/m^2. The values for irradiance and the temperature inside the bottle were recorded at an interval of an hour.

Microbial analysis

The microbial water quality analysis was carried out by Standard Methods in Soil and Water Quality Laboratory, Seoul National University (SNU) at room temperature 25°C. The detection of *P. aeruginosa* was made by multiple tube method. Asparagine broth (Titan Biotech Ltd.) for the presumptive stage of *P. aeruginosa* was used in a series of fifteen test tubes for all three dilutions of 10, 1 and 0.1 ml and thus 5 tubes per each dilution.

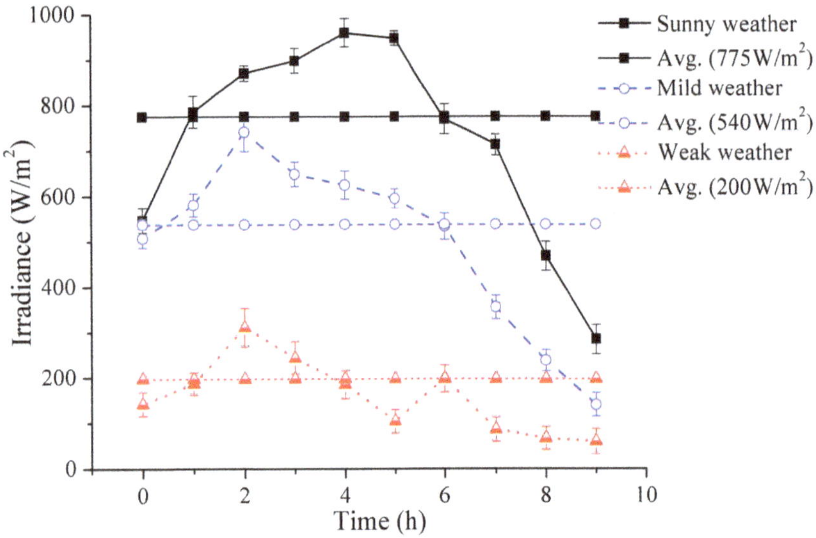

Figure 2. Irradiance change under different weather conditions.

The tubes were than incubated at 35°C for 24 hours. To check the number of *P. aeruginosa*, the tubes were examined under black light after 24 and then after 48 hours. The tubes with green fluorescent pigments were selected and then 0.1 ml of culture was further incubated at 35°C into acetamide broth (Titan Biotech Ltd.) to complete the confirmation stage [26]. For *E. coli* detection and measurements, DifcoTM Lauryl Tryptose Broth (Becton, Dick-insonandCompany) was used for the presumptive phase while the confirmation was done by using the BactoTM EC Medium with Mug (Becton Dickinson France S.A.).

Miscellaneous measurements

A dark control was separately kept in 1 PET bottle wrapped in aluminium foil, over the same time period for each experiment. Nearly 5–6 repetitions were carried out for most of the analysis to avoid any experimental error and the error bars are shown in almost all the time graphs.

For the effects of different initial pH values, four pH values were chosen for both SODIS and SOCODIS system and the comparison was made only at weak weather conditions. For initial pH adjustment in stored rainwater, diluted HCl and NaOH were used. To check the effect on *P. aeruginosa* inactivation at different turbidity values, kaolin was introduced in raw rainwater samples. The turbidity values were set at 20NTU and 100 NTU and the effects were checked and compared for best suited case $SODIS_{(Abs.)}$ and for SOCODIS system, separately.

The behavior of inactivation curves on experimental data was investigated with GInaFit (Geeraerd and Van Impe Inactivation Model Fitting Tool) using Geeraerd Log-linear plus shoulder model [27]. Geeraerd model has been widely used in many SODIS scientific contributions to fit experimental results [28].

Results and Discussions

Table 1 shows the average initial values of physicochemical parameters (turbidity, pH, EC & DO) and the number of *P. aeruginosa* observed in stored rainwater samples. The physicochemical parameters were only used as reference values and the major emphasize of this study was on the microbial analysis. The concentration of *P. aeruginosa* in harvested rainwater before disinfection varied throughout the year as 350–900 CFU/ 100 ml depending upon both wet and dry seasons. No artificial method of *P. aeruginosa* or *E. coli* spiking was used.

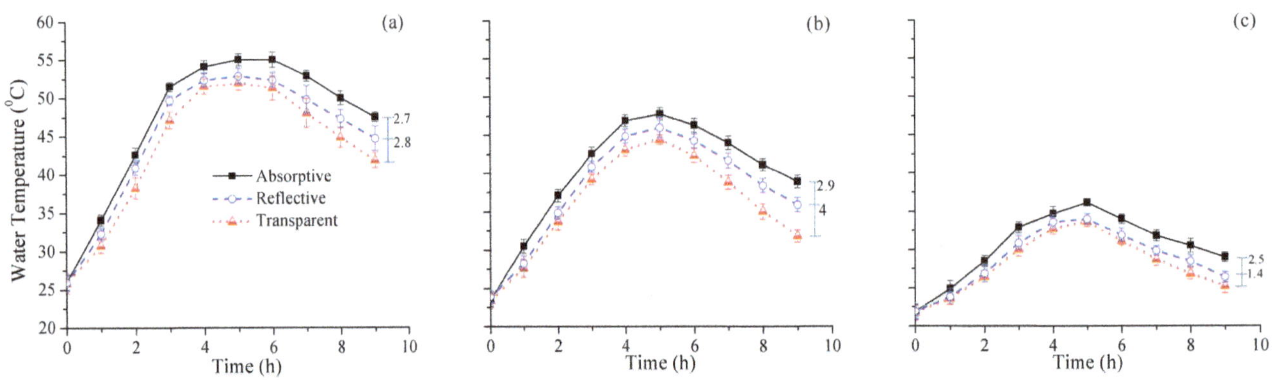

Figure 3. Temperature change in PET bottles with different rear surfaces against exposure time under (a) sunny, (b) mild, and (c) weak weather conditions.

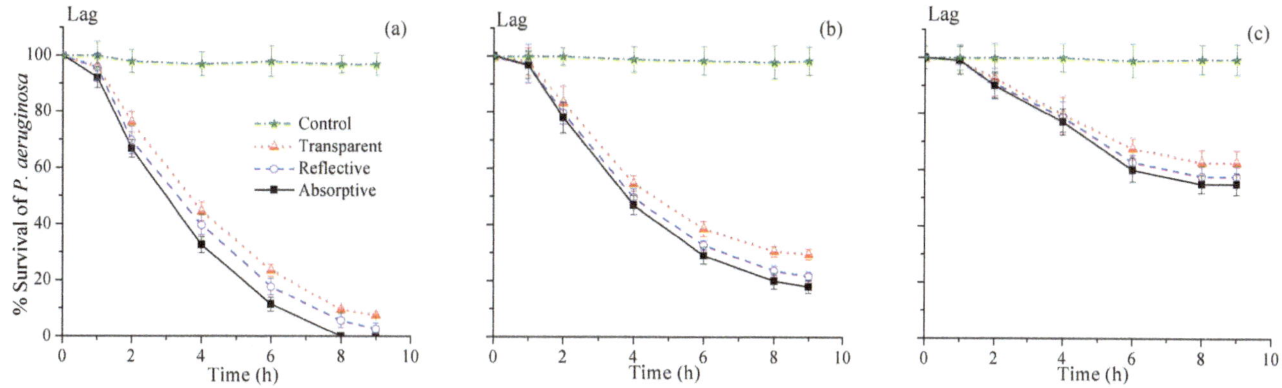

Figure 4. ***P. aeruginosa*** **inactivation with different rear surfaces of PET bottles under (a) sunny, (b) mild, and (c) weak weather conditions.**

Chemical and microbiological parameters fluctuated during the course of the study, with the highest levels of microbiological contamination in wet seasons when the sunlight irradiance were highest and sunny weather conditions prevailed. Overall, the collected rainwater had a relatively good physicochemical quality but did not meet the requirements for potable water. Most of the water-quality parameters including the bacterial counts were higher both at the beginning and ending of the wet season mainly due to dirty catchment surfaces and addition of first-flush into rainwater tanks after long dry periods. Initial pH was high in dry periods but decreased during the rainy season and increased again after the rains stops. Both *E. coli* and *P. aeruginosa* levels were lower during the wet season due to short water storage time despite the fact that the fresh contamination was added with the rains. Due to filtration of the rainwater through the filters attached to the weirs before entering into the rainwater storage tank, a reduction in the concentrations of *P. aeruginosa* was expected, but this was not always the case. This was evident by observing the microbial load even after the filtrations step and by observing the good numbers of microorganisms even after the first-flush.

Evaluation of solar radiation and temperature under different weather conditions

Solar irradiance plays the most important role in solar based disinfection systems. For efficient SODIS, a geographical area should receive radiation above 500 W/m^2 for 3–5 hours [19,29]. Fig. 2 represents the pattern of sunlight intensity by using the hourly average values of irradiance for categorizing the weather condition into sunny, mild and weak weathers. The time 0h corresponds to 8am or 9am in the morning while 9h corresponds to 5pm or 6 pm in the evening. The irradiance values varied from 700–950 W/m^2 for sunny weather with an average of 775 W/m^2 (usually from May to August), 500–700 W/m^2 for mild weather with an average of 540 W/m^2 (usually from September to October and March to April) and 140–300 W/m^2 with an average of 200 W/m^2 for weak weather (usually from November to February).

The temperature difference at 9th hours between absorptive and reflective rear surfaces container was 2.5–3°C under all weather conditions. Temperature comparison was also made using the same rear surfaces as mentioned earlier under different weather conditions. The water temperature was measured at regular intervals of 1h by keeping the device inside the bottles while the ambient temperature was not recorded. The average temperature values drawn for sunny, mild and weak weather conditions are shown in Fig. 3(a, b, & c). Absorptive container used the mechanism to enhance the water temperature by absorbing radiation while the irradiance affect was dominant in reflective container.

There was almost an exponential rise in water temperature reaching maximum value in the afternoon after which it drops with decreasing sun irradiance. Temperature was affected by weather conditions and was directly proportional to irradiance. This affect can be affirmed by comparing Fig. 2 with Fig. 3. The effective temperature for microbial inactivation was considered above 50°C with dominant thermal effect of sunlight [30]. The results of the current study were comparable with the previous findings as the temperature was raised above 50°C, during sunny weather conditions, in almost all containers shown in Fig. 3(a). However, mild weather conditions may not be fit for SODIS if only thermal rather than synergistic effects of thermal and UV radiations are considered since the maximum temperature attained was less than 50°C. Low temperature under weak sunlight conditions may have adverse effects on the efficiency of SODIS because of less or no thermal and/or synergistic effects of thermal and UV radiations. The observed values of temperature and irradiance were according to the weather pattern of South Korea (Latitude: 37°35' North, Longitude: 127°03' East). Similar pattern of temperature and irradiance can be observed for exposure time and can be generalized for any location in the world.

Effects of different rear surfaces of PET bottles under different weather conditions in SODIS

The optimization of system by selecting the best type of container suited under different weather conditions was one of the main objectives of this study. For this purpose, transparent, absorptive and reflective containers were used to compare the efficacy of SODIS at sunny, mild and weak weather conditions as shown in Fig. 4 (a, b & c). Fig. 4(a) shows microbial inactivation in all three containers with different rear surfaces under sunny weather conditions. The complete inactivation of *P. aeruginosa* in absorptive container confirmed the dominant effect of thermal inactivation. Also the higher inactivation of *P. aeruginosa* (about 97%) in reflective container showed that synergistic effect of temperature and radiation was available. The inactivation rate under sunny weather condition with different rear surface was in order as absorptive>reflective>transparent containers. In transparent container the temperature was raised above 50°C under sunny weather condition but the complete inactivation of *P.*

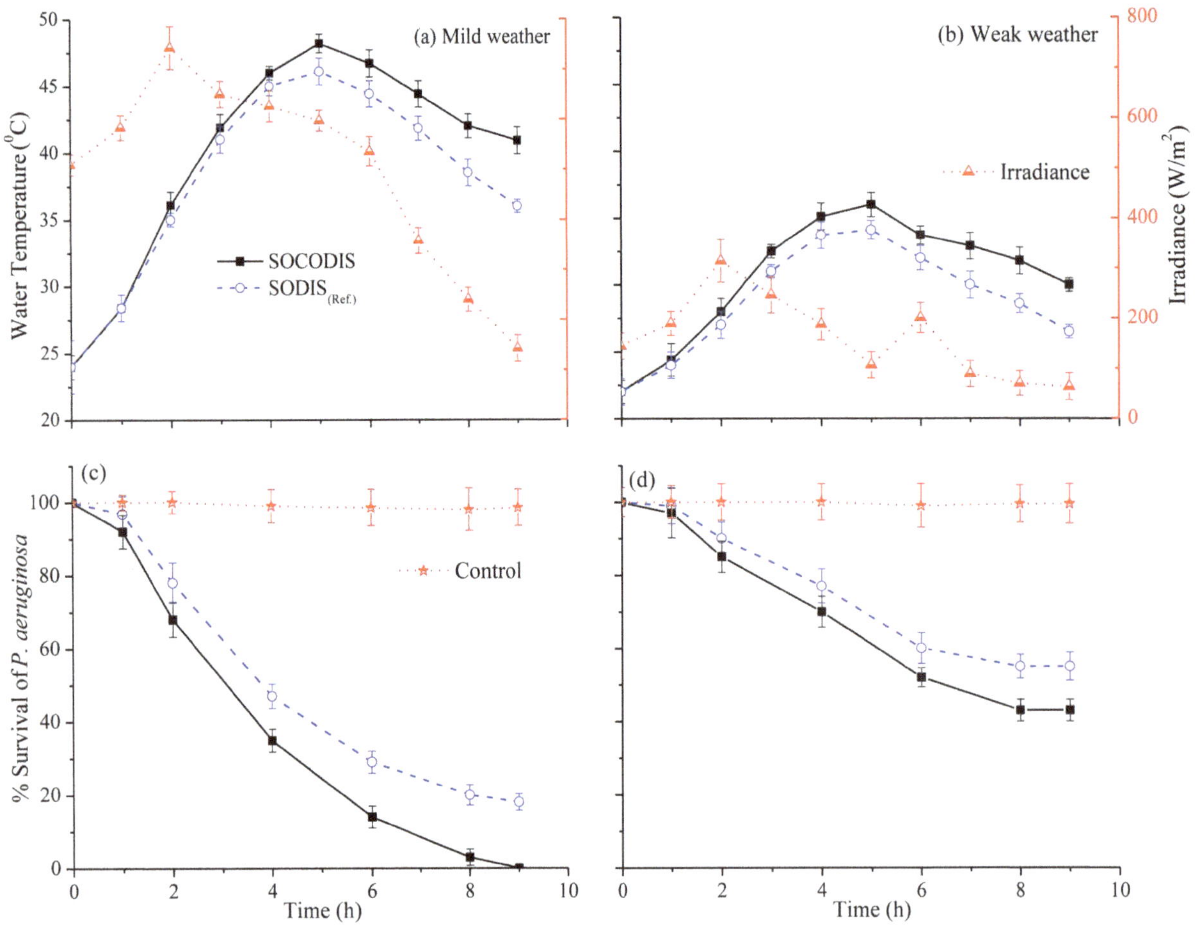

Figure 5. Comparison of temperature and *P. aeruginosa* inactivation in $SODIS_{(Ref.)}$ and SOCODIS system under mild (a & c), and weak (b & d) weather conditions.

aeruginosa was not achieved even after 9h of exposure. This observation showed the resistivity of *P. aeruginosa* to SODIS and was contrary with the previous studies [22, 31, 32]. Under mild weather conditions as shown in Fig. 4(b) however, the inactivation rate of *P. aeruginosa* was unexpectedly higher in reflective containers compared to absorptive containers but the complete inactivation of *P. aeruginosa* was not achieved in any case. The exact reason of this slightly higher inactivation in reflective container was unknown at least in this study.

The difference in inactivation of *P. aeruginosa* with transparent rear surface under mild conditions was about 10%. As none of the water sample, with any rear surface, under mild weather condition could reach to the 50°C so the irradiance, other dominant factor under SODIS, may be the possible reason for *P. aeruginosa* inactivation. The effects of irradiance looked to be more significant with reflective rear surface as indicated in Fig. 4(b). Transparent containers showed poor performance where by both effects of radiation and temperature were not concentrated as was the case in reflective and/or absorptive containers. Weak weather as displayed in Fig. 4(c) showed bad results for similar effects of different rear surfaces. Only 37–45% inactivation of *P. aeruginosa* after 9 hours of exposure to direct sunlight showed that least effect of temperature and/or irradiance was available.

In all above mentioned conditions, except $SODIS_{(Abs.)}$ under sunny weather condition, the inactivation rate of *P. aeruginosa* was almost negligible for first hour. This duration represents the lag phase in which the bacteria showed some resistance to solar inactivation. After lag phase, exponential inactivation of *P. aeruginosa* was observed for further 3–5 hours which was the peak hours of sunlight irradiation on any day and under all weather conditions. For last 2–3 hours of exposure, there was almost negligible inactivation at least under weak sunlight intensity.

Comparison of $SODIS_{(Ref.)}$ and SOCODIS systems for radiation and temperature effects on microbial inactivation

The efficiency of SODIS and SOCODIS systems was investigated under different weather conditions based upon the irradiance range and temperature. During 9 hours exposure to sunlight, the inactivation rate of *P. aeruginosa* was initially measured after 1h and then after every 2h interval. Finally the comparison of $SODIS_{(Ref.)}$ was made with SOCODIS system. All the results exhibited almost the same trend observing a close correlation between irradiance and time required to inactivate *P. aeruginosa*. At first, for the comparison purposes of $SODIS_{(Ref.)}$ and SOCODIS, the temperature changes under mild and weak weather conditions were measured at regular intervals of 1h. The average temperature values of all four bottles, kept inside the solar collector, for mild and weak weather conditions are shown in Figs. 5(a) and 5(b), respectively. Comparing the maximum temperature reached in SODIS and SOCODIS system, difference between peak values

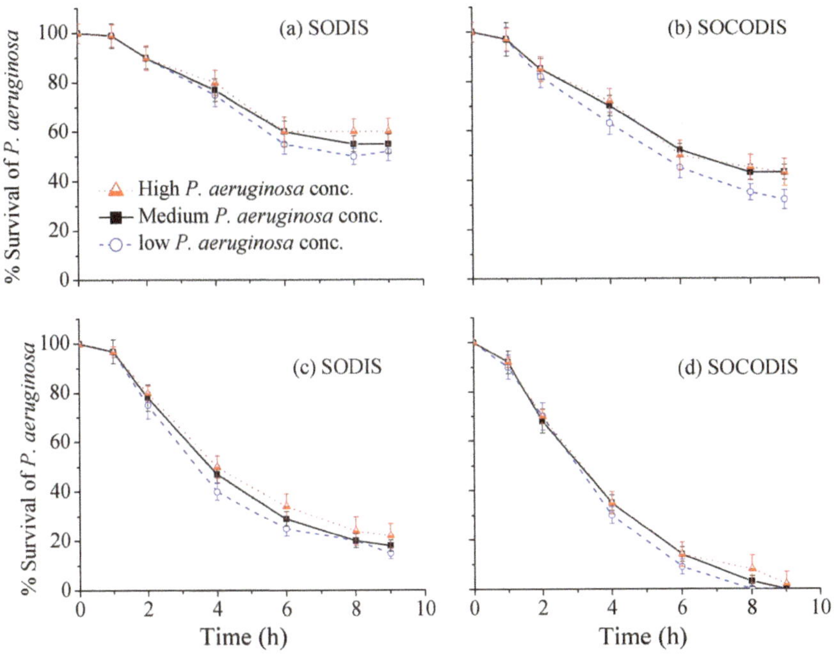

Figure 6. Effects of different initial *P. aeruginosa* concentratinos on disinfection efficiency in both SODIS and SOCODIS system under weak (a & b) and mild (c & d) weather conditions.

under mild and weak weather conditions was about 2–3°C. The relatively greater temperature increase in SOCODIS system was observed because the solar irradiation was reflected back onto the PET bottles from all four sides due to the open side wings of solar box. This effect was achieved only from one side i. e. base of the PET bottle, in SODIS $_{(Ref.)}$ case. The synergistic effects of temperature and irradiance were also reported in earlier studies [31] which were evident under mild weather conditions as shown in Fig. 5(c) where the complete inactivation of *P. aeruginosa* was achieved.

However, the SOCODIS system remained inefficient under weak weather conditions, mostly because the synergistic effect of radiation and temperature was absent shown in Fig. 5(d). The difference in disinfection efficiency between the two systems can be summarized as follows: the SOCODIS system was about 20–30% more efficient than the simple SODIS system. It is well known previously that the effectiveness of SODIS improves when a solar collector is used, enhancing the irradiance effect by reflecting and returning the solar energy to the water samples, as support for bottles [31].

The exact mode of action by which SODIS or SOCODIS systems achieved *P. aeruginosa* inactivation is not fully understood. It has recently been demonstrated that solar photons attack proteins in bacteria, directly or indirectly via reactive oxygen species (ROS), causing protein oxidation. This leads to the disruption of metabolic function and damage to the cell membrane losing the bacterial cultivability [33]. *P. aeruginosa* inactivation is directly related to the sunlight intensity and UV dose, especially in the UV-A range. According to some previous studies, after getting the required UV dose, the disruption of bacterial cells continues even the exposed water samples are removed from sun and stored in dark conditions [33,34].

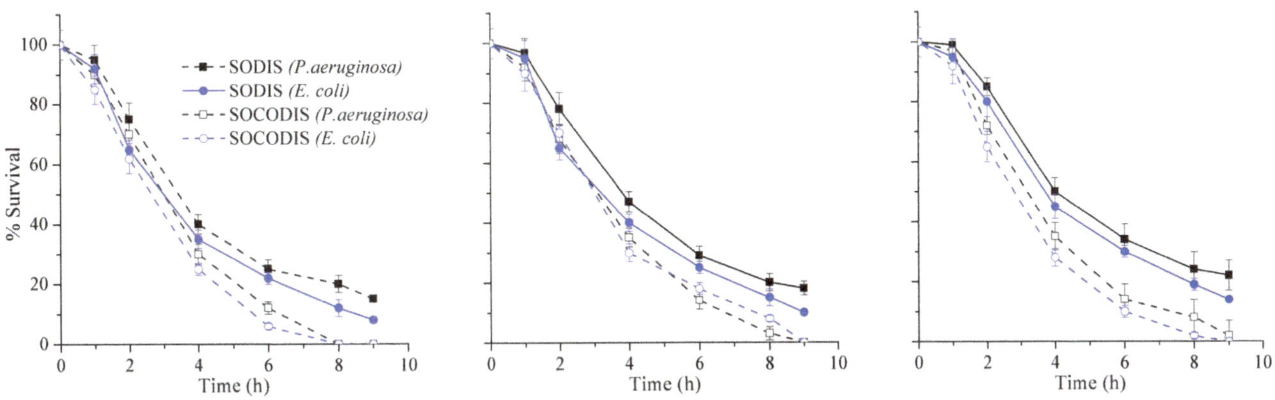

Figure 7. Effects on *P. aeruginosa* inactivation in both SODIS and SOCODIS system at mild weather at (a) low, (b) medium, and (c) high concentrations of *E. coli*.

Table 2. Comparison of k_{max} (1/min) for *P. aeruginosa* disinfection between SODIS and SOCODIS system under different weather conditions (Rainwater with low turbidity (2-5NTU) and neutral pH values).

Weather conditions	SODIS			SOCODIS
	Transparent	Absorptive	Reflective	
Sunny	0.47	0.63	0.56	ND*
Mild	0.17	0.22	0.26	0.71
Weak	0.07	0.1	0.11	0.26

* Not determined.

UV-A is highest as a proportion of solar irradiance when the sun is at its highest altitude near solar noon, as is clear from an increase in the inactivation rate at this time, from 11 am to 3 pm in conducted experiments. All the results showed same tendency signifying a close relationship between UV dose and time required to inactivate microorganisms. One suspicion on solar disinfection by using PET bottles is the leaching of carcinogenic degradation agents from plastic bottles [35]. However, this suspicion was removed in another sunlight exposure study, PET degradation products such as terephthalate monomers and dimers are primarily formed at the outer surface of the bottles [36]. Some compounds including the carbonyls and plasticizers are found in the water but within the safe limits set for drinking water quality [37].

Effects of different initial concentrations of *P. aeruginosa* on disinfection efficiency. Since the initial concentrations of *P. aeruginosa* varied throughout the year depending upon the catchment conditions which in turn is influenced by the wet and dry seasons, therefore, the effects of this different initial concentrations on the disinfection efficiency was investigated at mild weather and the results are presented in Fig. 6 for both SODIS and SOCODIS system. Three different initial concentrations of *P. aeruginosa* (350, 700 and 980 CFU/ml) were chosen at different times during the four-months span of each weak and mild weathers but mainly at the beginning and ending of the both weather conditions.

Low to high concentrations of *P. aeruginosa* were possible under weak weather conditions due to both wet and dry season for four months from November to February. Mild weather conditions spanning nearly over four month (March, April, September and October) yielded medium to high concentrations of *P. aeruginosa* during March and April, due to relatively dirty catchment surface as a result of dry seasons, and low to medium concentrations, during relatively wet months and resulted clean catchment surface, in September and October. As shown in Fig. 6, there is very little effect of higher initial *P. aeruginosa* concentrations on disinfection efficiency especially in SODIS under both weak and mild weather conditions. The inactivation of *P. aeruginosa*, however, increased slightly about 5–10% by lowering the initial concentrations in SOCODIS system and this could be due to the concentrated effects of sunlight radiations.

Effects of the presence of *E. coli* on disinfection of *P. aeruginosa*. The effects of the presence of other bacterial organisms like *E. coli* on disinfection efficiency of *P. aeruginosa* was investigated and for this purpose, three different initial concentrations of *E. coli* (210, 320 and 450 CFU/100 ml as low, medium and high, respectively) were chosen under mild weather conditions. It can be concluded from the results presented in Fig. 7 that the presence of other bacterium like *E. coli* had little effects on the disinfection performance of SODIS or SOCODIS system for *P. aeruginosa*. Higher *E. coli* concentrations, however, slightly decreased the inactivation of *P. aeruginosa* in addition to longer lag period, as is shown in Fig. 7 and this might be due to the increasing competition of different organisms for absorbing sunlight radiations or due to the bacterium selectivity of the UV-A radiations especially during early hours of exposure.

Reaction kinetics

Inactivation kinetics of *P. aeruginosa* was investigated by using the Geeraerd Log linear and shoulder model in GInaFit and inactivation rate constant (k_{max}) in all cases of SODIS and SOCODIS systems was calculated and compared. The model was originally defined by coupling two differential equations [27].

$$\frac{dN}{dt} = -k_{\max}.N.(\frac{1}{1+C_c}).(1-\frac{N_{res}}{N}) \quad (1)$$

$$\frac{dC_c}{dt} = -k_{\max}.C_c \quad (2)$$

Where C_c is related to the physiological state of the cells, k_{max} is the specific inactivation rate [1/time unit] and N_{res} is the residual population density [CFU/mL]. Following format of the model is used when actually applying to the experimental data in GInaFit.

Table 3. Comparison of k_{max} (1/min) for *P. aeruginosa* disinfection between $SODIS_{(Abs.)}$ (suny weather) and SOCODIS system (mild weather) at different pH and turbidity values.

pH (under Turbidity~5NTU)	Sunny weather	Mild weather	Turbidity (under pH~7)	Sunny weather	Mild weather
	$SODIS_{(Abs.)}$	SOCODIS	NTU	$SODIS_{(Abs.)}$	SOCODIS
3	0.83	0.98	~5	0.63	0.71
5	0.72	0.83	20	0.48	0.55
7	0.63	0.71	50	0.39	0.36
10	0.59	0.62	100	0.28	0.22

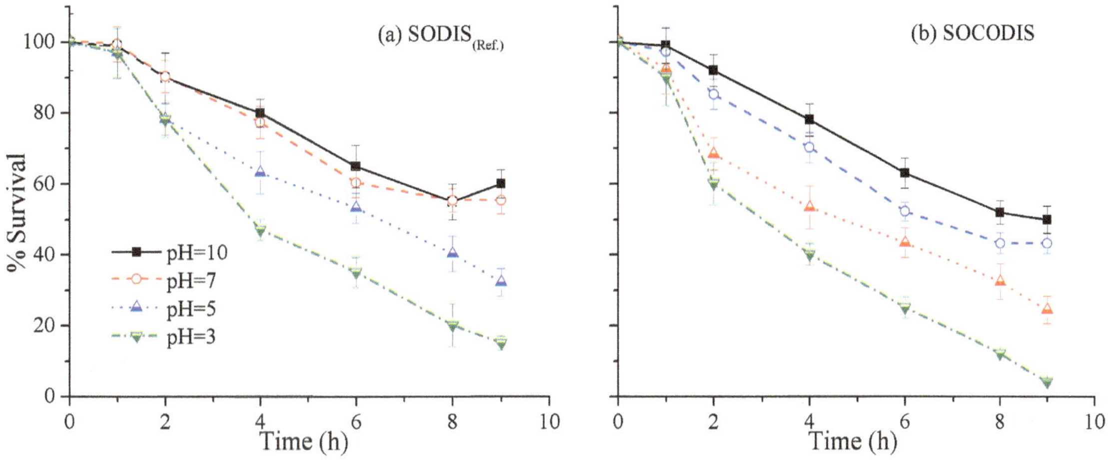

Figure 8. Effects of different initial pH values on *P. aeruginosa* inactivation in (a) SODIS$_{(Ref.)}$, and (b) SOCODIS system at weak weather conditions.

$$\log 10(N) = \log 10\left[(10^{\log 10(N(0))} - 10^{\log 10(N res)}).e^{-k\max t}.\left(\frac{e^{k\max S_1}}{1+(e^{k\max S_1}-1).e^{-k\max t}}\right) + 10^{\log 10(N res)}\right] \quad (3)$$

The first two mentioned equations (eq.1&2) can be obtained by substituting S_L, a parameter represents the shoulder (time unit), or N_{res} equal to zero. Table 2 presents a brief overview of *P. aeruginosa* inactivation based on the inactivation rate constants (k_{max}), calculated using the Geeraerd model.

The k_{max} (1/min) values in Tables 2 and 3 were calculated based on the Geeraerd model, as described earlier, and represent the inactivation rate constants for *P. aeruginosa*, with a corresponding coefficient of determination (R^2) of 1, except where mentioned. A complete application of the model, i.e. with tailing and shoulder, was not observed in most of the cases and modeling with tailing was unlikely for the observed data probably due to the synergistic effects of UV radiations and temperature. For mild and weak weather conditions, temperature less than 50°C but high k_{max} values validated the assumption that irradiance effects were significant in case of SOCODIS unlike SODIS for the inactivation of *P. aeruginosa*. This finding is in compliance with previously reported results [38].

Effects of different pH and turbidity values on disinfection efficiency

Both pH and turbidity of the stored rainwater are among the main factors that may change the efficacy of solar disinfection systems. So, the effects of different initial pH and turbidity values on the efficiency of the solar disinfection systems were tested and the results are presented in Figs. 8 and 9, respectively. As mentioned earlier, diluted HCl and NaOH were used for adjusting initial pH values while kaolin was added for different initial turbidity values. As shown in Fig. 8, four initial pH values i.e. 10, 7, 5 and 3 were selected with rainwater with a low initial turbidity value (<5 NTU) and the comparison was performed only under weak weather conditions mainly due to inefficiency of the SOCODIS system at this weather. The overall disinfection

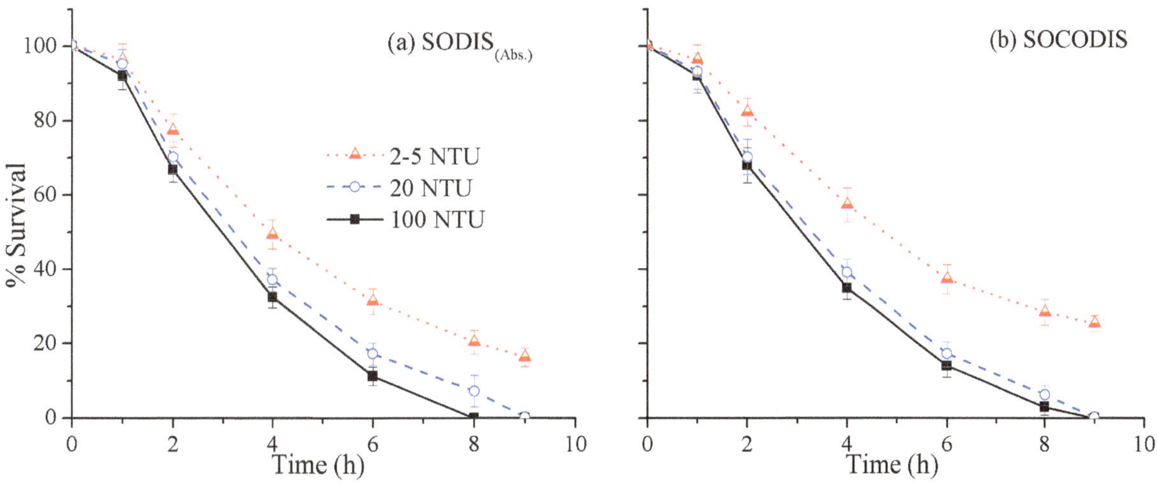

Figure 9. Effects of different turbidity values on *P. aeruginosa* inactivation in (a) SODIS$_{(Abs.)}$ (sunny weather), and (b) SOCODIS system (mild weather).

efficiency increased by about 15% by lowering the initial pH value from 10 to 3 and nearly complete inactivation of *P. aeruginosa* was achieved at the acidic conditions. Low pH may have increased the inactivation of *P. aeruginosa* by presenting significant additional stress to the cells [5,39].

Table 3 shows a comparison of the *P. aeruginosa* inactivation between SODIS and SOCODIS system under mild and sunny weather conditions at different initial pH values. These findings differed from the earlier studies stating no effects of pH values on microbial inactivation rates [40]. In this study lowering initial pH values had significant effect of *P. aeruginosa* inactivation as reported previously [41]. It is worth mentioning that there are no health-based guidelines for pH and it is reported that the consumption of food or beverages with low (2.5) or high (11) pH does not result in adverse health effects [42].

For investigating the effects of initial turbidity values with rainwater having nearly neutral pH values, best suited solar disinfection cases, $SODIS_{(Abs.)}$ under sunny weather and SOCODIS system under mild weather condition were selected for comparison as shown in Figs. 9(a) and 9(b), respectively. Rainwater samples collected from an underground storage tank had a turbidity value of less than 5 NTU. For comparison purposes, higher values (20 and 100NTU) were achieved by adding kaolin. Samples with low turbidity showed better results; the difference between inactivation of *P. aeruginosa* was insignificant up to about 20 NTU. However, the SOCODIS system showed poor performance in terms of *P. aeruginosa* inactivation at higher turbidity, and disinfection efficacy decreased by almost 10–15%. This could be because of a loss of UV radiations due to scattering and absorption by suspended particles in rainwater samples. Various microbiological studies have revealed that highly turbid waters do not allow the synergistic effect to occur [22,31,38,43]. However, the highly turbid water may undergo a substantial temperature increase because of the absorption of solar radiation by suspended particles, which may contribute to higher disinfection efficiencies [22]. The case of $SODIS_{(Abs.)}$ system under sunny weather conditions was in accordance with previously reported literature but contrarily the complete inactivation of *P. aeruginosa* was not achieved.

Table 3 summarizes the change in the inactivation rate constants of *P. aeruginosa* with the variable values of turbidity when Geeraerd model was used for best suited cases of $SODIS_{(Abs.)}$ and SOCODIS systems. As can be noted from the Table, the values of k_{max} were decreased with the increased turbidity level in all cases. Surprisingly, however, the difference between the k_{max} values for turbidity level 20 and 100 under $SODIS_{(Abs.)}$ was lesser (0.20) as compared with SOCODIS case (0.33). It can be due to the substantial temperature increase because of the absorption of solar irradiance by suspended particles [22] when the rear surfaces of PET bottles were painted black.

Conclusion

$SODIS_{(Abs.)}$ under, sunny weather conditions, was the most efficient case (100% inactivation) as compared to the $SODIS_{(Trans.)}$ (92%) and $SODIS_{(Ref.)}$ (97%) cases for the inactivation of *P. aeruginosa* in harvested rainwater. This could be due to the thermal and/or synergistic effect of temperature and irradiation. Generally, a 20% increase in the disinfection efficiency of the SOCODIS system, even under mild weather conditions, suggested that the simple solar collector/box is an efficient, cost effective and user friendly option especially for remote areas. However under weak weather conditions, a longer exposure time or other supplementary means with the SOCODIS system are required for the complete inactivation of *P. aeruginosa*.

There was very little effect of high initial *P. aeruginosa* or *E. coli* concentrations on disinfection efficiency but the inactivation of *P. aeruginosa* increased slightly about 5–10% by lowering the initial concentrations in SOCODIS system. Higher *E. coli* concentrations, on the other hand, slightly decreased the inactivation of *P. aeruginosa* in addition to longer lag period. Although, not completely inactivated under weak weather conditions, the overall disinfection efficiency increased by about 15% by lowering the initial pH value from 10 to 3 especially in SOCODIS system. Above 20 up to 100NTU, 16–25% decrease in activation of *P. aeruginosa* suggested that water should be filtered before applying solar disinfection. Overall, the inactivation of *P. aeruginosa* in harvested rainwater by solar disinfection systems can be considered a useful addition in making the stored rainwater available for potable purposes. The detailed investigation and disinfection of *P. aeruginosa* with sunlight based disinfection system under different weather conditions and variable parameters will help researchers to understand and further improve the innovative SOCODIS system. This can also be viewed as an easy approach for the poor communities living in the developing world in accessing the potable water using efficient and cost-effective point-of-use treatment methods.

Author Contributions

Conceived and designed the experiments: MTA MN MH. Performed the experiments: MTA MN. Analyzed the data: MN MNA. Contributed reagents/materials/analysis tools: MNA MH. Wrote the paper: MTA MN MNA.

References

1. WHO/UNICEF Joint Monitoring Program for Water Supply and Sanitation (2012) Progress on Drinking Water and Sanitation Update, USA.
2. Black RE, Cousens S, Johnson HL, Lawn JE, Rudan I, et al. (2010) Child Health Epidemiology Reference Group of WHO and UNICEF Global, regional, and national causes of child mortality in 2008: a systematic analysis. The Lancet 375: 1969–1987.
3. Amin MT, Han MY (2007) Scope/need of soft path water resource management in developing countries. Water Sci Technol: Water Supp 7(5–6): 185–192.
4. Amin MT, Han MY (2009) Roof-harvested rainwater for potable purposes: application of solar disinfection (SODIS) and limitations. Water Sci Technol 60(2): 419–431.
5. Amin MT, Han MY (2009) Roof-harvested rainwater for potable purposes: application of solar collector disinfection (SOCODIS). Water Res 43: 5225–5235.
6. Han MY, Mun JS (2008) Particle behavior consideration to maximize the settling capacity of rainwater storage tanks. Water Sci Technol 56 (11): 73–79.
7. Miguntanna NS, Egodawatta P, Goonetilleke A (2010) Pollutant characteristics on roof surfaces for evaluation as a storm water harvesting catchment. Desalin Water Treat 19 (1–3): 205–211.
8. Nawaz M, Han MY, Kim TI, Manzoor U, Amin MT (2012) Silver disinfection of Pseudomonas aeruginosa and *E. coli* in rooftop harvested rainwater for potable purposes. Sci Total Environ 431: 20–25.
9. Kerr KG, Snelling AM (2009) Pseudomonas aeruginosa: a formidable andever-present adversary. J Hosp Infect 73(4): 338–344.
10. Morais PV, Mesquita C, Andrade J, DaCosta M (1997) Investigation of persistent colonization by Pseudomonas aeruginosa-like strains in a spring water bottling plant. Appl Environ Microbiol 63(3): 851–856.
11. Uba BN, Aghogho O (2000) Rainwater quality from different roof catchments in the Port Harcourt district, Rivers State, Nigeria. J Water Supply Res T 49:281–288.
12. Albrechtsen HJ (2002) Microbiological investigations of rainwater and graywater collected for toilet flushing. Water Sci Technol 46:311–316.
13. Kaushik R, Balasubramanian R (2012) Assessment of bacterial pathogens in fresh rainwater and airborne particulate matter using Real-Time PCR. Atmos Environ 46:131–139.

14. Kaushik R, Balasubramanian R, de la Cruz AA (2012) Influence of air quality on the composition of microbial pathogens in fresh rainwater. Appl Environ Microbiol 78:2813–2818.
15. Vialle C, Sablayrolles C, Lovera M, Huau M-C, Jacob S, et al. (2012) Water Quality Monitoring and Hydraulic Evaluation of a Household Roof Runoff Harvesting System in France. Water Resour Manag 26:2233–2241.
16. Reed RH (1997) Solar inactivation of fecal bacteria in water: the critical role of oxygen. Lett Appl Microbiol 24(4): 276–80.
17. Sommer B, Marino A, Solarte Y, Salas ML, Dierolf C, et al. (1997) SODIS-an emerging water treatment process. J Water Supply Res T 46(3): 127–137.
18. Acra A, Raffoul Z, Karahagopian Y (1984) Solar disinfection of drinking water and oral rehydration solutions, UNICEF Guidelines for Household Application in Developing Countries.
19. Wegelin M, Canonica S, Mechsner K, Fleischmann T, Pesara F, et al. (1994) Solar water disinfection: scope of the process and analysis of radiation experiments. J Water Supply Res T 43(3): 154–169.
20. Rijal GK, Fujioka RS (2001) Synergistic effect of solar radiation and solar heating to disinfect drinking water sources. Water Sci Technol 43(12): 155–162.
21. Abdel Dayem AM, El-Ghetany HH, El-Taweel GE, Kamel MM (2011) Thermal performance and biological evaluation of solar water disinfection systems using parabolic trough collectors. Desalin Water Treat 36 (1–3): 119–128.
22. Joyce TM, McGuigan KG, Elmore-Meegan M, Conroy RM (1996) Inactivation of fecal bacteria in drinking water by solar heating. Appl Environ Microbiol 62(2): 399–402.
23. Conroy RM, Elmore-Meegan M, Joyce TM, McGuigan KG, Barnes J (1996) Solar disinfection of drinking water and incidence of diarrhea in Maasai children: a controlled field trial. The Lancet 348: 1696–1697.
24. Hobbins M, Mausezah ID, Tanner M (2003) Home-based drinking water purification: the SODIS Health Study. J Water Supply Res T 56 (4): 245–252.
25. Amin MT, Han MY (2011) Improvement of solar based rainwater disinfection by using lemon and vinegar as catalysts. Desalination 276(1-3): 416–424.
26. American Public Health Association, American Water Works Association and Water Environment Federation (2012) Standard methods for the examination of water and wastewater. American Public Health Association, Washington, D.C.
27. Geeraerd AH, Valdramidis VP, Van Impe JF (2005) GInaFiT, a freeware tool to assess non-log-linear microbial survivor curves. Int J Food Microbiol 102: 95–105.
28. Geeraerd AH, Herremans CH, Van Impe JF (2000) Structural model requirements to describe microbial inactivation during a mild heat treatment. Int J Food Microbiol 59(3): 185–209.
29. Oates PM, Shanahan P, Polz MF (2003) Solar disinfection (SODIS): simulation of solar radiation for global assessment and application for point-of-use water treatment in Haiti. Water Res 37: 47–54.
30. El Ghetany HH, Abdel Dayem A (2010) Numerical simulation and experimental validation of a controlled flow solar water disinfection system. Desalin Water Treat 20(1–3): 11–21.
31. McGuigan KG, Joyce TM, Conroy RM, Gillespie JB, Elmore-Meegan M (1998) Solar disinfection of drinking water contained in transparent plastic bottles: characterizing the bacterial inactivation process. Appl Environ Microbiol 84(6): 1138–1148.
32. Lonnen J, Kilvington S, Kehoe SC, Al-Touati F McGuigan KG (2005) Solar and photocatalytic disinfection of protozoan, fungal and bacterial microbes in drinking water. Water Res 39: 877–883.
33. Bosshard F, Riedel K, Schneider T, Geiser C, Bucheli M, et al. (2010) Protein oxidation and aggregation in UVA-irradiated *Escherichia coli* cells as signs of accelerated cellular senescence. Environ Microbiol 12(11): 2931–2945.
34. Ubomba-Jaswa E, Navntoft C, Polo-Lopez MI, Fernandez-Ibanez P, Mcguigan KG (2009) Solar disinfection of drinking water (SODIS): an investigation of the effect of UV-A dose on inactivation efficiency. Photoch Photobio Sci 8: 587–595.
35. Lilya D (2001) Analysis and risk assessment of organic chemical migration from reused PET plastic bottles, M.Sc. thesis, College of Graduate Studies, University of Idaho, Moscow.
36. Wegelin M, Canonica S, Alder A, Marazuela D, Suter M, et al. (2001) Does sunlight change the material and content of polyethylene terephthalate (PET) bottles?. J Water Supply Res T 50(3): 125–133.
37. Schmid P, Kohler M, Meierhofer R, Luzi S, Wegelin M (2008) Does the reuse of PET bottles during solar water disinfection pose a health risk due to the migration of plasticisers and other chemicals into the water?. Water Res 42: 5054–5060.
38. Kehoe SC, Joyce TM, Ibrahim P, Gillespie JB, Shahar RA, et al. (2001) Effect of agitation, turbidity, aluminum foil reflectors and container volume on the inactivation efficiency of batch-process solar disinfectors. Water Res 35(4): 1061–1065.
39. Foegeding EA, Lanier TC, Hultin HO (1996) Characteristics of edible muscle tissues. In: Owen Fennema R (Ed.). Food Chemistry. Chapter 15, vol. 3: 879–942.
40. Rincon A, Pulgarin C (2004) Effect of pH, inorganic ions, organic matter and H_2O_2 on *E. coli* k12 photocatalytic inactivation by TiO_2 – implications in solar water disinfection. Appl Catal B Environ 51 (4): 283–302.
41. Fisher MB, Keenan CR, Nelson KL, Voelker BM (2008) Speeding up solar disinfection (SODIS): effects of hydrogen peroxide, temperature, pH, and copper plus ascorbate on the photoinactivation of *E. coli.* J Water Health 6 (1): 35–51.
42. NHMRC (1996) Australian Drinking Water Guidelines. National Health and Medical Research Council. Commonwealth of Australia Sydney, Australia.
43. Gomez-Couso H, Fontan-Sainz M, Ares-Mazas E, (2010) Thermal contribution to the inactivation of Cryptosporidium in plastic bottles during solar water disinfection procedures. Am J Trop Med Hyg 82(1): 35–39.

Designing and Piloting a Program to Provide Water Filters and Improved Cookstoves in Rwanda

Christina K. Barstow[1], Fidele Ngabo[2], Ghislaine Rosa[3], Fiona Majorin[3], Sophie Boisson[3], Thomas Clasen[3,4], Evan A. Thomas[5]*

1 Civil, Environmental and Architectural Engineering, University of Colorado, Boulder, Colorado, United States of America, 2 Maternal and Child Health, Ministry of Health, Kigali, Republic of Rwanda, 3 Department of Infectious and Tropical Diseases, London School of Hygiene and Tropical Medicine, London, United Kingdom, 4 Department of Environmental Health, Rollins School of Public Health, Emory University, Atlanta, Georgia, United States of America, 5 Mechanical and Materials Engineering, Portland State University, Portland, Oregon, United States of America

Abstract

Background: In environmental health interventions addressing water and indoor air quality, multiple determinants contribute to adoption. These may include technology selection, technology distribution and education methods, community engagement with behavior change, and duration and magnitude of implementer engagement. In Rwanda, while the country has the fastest annual reduction in child mortality in the world, the population is still exposed to a disease burden associated with environmental health challenges. Rwanda relies both on direct donor funding and coordination of programs managed by international non-profits and health sector businesses working on these challenges.

Methods and Findings: This paper describes the design, implementation and outcomes of a pilot program in 1,943 households across 15 villages in the western province of Rwanda to distribute and monitor the use of household water filters and improved cookstoves. Three key program design criteria include a.) an investment in behavior change messaging and monitoring through community health workers, b.) free distributions to encourage community-wide engagement, and c.) a private-public partnership incentivized by a business model designed to encourage "pay for performance". Over a 5-month period of rigorous monitoring, reported uptake was maintained at greater than 90% for both technologies, although exclusive use of the stove was reported in only 28.5% of households and reported water volume was 1.27 liters per person per day. On-going qualitative monitoring suggest maintenance of comparable adoption rates through at least 16 months after the intervention.

Conclusion: High uptake and sustained adoption of a water filter and improved cookstove was measured over a five-month period with indications of continued comparable adoption 16 months after the intervention. The design attributes applied by the implementers may be sufficient in a longer term. In particular, sustained and comprehensive engagement by the program implementer is enabled by a pay-for-performance business model that rewards sustained behavior change.

Editor: Rachel A. Nugent, University of Washington, United States of America

Funding: All funding for this study was provided by DelAgua Health Limited www.delaguahealth.com) through a sponsored research grant to Portland State University and the London School of Hygiene and Tropical Medicine. The DelAgua Health Board and CEO had no role in study design, data collection and analysis, decision to publish, or preparation of the manuscript. All such responsibility was devolved to Author Thomas.

Competing Interests: DelAgua Health is the funder as well as implementer of the program described. Authors Thomas and Barstow are compensated by, and responsible to, DelAuga Health for the conceptualization, development, implementation, monitoring and reporting of the program described, in cooperation with the Rwanda Ministry of Health. Author Thomas is also the Principal Investigator at Portland State University, primarily responsible for designing and implementing the sensors described in the paper, and collaborating with other research partners including the London School of Hygiene and Tropical Medicine and Emory University. This does not alter the authors' adherence to all the PLOS ONE policies on sharing data and materials.

* E-mail: evan.thomas@pdx.edu

Introduction

Access to improved drinking water and clean burning stoves could benefit the millions who suffer from diarrheal disease and pneumonia, two of the leading causes of death around the world for children under five. Worldwide, of the 7.6 million deaths in children under 5 in 2010, 64% were associated with infectious diseases including 18% with pneumonia and 11% with diarrhea. Combined, pneumonia and diarrhea kill over 2 million children each year [1].

Some of these deaths may be avoided through interventions to improve indoor air quality and household water quality: pneumonia is often linked to indoor air pollution from biomass fuels [2,3] and diarrhea to deficiencies in water and sanitation, including poor water quality [4]. Many cookstove interventions have shown a reduction in indoor air pollutants such as carbon monoxide and fine particulate matter [5,6]. Similarly, interventions targeted at improving household water quality through the implementation of water treatment strategies such as chemical treatment, boiling, solar disinfection or filtration have been shown to reduce diarrheal disease [7,8].

Even with the fastest annual reduction in child mortality in the world, the Republic of Rwanda still faces challenges related to pneumonia and diarrhea: among deaths of children under 5, pneumonia accounts for 18% and diarrhea for 8% [9]. Cooking practices in a rural Rwandan household may contribute to this pneumonia burden since the predominate fuel and cookstove pairing is wood on a three stone fire [10]. Additionally, while Rwanda has demonstrated significant progress towards the Millennium Development Goals, almost 30% of households do not have access to an improved water source [11], and the improved water sources may become contaminated during collection, transport or storage within the home [12,13]. Once water is in the households, less than half (46.1%) of rural Rwandan families report treating their drinking water, with boiling as the leading treatment method (38.1%) [11], which again can become recontaminated after treatment [14]. The Rwanda Standard for Potable Water states that the microbiological limits for potable water for total CFU/100 ml of total coliforms should be 0 [15]. A baseline water quality assessment of 230 improved water sources, 78 unimproved water sources, and stored water in 468 households across all 30 districts in Rwanda indicated that 27.8% of improved water sources, 80.2% of unimproved water sources, and 58.3% of stored household water supplies exceed this standard [16], falling into the "intermediate", "high" or "very high" risk World Health Organization categories [17] for biological contamination of drinking water supplies. Another study of households within the other 11 districts of the project area prior to the start of the program implementation indicated that 81.1% of households exceed this standard, with 59.1% falling into the "intermediate" or "high" risk categories [18].

DelAgua Health, a for-profit social enterprise, was established to combine household technologies that address environmental health issues with market-based mechanisms. DelAgua Health participates in the United Nations Clean Development Mechanism (CDM) to earn carbon credits associated with the reduced use of, and demand for, fuel wood associated with water treatment and cooking, and then sell those credits to buyers as a way to recover costs and profit [19].

Carbon finance markets facilitate the reduction of greenhouse gas emissions worldwide through economic incentives, while allowing cleaner economic development to take place. Each emission reduction credit represents the non-emission of one tonne of carbon dioxide into the atmosphere. The carbon credits generated under the CDM help Kyoto Protocol Annex I countries to meet their binding targets, and can be traded in the marketplace. However, the carbon markets have yet to be well utilized to finance the distribution of humanitarian technologies in the least developed countries, particularly in Africa. Although the CDM is a multi-billion dollar industry, fewer than 2 percent of projects are registered in African nations [20].

Depending on the project location, structure, methodology and registration mechanism employed, a water treatment and/or cookstove program can earn between approximately 1/2 and 5 carbon credits per household, per year. The carbon credits earned are a function of the approved methodology, referenced to a baseline condition and the current performance of the program, as audited by independent firms. The reported reductions are then issued by the registration authority and are then sold to buyers. Carbon credit buyers may be banks, energy companies, brokers, or sovereign nations who require credits for either regulatory compliance or voluntary social responsibility efforts, or both. Because the carbon credits are issued in proportion to the present adoption and proper use of intervention technologies, this encourages sustained engagement by the program implementer and creates a pay-for-performance model.

In Rwanda, DelAgua Health is partnered with the Ministry of Health since 2012 to distribute free of charge household water treatment and high efficiency cookstoves to approximately 600,000 households (about 3 million people), throughout the country's 30 districts. The project will target *Ubudehe* categories 1 and 2, the government-recognized poorest 30% of the country. *Ubudehe* category is determined by community members based on classifications outlined by the Rwandan Ministry of Local Government [21]. Households categorized as *Ubudehe* 1 and 2 already receive free medical and other assistance through government programs.

A pilot program was initiated in October 2012 to provide input for the full effort, scheduled to start in mid-2014. This pilot was conducted after findings from a preliminary study of 100 households in July 2012. This effort was judged by the implementers to be sufficiently promising for testing at a larger scale. This paper discusses the design, development, implementation, monitoring and periodic modification of the October 2012 pilot program. We summarize the results of surveys collected to evaluate key outputs including intervention uptake and use. Other aspects of the pilot are described elsewhere, including a novel method for assessing intervention use with remotely reporting sensors [22] and a randomized controlled trial to study the impact of the intervention on drinking water quality and household air pollution [23].

Materials and Methods

Design Objectives

The objective of this study was to identify if certain design criteria, integrated together and applied to environmental health technologies, could result in a meaningful proportion of continued use of stoves and water filters. These design attributes are evaluated as a whole, though estimates of relative value are provided in the discussion. The three program design choices considered fundamental were:

1. **Free Distribution.** Free provisioning of high quality stoves and water filters under the authority of the Government of Rwanda and through established community mechanisms including community meetings and community leadership.
2. **Behaviour Change.** A behaviour change messaging and monitoring effort that prioritizes consistent and correct adoption of the stoves and filters through community and household level activities, focusing on both health and non-health benefits.
3. **"Pay for Performance" Public-Private Partnership**. A public-private partnership with the Rwanda Ministry of Health enabled by anticipated carbon credit revenues, which allows sustained, comprehensive community engagement by virtue of future anticipated "pay for performance" carbon credit revenues.

Program Setting and Population

The pilot was conducted in a convenience sample of 15 non-randomly selected villages spread across 11 districts in Western Rwanda (Figure 1). The 15 villages were selected to have at least one village per the 11 districts and the remaining four villages in districts with the largest populations. Additional inclusion criteria included ensuring that no villages were in adjacent sectors (district subdivisions), less than 20% of households in each village served by piped water; less than 60% of households in each village using any water treatment other than boiling; less than 20% of households in each village using cooking fuel sources other than biomass or charcoal; and less than 20% of households in each village using any stove other than a 3-stone fire or two other locally made unimproved stoves (known as *Rondereza* and *Imbabura* stoves). These inclusion criteria were selected by program staff to be representative of typical rural villages in Rwanda, based on rural water service and energy use characteristics identified in the Rwanda 2011 Demographic and Health Survey. Program staff visited each candidate village in advance to confirm with village officials that it met eligibility criteria. All 1,943 households who were registered as members of the 15 villages were eligible to participate in the study. While the full program will consist of distribution to only *Ubudehe* 1 and 2 households, this pilot program consisted of all households, of any *Ubudehe* category, in the 15 villages. The full program originally consisted of distribution to all households in the Western province of Rwanda but was later revised to be a country wide program of *Ubudehe* 1 and 2 households. This program change was directed by the Government of Rwanda Ministry of Health.

Intervention Hardware

The water filter used in this program, the Vestergaard Frandsen LifeStraw Family 2.0 is a point-of-use microbial water treatment system intended for routine use in low-income settings. The system is a table-top unit where the user pours untreated water through a 20 micron pre-filter into a six liter influent water tank. Water is then gravity-filtered through a 0.20 micron hollow-fiber ultrafiltration membrane into a 5.5 liter safe storage container. Water can be dispensed from the safe storage container through a plastic tap, limiting recontamination. The filter is backwashed by squeezing a plastic bulb located on the opposite side of the tap. The membrane can filter up to 18,000 liters of water [24], enough to supply a family of five with microbiologically clean drinking water for three to five years. The system exceeds the 'highly protective' World Health Organization Standard for household water treatment technologies [25,26]. In a recent study, an earlier version of this filter was shown to be highly effective in improving water quality

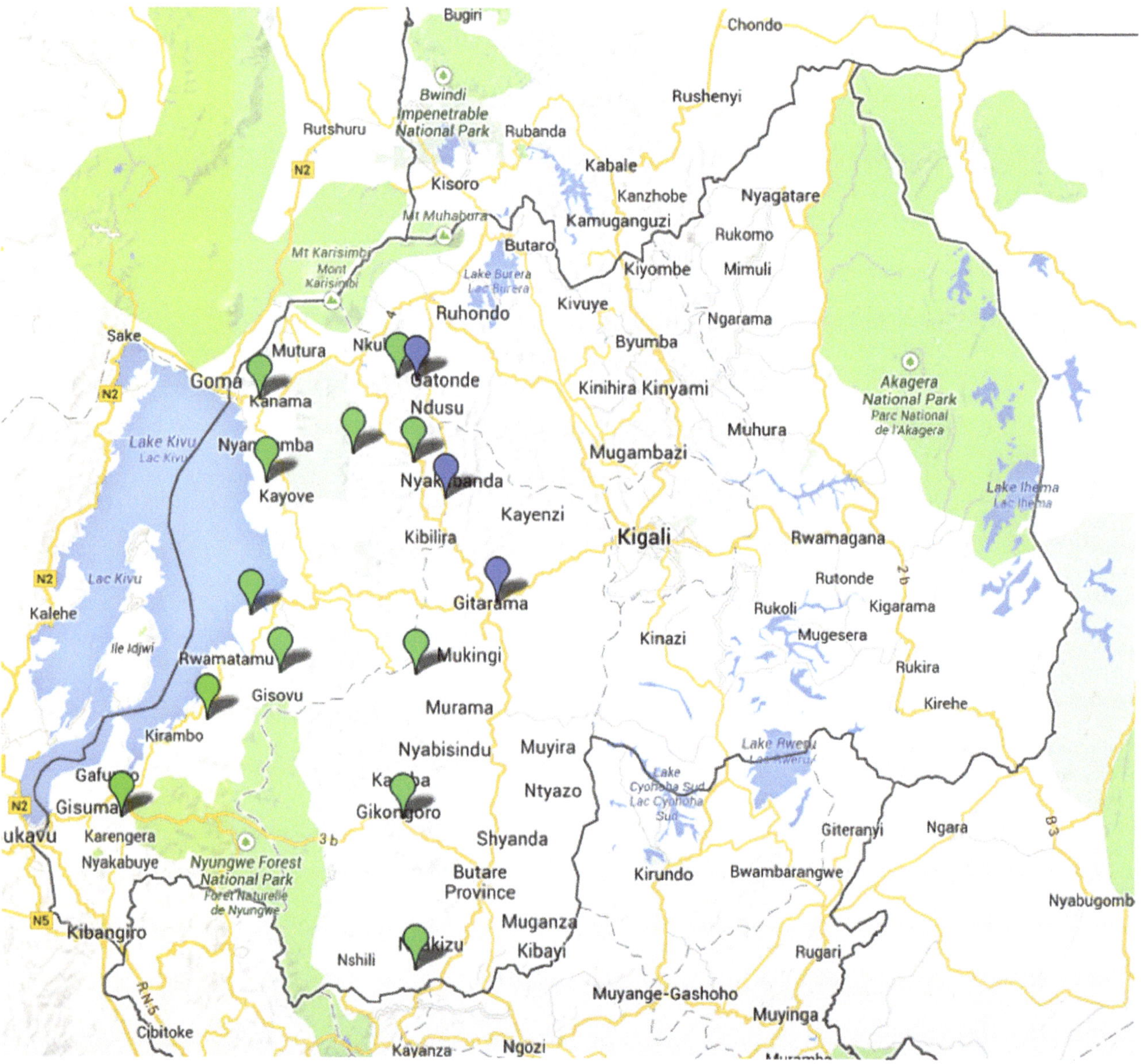

Figure 1. Program villages. RCT villages shown with blue pins.

and was protective against diarrhea among HIV positive individuals, reducing longitudinal prevalence by over 50% [27].

The cookstove used in this program, the EcoZoom Dura, is based on the rocket-stove concept that is designed to concentrate the combustion process while channeling air flow to create a more complete burn. A complete burn of carbon rich material will also result in little to no smoke. Included with the stove are a "stick support" on which fuelwood is placed to promote air flow and a "pot skirt" which increases thermal efficiency. In the field, performance is variable but when properly used, a rocket stove will significantly reduce fuelwood use by at least 50%, although reductions in indoor air pollution vary between designs, fuel types and use [28]. The thermal efficiency of this stove is 38% [29].

Intervention Design

The program is designed leveraging established behavior change theories, including the Diffusion of Innovation theory [30] and the Health Belief Model (HBM) [31]. In particular, the program design assumes that continued, comprehensive engagement is critical in order to effect positive behavior change. The program design takes a hybrid approach, integrating pieces of these theories that apply in a Rwandan context to shape the communication strategy of the program.

Several components of the diffusion of innovation theory are applied to the program. At the initial distribution meeting, community members are informed about the potential health and other benefits of the water filter and cookstove creating 'initial knowledge' around the technologies. The 'persuasion stage' is initiated through demonstrations at both the community meeting and the household. The stove demonstration includes assembling the stove, how to adjust a pot skirt which can be fitted to different sized pots and finally a fire being started in the stove with some demonstrations including boiling a pot of water to show the rapidity of the cooking process. The water filter includes a demonstration of filtering visually dirty water with clear water coming out of the tap and the maintenance procedure, which includes backwashing the filter. Progressing to the 'decision stage' the household is then asked to demonstrate use and maintenance of the technology, allowing them to trial the technology. Households then move into the 'implementation stage' where they can choose to adopt or reject the technology. About a month later, the program implements the 'confirmation stage' where households who have chosen to partially adopt or reject the technology are given additional training and messaging to hopefully reverse their decision.

Village chiefs are promoted as 'early adopters' because of their high degree of influence and respect within their villages. In this program the change agents are Community Health Workers (CHWs). The CHW system in Rwanda includes three CHWs per village who are part-time volunteers of sector health centers and are compensated with a stipend. They provide basic services such as maternal and newborn health monitoring, vaccination advocacy, family planning, treatment of malaria, and sanitation and hygiene education. Through this program CHWs play several important roles including informing households of the need for the devices, encouraging adoption and analyzing potential problems with the technologies. The CHWs play an especially important role with the 'late adopters' and 'laggards' as more effort is needed to change the household's old habit and promote the new behaviors [32].

The Health Belief Model is used to shape messaging. The belief that there is a health threat is compelled by messaging related to clean drinking water and clean indoor air. Households are educated about the reduced risk of diarrheal disease from water borne diseases and the reduced risk of respiratory problems from breathing indoor air pollutants. Additionally an important concept in social cognitive theory is often added to the health belief model, self-efficacy, which states that the user must believe that they can adopt the new behavior [33]. This is facilitated through households gaining confidence in the use of the technologies by having members of the household demonstrate proper use.

While the health belief model provides relevant guidance on behavior change theory it is important that the program also express non-health benefits to users. Previous interventions related to both water quality and improved cookstoves emphasize the need to highlight non-health benefits such as those related to economic and social benefits [32,34]. Thus CHWs educate households on additional benefits such as reduction in medical costs from the water filter and a reduction in cooking time and expenditure on fuel costs for the cookstove.

Adoption and Monitoring Survey

Households were assigned to receive the adoption and monitoring survey in one of six rounds by a random number generator. Approximately 325 households were surveyed each month with the exception of month five where approximately twice as many households were surveyed as rounds 5 and 6 were combined because of time constraints. Environmental Health Officers (EHOs) were responsible for conducting the surveys. Environmental Health Officers (EHOs) are full time employees of the Ministry of Health at sector health centers who are responsible for a range of health interventions, including, food safety, waste management, water, sanitation and hygiene inspection, indoor air pollution and environmental emergency health interventions [21]. Each household was surveyed by EHOs only once during the five month period.

The survey consisted of about 100 questions and was administered using a smartphone in Kinyarwanda by an EHO. Information included household identifying information, demographics, cooking practices, and water treatment and collection practices. The survey included both self-reported use questions to be answered by members of households and observational questions which EHOs answered based on their observations. Observational use of the water filter was measured by checking if water was present in the filter at the time of the visit while observational use of the stove was only confirmed if the stove was actually being used at the time of the visit.

Survey Data Analysis

All survey data was uploaded to the doForms database where it was analyzed using T-SQL and R-Project. Only surveys that fell within 15 to 90 minute survey duration were included in the analysis. All numerical outcomes were additionally analyzed using an outlier analysis where only 1.5 times the upper and lower interquartile range were included in that particular outcome. This outlier exclusion was chosen to be consistent with the program's carbon credit monitoring requirements [35][. Analysis of variance was used to compare group means. Additionally any missing data was excluded from the analysis.

Focus Group Discussions

Three focus group meetings were conducted concurrently with EHOs and CHWs to assess qualitative aspects of the program. A total of 30 participants attended the meetings including one CHW from each of the pilot villages and all EHOs. CHWs were chosen by DelAgua staff as the highest performing CHWs within each village. Topics covered included perceived adoption of technologies within their villages, problems with filter and stove hardware, effectiveness of household messaging and boundaries to exclusive adoption of the filter and stove.

Ethics and Consent

The study was reviewed and approved by the Rwanda National Ethics Committee (IRB #328/RNEC/2012), University of Colorado Institutional Review Board (Protocol #12-0564), and the London School of Hygiene and Tropical Medicine Ethics Committee (Trial registration: Clinicaltrials.gov NCT01882777). Each participating household gave informed, verbal consent after having received complete details regarding the purpose of the survey as well as information regarding privacy of personal information. Enumerators were required to confirm electronically, on their smartphone surveys, if a.) the respondent was over 18, and b.) if they gave verbal consent, before the smartphone allowed the survey to continue. These consent records are kept on a password protected server. Verbal consent was requested and approved by the approving ethics committees, based on the high percentage of illiteracy within the study population. Rwandan residents are often asked about their water and energy habits by community health workers, and the signing of a document adds a level or formality that may mislead participants. Participants were given the opportunity to ask any questions before agreeing to participate. All households were entitled to retain their filters and stoves at the conclusion of the study.Participation in the study was not a prerequisite to receiving the filters and stoves.

Results

Program Delivery

All households that were registered in the 15 villages according to the village chief's list were distributed a stove and filter. The model of distributing at a central location allowed for the implementer to transport the technologies to a location that could be reached by vehicles of which many households could not. It was then members of the household's responsibility to get the stove and filter to their homes.

Household Characteristics

A total of 1943 households participated in the study, from which 1634 (84.1%) valid surveys were included in the analysis. Selected household characteristics are summarized in Table 1. The mean household size was 4.55, consistent with Rwanda's 2011 Demographic and Health Survey household size of 4.5. Approximately one-third (29.3%) of households reported being categorized as *Ubudehe* 1 or 2, 70.5% were classified as 3 or 4 and 0.2% as *Ubudehe* 5 or 6. *Ubudehe* 1 and 2 household size was significantly lower than the entire study population with 3.85 persons per household. However, fuel type and water source were similar with the majority of all households (92.4%) and *Ubudehe* 1 and 2 households (89.1%) using wood as their primary fuel source and all households (41.1%) and *Ubudehe* 1 and 2 households (42.9%) reporting a public tap as their primary drinking water source.

Filter Adoption and Use

Adoption of the LifeStraw filter was measured at approximately 90% or greater by several metrics. Households reported use of the filter had the highest adoption rate with 96.5% of households reporting the water filter as the treatment method for the last water they drank. An observational measure of use through presence of water in the filter showed a slightly lower adoption rate with 9 out of 10 households having water in their filter at the time of household visit (Table 2). Similar adoption rates as measured observationally were seen over the five follow up visits with the first follow up visit having the highest adoption rate of 92.6% and the lowest adoption rate reported as 86.4%. No longitudinal trend in adoption was observed through the five months of the study.

Adoption rates were further analyzed to understand any differences between all households in the study and those who were identified as *Ubudehe* 1 and 2 as well as any differences across the 15 villages. *Ubudehe* 1 and 2 adoption rates were similar to those seen by the full program with an *Ubudehe* 1 and 2 reported adoption rate of 95.6% and an observational adoption rate of 88.5% (Table 2). Observed filter adoption across all 15 villages varied between 74.5% and 98.1% (Table 3) with the lowest adoption rate occurring in one village (Mara) by almost 10% below all other villages. This is likely due to rodents destroying the filter tubes with 21.8% of filter repair of this problem in this village.

Households reported treating an average of 5.06 liters per day in all households and 5.11 liters per day in *Ubudehe* 1 and 2 households with no significant difference between monthly survey rounds. This equates to an average of 1.27 liters per person per day for all households and 1.11 liters per person for *Ubudehe* 1 and 2 households, possibly because of the smaller household sizes (Table 2). Regardless, water quantity consumption is lower than advised by CHWs at 2 liters per person per day. Similar consumption rates of 1 to 1.5 liters per person per day were reported at focus group discussions conducted with community health workers. Primary reasons discussed at the focus group for not consuming more water included not having a container to carry water when leaving the household, an inability to drink 2 liters per day, and a preference for drinking other beverages.

12.8% of households that reported using the filter also reported doing so for purposes other than drinking water. The most common uses were cooking (28.4%), hand washing (28.4%) and washing dishes (27.4%) (Table 2).

Almost two thirds (63.7%) of households reported backwashing their filter every time they treated water as advised in the household visit (Table 2). Not backwashing the filter frequently enough may be the cause of the most common reported problem with the filter, which was that it was clogged and wouldn't filter water (N = 45). The next most common problem reported was damage to rubber tubes because of rodents (N = 33). Overall 11.1% of households over the 5-month period reported any problems with the filter during the household visits. 57 (2.9%) filters were replaced and 366 (18.8%) filters were repaired with the same primary reported problems of tubes being damaged by rodents (N = 119) and filter clogged (N = 124) (Table 4).

The majority of filter problems were addressed through repair and replacement by program staff. Households contacted program staff through phone numbers on informational posters which were provided during the initial household visits. Common repairs included replacing tubes eaten by rodents or power backwashing the filters using a hand pump pressurized canister.

Stove Adoption and Use

As seen with filter adoption, reported primary use of the EcoZoom stove was around 90% for the entire population and *Ubudehe* 1 and 2 households (Table 5). Primary reasons given for stove adoption during focus group meetings included cost savings, time savings and cleanliness of the cook and kitchen when using the EcoZoom. However, 71.5% of these households reported continuing to use their traditional stove as well as their EcoZoom stove. Of households cooking at the time of the follow up visit (20.9%), about two thirds (63.7%) were using the EcoZoom stove, 21.9% a traditional 3-stone fire, and 11.4% cooking on a different

Table 1. Selected demographics and characteristics regarding water and energy practices.

	All households		*Ubudehe* 1 and 2	
	N	%	N	%
Number of Households	1634		478	
Household size, mean (95%CI)	4.55 (4.46–4.65)		3.85 (3.70–4.00)	
***Ubudehe* Category**				
1 or 2	478	29.3%		
3 or 4	1152	70.5%		
5 or 6	4	0.2%		
Fuel Type				
Wood	1510	92.4%	426	89.1%
Straw/Shrubs/Grass	93	6.3%	42	8.8%
Charcoal	28	1.9%	10	2.1%
LPG/Natural Gas/Biogas	1	0.1%	0	0.0%
Other	2	0.1%	0	0.0%
Drinking water source*	1576		457	
Public Tap	647	41.1%	196	42.9%
Protected Spring	592	37.6%	149	32.6%
Unprotected Spring	184	11.7%	62	13.6%
Surface Water	56	3.6%	26	5.7%
Hand Pump	37	2.3%	6	1.3%
Piped Water in Home	31	2.0%	8	1.8%
Unprotected Well	24	1.5%	8	1.8%
Protected Well	4	0.3%	1	0.2%
Rainwater	1	0.1%	1	0.2%

*Missing 58 (3.6%) answers.

traditional stove. Ten households (2.9%) were also observed using both the EcoZoom stove and a traditional stove.

Using the same metric of observed cooking use, EcoZoom use at the time of the household visit varied from 35.5% – 100.0% across the 15 pilot villages though the sample sizes were low in some villages with only 342 total observed cooking events. The three villages with the lowest observed use were Buhunde, Gisoro and Burorero (Table 3).

The two primary reasons reported during household surveying for not using the EcoZoom stove were inability to use wet wood in the EcoZoom stove (N = 46) and difficulty in using the stove (N = 31). This reported difficulty may refer to the required increased frequency of fire tending while cooking on the EcoZoom stove as 67.1% of households reporting tending the fire more with the EcoZoom stove than with their old stove (Table 4). Additionally the food most frequently reported cooked was beans (53.9%), requiring cooks to tend the fire frequently over a long period of time. Focus group discussions further emphasized these problems with the primary reasons for not using the stove as high frequency of fire tending, difficulty in burning wet wood when dry wood was unavailable and the inability to warm the house. Tending the fire was expressed most frequently as the primary issue since cooks use smaller pieces of wood to keep the fire going.

To assess the degree of EcoZoom stove use compared to other traditional stoves, households were asked the number of times per week they used each stove in their home. The EcoZoom stove was reported being used on average 1.37 times per day for 71.2% of cooking events in a household with a significant difference between monthly survey rounds (Table 5).

Primary reasons for continued use of a traditional stove included needing more than one stove at a time (N = 649) and the inability to cook on the EcoZoom stove when only wet wood was available (N = 330) (Table 4).

Cooking location was assessed because of the program emphasis on cooking outdoors during education and training activities. 342 observations of cooking location at the time of follow up visit were made, where 42.4% were cooking outdoors, 36.0% were cooking indoors and 21.6% in a separate kitchen. Slightly higher rates of cooking indoors were observed in the *Ubudehe* 1 and 2 households with 37.3% cooking outdoors, 44.5% cooking indoors, and 18.2% cooking in a separate kitchen (Table 5).

To quantify wood savings households were asked to report the number of wood bundles they collected or purchased before and after receiving the EcoZoom stove. Of the 1551 valid responses, an average wood reduction of 65.8% was reported across all rounds with no significant difference between the five rounds (Table 5).

A total of 73 stove problems were reported with the two most common problems being the pot skirt screw missing (N = 22) and the pot skirt degrading (N = 8). These were also the two most common reasons for stove repair with 48 pot skirts (2.5%) being replaced due to melting and 11 pot skirt replacements due to missing adjustment screws (Table 4). No stoves were replaced during the five months following distribution.

Table 2. Primary filter adoption and use outcomes.

	All households				*Ubudehe* 1 and 2			
	N	**%**	**Mean (95% CI)**	**ANOVA p-value**	**N**	**%**	**Mean (95% CI)**	**ANOVA p-value**
Reported Filter Use	1576	96.5%			457	95.6%		
Observed Filter Use	1471	90.0%			423	88.5%		
Other Purposes for Filtered Water	201	12.8%			74			
Cooking	57	28.4%			24	32.4%		
Hand Washing	57	28.4%			20	27.0%		
Washing Dishes	55	27.4%			18	24.3%		
Other	32	15.9%			12	16.2%		
Cleaning Frequency	1576				457			
Everytime	1004	63.7%			265	58.0%		
Daily	234	14.8%			71	15.5%		
Every Other Day	53	3.4%			20	4.4%		
2 Times per Week	236	15.0%			83	18.2%		
Once per Week	48	3.0%			18	3.9%		
Once per Month	1	0.1%			0	0.0%		
Liters per household per day	1210*		5.06 (4.99–5.12)	0.193	304*		5.11 (5.04–5.18)	0.194
Liters per person per day	1436*		1.27 (1.25–1.29)	0.090	304*		1.11 (1.10–1.13)	0.194

*Only records within 1.5 IQR were included.

Table 3. Filter and stove use by village.

	Observed Filter Use			Cooking on EcoZoom only at time of visit		
Village	**Valid observations**	**N**	**%**	**Valid observations**	**N**	**%**
Mara	110	82	74.5%	35	24	68.6%
Kigaga	90	82	91.1%	23	19	82.6%
Buhunde	117	103	88.0%	31	11	35.5%
Rushishi	147	137	93.2%	26	15	57.7%
Nyarubuye	116	102	87.9%	16	10	62.5%
Karambo	177	149	84.2%	34	23	67.6%
Rubona	65	62	95.4%	21	13	61.9%
Burorero	214	191	89.3%	18	7	38.9%
Nyabivumu	61	58	95.1%	22	22	100.0%
Rambura	116	107	92.2%	12	6	50.0%
Kabuga	111	106	95.5%	23	9	39.1%
Rupango	106	102	96.2%	31	27	87.1%
Gisoro	54	53	98.1%	8	3	37.5%
Nyarutovu	81	77	95.1%	31	21	67.7%
Gasumo	69	60	87.0%	11	8	72.7%
All villages	1634	1471	90.0%	342	218	63.7%

Discussion and Conclusions

In the pilot program described here high levels of uptake and continued use of water filters and improved cookstoves were found. The rigorous five-month follow up study was complemented by qualitative assessments by the implementation team periodically over an additional 11 months. 16 months after the intervention, adoption rates of the water filters and cookstoves were assessed to being comparable to those observed during the detailed 5-month study. This outcome may be described through the three design choices outlined previously. These design choices are intended to be applied in the full-scale program scheduled for deployment in 2014 and 2015. A key design difference in the planned full-scale program is that it will reach only the poorest households in a given village.

Behavior Change

The primary purpose of the technologies provided is to realize a health benefit. As a first step, communicating these potential health benefits to a user is often seen as an appropriate prerequisite to adoption. A lack of knowledge of potential health benefits has been shown to result in poor adoption of products like stoves and filters [36]. It has also been demonstrated that knowledge of health benefits alone is not sufficient to result in sustained behavior change in an individual or household [34]. The program studied here uses theories of behavior change such as diffusion of innovation and the health based model with both health based messaging as well as economic and social messaging to promote behavior change within the program.

In the case of the filter, adoption was measured around 90% for the five months of the study. A high adoption outcome has been seen previously with earlier versions of the LifeStraw Family filter with 96% adoption in Zambia [27] after 12 months and 68% adoption in the Democratic Republic of Congo after eight months [37]. Additionally, compared to other point-of-use water methods, filtration often has higher adoption rates [38] possibly because it is seen as easier to use [39] and doesn't result in a change in taste and odor [34]. However, while adoption of the filter was high, the recommended water consumption of 2 liters per person per day was not reached in most cases, suggesting that households may be drinking untreated water at times, a behavior also seen in the study conducted in the Democratic Republic of Congo [37]. Even occasional consumption of untreated water can greatly reduce the potential health benefits from water quality interventions [40,41]. Reasons that households gave for not exclusively drinking clean water included not having clean water when they were away from home and not having any filtered water at the time they were thirsty. Similar reasoning has been found in other studies where water treatment needed to be integrated into the everyday lifestyle of the family [34]. However, guidance from the existing literature is limited to help guide behavior change program development [42] so additional qualitative research is needed to understand people's behavior and preferences for exclusive drinking of treated water to adjust messaging and education activities to be more effective. In collaboration with the manufacturer, the product has been updated based on recipient feedback to protect the soft tubes and the backwash bulb, and to allow for a separable safe water storage container for ease of cleaning.

Adoption of the stove was also around 90%. Non-health benefits such as a cleaner appearance and cooking environment were more highly valued than health or environmental impacts, as observed in other studies. Exclusive use of the EcoZoom stove was only reported in 28.5% of households with most continuing to use their old stove with the EcoZoom stove. The earliest models for stove adoption assumed a "fuel switch" wherein behavior is switched over a short period of time from one stove/fuel combination to another. More recently, continued "stove stacking", where the use of multiple stoves for varying purposes, has been shown to be a more stable behavior, and can result in as high as 90% of stove usage events on the improved stove [43]. Studies have shown that households do not move from older existing methods of cooking such as a 3-stone fire to exclusive use of an improved stove. In

Table 4. Reasons for stove and filter problems, repairs and replacements.

	N	%
Reported reasons for not using EcoZoom	168	
Don't have dry wood	46	27.4%
Difficult to use	31	18.5%
Doesn't warm the house	20	11.9%
Use of a different fuel	13	7.7%
Other (20)	76	45.2%
Reported stove problems	73	
Pot skirt missing screw	22	30.1%
Pot skirt damaged	8	11.0%
Stove too small	6	8.2%
Difficulty in moving pot skirt to another pot	6	8.2%
Ceramic chamber cracked	4	5.5%
Stick support damaged	4	5.5%
Other (12)	23	31.5%
Reported reasons for continued use of old stove	1386	
More than one stove needed	649	46.8%
Don't have dry wood	330	23.8%
Need to warm house	126	9.1%
Pot is too big for stove	73	5.3%
Other (28)	208	15.0%
Reasons for stove repair	67	
Skirt replaced/Skirt damaged	48	71.6%
Skirt replaced/adjustment screws missing	11	16.4%
Stick support replaced/Broken	3	4.5%
Stick support replaced/Missing	1	1.5%
Other	4	6.0%
Reported filter problems	182	
Filter broken or clogged	45	24.7%
Tubes damaged/eaten by rodents	33	18.1%
Tubes are kinked	26	14.3%
Difficulty in backwashing	22	12.1%
Tap is leaking or broken	18	9.9%
Backwash bulb is damaged	17	9.3%
Other (6)	21	11.5%
Reasons for filter repair*	391	
Filter cartridge clogged	124	31.7%
Tubes replaced from rodent damage	119	30.4%
Backwash water not going into container	62	15.9%
Broken tap handle	39	10.0%
Broken tap - leaking	10	2.6%
Backwash leaking	5	1.3%
Backwash bulb replaced	2	0.5%
Other	30	7.7%

*366 total filters repaired - 25 had multiple problems.

order to realize the potential health benefits of improved stoves, exclusive use will need to be further promoted within the program [44,45,46]. Working with the stove manufacturer, the pot skirt has been updated to reduce degradation.

This stove stacking behavior suggests that the true innovation being introduced is not necessarily the stove itself, but the modified cooking practices required to realize the health and other benefits. Previous studies have found that an improved stove must meet the user's traditional cooking needs in order for adoption to occur.

Table 5. Primary stove adoption and use outcomes.

	All households				*Ubudehe* 1 and 2			
	N	**%**	**Mean (95% CI)**	**ANOVA p-value**	**N**	**%**	**Mean (95% CI)**	**ANOVA p-value**
Reported Primary Stove Use	1634				478			
EcoZoom	1479	90.5%			430	90.0%		
Traditional 3-stone Fire	68	4.2%			23	4.8%		
Other Traditional Stove	78	4.8%			22	4.6%		
Other Stove	9	0.6%			3	0.6%		
Stove Using at Time of Visit	342	20.9%			110	23.0%		
EcoZoom Only	218	63.7%			70	63.6%		
Traditional 3-Stone Fire	75	21.9%			28	25.5%		
Other Tradional Stove	39	11.4%			11	10.0%		
EcoZoom and Other Stove	10	2.9%			1	0.9%		
Cooking Location at Time of Visit								
Outdoor	145	42.4%			41	37.3%		
Indoor	123	36.0%			49	44.5%		
Separate Kitchen	74	21.6%			20	18.2%		
Continuing to Use Old Stove	1169	71.5%			316	66.1%		
# of EcoZoom uses per week	1517*		1.37 (1.34–1.40)	<0.000001	438*		1.38 (1.32–1.44)	0.869
% EcoZoom Use	1625*		71.2% (69.8%–72.7%)	0.052	474*		73.7% (71.0%–76.4%)	0.896
% Wood Reduction	1551*		65.8% (65.0%–66.6%)	0.593	443*		65.9% (64.3%–67.4%)	0.082

*Only records within 1.5 IQR were included.

The primary barriers to adoption or exclusive use of the EcoZoom stove in this study center around modifying current practices such as additional fire tending. While the EcoZoom stove is likely to decrease the cooking time of an individual cooking event, additional fire tending compared to a traditional 3-stone fire is often necessary, as reported in the focus group convened for this study. There may be an increased emissions exposure risk associated with greater fire tending that has not yet been characterized. When cooking meals on a 3-stone fire that requires long cooking times, such as beans, the cook will often prepare a large fire and perform other household tasks while the beans are cooking. Use of the EcoZoom stove requires a behavior modification to persuade the cook to stay by the fire while the meal is cooking. Additionally burning of wet wood in an improved stove can be more difficult than a 3-stone fire, so a cook often prefers to use the easier cooking method and therefore it is suggested that a careful examination of cooking practices, and focusing on those practices rather than the intrinsic benefits of the technology may result in higher adoption rates [43].

42.4% of observed cooking events occurred outdoors as instructed through education and training activities. While the EcoZoom stove has the potential to reduce indoor air pollution compared to traditional stoves, the primary reduction may likely come from moving cooking out of the home. As less than half (42.4%) of observed cooking events occurred outdoors as instructed through education and training activities, further messaging targeted at cooking location will need to be performed to increase outdoor usage.

A 65.8% reduction in wood usage by users of the EcoZoom stove is likely to significantly reduce time to collect wood and expenditure on fuelwood. Much of the wood use in Rwanda is with small sticks and branches which burn fast on three stone fires. Wood reduction was calculated through the ratio of wood bundles used before and after receiving the EcoZoom stove. In order to better quantify wood savings, additional methods will need to be employed to better understand wood usage such as the kitchen performance test which can evaluate stove performance in real-world settings [47].

With respect to both the filter and stove, behavior in *Ubudehe* categories 1 and 2 was similar to the overall population. This suggests that similar results could be expected during a large distribution of only *Ubudehe* 1 and 2 households. However, the effect of distributing to only a part of a village while the other households do not receive the technologies is unknown.

Free Distribution

Recent studies have examined cost-sharing for bed-nets, cook-stoves and water treatment systems and have found that there is no correlation between free distributions and low adoption rates [48]. Meanwhile, a study examining point-of-use chlorination through marketing campaigns and coupon schemes found these to be ineffective strategies but found free chlorination dispensed at water sources along with community providers as the most effective strategy in potentially preventing diarrheal incidence in areas like rural Kenya [49]. Furthermore, Bensch and Peters determined that a free stove program in Senegal resulted in high uptake of almost 100% of households [50].

These studies suggest that adoption and price are not fundamentally correlated, and that other factors including community engagement, government support and education are worth more careful study. With respect to the private-public partnership described here, the free giveaway nature of the program did not appear to adversely affect technology adoption on a community level, and resulted in a broader population exposure to the interventions than would have been possible via a retail effort over the same time period. The high rate of exclusive use of filters suggest that free distribution did not impact filter use, though it may have impacted intervention stove use.

Public-Private Partnership

The extensive logistical and behavior change messaging components of this program require sustained funding. Rwanda is not yet able to finance all health service activities directly; it relies both on direct donor funding to government programs, as well as careful coordination of programs managed by international non-profits and health sector businesses. By 2002, the government was spending 8.6% of its revenue on health care, which was only a third of the total costs, the remainder covered by donors [47]. Donation based non-profits are not providing services to the target populations serviced by this program. The business model anticipated by the for-profit implementer is designed to recuperate invested costs by the generation and sale of carbon credits associated with the proportion of the intervention that continues to demonstrate successful behavior change. The outcomes observed to-date support the business model in that high adoption rates will correlate to carbon credit generation sufficient to generate sustainable revenue that will allow continued program investment.

Study Limitations

Many of the results described here are from self-reported survey data that may result in over-reporting because of courtesy bias [51]. Over-reporting was measured in this study through the use of remote sensors which revealed over-reporting in frequency of use of both the water filter and cookstove [22]. This contributes to existing evidence of courtesy bias in self-reported outcomes of product distributions. Additionally while the survey directly asked households about their use of other cookstoves, it did not ask about households about supplementing their drinking water from other sources. To fully understand this issue, additional surveying and analysis is necessary. Respondent fatigue may also have been an issue throughout the study as some households were visited several times during a single month. Additionally the short duration of this study (five months) with less rigorous follow up through at least month 16 doesn't allow for complete characterization of the technologies or long-term adoption.

Acknowledgments

The authors thank the important contributions from colleagues and students at the Rwanda Ministry of Health, DelAgua Health, Vestergaard Frandsen, EcoZoom, Portland State University, the University of Colorado at Boulder and the London School of Hygiene and Tropical Medicine.

Author Contributions

Conceived and designed the experiments: CB FN GR SB TC ET. Performed the experiments: CB GR FM ET. Analyzed the data: CB ET. Contributed reagents/materials/analysis tools: CB ET. Wrote the paper: CB TC ET.

References

1. Liu L, Johnson HL, Cousens S, Perin J, Scott S, et al. (2012) Global, Regional, and National Causes of Child Mortality: An Updated Systematic Analysis for 2010 with Time Trends Since 2000. Lancet, 379: 2151–2161.
2. Smith K, Samet J, Romieu I, Bruce N (2000) Indoor Air Pollution in Developing Countries and Acute Lower Respiratory Infections in Children. Thorax 55: 518–532.

3. Ezzati M, Kammen D (2001) Indoor Air Pollution from Biomass Combustion and Acute Respiratory Infections in Kenya: An Exposure-Response Study. Lancet 358: 619–24.
4. Black R, Morris SS, Bryce J (2003) Where and Why are 10 Million Children Dying Each Year. Lancet 361: 2226–2234.
5. Roden C, Bond T, Conway S, Pinel AB, MacCarty N, et al. (2009) Laboratory and Field Investigations of Particulate and Carbon Monoxide Emissions from Traditional and Improved Cookstoves. Atmospheric Environment 43: 1170–1181.
6. Masera O, Edwards R, Arnez CA, Berrueta V, Johnson M, et al. (2007) Impact of Patsari Improved Cookstoves on Indoor Air Pollution in Michoacan, Mexico. Energy for Sustainable Development 11: 45–56.
7. Clasen T, Schmidt WP, Rabie T, Roberts I, Cairncross S (2007) Interventions to improve water quality for preventing diarrhoea: systematic review and meta-analysis. BMJ 334: Epub.
8. Fewtrell L, Kaufmann RB, Kay D, Enanoria W, Haller L, et al. (2005) Water, Sanitation and Hygiene Interventions to Reduce Diarrhoea in Less Developed Countries: A Systematic Review and Meta-Analysis. Lancet Infectious Diseases 5: 42–52.
9. UNICEF (2013). Committing to Child Survival: A Promise Renewed. United Nations Children's Fund.
10. European Union Energy Initiative (2009) Biomass Energy Strategy (BEST), Rwanda - Volume 3 Rural Supply and Demand. Available: http://www.cleancookstoves.org/resources_files/best-volume-3rural-supply.pdf. Accessed 2014 March 1.
11. Republic of Rwanda (2011) Rwanda Demographic and Health Survey 2010. National Institute of Statistics of Rwanda. Kigali.
12. Gundry S, Wright J, Conroy R (2004) A Systematic Review of the Health Outcomes Related to Household Water Quality in Developing Countries. Journal of Water and Health 1: 1–13.
13. Onda K, LoBuglio J, Bartram J (2012) Global Access to Safe Water: Accounting for Water Quality and the Resulting Impact on MDG Progress. Int. J. Enviorn. Res. Public Health 9: 880–894.
14. Wright J, Gundry S, Conroy R (2004) Household drinking water in developing countries: a systematic review of microbiological contamination between source and point-of-use. Tropical Medicine and International Health 9: 106–117.
15. Rwanda Bureau of Standards (2011) Potable Water Specification,RS 435:2009, Rev 1. Kigali.
16. Kirby M (2013) Water Quality Testing in 30 Districts of Rwanda. Unpublished.
17. WHO (1997) Guidelines for drinking water quality, 2nd edition. Volume 3: Surveillance and control of community supplies. Geneva. Available: http://www.who.int/water_sanitation_health/dwq/gdwqvol32ed.pdf. Accessed 2014 March 1.
18. Rosa G (2012) Water Quality Testing in 11 Districts in Rwanda. Unpublished.
19. Thomas E (2012) Leveraging carbon financing to enable accountable water treatment programs. Global Water Forum. Available: http://www.globalwaterforum.org/2012/09/23/leveraging-carbon-financing-to-enable-accountable-water-treatment-programs. Accessed 2014 March 1.
20. Williams M, Murthy S (2013) Reconciling the Carbon Market and the Human Right to Water: The Role of Suppressed Demand Under Clean Development Mechanism and the Gold Standard. Environmental Law 43: 517–562.
21. Rwanda Ministry of Health (2012) Third Health Sector Strategic Plan July 2012- June 2018. Ministry of Health. Kigali.
22. Thomas E, Barstow C, Rosa G, Majorin F, Clasen T (2013) Use of remotely reporting electronic sensors for assessing use of water filters and cookstoves in Rwanda. Environmental Science & Technology 47: 13602–13610.
23. Rosa G, Majorin F, Boisson S, Barstow C, Johnson M, et al. (2014) Assessing the impact of water filters and improved cookstoves on drinking water quality and household air pollution: A randomised controlled trial in Rwanda. PLOS One In Press.
24. Clasen T, Naranjo J, Frauchiger D, Gerba C (2009) Laboratory assessment of a gravity-fed ultrafiltration water treatment device designed for household use in low-income settings. American Journal of Tropical Medicine and Hygiene 80: 819–823.
25. WHO (2011) Evaluating household water treatment options: Health-based targets and microbiological performance specificiations.
26. Narajo J, Gerba C (2011) Assessment of the LifeStraw Family Unit using the World Health Organization Guidelines for Evaluating Household Water Treatment Options: Health-based Targets and Performance Specifications. University of Arizona.
27. Peletz R, Simunyama M, Sarenje K, Baisley K, Filteau S, et al. (2012) Assessing water filtration and safe water storage in households with young children of HIV-positive mothers: A randomized, controlled trial in Zambia. PLOS One doi:10.1371/journal.pone.0046548.
28. Johnson M, Pilco V, Torres R, Joshi S, Rajeev M, et al. (2013) Impacts on household fuel consumption from biomass stove programs in India, Nepal, and Peru. Energy for Sustainable Development 17: 403–411.
29. Aprovecho Research Center (2012) Results of Water Boiling Test 41 Testing of the EcoZoom Dura Stove.
30. Rogers E (2003) Diffusion of Innovations 5th Edition. New York: Free Press.
31. Becker M (1974) The Health Belief Model and Personal Health Behavior. Health Education Monographs 2: 324–473.
32. Aboud FE, Singla DR (2012) Challenges to Changing Health Behaviors in Developing Countries: A Critical Overview. Social Science and Medicine 75: 589–594.
33. Rosenstock I, Stretcher V (1988) Social Learning Theory and the Health Belief Model. Health Education Behavior 15(2): 175–183.
34. Figueroa M, Kincaid D (2010) Social, cultural and behavioral correlates of household water treatment and storage. USAID.
35. United Nations Clean Development Mechanism (2013) DelAgua Public Health Program in Eastern Africa, 2012. Available: http://cdm.unfccc.int/ProgrammeOfActivities/Validation/DB/32Z3TT7LC70PMDWOEW1706HJ0ZJLN7/view.html. Accessed 2014 March 1.
36. Lewis J, Pattanayak S (2012) Who adopts improved fuels and cookstoves? A systematic review. Environmental Health Perspectives 120(5): 637–645.
37. Boisson S, Kiyombo M, Sthreshley L, Tumba S, Makambo J (2010) Field Assessment of a Novel Household-Based Water Filtration Device: A Randomized, Placebo-Controlled Trial in the Democratic Republic of Congo. PLoS ONE doi:10.1371/journal.pone.0012613.
38. Hunter P (2009) Household Water Treatment in Developing Communities: Comparing Different Intervention Types Using Meta-Regression. Environmental Science and Technology 43(23) 8991–8997.
39. Albert J, Luoto J, Levine D (2010) End-User Preferences of Competing POU Water Treatment Technologies Among the Rural Poor of Kenya. Enviornmental Science and Technology 44(12) 4426–4432.
40. Brown J, Clasen T (2012) High Adherence is Necessary to Realize Health Gains from Water Quality Interventions. PLoS One doi:10.1371/journal.-pone.0036735.
41. Enger KS, Nelson KL, Rose JB, Eisenberg JN (2013) The Joint Effects of Efficacy and Compliance: A Study of Household Water Treatment Effectiveness Against Childhood Diarrhea. Water Research doi:10.1016/j.wa-tres.2012.11.034.
42. Fiebelkorn P, Person B, Quick RE, Vindigni SM, Jhung M, et al. (2012) Systematic Review of Behavior Change Research on Point-of-Use Water Treatment Interventions in Countries Categorized as Low-to Medium-Development on the Human Development Index. Social Science and Medicine doi:10.1016/j.socscimed.2012.02.011.
43. Ruiz-Mercado I, Masera O, Zamora H, Smith KR (2011) Adoption and sustained use of improved cookstoves. Energy Policy 39: 7557–7566.
44. Geary CW, Prabawanti C, Aristani C (2012) A Field Assessment of Adoption of Improved Cookstove Practices in Yogyakarta, Indonesia: Focus on Structural Drivers. Fhi360 Available: http://www.fhi360.org/sites/default/files/media/documents/CookstovePracticeFieldAssessment101512.pdf. Accessed 2014 March 1.
45. Masera OR (2000) From Linear Fuel Switching to Multiple Cooking Strategies: A Critique and Alternative to the Energy Ladder Model. World Development 28(12) 2083–2103.
46. Redman A (2011) Stove Stacking: The Integration of the Justa Cookstove in One Rural El Salvadorian Community. EHOS Conference 2011. Kirkland.
47. Shell Foundation (2007) Kitchen Performance Test.
48. Jameel Poverty Action Lab (2011) The price is wrong: Charging small feels dramatically reduces access to important products for the poor. Jameel Poverty Action Lab. Availible: http://www.povertyactionlab.org/publication/the-price-is-wrong. Accessed 2014 March 1.
49. Kremer M, Miguel E, Mullainathan S (2008) Source dispensers and home delivery of chlorine in Kenya. Innovations for Poverty Action. Availible: http://www.povertyactionlab.org/evaluation/source-dispensers-and-home-delivery-chlorine-kenya. Accessed 2014 March 1.
50. Bensch G, Peters J (2012) A recipe for success? Randomized free distribution of improved cooking stoves in Senegal. Ruhr-Universität Bochum (RUB). Availible: http://repec.rwi-essen.de/files/REP_12_325.pdf. Accessed 2014 March 1.
51. Wood L, Egger M, Gluud LL, Schulz K, Juni P, et al. (2008) Empirical Evidence of Bias in Treatment Effect Estimates in Controlled Trials with Different Interventions and Outcomes: Meta-Epidemiological Study. BMJ. http://dx.doi.org/10.1136/bmj.39465.451748.AD. doi:10.1136/bmj.39465.451748.AD.

10

Short-Term Effect of Nutrient Availability and Rainfall Distribution on Biomass Production and Leaf Nutrient Content of Savanna Tree Species

Eduardo R. M. Barbosa[1,2]*, Kyle W. Tomlinson[1,3], Luísa G. Carvalheiro[4,5], Kevin Kirkman[6], Steven de Bie[1], Herbert H. T. Prins[1,6], Frank van Langevelde[1]

1 Resource Ecology Group, Wageningen University, Wageningen, The Netherlands, **2** Departamento de Botânica, Laboratório de Termobiologia, Instituto de Ciências Biológicas, Universidade de Brasília, Brasília, DF, Brazil, **3** Community Ecology & Conservation Group, Xishuangbanna Tropical Botanical Garden, Chinese Academy of Sciences, Yunnan, China, **4** School of Biology, University of Leeds, Leeds, the United Kingdom, **5** Terestrial Zoology, Naturalis Biodiversity Center, Leiden, The Netherlands, **6** School of Life Sciences, University of KwaZulu-Natal, Scottsville, South Africa

Abstract

Changes in land use may lead to increased soil nutrient levels in many ecosystems (e.g. due to intensification of agricultural fertilizer use). Plant species differ widely in their response to differences in soil nutrients, and for savannas it is uncertain how this nutrient enrichment will affect plant community dynamics. We set up a large controlled short-term experiment in a semi-arid savanna to test how water supply (even water supply vs. natural rainfall) and nutrient availability (no fertilisation vs. fertilisation) affects seedlings' above-ground biomass production and leaf-nutrient concentrations (N, P and K) of broad-leafed and fine-leafed tree species. Contrary to expectations, neither changes in water supply nor changes in soil nutrient level affected biomass production of the studied species. By contrast, leaf-nutrient concentration did change significantly. Under regular water supply, soil nutrient addition increased the leaf phosphorus concentration of both fine-leafed and broad-leafed species. However, under uneven water supply, leaf nitrogen and phosphorus concentration declined with soil nutrient supply, this effect being more accentuated in broad-leafed species. Leaf potassium concentration of broad-leafed species was lower when growing under constant water supply, especially when no NPK fertilizer was applied. We found that changes in environmental factors can affect leaf quality, indicating a potential interactive effect between land-use changes and environmental changes on savanna vegetation: under more uneven rainfall patterns within the growing season, leaf quality of tree seedlings for a number of species can change as a response to changes in nutrient levels, even if overall plant biomass does not change. Such changes might affect herbivore pressure on trees and thus savanna plant community dynamics. Although longer term experiments would be essential to test such potential effects of eutrophication via changes in leaf nutrient concentration, our findings provide important insights that can help guide management plans that aim to preserve savanna biodiversity.

Editor: Ben Bond-Lamberty, DOE Pacific Northwest National Laboratory, United States of America

Funding: Funding and research assistance for this project were provided by Shell Research Foundation (http://www.shellfoundation.org/). The funders had no role in study design, data collection and analysis, decision to publish, or preparation of the manuscript.

Competing Interests: The authors have declared that no competing interests exist.

* E-mail: eduardormbarbosa@gmail.com

Introduction

Recent studies predict an increase in nitrogen deposition over southern Africa during the next few decades [1], due to rising industrial emissions and changes in land use [2]. Soil nitrogen enrichment can lead to soil acidification, which reduces soil fertility by promoting leaching of certain nutrients (such as calcium and magnesium) [3]. Moreover, increased nitrogen availability might also affect the carbon flux from soils of natural ecosystems [4] through changes in plant and soil microbial communities [5]. Such environmental changes can have important impacts for African savannas, especially on the species composition and abundance. Furthermore, alterations in rainfall patterns are also expected in the region where savannas occur [6]. However, little information on the effects of changes in soil nutrient and water availability on the leaf nutrient concentration of savanna trees is found in the literature [7].

Plant productivity and above-ground biomass are thought to increase with higher soil resource availability (e.g. nitrogen, water, phosphorus) [8–10]. In drier regions (such as semi-arid savannas), highly variable rainfall may negatively affect plant nutrient uptake and storage [11,12], potentially limiting plant growth during the growing season [13]. Indeed, performance of savanna tree seedlings is suggested to be worse when grown in nutrient-rich soils than in nutrient-poor soils [14,15]. This effect may be caused by the intensification of herbaceous competition for water and not by direct negative effects of high nutrient availability on tree seedlings [15]. Moreover, increased amounts of nutrients in plant leaves might increase their quality as food for herbivores [16–18], whereas increased water availability may increase biomass but decrease leaf nutrient concentration [18]. Tree seedling recruitment is a critical stage in the regeneration of trees and overall plant population dynamics [19–22]. However, there is a lack of empirical studies involving multiple plant species [22]. Most experimental studies evaluating the growth of tree species in response to resource supply and disturbance, with and without grass competitors, focus on single species [23,24], and there are few comparative investigations on seedlings of savanna tree species either within or across communities. This lack of empirical knowledge critically limits our ability to understand how seedlings of different species in a community perform under different environmental condition, and consequently, how plant community dynamics might change under modifications in the land use and climate conditions [25,26].

As plant species differ widely in their response to differences in soil nutrients [27], changes in soil resource availability (water and nutrient availability) may change structural heterogeneity in tree cover [28] or leaf nutrient concentration, and thereby influence primary productivity [29,30]. Differences in functional traits can mechanistically explain why species differ in their performance across resource and disturbance gradients [31,32]. Qualitative trait differences between species which are associated with nutrient and water gradients have been recognised [33]. Notably, within African savannas, dystrophic or humid savannas are dominated by broad-leafed species that are also non-spinescent, non-N-fixing species, whereas eutrophic or arid savannas are dominated by fine-leafed species which may additionally be spinescent or N-fixing [33–36]. These two groups can also be distinguished on the basis of their leaf chemistry, physiology and morphology [32]. As these functional traits are already found in tree seedlings during their first season of growth (e.g. N-fixing associations can be established early as two/three weeks after planting) [37], there is reason to believe that seedlings of tree species representing these functional species groups respond differently to changes in supply rates of resources, and that these differences may in part explain why they dominate in different environments.

Savannas are often characterized by water-limited plant growth during the growing season [38]. However, the amount of rainfall within the wet season is highly unpredictable, especially in semi-arid savannas [39]. Dry periods during the wet season can have an important impact limiting tree seedling survival [28]. Such dry periods may become more frequent in the near future, as global climate models indicate rising temperatures and increasingly erratic rainfall patterns across Southern African regions [40,41]. Climatic changes may also lead to slightly extended later summer rainfall over eastern South Africa [42]. Here we evaluated the short-term effects of water variability (even water supply vs. natural rainfall) and soil nutrient availability (no fertilisation vs. NPK addition) on above-ground biomass production and leaf nutrient concentrations of seedlings of two important functional groups of semi-arid savanna trees: broad-leafed (4 species) and fine-leafed (4 species) species (Table 1). We focus on the leaf concentrations of nitrogen (N), phosphorus (P) and potassium (K) because these nutrients are important in many plant metabolism processes [43], and in the diets of herbivores [16,18].

As increases in soil resources and reduction of periods of soil moisture deficiency are thought to increase plant productivity [8–11], we expected that all species respond positively to even water availability (no dry periods during the wet season) and to increased nutrient supply by increasing above-ground plant biomass (Hypothesis 1). As fine-leafed species are the dominant tree species in nutrient-rich savannas [35], are N-fixing and may have greater photosynthetic rates [44] than broad-leafed species, we expected fine-leafed species to have always higher leaf nutrient concentrations than broad-leafed species (Hypothesis 2). However, during growth, most nutrients (50–75%) are thought to be located in the leaves (e.g. [45,46], their concentration depending mostly on soil nutrient availability [47,48] and soil moisture [49]. Longer periods of soil moisture availability during the growing season may decrease leaf nutrient concentration, due to dilution effects of increased plant growth [18,50]. Therefore, we expected that the two species groups would increase leaf nutrient concentration with increasing soil nutrient availability, and that it would decrease with constant water supply (Hypothesis 3).

Methods

To test whether the two functional species groups differed in their response to variation in the growth conditions, we set up a large controlled, short-term field experiment in the Lowveld savanna region [35]. The study was carried out on private land of the Southern African Wildlife College (SAWC), Limpopo Province, South Africa (24°15′20.23″S, 31°23′23.63″E). For future permissions for fieldwork at the SAWC please contact Mrs. Theresa Sowry (tsowry@sawc.org.za) or Mr. Francois Nel (fnel@sawc.org.za). The experiments were run during the wet season of 2009–2010 (November–May), in a fenced area that excluded large herbivores. The mean rainfall during the growing season (from October till April) of the previous 10 years (2000–2010) is *ca.* 456 mm (Satara Camp, Kruger Park around 40 km northeast of the research site). The mean maximum temperature during January (hottest month) is 33.7°C and the mean minimum temperature for June (coolest month) is 9.4°C [42]. The vegetation is described as Granite Lowveld [35], and the area is classified as semi-arid under the Köppen-Geiger System [36]. Soils in the experimental site were shallow (ca. 1.5 m depth) and mainly derived from granite [34] with occasional gabbro extrusions. Soils derived from granite tend to be coarse-textured and nutrient-poor (i.e., low availability of N and P) on crests and mid-slopes [51], but nutrient availability may be elevated in bottom positions in the landscape, and very locally such as on termitaria or underneath large *Acacia* trees [51].

Species

We selected eight locally abundant tree species that make up a large proportion of vegetation cover in the Lowveld savanna region in South Africa. Although most of the selected species belong to the Fabaceae family (with the exception of *Combretum apiculatum*), these species are classified into two different sub-families: Mimosoideae (fine-leafed species) and Caesalpinioideae (broad-leafed species). In African semi-arid savannas, broad-leafed and medium-leafed species are found on dystrophic soils, characterised by high fire frequency (annual to triennial) and MAPs from 600–1500 mm [14]. Fine-leafed species (largely Mimosoideae) are found on eutrophic soils or skeletal soils with low fire frequency (quintennial or longer) and MAPs of 300–800 mm [35]. The study species were separated in two different functional species groups: four species with characteristic small leaves, spines, and N-fixing associations (hereafter termed 'fine-leafed species'), and four species with characteristic broad leaves, no spines, and lacking N-fixing associations (hereafter termed 'broad-leafed species') (see Table 1). All the seeds used in the experiment were collected in areas surrounding the experimental site. Since these species are abundant in the savannas of Southern African region, changes in their populations due to varying environmental conditions will likely have substantial effects on the local vegetation structure.

Treatments

The study site was ploughed (about 20 cm deep) to homogenize the soil and to give all treatments the same starting conditions. Five

Table 1. Functional trait data for tree species used in the experiment.

Species	Family	Sub-family	N_2-fixing[‡]	Leaf type[‡‡]	Leaf size (cm^2)[††]	Spinescence
Fine-leafed species						
Acacia nigrescens Oliv.	Fabaceae	Mimosoideae	Yes	Bipinnate	16.0 (±3.6)	Yes
Acacia nilotica Willd.	Fabaceae	Mimosoideae	Yes	Bipinnate	12.0 (±2.1)	Yes
Acacia tortilis Hayne	Fabaceae	Mimosoideae	Yes	Bipinnate	12.5 (±7.0)	Yes
Dichrostachys cinerea Wight and Arn	Fabaceae	Mimosoideae	Yes	Bipinnate	31.7 (±28.4)	Yes
Broad-leafed species						
Colophospermum mopane J. Léonard	Fabaceae	Caesalpinioideae	No	Pinnate	47.2 (±21.6)	No
Combretum apiculatum Sond.	Combretaceae	–	No	Simple	25.3 (±5.3)	No
Schotia brachypetala Sond.	Fabaceae	Caesalpinioideae	No	Pinnate	42.6 (±17.6)	No
Peltophorum africanum Sond.	Fabaceae	Caesalpinioideae	No	Bipinnate	99.5 (±81.8)	No

For continuous values the standard deviation is indicated between brackets. Sources of data are indicated in postscripts.
‡[51], [70].
††Obtained from the experimental seedlings from the treatment W0N0 (Natural rainfall-No nutrient addition).
‡‡[70].

blocks were laid out in a restricted area (90×90 m) in the experimental area (maximum distance between the blocks was 40 m). Inside each block, four 4-m^2 plots were located, separated by a 2 m gap between the plots (Figure 1) Seedlings were subjected to two different water regimes: one covered with a rain-out shelter (W1 - even water supply) and another exposed to natural rainfall conditions (W0 - natural rainfall or uneven water supply). The nutrient treatment was separated in two different nutrient applications (N0 - no nutrient supply vs. N1 - nutrient supply), leading to a total of 20 experimental plots.

Three weeks before the experiment all seeds were sown in nursery bags (using the same soil of the experimental area). At four weeks after germination, 20 seedlings per species were randomly transplanted in treatment combination plots (five replicate blocks, each with four seedlings per plot). The seedling positions in the plots were randomly selected. The seedling density inside of the plots (20 seedlings per m^2) was lower than the normal early seedling density in savannas (more than 50 seedlings per m^2 in the seed/seedling bank [52]. The seedlings were then followed for six months (November 2009 to May 2010). Although our experiments were performed during a short period of time (six months), this period of time is equivalent to a growing season in the area where the study was conducted. As savanna tree species show great differences in growth strategies, which allow them to cope with the high unpredictability of the amount of annual rainfall [39], we expected differences in the responses to the variation in resource avaliabiity between species even in short-term experiments.

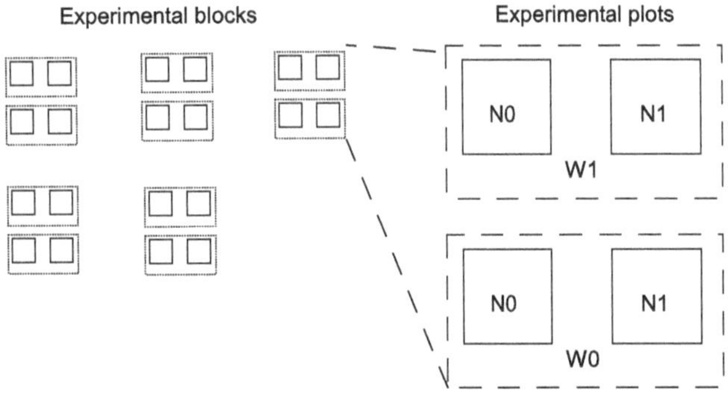

Figure 1. The experimental design. W0 - natural rainfall treatment, W1– even watering treatment, N0– no addition of nitrogen-phosphorus-potassium fertilizer, N1 addition of nitrogen-phosphorus-potassium fertilizer.

Plots within the uneven water supply treatment (W0) received 623 mm of water from natural rainfall during the period of the experiment, which was higher than the mean rainfall of the previous 10 years for the area (456 mm) (Figure 2). The distribution was uneven during the experiment: 206 mm in November, 114 mm in December, 55 mm in January, 31 mm in February, 57 mm in March and 160 mm in April. For the even water treatment (W1), natural rainfall was excluded from the treatments by rain-out shelters (200 μm clear greenhouse polyethylene film, allowing around 95% of sun light irradiation) and we supplied a fixed amount of 46.3 mm (185 l per 2×2 m plots) of water to the seedlings every two weeks for the six months of the experiment, yielding a total of 556 mm water over the experiment, using sprinkler irrigation systems. Due to the lack of the rainfall data from the research site, the amount of water applied in W1 was based on the water deficit rules as defined in the Köppen-Geiger climate classification (550 mm per season), based on a recent update of these regional classifications [36].

To increase the nutrient availability for the tree seedlings in the high nutrient treatment (treatment N1), we used a granular slow-release inorganic fertilizer containing nitrogen (N), phosphorus (P) and potassium (K) in the ratio 3:1:2 (Osmocote Exact Standard 15:9:11, Scotts International, The Netherlands). The fertilizer was added once before the seedlings being transplanted in treatment seedlings at a rate of 4 g N m^{-2} (640 g per plot), following rates previously applied [15]. Normal annual amount of nitrogen mineralized in the study region was estimated at 5.8 g N m^{-2} [53], so N1 treatment increased local nitrogen availability ca.1.7 times.

Shoot Foliar Nutrient Concentration and Biomass

The shoots of seedlings were harvested six months after planting in May 2010. These were oven-dried at 70°C for at least 48 h, and their dry weights were measured. To quantify the concentration of the elements N, P and K in leaves, the leaf material was digested with a mixture of H_2SO_4, Se and salicylic acid [54]. The concentrations of N and P in the leaves were measured with a Skalar San-plus auto-analyzer, and K was measured with an

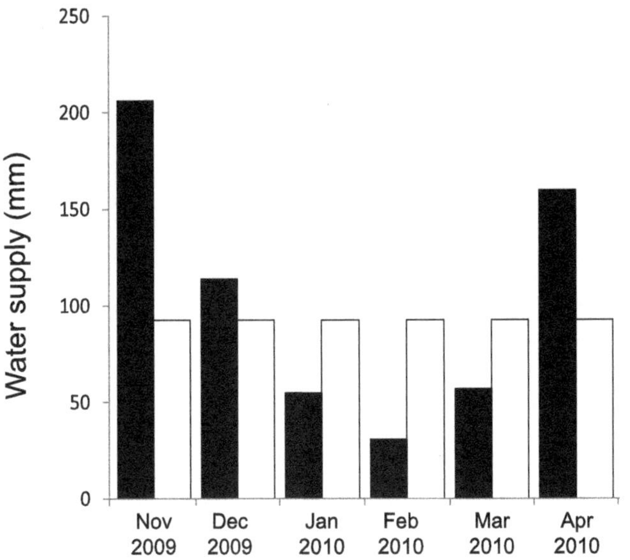

Figure 2. Monthly water availability (mm) during the experimental period. The black bars represent the monthly rainfall in the area during the wet season of 2009–2010 (natural rainfall treatment-W0). The white bars represent the monthly water supplied in the treatment W1. The rainfall data are from Satara Camp, Kruger Park around weather station (40 km northeast from the research site).

Atomic Absorption Spectrometer (AAS) from Varian (Palo Alto, CA, USA).

Data Analysis

To test how the different treatments affected leaf nutrient concentrations and biomass production, we used general linear mixed models (GLMM), using maximum likelihood [55]. Water regime (W) and nutrient addition (N) and functional species group (FG) were included as fixed variables. To account for inter-specific variability, species was treated as a random factor in the model (Species, 8 levels), and plot within experimental block. As the inclusion of block and plot position did not significantly improve the model (all plots were very close to each other), these two random factors were dropped from the final model. The individual species analyses are provided in Table S1.

Mixed model analyses were conducted in R (R Development Core Team, 2013 - version 3.0.2) using the *lmer* function of the package lme4 [55]. To test the significance of the terms in the statistical model we ran Monte Carlo Markov Chain simulations (100,000 iterations) using the LanguageR package (http://cran.r-project.org/web/packages/languageR/languageR.pdf) to analyse the seedling biomass production and leaf nutrient concentrations.

The data used for this manuscript is made available via SANParks Data Repository website (https://knb.ecoinformatics.org/knb/style/skins/sanparks/index.jsp) and can also be obtained from the corresponding author.

Results

Contrary to our expectations (Hypothesis 1), short-term changes in soil resource availability (water and nutrient availability) did not affect above-ground biomass production of any of the tree species (Table 2 and Figure 3).

In relation to leaf quality, we expected that fine-leafed species would present higher leaf nutrient concentrations than broad-leafed species (Hypothesis 2). Indeed, overall leaf N concentration

Table 2. The effect of functional species groups (FG) (fine-leafed vs. broad-leafed) of water (regular water supply vs. natural rainfall) and nutrient (no addition vs. NPK addition) treatments on leaf nutrient concentrations and above-ground biomass.

	Leaf nitrogen concentration			Leaf phosphorus concentration			Leaf potassium concentration			Total Biomass		
	Post mean	effective samples	p- MCMC	Post mean	effective samples	p- MCMC	Post mean	effective samples	p- MCMC	Post mean	effective samples	p- MCMC
FG	**−0.49**	**−0.14**	**0.011**	0.004	0.024	0.96	−0.01	0.20	0.93	−0.24	0.23	0.27
Nutrients	**−0.13**	**−0.01**	**0.029**	**−0.01**	**−0.003**	**0.015**	−0.03	0.05	0.44	−0.07	0.10	0.45
Water	**−0.24**	**−0.12**	**<0.0001**	**−0.01**	**−0.003**	**0.015**	**−0.10**	**0.00**	**0.040**	−0.07	0.10	0.46
FG × Nutrients	−0.09	0.09	0.31	−0.01	0.008	0.42	0.02	0.15	0.73	0.02	0.28	0.85
FG × Water	−0.01	0.17	0.92	0.005	0.020	0.49	−0.07	0.06	0.28	−0.07	0.18	0.58
Water × Nutrients	**0.20**	**0.37**	**0.024**	**0.03**	**0.045**	**<0.0001**	0.12	0.25	0.06	0.09	0.34	0.47
FG x Nutrients × Water	−0.07	0.17	0.60	0.01	0.026	0.56	−0.03	0.15	0.73	−0.08	0.26	0.68

Species was used as a random variable to correct for the variation among the different species. P values were obtained with Monte Carlo Markov Chain simulations (100000 iterations), using the MCMCglmm package and LanguageR package for R software (R Development Core Team, 2013, version 3.0.2). The significant values are represented in bold.

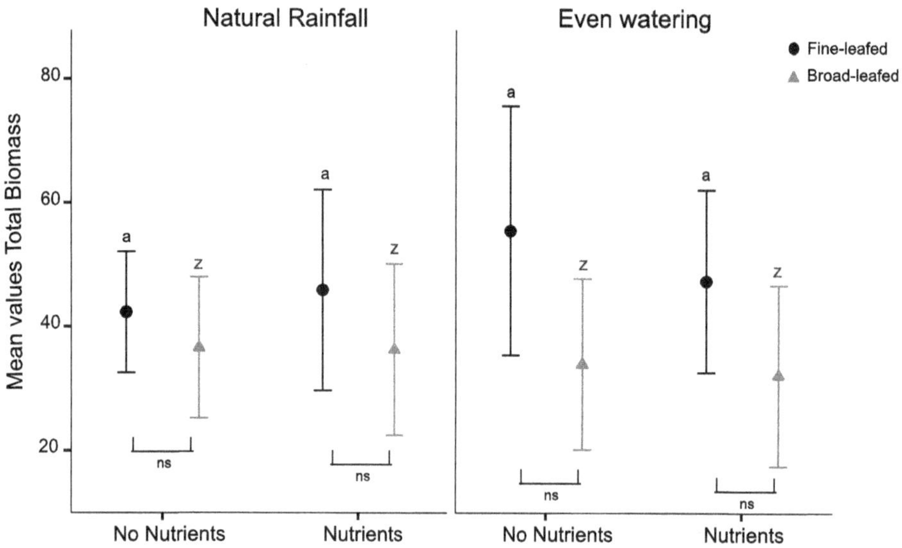

Figure 3. Effect of soil water and nutrient supply on the mean above-ground dry biomass (g). Bars represent 95% confidence intervals. The black circles represent the fine-leafed species, and the grey triangles represent the broad-leafed species. Nutrients represents the where Nitrogen-Phosphorus-Potassium (NPK) fertilazer was added, and No nutrients represents the plots where no NPK fertilazer was added. Statistical details are presented in Table 2. The letters represent differences between the treatments. Results of pairwise comparisons between the species groups within a treatment combination are indicated by the brackets (*: $p<0.05$; **: $p<0.01$; ***: $p<0.001$; ns: non-significant).

was lower in broad-leafed species, but no significant differences were found between the two groups for leaf K and P concentration. In relation to the responses of leaf quality to treatments, we expected that leaf nutrient concentration would increase with increasing soil nutrient availability, and that it would decrease with constant water supply (Hypothesis 3). However, leaf P concentration only increased with NPK fertilizer input under even water supply in both species groups (Figure 4). A non-significant positive trend in leaf K concentration was also apparent. In contrast, under uneven water supply (natural rainfall), foliar concentrations of P and N were lower under the nutrient addition treatment (Table 2 and Figure 4). This trend was more accentuated for broad-leafed species with respect to leaf N concentration. Moreover, leaf K concentration of broad-leafed species was significantly lower than fine-leafed species when grown under constant water supply.

Discussion

Plant productivity and above-ground biomass are thought to increase with higher soil resources (e.g. nitrogen, phosphorus) and water availability [8–10]. However, our results show that changes in leaf nutrient concentration varied with changes in the soil resource availability, even when biomass is not affected. Here we discuss the variability of responses to nutrient and water supply of two functional groups of tree species that are representative for the African savanna biome as a whole.

Effect of Water and Nutrient Availability on Shoot Growth and Leaf Nutrient Concentration

Contrary to our expectations (Hypothesis 1), increased short-term nutrient input and water availability did not significantly influence above-ground biomass production of the two functional species groups studied. Three plausible explanations arise. Firstly, the nutrient additions may have been insufficient to cause a difference in growth between the unfertilised and fertilised plots (raised the available N in the soil by at most 70%). Indeed, as we used a slow release fertilizer, nutrients added may not have been immediately available. However, nutrient addition had a strong effect on leaf chemical composition, suggesting that nutrient additions did increase nutrient uptake by the tree seedlings. A second explanation is that seedlings of the considered species are not limited by soil nutrient availability in the study, savanna tree species being able to cope with low resource conditions. While our study included fine-leafed species, which can be found in regions with high nutrient soils (e.g. *Acacia* species), the seeds used in this study were collected in areas with relative low soil nutrients. It is, therefore, possible that the source populations of the seeds used in this study are adapted to grow in relatively infertile soils exhibiting lower maximum potential growth rates, and responding less to nutrient addition [56]. Moreover, under frequent water supply and high nutrient availability, it is possible that plant species of semi-arid environments allocate more resources to the root system [57] while above ground biomass remains constant.

The lack of growth response to improved fertility has been observed previously for tree seedlings growing in low nutrient environments [32,58]. While the application of nutrient fertilizers may mitigate the adverse effects of water stress on plant development [59], the potential effect of nutrient addition on plants depends on their growth potential [58]. Therefore, changes in tree species composition are gradual, potentially taking a long time to be noticeable [60]. A longer term experiment (e.g. several years) would be essential to verify if nutrient enrichment and changes in water supply have delayed effects on savanna tree growth and biomass.

In relation to leaf-nutrient concentration, as expected fine-leafed species had higher leaf nutrient concentrations than broad-leafed species (Hypothesis 2). Fine leafed species are dominant on eutrophic soils [33–36], and hence are likely to be adapted to this high nutrient availability. Indeed all fine-leafed species studied here are able to fix atmospheric nitrogen [51], leading to a higher access to nitrogen. This extra N may be stored in leaves for future use, explaining the higher nitrogen values found in this study. Such storage might be important for shoot biomass recovery after

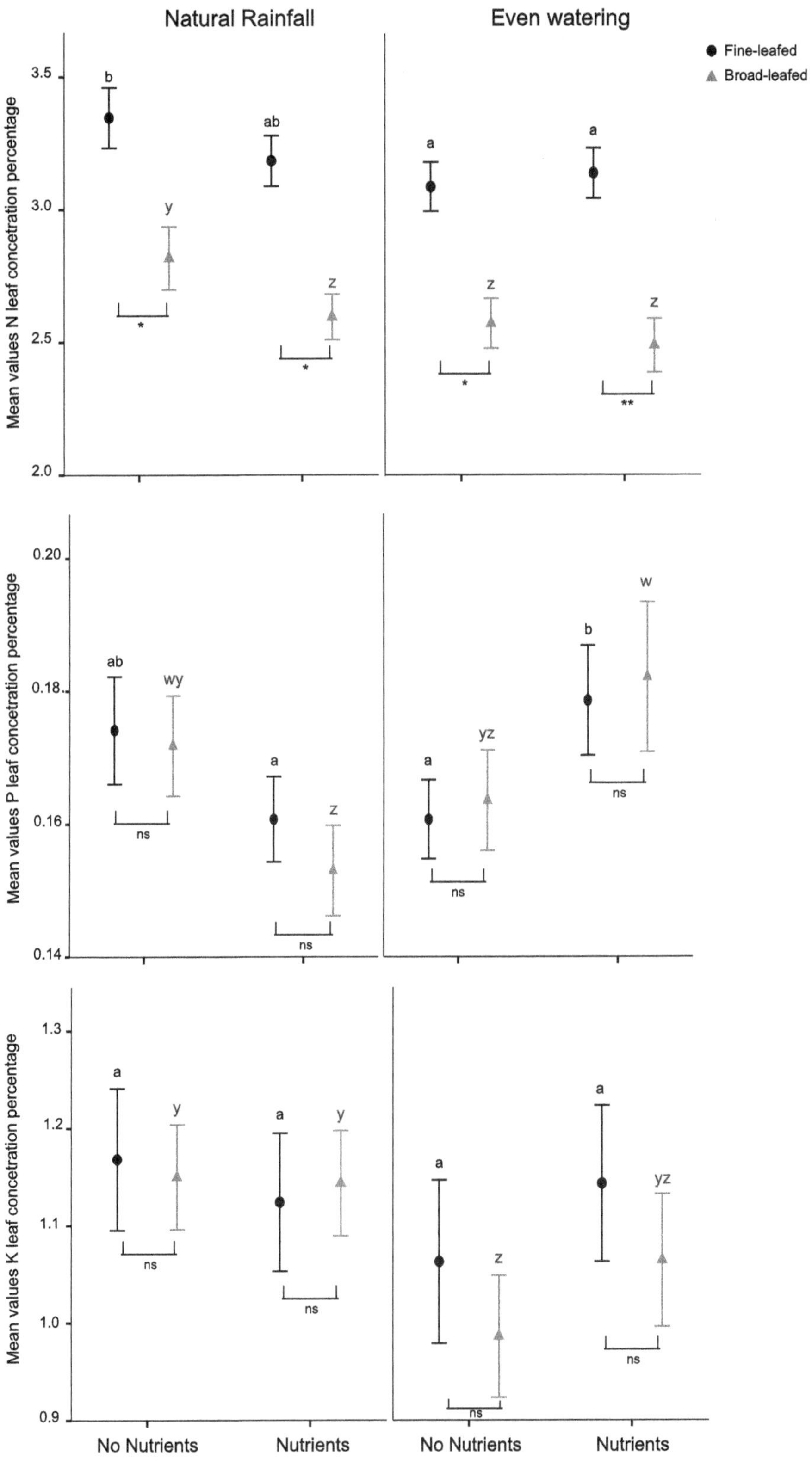

Figure 4. Effect of soil water and nutrient supply on the mean leaf nutrient concentration. Bars represent 95% confidence intervals. N = leaf nitrogen concentration, P = leaf phosphorus concentration and K = leaf potassium concentration. The black circles represent the fine-leafed species, and the grey triangles represent the broad-leafed species. "Nutrients" indicates where Nitrogen-Phosphorus-Potassium (NPK) fertilizer was added, and "No nutrients" represents the plots where no NPK fertilizer was added. Statistical details are presented in Table 2. The letters represent differences between the treatments. Results of pairwise comparisons between the species groups within a treatment combination are indicated by the brackets (*: $p<0.05$; **: $p<0.01$; ***: $p<0.001$; ns: non-significant).

intense fire or herbivory, as the N stored in the leaves not lost in the defoliation event can enhance shoot and leaf growth rates [27], which characterize nutrient rich savannas [35,36]. Fine leafed species could, hence, show high growth rate [61]. Indeed, in all treatments the average biomass of the fine-leafed species was higher than for broad-leafed species, which could be due to this extra nitrogen availability. However, these differences between the two groups were not significant, which possibly due to the short period of the experiment.

In contrast to plant productivity, plant quality (measured as the nutrient concentration in leaves) did significantly vary with nutrient input and water availability (Figure 4). A previous study of leaf nutrient concentration of grasses [18] suggested that plant quality (i.e. leaf nutrient concentration) increases with soil nutrient concentration and decreases with water availability (similar to Hypothesis 3). Indeed, for both functional groups, leaf nitrogen concentrations were higher under natural conditions (natural rainfall/no NPK input) than in other treatments (Fig. 4). For plants growing under high N levels, plants may invest mostly in growth, leading to a dilution effect of N content of leaves [18,50], potentially explaining the lower N content found in treatments with NPK addition. At low soil N availability, most N is stored in the leaves as amino acids, amides, or protein (enzymes such as Rubisco) [58]. The total Rubisco in the leaves increases linearly with increase of leaf N content, being essential for photosynthesis [62]. However, the activation of the Rubisco is regulated by CO_2 levels in the leaves [63]. As leaf N content and CO_2 assimilation by the leaves have a non-linear relationship, and hence most of the Rubisco in the leaves is inactive [64], not being used in photosynthesis. Such storage of N (e.g. as amino acids, or inactive Rubisco) can, however, be exported to support growth of other parts of the plant, whenever is needed [65].

In contrast with our expectations, our empirical results with savanna tree seedlings show that nutrient input increased leaf phosphorus concentration only when water input was regular, whereas decreases in leaf concentrations for this element occurred when water availability was uneven. Seedling dependence on water for a positive effect of nutrient (K and P) availability in leaves can be explained by the fact that nutrient uptake depends on water movement within plants [66]. Furthermore, the negative effect of nutrient input under the uneven water supply treatment (natural rainfall) also suggests that such irregularity in water provision stimulates allocation of resources away from leaves towards other organs, such as roots, that can support growth and survival when soil reserves are unavailable [67] for example during the dry (non-growing) season. This re-allocation is more likely to be noticeable at the end of the growing season, when our measurements were taken. Further studies on root production in tree seedlings across soil nutrient and moisture gradients would help to confirm where absorbed nutrients are allocated.

Implications for Herbivore-plant Relationships

The availability of soil nutrients is influenced by herbivore density through dung and urine [7]. The results of this short-term experiment suggests that when combined with the natural (i.e. uneven) rainfall patterns, high soil nutrient availability may lead to a decrease of the leaf quality of tree seedlings as forage for herbivores (due to lower nutrient concentration in leaves, in this study mostly nitrogen and phosphorus), even when overall biomass does not change (Figure 3). Tree seedlings are a common food source for herbivores, especially due the high nutrient levels, low levels of defensive structures, and secondary defensive compounds [68]. This decline in leaf quality might, hence, increase the need of consumption by herbivores to acquire the amounts of nutrients needed by them, magnifying the effects of high herbivore density. This increase in browsing may affect tree recruitment, potentially impacting long-term dynamics and vegetation structure in savannas [22,69]. Further changes in soil resource levels (e.g. higher N deposition, changes in wet season rainfall patterns) can lead to further accentuation of impacts of increased herbivore density for plant community dynamics in savannas. As several herbivore species are limited by the nutrient concentration of tree leaves, especially in pregnant and lactating animals [43], such changes may affect their population dynamics (e.g., reproduction, breeding times, and foraging range [14,16,18,43]. Future studies would be needed to test such potential effects of soil nutrient input on herbivores, via change in tree seedlings nutrient content.

Concluding Remarks

The results of our short-term multi-species experiment show that differences in soil resource availability lead to changes in leaf quality (leaf nutrient concentration). The effect of nutrient input on leaf quality (especially nitrogen and phosphorus concentrations) depends on water availability. Under more uneven water availability, leaf nutrient concentration decreases, while under regular rainfall it increases. While changes in the soil conditions might not directly affect plant species distribution [7], the changes in leaf quality may affect browsing pressures, and consequently affect overall vegetation structure. Our results hence suggest that, in response to the predicted changes in the rainfall distribution during the wet season in Southern African savannas (which is expected to become more erratic, with increases of the interval between each rainfall event [3,41], leaf quality of tree seedlings for a large number of species will change, potentially affecting vegetation communities and herbivore population dynamics. Long-term experiments across multiple growing seasons are essential to confirm the robustness of the results obtained in this study. Moreover, close monitoring of how vegetation and herbivore communities will change in response to climate and land-use changes is essential both to understand the full extent of the ecological and consequences and to contributing to the development of adequate policies and management plans that aim to preserve biodiversity.

Supporting Information

Table S1 Effect of water (regular water supply vs. natural rainfall) and nutrient (no addition vs. NPK addition) treatments on leaf nutrient concentrations and above-ground biomass in all study species used in this research. P values were obtained with Monte Carlo Markov Chain simulations (100000 iterations), using the MCMCglmm package and LanguageR package for R software (R Development Core Team, 2013).

Acknowledgments

We are thankful to the staff of Southern African Wildlife College, SANParks, School of Biological and Conservation Sciences (University of KwaZulu-Natal) and the Resource Ecology Group (Wageningen University) for technical support, especially A. van den Driessche, C. Y.Ravenstein, C. Mthabine, P. Mthabine, L. Mthabine, D. Mlambo and P. Ngomana.

Author Contributions

Conceived and designed the experiments: ERMB KWT FVL. Performed the experiments: ERMB. Analyzed the data: ERMB LGC KWT. Wrote the paper: ERMB. Provided comments in several stages of the project and manuscript writing: FVL KWT LGC KK SDB HHTP.

References

1. Miyazaki K, Eskes HJ, Sudo K (2012) Global NOx emission estimates derived from an assimilation of OMI tropospheric NO2 columns. Atmos Chem Phys 12: 2263–2288.
2. Adams M, Ineson P, Binkley D, Cadlsch G, Tokuchi N, et al. (2004) Soil function responses to excess nitrogen inputs at global scale. Ambio 33: 530–536.
3. Allison SD, Treseder KK (2008) Warming and drying suppress microbial activity and carbon cycling in boreal forest soils. Glob Change Biol 14: 2898–2909.
4. Neff JC, Townsend AR, Gleixner G, Lehman SJ, Turnbull J, et al. (2002) Variable effects of nitrogen additions on the stability and turnover of soil carbon. Nature 419: 915–917.
5. Xu LK, Baldocchi DD, Tang JW (2004) How soil moisture, rain pulses, and growth alter the response of ecosystem respiration to temperature. Glob Biogeochem Cycles: 18 GB4002.
6. IPCC (Intergovernmental Panel on Climate Change) (2007) Climate Change 2007: The Physical Science Basis. Cambridge: Cambridge University Press582 p.
7. Van der Waal C, Kool A, Meijer SS, Kohi E, Heitkonig IMA, et al. (2011) Large herbivores may alter vegetation structure of semi-arid savannas through soil nutrient mediation. Oecologia 165: 1095–1107.
8. Tilman D (1984) Plant dominance along an experimental nutrient gradient. Ecology 65: 1445–1453.
9. Gower ST, Vogt KA, Grier CC (1992) Carbon Dynamics of Rocky Mountain Douglas-Fir: Influence of Water and Nutrient Availability. Ecol Monogr 62: 43–65.
10. Polis GA (1999) Why Are Parts of the World Green? Multiple Factors Control Productivity and the Distribution of Biomass. Oikos 86: 3–15.
11. Parks SE, Haigh AM, Creswell GC (2000) Stem tissue phosphorus as an index of the phosphorus status of *Banksia eri cifolia* L. Plant Soil 227: 59–65.
12. Shane MW, Cramer MD, Funayama-Noguchi S, Cawthray GR, Millar HA, et al. (2004) Developmental physiology of cluster-root carboxylate synthesis and exudation in *Hakea prostrata* R. Br. (Proteaceae): expression of PEP-carboxylase and the alternative oxidase. Plant Physiol 135: 549–560.
13. Kambatuku JR, Cramer MD, Ward D (2011) Savanna tree-grass competition is modified by substrate type and herbivory. J Veg Sci 22: 225–237.
14. Kraaij T, Ward D (2006) Effects of rain, nitrogen, fire and grazing on tree recruitment and early survival in bush-encroached savanna, South Africa. Plant Ecol 186: 235–246.
15. Van der Waal C, De Kroon H, de Boer F, Heitkönig IMA, Skidmore AK, et al. (2009) Water and nutrients alter herbaceous competitive effects on tree seedlings in a semi-arid savanna. J Ecol 97: 430–439.
16. Ahrestani FS, Heitkönig IMA, van Langevelde F, Vaidyanathan S, Madhusudand MD, et al. (2011) Moisture and nutrients determine the distribution and richness of India's large herbivore species assemblage. Basic Appl Ecol 12: 634–642.
17. Du Toit JT, Bryant JP, Frisby K (1990) Regrowth and palatability of Acacia shoots following pruning by African savanna browsers. Ecology 71: 149–154.
18. Olff H, Ritchie ME, Prins HHT (2002) Global environmental controls of diversity in large herbivores. Nature 415: 901–04.
19. Higgins SI, Bond WJ, Trollope WSW (2000) Fire, resprouting and variability: a recipe for grass-tree coexistence in savanna. J Ecol 88: 213–229.
20. Kitajima K, Fenner M (2000) Ecology of seedling regeneration. In: Fenner M, editor. Seeds: the ecology of regeneration in plant communities. Wallingford: CAB, International. 331–360.
21. Wiegand K, Saltz D, Ward D (2006) A patch dynamics approach to savanna dynamics and bush encroachment – Insights from an arid savanna. Perspect Plant Ecol Evol Syst 7: 229–242.
22. van Langevelde F, Tomlinson KW, Barbosa ERM, de Bie S, Prins HHT, et al. (2011) Understanding tree-grass coexistence and impacts of disturbances and resource variability in savannas. In: Hill M, Hanan N editors. Ecosystem function in savannas. Boca Raton: CRC Press. 257–271.
23. Kambatuku JR, Cramer MD, Ward D (2011) Savanna tree–grass competition is modified by substrate type and herbivory. J Veg Sci 22: 225–237.
24. Ward D, Esler KJ (2011) What are the effects of substrate and grass removal on recruitment of *Acacia mellifera* seedlings in a semi-arid environment? Plant Ecol 212: 245–250.
25. Ward D (2005) Do we understand the causes of bush encroachment in African savannas? Afr J Range Forage Sci 22: 101–105.
26. Lehmann CER, Ratman J, Hetley LB (2009) Which of these continents is not like the other? Comparisons of tropical savanna systems: key questions and challenges. New Phytol 181: 508–511.
27. Lambers H, Chapin FS, Pons TL (2008) Plant physiological ecology. New York: Springer. 610 p.
28. Wijesinghe DK, John EA, Hutchings MJ (2005) Does pattern of soil resource heterogeneity determine plant community structure? An experimental investigation. J Ecol 93: 99–112.
29. Sankaran M, Ratnam J, Hanan NP (2008) Woody cover in African savannas: The role of resources, fire and herbivory. Glob Ecol Biogeogr 17: 236–245.
30. Van Langevelde F, Van De Vijver C, Kumar L, Van De Koppel J, De Ridder N, et al. (2003) Effects of fire and herbivory on the stability of savanna ecosystems. Ecology 84: 337–350.
31. Chapin FS, Autumn K, Pugnaire F (1993) Evolution of suites of traits in response to environmental stress. Am Nat 142: S78–S92.
32. Barbosa ERM, Van Langevelde F, Tomlinson KW, Carvalheiro LMGR, Kirkman K, et al. (2013) Tree species from different functional groups respond differently to environmental changes during establishment. Oecologia. "In press". DOI: 10.1007/s00442-013-2853-y.
33. Scholes RJ (1997) Savanna. In: Cowling RM, Richardson DM, Pierce SM, editors. Vegetation of southern Africa. Cambridge: Cambridge University Press. 258–277.
34. Venter FJ, Scholes RJ, Eckhardt HC (2003) The abiotic template and its associated vegetation pattern. In: du Toit JT, Rogers KH, Biggs HC, editors. The Kruger Experience. Washington DC: Island Press. 83–129.
35. Mucina L, Rutherford MC (2006) The Vegetation Map of South Africa, Lesotho and Swaziland. Pretoria: SANBI. 807 p.
36. Kottek M, Grieser J, Beck C, Rudolf B, Rubel F (2006) World map of the Köppen-Geiger climate classification updated. Meteorol Z 15: 259–263.
37. Bohrer G, Kagan-Zur V, Roth-Bejerano N, Ward D (2001) Effects of environmental variables on vesicular-arbuscular mycorrhizal abundance in wild populations of *Vangueria infausta*. J Veg Sci 12: 279–288.
38. Kambatuku JR, Cramer MD, Ward D (2012) Overlap in soil water sources of savanna woody seedlings and grasses. Ecohydrology 6: 464–473.
39. Ward D (2009) The biology of deserts. Oxford: Oxford University Press. 304 p.
40. IFPRI (International Food policy Research Institute) (2013) African Agriculture and Climate Change Country Summary – South Africa report. Johnston P, Hachigonta S, Sibanda L, Thomas M, Timothy S, editors. Washington DC: IFPRI. 38 p.
41. IPCC (Intergovernmental Panel on Climate Change) (2012) Managing the Risks of Extreme Events and Disasters to Advance Climate Change Adaptation. Field CB, Barros V, Stocker TF, Qin D, Dokken DJ, Ebi KL, et al., editors. Cambridge: Cambridge University Press. 582 p.
42. Marschner H (1995) Mineral Nutrition of Higher Plants. London: Academic Press. 889 p.
43. Prins HHT, Van Langevelde F (2008) Assembling a diet from different places. In: Prins HHT, van Langevelde F, editors. Resource Ecology: Spatial and Temporal Dynamics of Foraging. Berlin: Springer-Verlag. 129–155.
44. Kgope BS (2004) Differential photosynthetic responses of broad- and fine-leafed savanna trees to elevated temperatures. S Afr J Bot 70: 760–766.
45. Chapin FS, Kedrowski RA (1983) Seasonal changes in nitrogen and phosphorus fractions and autumn retranslocation in evergreen and deciduous taiga trees. Ecology 64: 376–391.
46. Pregitzer KS, Dickmann DI, Hendrick R, Nguyen PV (1990) Whole-tree carbon and nitrogen partitioning in young hybrid poplars. Tree Physiol 17: 79–93.
47. Paquin R, Margolis HA, Doucet R, Coyea MR (2000) Physiological responses of black spruce layers and planted seedlings to nutrient addition. Tree Physiol 20: 229–237.
48. Prior SA, Runion GB, Mitchell RJ, Rogers HH, Amthor JS (1997) Effects of atmospheric CO2 on longleaf pine: Productivity and allocation as influenced by nitrogen and water. Tree Physiol 17: 397–405.
49. Scholes RJ, Walker BH (1993) An African Savanna: Synthesis of the Nylsvley Study. Cambridge: Cambridge University Press. 306 p.
50. Rittenhouse LR, Roath LR (1987) Forage quality: primary chemistry of grasses. In: Capinera JL, editor. Integrated pest management on rangeland a shortgrass prairie perspective. Boulder: West-view. 25–37.
51. Treydte AC, Heitkonig IMA, Prins HHT, Ludwig F (2007) Trees improve grass quality for herbivores in African savannas. Perspect Plant Ecol Evol Syst 8: 197–205.
52. Tefera SB (2011) Soil seed bank dynamics in relation to land management and soil types in the semi-arid savannas of Swaziland. Afr J Agric Res 6: 2494–2505.
53. Scholes RJ, Bond W, Eckhardt H (2003) Vegetation dynamics in the Kruger ecosystem. In: du Toit J, Rogers K, Biggs H, editors. The Kruger experience. Ecology and Management of savanna heterogeneity. Washington DC: Island Press.pp 131–148.
54. Novozamsky I, Houba VJG, Eck RV, Vark VW (1983) A novel digestion technique for multi-element analysis. Commun Soil Sci Plant Nutr 14: 239–249.
55. Zuur AF, Ieno EN, Walker NJ, Saveliev AA, Smith GM (2009) Mixed Effects Models and Extensions in Ecology with R. New York: Springer-Verlag. 596 p.
56. Chapin FS (1980) The mineral nutrition of wild plants. Annu Rev Ecol Syst 11: 233–260.
57. Shipley B, Meziane D (2002) The balanced-growth hypothesis and the allometry of leaf and root biomass allocation. Funct Ecol 16: 326–331.
58. Chapin FS, Vitousek PM, Van Cleve K (1986) The Nature of Nutrient Limitation in Plant Communities. Am Nat 127: 48–58.
59. Garge BK, Burman U, Kathju S (2004) The influence of phosphorus nutrition on the physiological response of moth bean genotypes to drought. J Plant Nutr Soil Sci 167: 503–508.

60. Christensen N L, Peet RK (1981) Secondary forest succession on the North Carolina Piedmont. In: West D C, Shugart HH, Botkin DB, editors. Forest succession: concepts and applications. New York: Springer-Verlag. 230–245.
61. Tilman D (1986) Nitrogen-limited growth in plants from different successful stages. Ecology 67: 555–563.
62. Pessarakli M (2005) Handbook of photosynthesis, 2nd edition. Boca Raton: CRC Press. 928 p.
63. Portis AR (1992) Regulation of ribulose 1,5-bisphosphate carboxylase/oxygenase activity. Annu Rev Plant Physiol Plant Mol Biol. 43: 415–437.
64. Cheng L, Fuchigami LH (2000) Rubisco activation state decreases with increasing nitrogen content in apple leaves. J Exp Bot 51: 1687–1694.
65. Chapin FS, Schultz E, Mooney H (1990) The ecology and economics of storage in plants. Ann Rev Ecol Syst 21: 423–447.
66. Hu Y, Schmidhalter U (2005) Drought and salinity: A comparison of their effects on mineral nutrition of plants. J Plant Nutr Soil Sci 168: 541–549.
67. Grime JP (1979) Plant strategies and vegetation processes. New York: John Wiley and Sons. 222 p.
68. Fornara DA, Du Toit JT (2008) Responses of woody saplings exposed to chronic mammalian herbivory in an African savanna. Ecoscience 15: 129–135.
69. Sankaran M, Ratnam J, Hanan NP (2004) Tree-grass coexistence in savannas revisited - insights from an examination of assumptions and mechanisms invoked in existing models. Ecol Lett 7: 480–490.
70. Tomlinson KW, Poorter L, Sterck F, Borghetti F, Ward D, et al. (2013) Leaf adaptations of evergreen and deciduous trees of semi-arid and humid savannas on three continents. J Ecol 101: 430–440.

Drinking Water in Transition: A Multilevel Cross-sectional Analysis of Sachet Water Consumption in Accra

Justin Stoler[1,2]*, John R. Weeks[3], Richard Appiah Otoo[4]

1 Department of Geography and Regional Studies, University of Miami, Coral Gables, Florida, United States of America, **2** Department of Public Health Sciences, Miller School of Medicine, University of Miami, Miami, Florida, United States of America, **3** Department of Geography, San Diego State University, San Diego, California, United States of America, **4** Ghana Urban Water Limited, Head Office Operations, Accra, Ghana

Abstract

Rapid population growth in developing cities often outpaces improvements to drinking water supplies, and sub-Saharan Africa as a region has the highest percentage of urban population without piped water access, a figure that continues to grow. Accra, Ghana, implements a rationing system to distribute limited piped water resources within the city, and privately-vended sachet water–sealed single-use plastic sleeves–has filled an important gap in urban drinking water security. This study utilizes household survey data from 2,814 Ghanaian women to analyze the sociodemographic characteristics of those who resort to sachet water as their primary drinking water source. In multilevel analysis, sachet use is statistically significantly associated with lower overall self-reported health, younger age, and living in a lower-class enumeration area. Sachet use is marginally associated with more days of neighborhood water rationing, and significantly associated with the proportion of vegetated land cover. Cross-level interactions between rationing and proxies for poverty are not associated with sachet consumption after adjusting for individual-level sociodemographic, socioeconomic, health, and environmental factors. These findings are generally consistent with two other recent analyses of sachet water in Accra and may indicate a recent transition of sachet consumption from higher to lower socioeconomic classes. Overall, the allure of sachet water displays substantial heterogeneity in Accra and will be an important consideration in planning for future drinking water demand throughout West Africa.

Editor: Lawrence Kazembe, Chancellor College, University of Malawi, Malawi

Funding: This research was funded in part by grant number R01 HD054906 from the Eunice Kennedy Shriver National Institute of Child Health and Human Development ("Health, Poverty and Place in Accra, Ghana," John R. Weeks, Project Director/Principal Investigator). The content is solely the responsibility of the authors and does not necessarily represent the official views of the National Institute of Child Health and Human Development or the National Institutes of Health. Additional funding was provided by Hewlett/PRB ("Reproductive and Overall Health Outcomes and Their Economic Consequences for Households in Accra, Ghana," Allan G. Hill, Project Director/Principal Investigator). The 2003 Women's Health Study of Accra was funded by the World Health Organization, the United States Agency for International Development, and the Fulbright New Century Scholars Award (Allan G. Hill, Principal Investigator). The funders had no role in study design, data collection and analysis, decision to publish, or preparation of the manuscript.

Competing Interests: The authors have declared that no competing interests exist.

* E-mail: stoler@miami.edu

Introduction

Population growth in the developing world continues to put a strain on drinking water supplies, even amid declining fertility. As of 2010, approximately 884 million people–over a third of whom live in sub-Saharan Africa–still did not have access to an improved drinking water source [1]. Sub-Saharan Africa is the only region not on track to meet the Millennium Development Goal target of halving the proportion of the population without sustainable access to safe drinking water [1]; the percentage of individuals with access to piped water in the dwelling, yard, or plot stagnated from 1990–2010 and fell in urban areas as urban populations grew by over 30% [2,3]. Despite international efforts to extend access, morbidity and mortality attributable to inadequate water and sanitation remain high, particularly for children under five [4]. In Ghana, rapid urbanization continues to erode the government's ability to provide municipal water to its urban centers. The percentage of the urban population with access to an improved water source increased from 84% in 1990 to 90% in 2008, yet the percentage with access to piped water decreased steadily from 41% to 30% [1].

In Accra, Ghana's coastal capital, drinking water shortages are not driven by lack of surface or ground water *per se*, but are historically attributable to poor governance and improper water resource management [5]. Population growth and urban water mismanagement have resulted in water demand in Accra that far exceeds the water production capabilities of the two local water treatment plants. Ghana Urban Water Ltd. (GUWL; a subsidiary of the Ghana Water Company Ltd.) has instituted a rationing program for water distribution within city limits [6], and by one estimate 75% of Accra lacks 24-hour water access while another 10% has no access at all [7]. Municipal water rationing varies both geographically and socioeconomically by neighborhood in Accra [8], and these service gaps have been linked to the recent ubiquity of packaged "sachet water" sold in sealed 500 ml plastic sleeves. Over the last five years, sachet water has become an important primary drinking water source for the urban poor, and may even confer a health benefit when sachets replace the consumption of improperly stored water in the home [8]. As piped water access in Accra continues to decline, it remains unclear whether sachet water is strictly a phenomenon of the poor, or if sachets are being consumed by a more socioeconomically diverse population and with similar protective health effects.

The most recent Joint Monitoring Report from WHO/UNICEF [3] indicates that urban access to piped drinking water is on the decline throughout sub-Saharan Africa, a trend previously observed over three decades in East Africa [9], but the report attributes these service losses solely to population growth. International development agencies continue to ignore West Africa's transition to sachet water with no mention of it in several recent regional reports [1,3,10–12]. WHO's recent report on household water treatment and storage [13] discusses routine and emergency distribution of flocculent-disinfectant sachets for household use, but there is no discussion of pre-treated sachet water in the form addressed here. Because urban drinking water options have evolved over just the last few years, a fuller understanding of West Africa's urban drinking water patterns is crucial for catching up on missed Millennium Development Goals for sustainable drinking water and sanitation in the region.

This study utilizes survey data from the Women's Health Study of Accra (WHSA) to analyze the socioeconomic and geographic patterns associated with sachet water consumption and the effect of water rationing on sachet consumption. We present data on primary drinking water sources and socioeconomic characteristics for a sample of 2,814 women in Accra. Because GUWL water rationing is managed by local water districts, we expect that a household's drinking water options are influenced by the degree of neighborhood rationing after accounting for individual differences. We specifically test three hypotheses: (1) that residents of lower socioeconomic status are more likely to consume sachet water as their primary water source, (2) that residents enduring a greater number of days of water rationing in the neighborhood are more likely to consume sachets, and (3) that the interaction of higher rationing and lower socioeconomic status will produce the highest rates of sachet use. In light of sachet water's massive appeal throughout West Africa, we comment on the sustainability of sachets as an urban drinking water solution, as well as how the phenomenon might better inform water provision projects in the region.

Methods

Study Population

Accra is typical of many developing urban areas in West Africa both for its progress in health care delivery and its overall epidemiologic transition. The Accra metropolitan area, with its southern border on the Gulf of Guinea coast, extends about 11 km north just beyond the University of Ghana at Legon, and is roughly 20 km east to west. The metropolitan area contained 1.66 million people and 373,540 households according to the March 2000 Accra census, and is approaching 2.4 million people as of the 2010 census, while still growing at 3% annually [2]. Local governance of this population rests with the Accra Metropolitan Assembly, and logistical coordination with this single administrative body facilitated implementation of the study instrument.

Study Instrument

The WHSA is a community-based longitudinal population study conducted in Accra by the Harvard School of Public Health, the University of Ghana at Legon, and San Diego State University. The first round of the WHSA was conducted in 2003, with a follow-up survey in 2008–2009. The survey provides a comprehensive snapshot of health, disease, and related risk factors in each study year among a representative sample of women ≥18 years of age in the Accra metropolitan area [14,15]. Participants were originally selected through a two-stage cluster probability sample stratified by socioeconomic status (SES) quartiles, as determined by the 2000 census, and by four age groups. The sample was drawn from a population-weighted sample of 200 of Accra's 1,731 enumeration areas (EAs), and study women were selected according to the probability of being in one of 16 cells formed by the stratification of SES and age group [14]. To improve our understanding of neighborhood context, EAs are aggregated into 108 neighborhood units called *field modified vernacular neighborhoods* (FMVN) [16], which are field-validated configurations of previously-defined *vernacular neighborhoods* [17]. The study sample and spatial hierarchy are depicted in Figure 1.

The study interviewed 3,172 women in 2003, but due to significant challenges tracking women between surveys (primarily the lack of a formal address system to trace those who moved, and

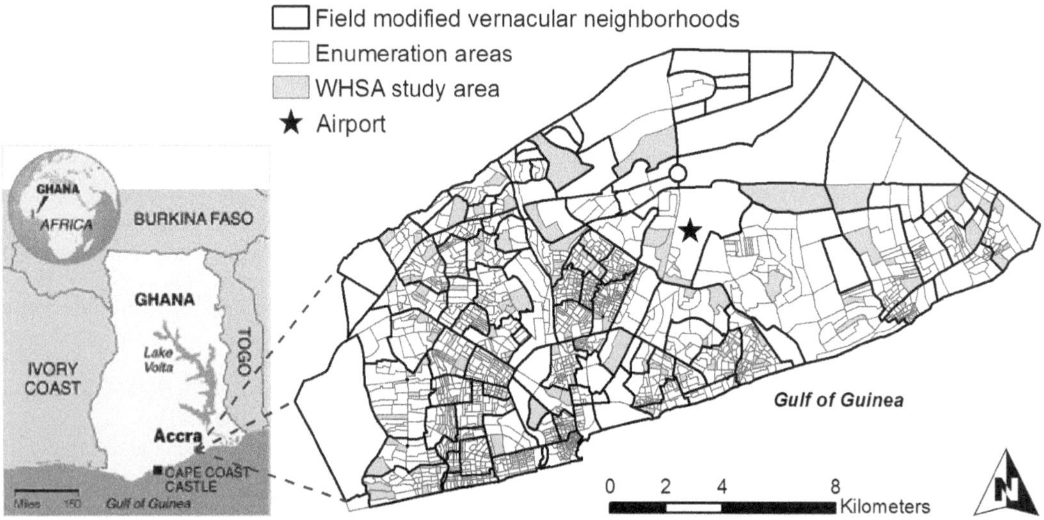

Figure 1. Digital layer of the study site in Accra, Ghana, depicting WHSA-II study sample with enumeration area (EA) and field modified vernacular neighborhood (FMVN) levels of analysis.

also Institutional Review Board prohibition of collecting original household location data via Global Positioning System receivers in 2003), 39% of women were lost to follow-up in 2009, in addition to 5% that were deceased. The 2008–2009 round of the WHSA (henceforth referred to as WHSA-II) interviewed 2,814 women, 995 of whom were replacements–matched by age and enumeration area–for those lost to follow-up. This paper explores the recent sachet water phenomenon, which was not an important factor in 2003 when 97.5% of WHSA respondents reported piped water access (either in or outside the home) as their primary drinking water source. In order to analyze the new dynamics of sachet water using the broadest sample possible, we restrict ourselves to a cross-sectional application using only the WHSA-II data set with an explicit acknowledgment of limitations on interpreting causality from the results.

All WHSA-II participants provided written informed consent; literate women provided their personal signature, while illiterate women provided an ink fingerprint in lieu of a signature. Institutional Review Board approval for all consent procedures, data collection, and analysis was granted by the respective boards of Harvard University, University of Ghana-Legon, and San Diego State University (SDSU).

Statistical Analyses

To test hypothesis 1, we use X^2 and F tests to compare socioeconomic and demographic characteristics between sachet drinkers and those using all other drinking water sources. To test hypothesis 2, we implement a series of iterative multilevel logistic regression models to separate compositional and contextual effects associated with sachet consumption as a primary drinking water source, with neighborhood factors treated as the exposures of interest. We use exploratory forward and backward stepwise regression models to advise the introduction of independent measures in the multilevel model-building process. The full model parsimoniously maximizes higher-level covariance parameters while minimizing the model fit statistic (−2 restricted log pseudo-likelihood). To test hypothesis 3, we introduce cross-level interactions with rationing into the full model, as well as individual-level interactions for other significant measures. We implement a random intercept model using the GLIMMIX procedure in SAS 9.2 with parameters fitted to the FMVN, EA, and/or woman as described below.

The theoretical underpinnings of measuring health and poverty through socioeconomic indicators have long been established in the social epidemiology and international development literature [18–21]. In this analysis, the individual effects modeled include respondent demographics such as age, ethnicity, and education; household characteristics such as dwelling type, and access to toilets and waste disposal services; and an individual-level measure of water pipe density within 500 m of the household (generated from a kernel density surface estimation of neighborhood water pipe penetration). We control for SES quartiles at the EA level, then introduce neighborhood, i.e. FMVN-level, factors that have been previously linked to adverse health outcomes: infrastructure conditions such as pipe density and days of water rationing [8], the proportion of vegetated land cover as a proxy for socioeconomic status [22], and a slum index and housing quality index constructed from 2000 census data to infer socioeconomic status at a finer level [16,22]. The proportion of vegetated land cover was summarized for each FMVN and EA from a classification of a high spatial resolution (2.4 m) QuickBird image captured in 2010. GUWL rationing metrics were recorded in July 2009 and are drawn from schedules of days per week of water service as implemented by local water districts in 2009. Rationing data are managed by GUWL as a GIS point layer and mapped as a surface of Theissen polygons; mean rationing values are summarized at the FMVN scale by an area-weighted algorithm. All FMVN-level measures were also calculated (or disaggregated) at the EA-level for comparison purposes, and the multilevel models were computed as 3-level (FMVN, EA, individual) and 2-level (EA, individual) models.

Results

Table 1 summarizes the individual characteristics of women from the WHSA-II interviews. Among 2,814 women in the study, 6.8% named sachet water as their primary drinking water source. The demographic profile of these women, compared with non-sachet users, was about 3 years younger on average, disproportionately more likely to live in an EA that ranks in the lowest SES quartile, and more likely to be any ethnicity other than Ga, particularly ethnic groups such as Mole-Dagbani that are grouped into the *other* category. There were no statistically significant differences observed between sachet drinkers and their counterparts in education, though sachet drinkers were significantly more likely to report lower levels of overall health. A profile of sachet consumers as younger, lower-income ethnic minorities begins to emerge, but additional socioeconomic details in Table 1 offer conflicting results.

Women relying on sachet water were less likely to be connected to a sewage system for liquid waste disposal (11.6% vs. 16.7%), an observation that is consistent with the "low income" profile, but was not statistically significant. Conversely, sachet drinkers were significantly more likely to report using a collection service for solid waste disposal (43.7%) than those drinking from other water sources (34.1%), yet access to a waste collection service is generally associated with higher-income populations in Accra. Solid waste disposal was the sole household socioeconomic variable that yielded statistically significant differences between sachet drinkers and the rest of the study population ($p = 0.009$). There were other non-significant hints of better amenities among sachet drinkers in Table 1: sachet drinkers were more likely than their counterparts to use their own bathroom for bathing (39.5% vs. 33.3%), and rated higher on a wealth score of durable goods ownership (mean 0.11 vs. -.01), a difference that translates to the 62nd vs. 59th percentile. There were no significant differences in dwelling type, rooms per dwelling, type of bathing facility, and toilet access. All of these variables except toilet access have previously differentiated sachet drinkers as being of lower means than other women in Accra [8]. The density of GUWL pipe infrastructure was calculated within varying buffers from each household, and sachet users had lower pipe density values within 500 m than non-sachet users (819 vs. 920 mm/km^2, $p = 0.03$).

Table 2 summarizes the overall mean FMVN-level characteristics and Pearson's correlations for 71 neighborhoods that encompass the WHSA-II study population, and selected characteristics are mapped in Figure 2. The neighborhood mean GUWL pipe density was 1,074 mm/km^2, a figure higher than the household means because major pipelines tend to co-locate with major roads that are not necessarily residential centers; the kernel density surface of water pipe density is depicted in Figure 2A. Each FMVN averaged about 5 days per week of running water service due to the GUWL rationing regime as shown in Figure 2B. Neighborhood pipe density was weakly correlated with days of running water, implying that water service was truly more dependent on a local water district's rationing policy than on local infrastructure capacity. The raw neighborhood average for sachets as the primary drinking water source was 7.5%; to account

Table 1. Women's individual characteristics from the WHSA-II.

	Sachet Water		Other Water Source	
Characteristic	**Freq.**	**% or mean (95% CI)**	**Freq.**	**% or mean (95% CI)**
Individual characteristics (%, n = 2,814)	190	6.8	2,624	93.2
Age (years) *		43.5 (41.1–46.0)		46.5 (45.8–47.2)
Major Ethnic Group (%) ***				
Akan	65	34.2	850	32.4
Ewe	33	17.4	358	13.6
Ga	53	27.9	1,085	41.3
Other	39	20.5	331	12.6
Education (%) ~				
None, other, religious	43	22.8	568	21.8
Primary	29	15.3	306	11.7
Middle	57	30.2	1,043	40.0
Secondary	36	19.0	435	16.7
Higher	24	12.7	255	9.8
Self-reported overall health (%) ***				
Excellent	5	2.6	319	12.3
Very good	33	17.5	650	25.0
Good	108	57.1	1,238	47.6
Fair or poor	43	22.8	396	15.2
Socioeconomic status quartile of EA ***				
Lower class	78	41.1	707	26.9
Lower middle class	36	18.9	605	23.1
Upper middle class	36	18.9	681	26.0
Higher class	40	21.1	631	24.0
Type of dwelling (%)				
House, semi-detached, flat	63	33.2	820	31.3
Compound house	125	65.8	1772	67.6
Hut, tent, kiosk, business, other	2	1.1	28	1.1
Number of rooms in dwelling		2.5 (2.3–2.8)		2.5 (2.4–2.6)
Solid waste disposal (%) **				
Collection service	83	43.7	894	34.1
Public dump	88	46.3	1,516	57.8
Burnt, buried, dumped elsewhere, other	19	10.0	214	8.2
Liquid waste disposal (%) ~				
Sewage system	22	11.6	438	16.7
Thrown in street, gutter, compound, other	168	88.4	2,186	83.3
Type of toilet access (%)				
WC or another house	73	38.4	978	37.3
KVIP or public toilet	88	46.3	1268	48.3
Pit latrine, bucket/pan, other, none	29	15.3	378	14.4
Type of bathing facility (%)				
Own bathroom	75	39.5	873	33.3
Shared with other households	109	57.4	1,635	62.3
Cubicle, open space, other	6	3.2	116	4.4
Wealth score ~		0.11 (−0.02–0.24)		−0.01 (−0.05–0.03)
Pipe density (mm/km^2) within 500 m *		819 (732–906)		920 (892–947)

~$p<0.10$;
*$p<0.05$;
**$p<0.01$;
***$p<0.001$.
Note: p-values for categorical measures are from X^2 test; p-values for continuous measure are from Welch F test of equality of means to account for variance heterogeneity.

Table 2. Characteristics and Pearson's correlations of 71 Field Modified Vernacular Neighborhoods (FMVN) comprising the WHSA-II study population.

Characteristic	Mean	SE	Range	Pearson's Correlation						
				WPD	H2O	SAC1	SAC2	VEG	SI	HQI
GUWL water pipe density (mm/km^2) (WPD)	1074	73	207–3018	1.00						
Days per week of running water (no rationing) (H2O)	4.92	0.20	1–7	0.16	1.00					
Sachets as primary water source (raw %) (SAC1)	7.53	1.81	0–100	0.16	0.09	1.00				
Sachets as primary water source (smoothed %) (SAC2)	6.59	0.75	0.62–35.57	0.09	−0.04	0.73*	1.00			
Vegetated land cover in 2010 (%) (VEG)	19.86	2.00	0–83.38	−0.14	0.05	0.12	0.12	1.00		
Slum index (SI)	2.00	0.04	1.13–2.57	−0.16	−0.07	−0.04	0.00	−0.65*	1.00	
Housing quality index (HQI)	2.44	0.07	1.58–3.51	0.06	0.01	0.04	0.02	0.79*	−0.92*	1.00

*$p<0.01$ (2-tailed).

for small sample denominators in several neighborhoods, we applied an empirical Bayes smoothing algorithm and mapped the smoothed sachet rates in Figure 2C (smoothed mean = 6.6%). As shown in Table 2, neither neighborhood-level sachet metric was correlated with any of the other neighborhood measures. Three additional FMVN-level measures of socioeconomic variation were considered: the proportion of vegetated land cover (mean 19.86%, SE 2.00), a slum index (2.00, 0.04), and a housing quality index (2.44, 0.07). While these socioeconomic proxies were statistically significantly correlated with each other, none yielded significant correlations with the water-related neighborhood-level measures in Table 2.

Table 3 contains the empty and full 2-level multilevel models. The empty model is the simplest form of the multilevel model (no covariates), and we estimated this form as a baseline for assessing the change in higher-level variance as we added independent measures. We began by fitting 3-level models with the FMVN atop the spatial hierarchy, but the introduction of interaction terms resulted in unstable models with an unexpected inflation of variance. This lack of fit may indicate that the FMVN scale is not optimal for assessing the role of rationing, despite a significant relationship between rationing and sachet use in these models (not shown). In lieu of "overfitting the data," the model was simplified to two levels with neighborhood parameters fitted at the EA level.

The total variance at the EA level was 2.040 in the empty model (Model 1). Model building began with the EA-level rationing and SES measures, and we proceeded to iteratively test combinations of covariates and first-order interaction terms until arriving at the most parsimonious model shown in Table 3. After adjusting for covariates, the full model without interactions (Model 2) minimized EA variance at 1.691, a decrease of 17.1%. Substantial variance remains, and this result suggests that living in a particular neighborhood strongly influenced the likelihood that a woman relied on sachet drinking water.

Table 3 gives the restricted pseudo-likelihood estimates for determinants of sachet use. We observed no support for our second hypothesis that neighborhood rationing was driving sachet consumption; the rationing variable was not statistically significant, though it did contribute to minimizing higher-level variance in the model. There is, as seen in Table 1, mixed support for our first hypothesis. At the individual level, lower self-reported overall health was strongly associated with sachet consumption: we observed a trend of increasing odds of sachet use for each category of overall health, with women reporting fair or poor health being significantly more likely to have used sachets than women reporting excellent health after adjusting for covariates ($p<0.001$). Age was also statistically significantly associated with sachet use, and the negative coefficient suggests that women were slightly less likely to use sachet water as their primary water source for each additional year of age, after controlling for covariates ($p<0.01$). The EA-level SES measure was also statistically significant; women in the lower class SES quartile were more likely ($p<0.01$) to use sachets than women in the upper class quartile, and we observed a similar trend of increasing odds of sachet use for each progressively lower SES quartile as seen with the *overall health* measure. Women without access to a sewer connection were also more likely to rely on sachets, but this difference was not statistically significant. These results affirm the profile of the "younger, less-well-off" sachet consumer, except that the wealth score was again positively associated with sachet use: women were more likely to use sachets ($p<0.01$) for each unit-increase in wealth score, which translated to a 20-percentile change when the score is centered on zero. This implies a massive relative increase in wealth score to produce a modest increase in the odds of sachet consumption, but it does indicate that even after controlling for covariates–particularly SES quartile–relative wealth still did impact the use of sachet water. Evidence of a possible interaction between SES and wealth appears in Table 4 where we stratified the mean wealth score for sachet users and non-users by SES quartile. Only the lower class quartile yielded a statistically significant difference in wealth score, as sachet users scored higher than expected (−0.15 vs. −0.58)–a difference that translated to the 54th vs. 35th percentile–and higher than sachet users in the lower-middle quartile (−0.34).

Models 3–5 in Table 3 summarize the beta coefficients and standard errors for three additional models that introduce interaction terms to the full model. Model 3 contains cross-level interactions between rationing and proportion of vegetated land cover, and between rationing and SES, but these relationships were not statistically significant. We added cross-level interactions in Model 4 to test if rationing and SES interacted with overall health and wealth score. Only the interaction between lowest-quartile SES and wealth score–the relationship depicted in Table 4–was statistically significantly associated with higher sachet use ($p<0.01$). While most of the interaction terms in Model 4 were not statistically significant, their presence in the model rendered the beta for *days of rationing* marginally significant ($p<0.10$) and *proportion of vegetated land cover* statistically significant ($p<0.01$), thus lending initial, though weak, support to hypothesis 2 that increased rationing was associated with increased sachet use. At the

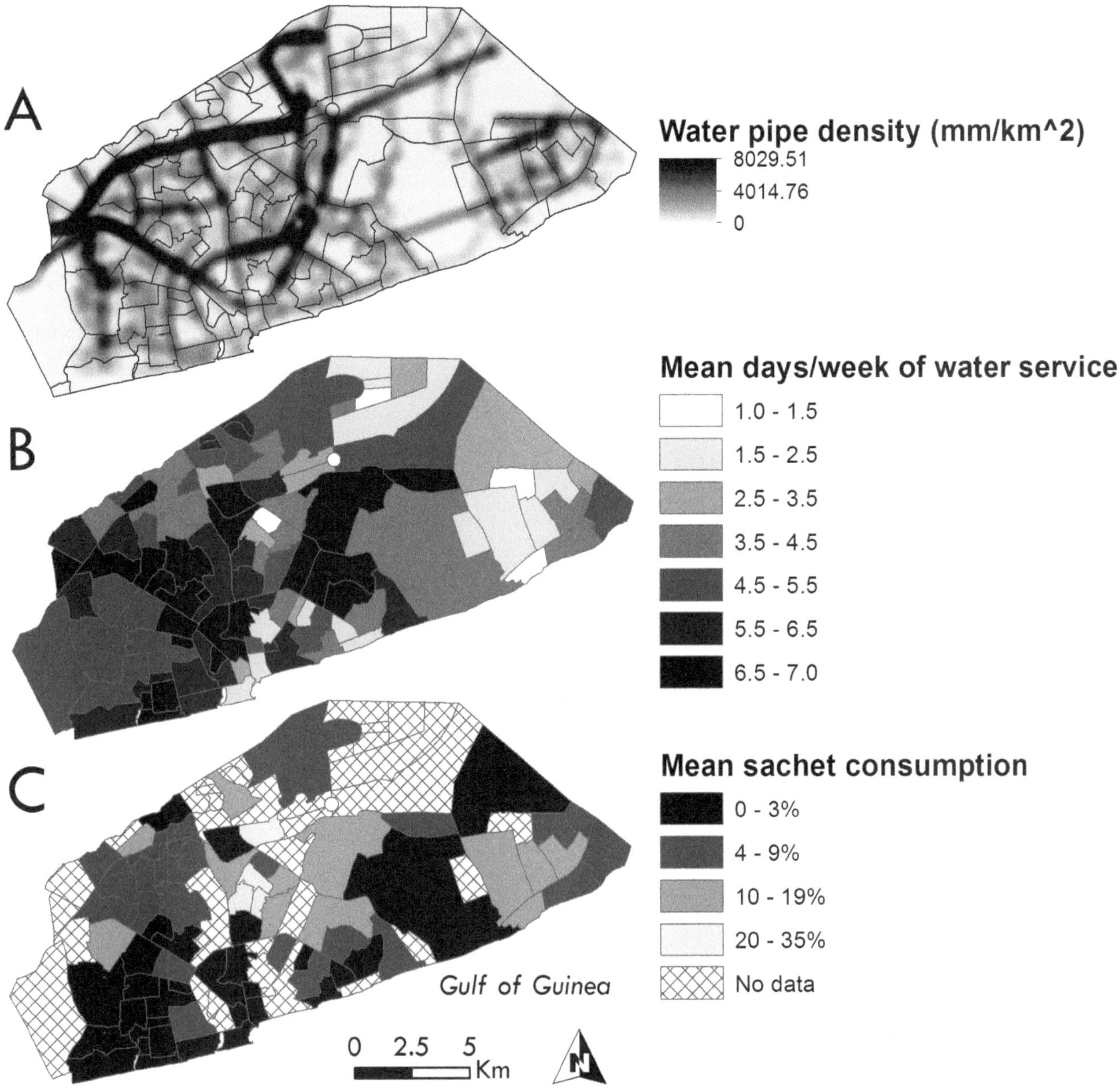

Figure 2. Accra metropolitan area neighborhoods (FMVN) depicting (A) mean density of 2009 GUWL water pipe infrastructure (mm/km2) using a 500 m kernel, (B) mean days per week of running water service according to the 2009 GUWL rationing regime, and (C) smoothed mean percentage of women reporting sachet water as the primary drinking water source for 71 FMVN sampled in the 2008–2009 WHSA-II.

individual level, living in the lower class SES quartile, worse overall health, and younger age were all associated with higher sachet use just as in Model 2. There was no strong evidence in Model 4 for hypothesis 3, which posits that the interaction between higher rationing and lower SES quartile–or any proxy for lower SES such as worse overall health or lower wealth–was associated with higher sachet use.

Model 5 contains additional individual-level interactions between overall health and wealth score, and between overall health and age. The interaction between overall health and age was statistically significant: for each additional year of age, women with fair or poor health ($p<0.01$), or very good health ($p<0.10$), were less likely to use sachets than women with excellent health, and the beta coefficients revealed a stronger effect for age than in Models 2–4. Just as in Models 3 and 4, worse overall health, living in a lower class EA, higher rationing, and a higher proportion of vegetated land cover were predictive of sachet use.

Models 3–5 present a more nuanced picture of the predictors of sachet use than Model 2, though the general conclusion is similar: younger, poorer women (or women living in poorer EAs) in neighborhoods experiencing higher rationing were most likely to rely on sachet water. The sole factor with a relationship contrary to hypothesis 1 was the significance of vegetated land cover at the EA level: a higher percentage of vegetation–which was statistically significantly associated with higher-SES quartiles in Table 5–was also predictive of sachet use. Previous research has shown a higher

Table 3. Beta coefficients (SE) for random intercept models assessing the association between rationing and sachet use among Accra women.

Characteristic	Model 1		Model 2		Model 3		Model 4		Model 5	
	Empty (no covariates)		Full model of EA- and individual-level factors		Full model with level-2 interactions		Full model with level-2 and cross-level interactions		Full model with level-1 interactions	
Intercept	−3.077***	(0.149)	−5.133***	(0.825)	−5.830***	(0.942)	−6.061***	(1.138)	−6.678***	(1.807)
Interactions: Level-2 (EA)										
Rationing × Vegetation					−1.420	(0.875)	−1.648~	(0.889)		
Rationing × Upper middle class SES					−0.404~	(0.241)	−0.389	(0.243)		
Rationing × Lower middle class SES					−0.072	(0.230)	−0.047	(0.238)		
Rationing × Lower class SES					−0.176	(0.230)	−0.109	(0.236)		
Interactions: Cross-level										
Rationing × OH Very good							−0.106	(0.216)		
Rationing × OH Good							−0.150	(0.205)		
Rationing × OH Fair or poor							−0.151	(0.217)		
Rationing × wealth score							0.084~	(0.050)		
Upper middle class SES × wealth score							0.156	(0.218)		
Lower middle class SES × wealth score							−0.405	(0.309)		
Lower class SES × wealth score							0.627**	(0.231)		
Interactions: Level-1 (Individual)										
Wealth score × OH Very good									0.213	(0.634)
Wealth score × OH Good									0.612	(0.606)
Wealth score × OH Fair or poor									0.896	(0.628)
Age × OH Very good									−0.035~	(0.019)
Age × OH Good									−0.007	(0.007)
Age × OH Fair or poor									−0.029**	(0.011)
Level-1 (Individual) variables										
Self-reported overall health										
Excellent†										
Very good			1.264*	(0.502)	1.259*	(0.502)	1.593~	(0.822)	2.955~	(1.745)
Good			1.853***	(0.483)	1.855***	(0.484)	2.312**	(0.796)	2.610	(1.646)
Fair or poor			2.151***	(0.516)	2.139***	(0.517)	2.633**	(0.833)	3.988*	(1.711)
Solid waste disposal										
Burnt, buried, dumped elsewhere, other†										
Public dump			−0.355	(0.303)	−0.326	(0.304)	−0.391	(0.305)	−0.428	(0.307)

Table 3. Cont.

	Model 1		Model 2		Model 3		Model 4		Model 5	
Characteristic	**Empty (no covariates)**		**Full model of EA- and individual-level factors**		**Full model with level-2 interactions**		**Full model with level-2 and cross-level interactions**		**Full model with level-1 interactions**	
Collection service			0.104	(0.323)	0.109	(0.325)	0.078	(0.327)	0.038	(0.327)
Liquid waste disposal										
Sewage system†										
Thrown in street, gutter, compound, other			0.592~	(0.333)	0.578~	(0.334)	0.497	(0.344)	0.484	(0.354)
Major ethnic group										
Other										
Ewe			0.024	(0.300)	0.015	(0.310)	0.004	(0.304)	−0.019	(0.309)
Ga			−0.255	(0.290)	−0.275	(0.291)	−0.312	(0.293)	−0.292	(0.296)
Akan			0.012	(0.269)	−0.007	(0.270)	−0.011	(0.273)	0.004	(0.275)
Age (years)			−0.016**	(0.005)	−0.016**	(0.005)	−0.015**	(0.005)	-	(-)
Wealth score			0.380**	(0.125)	0.392**	(0.126)	-	(-)	-	(-)
Level-2 (EA) variables										
SES quartile										
Upper class†										
Upper middle class			−0.057	(0.449)	0.827	(0.645)	0.718	(0.659)	0.720	(0.664)
Lower middle class			0.523	(0.482)	0.674	(0.695)	0.431	(0.712)	0.381	(0.717)
Lower class			1.354**	(0.489)	1.776*	(0.712)	1.703*	(0.724)	1.704*	(0.728)
% Vegetated land cover			2.069	(1.290)	4.881*	(1.964)	5.452**	(1.988)	5.464**	(2.010)
Days of rationing			0.087	(0.069)	0.395~	(0.233)	0.514~	(0.311)	0.531~	(0.312)
Random Effects										
Level-2 variance (EA)	2.040	(0.368)	1.691	(0.349)	1.636	(0.354)	1.668	(0.361)	1.711	(0.368)
Δ (%) in level-2 variance	-		−17.1%		−19.8%		−18.2%		−16.1%	
−2 res log pseudo-likelihood	15,945.99		15,791.81		15,905.96		16,083.04		16,285.13	
Generalized chi-square	1,427.79		1,429.90		1,483.29		1,504.83		1,527.98	

†Reference Category;
~ $p<0.10$;
*$p<0.05$;
**$p<0.01$;
***$p<0.001$.

Table 4. Differences in wealth score between sachet users and non-users stratified by SES quartile.

	Sachet Water		Other Water Source	
Characteristic	**Freq.**	**mean (95% CI)**	**Freq.**	**mean (95% CI)**
Individual characteristics (%, n = 2,814)	190	6.8%	2,624	93.2%
Wealth score ~		0.11 (−0.02–0.24)		−0.01 (−0.05–0.03)
By SES quartile				
Lower class *	78	−0.15 (−0.31–0.02)	707	−0.58 (−0.63– −0.53)
Lower middle class	36	−0.34 (−0.56– −0.11)	605	−0.27 (−0.34– −.021)
Upper middle class	36	0.35 (0.04–0.66)	681	0.16 (0.08–0.23)
Higher class	40	0.79 (0.47–1.11)	631	0.71 (0.62–0.79)

~ $p<0.10$;
*$p<0.001$.
Note: *p*-values are from Welch *F* test of equality of means to account for variance heterogeneity.

proportion of vegetated land cover to be associated with higher-status areas, not poverty [22], so this relationship may be an artifact of the transition in sachet consumption from higher to lower socioeconomic classes. The modeling of interaction terms in Table 3 lent weak support for hypothesis 2, as days of rationing persists as a marginally significant neighborhood-level influence on sachet use. Models 3 and 4 explain a slightly greater proportion of EA-level variance than Model 2, but the use of individual interaction terms in Model 5 does not improve overall results. There is no strong support for hypothesis 3; there are no significant interactions between rationing and any SES indicator.

Discussion and Conclusion

Over the last decade sachet water has risen from relative obscurity to become an important source of drinking water throughout West Africa. Previous research has explored sachet water in select neighborhoods of Accra [8,23] but this is the first study to analyze sachet water consumption in a West African metropolis using a socioeconomic and geographic cross-sectional approach. This study uses data collected in 2008–2009 to examine intra-urban variability of drinking water selection in Accra, and factors that influence these choices. We observe evidence of our first hypothesis, as sachet consumers tend to be younger, of poorer overall health, and of lower socioeconomic means, and yet those living in the economically poorest areas may be slightly better-off than their immediate neighbors. We also observe limited support for our second hypothesis that water rationing at the neighborhood scale positively influences sachet use, though rationing may be less important than individual-level factors. A smaller neighborhood unit was more appropriate for modeling sachet use, though there is still a considerable unexplained neighborhood effect. These neighborhood-level findings are consistent with recent reports from Accra [8]. The analysis does not support our third hypothesis regarding an interaction between rationing and lower socioeconomic status.

Two other recent population-based surveys, Measure DHS's Ghana Demographic and Health Survey (DHS) and the Harvard/SDSU Housing and Welfare Study of Accra (HAWS), each report higher rates of sachet consumption than those reported in the WHSA-II [8,23]. The last two Ghana DHS indicate that between 2003 and 2008 the percentage of urban households using sachet water as the primary water source increased from 6% to 37% nationally [24]. In Greater Accra, 87% of sachet-using households were in the top wealth quintile in the 2003 survey; by 2008 this rate had fallen to 71% and sachet use started trickling down into the middle and lower quintiles, but with scant evidence that sachets were a phenomenon of poverty [23,24]. The WHSA-II data were collected between October 2008 and March 2009 and yield only 10% (78/785) of lowest-SES quartile participants relying on sachets; this survey seems to have been administered amidst a widespread transition to sachet water by lower-income populations. The HAWS data, collected between September 2009 and March 2010 exclusively in Accra slum neighborhoods, report 50% of households depending on sachet water with lowest-income residents as the most likely consumers [8].

Table 5. Enumeration area (EA) differences in rationing, proportion of vegetated land cover, and sachet use stratified by SES quartile.

		Days of Rationing	Vegetation (%)*	Sachet Use (%)
SES Quartile of EA	***N***	**mean (SE)**	**mean (SE)**	**mean (SE)**
Lower class	48	2.18 (0.26)	5.03 (0.91)	12.06 (3.00)
Lower middle class	50	2.12 (0.25)	8.03 (1.27)	6.69 (1.95)
Upper middle class	53	1.75 (0.20)	15.17 (1.85)	5.62 (2.13)
Higher class	44	1.84 (0.23)	25.06 (2.42)	8.53 (2.73)
Total	195	1.97 (0.12)	13.24 (1.01)	8.14 (1.23)

*$p<0.001$ from Welch *F* test of equality of means to account for variance heterogeneity.

Both 2008 and 2010 were especially bad years for water service delivery as power outages and construction projects at both of Accra's water treatment plants led to multiple-week service interruptions in many neighborhoods. Sachet water was already a fairly ubiquitous product in Accra in 2008. It is plausible that sachet water's higher unit price prevented it from becoming a *primary* drinking water source for many low-income communities until severe piped water shortages finally made sachets a necessity, even if temporarily. This study's finding that sachet users in the lowest SES quartile score higher on a wealth index than sachet users in the lower-middle quartile underscores the reality that, in lieu of abject poverty, sachets may still be a discretionary, but increasingly attractive, choice to younger, poorer urban residents.

There are also several geographical discrepancies between the sachet patterns observed in the WHSA-II and those seen in the 2008 DHS and HAWS. In particular, the WHSA-II reports no sachet use among 270 women interviewed in the coastal Ga neighborhoods of Gbegbeyise, Chorkor, Korle Gonno, and Jamestown in western Accra. The DHS and HAWS data sets report significant sachet reliance in these and other neighborhoods where WHSA-II reports no sachet use. These coastal Ga neighborhoods in particular are known to have built much of their own water infrastructure (in a mix of legal and pirated networks), and the lack of maintenance by GUWL may increase the frequency of service disruptions. Given the geographic and temporal heterogeneity of water service in Accra, it is possible that some neighborhoods were interviewed in atypically water-abundant or water-scare periods, thus potentially biasing responses during any of these three surveys. It is also possible that there may have been some inadvertent bias in how the WHSA-II question about primary drinking source was framed to participants, despite all three surveys using the same DHS format. This mismatch between surveys may account for the lack of significant effects among neighborhood-level measures in this study, particularly days of rationing as observed in 2009–2010 [8]. More broadly speaking, it also highlights the difficulty of accuracy assessment for primary data collected in a developing urban setting. These survey discrepancies may soon be reconciled by the forthcoming release of 2010 Ghana Census results, as the 2010 questionnaire split the traditional household water question into separate items for drinking water sources and all other household water sources. The 2010 census may provide the most detailed picture yet of Ghana's evolving drinking water trends.

This analysis is limited by the cross-sectional nature of the WHSA-II data. Despite compiling a broad array of women's health outcomes, the WHSA-II did not measure diarrhea prevalence or other outcomes that might specifically be attributable to drinking water quality, thus it is difficult to assess the strong association between sachet use and overall health. Higher self-reported overall health has previously been positively associated with higher socioeconomic amenities in Accra [16,25,26], and ill-health has long been linked to chronic poverty [27], so the overall health measure can reasonably be interpreted as a SES proxy. Future research should directly target the relationship between sachet consumption and health outcomes suggested in the HAWS project [8].

Urban drinking water patterns in the developing world remain understudied, as urban areas are generally presumed to have sufficient piped water coverage. Accra is emblematic of many developing cities where this assumption is no longer valid [3,9,28–30]. Piped water access is lagging in urban sub-Saharan Africa [3], and residents are increasingly turning to privately vended water such as sachets. Problems with the unsustainable amount of plastic waste generated by sachet water consumption have been reported elsewhere [8,23,31], and have resulted in increased local media coverage of municipal water delivery issues faced by Ghana's Ministry of Water Resources, Works and Housing. Sachet water continues to spread throughout West Africa, and many cities are likely experiencing similar transitions in drinking water patterns without a complete understanding of potential health effects or an appreciation for the environmental sanitation consequences of all the plastic waste. Global safe water and sanitation goals face many challenges [32], but in Ghana improvements are primarily constrained by financial resources; the capital investments required for an adequate water supply and sanitation infrastructure are estimated at $1.3 billion for the rehabilitation and expansion of urban water infrastructure alone [33]. Local governments and non-governmental organizations interested in alternative sustainable water provision would be well-served to understand the demographic appeal of sachet water, as well as the geographic and economic realities that turn citizens into sachet customers.

Acknowledgments

The authors thank David Nunoo and Michael Nyoagbe at Ghana Urban Water Ltd., Head Office Operations, Accra, for providing important GIS data management and assistance.

Author Contributions

Conceived and designed the experiments: JS JRW. Performed the experiments: JS. Analyzed the data: JS. Contributed reagents/materials/analysis tools: JS JRW RAO. Wrote the paper: JS.

References

1. WHO/UNICEF (2010) Progress on sanitation and drinking water: 2010 update. Geneva: World Health Organization.
2. United Nations (2010) World Urbanization Prospects: The 2009 Revision. New York, NY: United Nations Population Division, Department of Economic and Social Affairs.
3. WHO/UNICEF (2011) Drinking water: Equity, safety and sustainability. Geneva: World Health Organization.
4. Fink G, Günther I, Hill K (2011) The effect of water and sanitation on child health: evidence from the demographic and health surveys 1986–2007. International Journal of Epidemiology 2011: 1–9.
5. Nsiah-Gyabaah K (2001) The looming national dilemma of water crisis in peri-urban areas in Ghana. Second Workshop on Peri-Urban Natural Resources Management Project at the Watershed Level. Kumasi, Ghana: DFID.
6. Van-Rooijen DJ, Spalthoff D, Raschid-Sally L (2008) Domestic water supply in Accra: How physical and social constraints to planning have greater consequences for the poor. Accra, Ghana. 5p.
7. WaterAid (2005) National Water Sector Assessment: Ghana. London, UK. 8 p.
8. Stoler J, Fink G, Weeks JR, Appiah Otoo R, Ampofo JA, et al. (2012) When urban taps run dry: Sachet water consumption and health effects in low income neighborhoods of Accra, Ghana. Health & Place 18: 250–262.
9. Thompson J, Porras IT, Wood E, Tumwine JK, Mujwahuzi MR, et al. (2000) Waiting at the tap: Changes in urban water use in East Africa over three decades. Environment and Urbanization 12: 37–52.
10. International Development Committee (2007) Sanitation and Water: Sixth Report of Session 2006–07, House of Commons. London, UK: The Stationery Office Limited.
11. United Nations (2008) The Millennium Development Goals Report 2008. New York, NY: United Nations.
12. United Nations Development Programme (2006) Human Development Report 2006: Beyond scarcity: Power, poverty and the global water crisis. New York, NY: United Nations.
13. WHO, Clasen TF (2009) Scaling up household water treatment among urban populations. Geneva: WHO. 84p.
14. Hill AG, Darko R, Seffah J, Adanu RMK, Anarfi JK, et al. (2007) Health of urban Ghanaian women as identified by the Women's Health Study of Accra. International Journal of Gynecology & Obstetrics 99: 150–156.

15. WHSA-II Writing Team (2011) Final report on the Women's Health Study of Accra, Wave II. Accra, Ghana: Institute for Statistical, Social and Economic Research, University of Ghana. 144 p.
16. Weeks JR, Getis A, Stow D, Hill AG, Rain D, et al. (2012) Connecting the dots between health, poverty and place in Accra, Ghana. Annals of the Association of American Geographers 102: 932–941.
17. Weeks JR, Getis A, Hill AG, Agyei-Mensah S, Rain D (2010) Neighborhoods and fertility in Accra, Ghana: An AMOEBA-based approach. Annals of the Association of American Geographers 100: 558–578.
18. Krieger N (1992) Overcoming the absence of socioeconomic data in medical records: validation and application of a census-based methodology. American Journal of Public Health 82: 703–710.
19. Krieger N, Williams D, Moss N (1997) Measuring social class in US public health research: concepts, methodologies and guidelines. Annual Review of Public Health 18: 341–378.
20. Lynch J, Kaplan G (2000) Socioeconomic Position. In: Berkman LF, Kawachi I, editors. Social Epidemiology. Oxford: Oxford University Press. pp. 13–35.
21. UNDP-UNEP Poverty and Environment Initiative (2008) Poverty and Environment Indicators. Cambridge: Capability and Sustainability Centre, St. Edmund's College.
22. Weeks JR, Hill AG, Stow DA, Getis A, Fugate D (2007) Can we spot a neighborhood from the ground? Defining neighborhood structure in Accra, Ghana. GeoJournal 69: 9–22.
23. Stoler J, Weeks JR, Fink G (2012) Sachet drinking water in Ghana's Accra-Tema metropolitan area: Past, present, and future. Journal of Water, Sanitation and Hygiene for Development 2: 223–240.
24. Macro International Inc (2011) MEASURE DHS STATcompiler.
25. Arku G, Luginaah I, Mkandawire P, Baiden P, Asiedu AB (2011) Housing and health in three contrasting neighborhoods in Accra, Ghana. Social Science & Medicine 72: 1864–1872.
26. Stoler J, Weeks JR, Getis A, Hill AG (2009) Distance threshold for the effect of urban agriculture on elevated self-reported malaria prevalence in Accra, Ghana. American Journal of Tropical Medicine and Hygiene 80: 547–554.
27. Sverdlik A (2011) Ill-health and poverty: A literature review on health in informal settlements. Environment and Urbanization 23: 123–155.
28. Collignon B, Vézina M (2000) Independent water and sanitation providers in African cities. Full report of a ten-country study. Washington, DC: Water and Sanitation Program.
29. Dominguez Torres C (2012) The future of water in African cities: Why waste water? Urban access to water supply and sanitation in sub-Saharan Africa, background report. Washington DC: World Bank. 35p.
30. Kessides C (2005) The urban transition in sub-Saharan Africa: Implications for economic growth and poverty reduction. Africa Region Working Paper Series No 97. Washington DC: World Bank. 84p.
31. Stoler J (2012) Improved but unsustainable: Accounting for sachet water in post-2015 goals for global safe water. Tropical Medicine & International Health 17: 1506–1508.
32. Moe CL, Rheingans RD (2006) Global challenges in water, sanitation and health. Journal of Water and Health 4: 41–57.
33. AfDB/OECD (2007) African Economic Outlook. Paris: AfDB/OECD. 618 p.

12

Spatial and Seasonal CH_4 Flux in the Littoral Zone of Miyun Reservoir near Beijing: The Effects of Water Level and Its Fluctuation

Meng Yang[1], Xuemeng Geng[1], John Grace[2], Cai Lu[1], Yi Zhu[1], Yan Zhou[1], Guangchun Lei[1]*

1 School of Nature Conservation, Beijing Forestry University, Beijing, China, **2** School of Geosciences, The University of Edinburgh, Edinburgh, United Kingdom

Abstract

Wetlands, and especially their littoral zones, are considered to be CH_4 emissions hotspots. The recent creation of reservoirs has caused a rapid increase in the area of the world's littoral zones. To investigate the effects of water depth and water level fluctuation on CH_4 fluxes, and how these are coupled with vegetation and nutrients, we used static closed chamber and gas chromatography techniques to measure CH_4 fluxes in the littoral zone of a large reservoir near Beijing, China, from November 2011 to October 2012. We found that CH_4 flux decreased significantly along a transect from open water to dry land, from 3.1 mg m^{-2} h^{-1} at the deep water site to approximately 1.3 mg m^{-2} h^{-1} at the shallow water site, and less than 0.01 mg m^{-2} h^{-1} in the non-flooded area. Water level influenced CH_4 flux by affecting soil properties including soil redox potential, soil carbon and nitrogen, and bulk density. The largest emission of all was from the seasonally flooded site after a flooding event (up to 21.1 mg m^{-2} h^{-1}), which may have been caused by vegetation decomposition. Submerged sites had greater emissions, while the driest site had lower emissions. Immediately after the monthly measurements had been made, we removed the aboveground vegetation to enable an assessment of the gas transportation per unit of biomass. Removal of biomass decreased emissions by up to 53%. These results indicated the dominant effect of water depth on CH_4 flux through effects of soil conditions, plant species composition and distribution. This study suggests that temporally flooded wetlands, including littoral zones, contribute significantly to the global CH_4 burden. However, the current challenge is to capture their spatial extent and temporal variation in the fluxes.

Editor: Ben Bond-Lamberty, DOE Pacific Northwest National Laboratory, United States of America

Funding: This research was supported by China National Forestry Public Welfare Fund (No. 200804005). The funders had no role in study design, data collection and analysis, decision to publish, or preparation of the manuscript.

Competing Interests: The authors have declared that no competing interests exist.

* E-mail: guangchun.lei@foxmail.com

Introduction

Methane is the second most important contributor to total greenhouse gas emissions, with a Global Warming Potential of 25 over a 100-year time-span [1]. Wetlands are considered to be the most important CH_4 source, thought to account for 24–39% of the total global emission of CH_4, albeit with a large degree of uncertainty [2]. Reservoirs are an important type of wetland, whose combined area across the globe has increased in recent years to now occupy approximately 5×10^5 km^2, which is 1/3 the size of all natural lakes [3]. One reason for the large development of reservoirs is their potential as a clean energy source, although uncertainties in their methane emissions have led some scientists to question whether they are as clean as people believe [4]. Several authors have insisted that the subject of greenhouse gas emissions from reservoirs requires more scientific study [5–7]. The Ministry of Water Resources of the People's Republic of China has stated that China has over 80,000 reservoirs, large and small [8]. However, work in China on their greenhouse gas emissions, and the extent to which these reservoirs contribute to total emissions, is limited [9].

The littoral zone of reservoirs features fast and complicated material cycles and is reported to be a hotspot for CH_4 emissions [10,11]. Furthermore, there are indications from elsewhere about the importance of the littoral zone: for example, in boreal systems, where about 70% of total CH_4 emissions comes from the littoral zone, even though its area is no more than 24% of the total wetland area [12]. Water depth and its diurnal and seasonal fluctuations are thought to be the main drivers determining the characteristics of the littoral habitat, including plant species composition and distribution, soil conditions, and CH_4 flux [11,13–16]. However, the interaction between water level fluctuation, CH_4 production, transportation, and emissions is still not fully understood [17,18], especially the effect of dynamic water fluctuation coupling with vegetation amd nutrients [19,20]. The rate of change of the water table, as distinguished from just differences in the water table, is known to be a major environmental factor controlling the CH_4 flux [21].

In summary, studies of the CH_4 flux from the littoral zone are essential for evaluating the atmospheric and climatic impacts of reservoirs as energy sources, as well as for a better understanding of the biogeochemical mechanism of CH_4 flux. Research on the mechanisms of how the water level and its fluctuations influence CH_4 flux is an important foundation for related work aimed at modeling and controlling carbon loss through ecosystem management. In this context, we report the seasonal and monthly variation of CH_4 flux of the littoral zone of Miyun Reservoir,

Beijing, China. We hypothesized that water depth and water level fluctuation would be the main factors driving the CH_4 flux variability, together with soil and vegetation.

Materials and Methods

This study was authorized by Beijing North Miyun Reservoir Eco-agriculture Co. Ltd. It did not harm any protected species, and thus no ethics committee was required to authorize the work. All data used in this paper were collected in the field by the research team, and the raw data are available upon request from the senior author. No data were downloaded from publicly available resources, and thus no permission was required.

Study area

The research was carried out at Miyun Reservoir (40°29′N, 116°50′E), which is located in the northern mountainous area near Beijing, China. It was built in 1960 with a maximum water area of 188 km^2. Its catchment is characterized by warm temperate semi-humid monsoonal climate with an annual average air temperature of 10.5°C, maximum air temperature of 38°C, and a minimum of −18°C. The reservoir is normally covered by ice from the middle of November to the end of March. The growing season is from April to November. The annual average precipitation is close to 600 mm, of which 80% is concentrated from July to August [22]. The annual change in the water level is 1–5 m because of rainfall and water supply for domestic use. The area between the highest and lowest water level from 1984 to 2005 was 84 km^2 [23]. In the summer of 2012, when the work was carried out, continuous heavy rain in July caused a sudden water level increase of one meter, and part of the littoral vegetation was inundated.

We divided the littoral zone into five areas based on water level (Figure 1). Sites were selected ranging from locations in open water to the dry area on higher ground, to provide five contrasting environments: (i) deep water area (DW); (ii) shallow water area (SW); (iii) seasonal (August and September) flooded area (SF); (iv) 'seasonally flooded control' (SFC) area, which was 500 m away from SF, had the same plant species and similar soil carbon/nitrogen content as SF, but escaped the flood in August and September because of its 1-m-higher elevation; and (v) permanent non-flooded area (NF). Details of the water depths in each of these areas are shown in Figure 2B. Three typical plant communities in each area were selected, and Table 1 shows the dominant species in different seasons in all zones. For more details on biomass and soil, see Figures 2 and 3.

CH_4 flux measurements

Methane flux was measured in November 2011, then May, July, August, September and October 2012. The experiment of three plots at site SFC was carried out just after the flooding and during the time when the water level dropped from August to October 2012. In order to reduce uncertainty in the average daily flux, sampling to capture any diurnal variation was performed at three-hourly intervals (local time: 6, 9, 12, 15, 18, 21 and 24 h). Each plot had four replicates located within three meters from each other. To eliminate disturbance to the soil, wooden access platforms were built.

The static opaque chamber technique was used to determine the CH_4 flux [24]. The chambers were made of stainless steel (volume: 125 L; surface area: 0.25 m^2) and covered with polyethylene foam to avoid any warming effect inside the chamber. An internal chamber (volume: 200 L; surface area: 0.25 m^2) could be added if plants were tall. Two fans were built

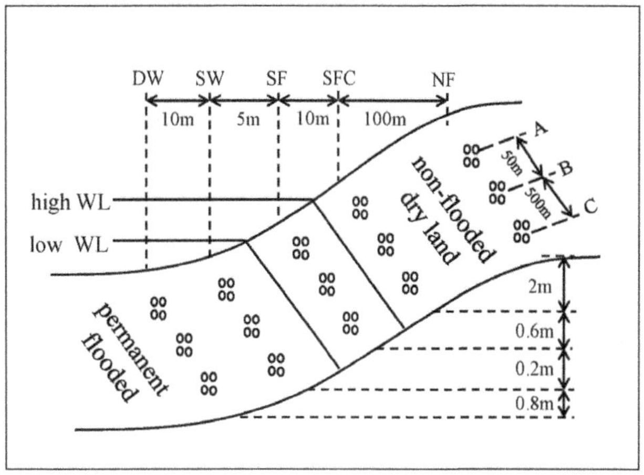

Figure 1. Experimental design. WL: water level. The sites are grouped at different heights. DW: deep water site; SW: shallow water site; SF: seasonally flooded site; SFC: 'control site' for the seasonally flooded site; NF: non-flooded site. A, B and C denote samples from different vegetation types within each height band (see Table 1 for information on plant species). There were four replicates in each case, repeatedly sampled six times (also repeatedly sampled seven times in a day) in the year. For more details on water depth and other environmental parameters, see Figures 2 and 3.
doi:10.1371/journal.pone.0094275.g001

into the chamber for air mixing. Four gas samples (200 mL each) were taken using 100-mL polypropylene syringes at 15-min intervals over a 45-min period after enclosure, and stored in 500-mL plastic and aluminum membrane gas sampling bags. The concentration of CH_4 was analyzed within one week by gas chromatography (7890A, Agilent, California, USA) equipped with a flame ionization detector (FID). Gases were separated with a column (3 m, 3.2 mm) packed with Porpak Q (80/100 mesh). The temperatures of the oven, injector, and detector were 70°C, 20°C, and 200°C, respectively. The flow rate of the carrier gas (N_2) was 25 mL min^{-1}. Standard CH_4 gas (2.03 ppm in air, China National Research Center for Certified Reference Materials, China) was used for precision verification for CH_4 concentrations. The coefficient of variation was below 1%. The flux of CH_4 was calculated following Chen et al. [25]. In order to determine the effect of aboveground vegetation on flux, one more flux measurement was taken at 9 am the following day (after seven times sampling for diurnal variation), with aboveground plant material removed. Chambers were reset into new positions near the old positions each sampling month. All positions at each site were within an area of 20 m^2, but not so close to each other to cause artifacts in the data through (for example) changes in the local hydrology.

Environmental factors

In order to analyze the effects of environmental parameters on fluxes, the following factors were taken into account: water level, dissolved oxygen (DO) in water, soil total carbon (TC) and nitrogen (TN), soil bulk density, biomass, and air temperature.

Water level was measured after gas sampling at DW, SW and SF (when SF had standing water in August and September 2012). At site SF (when there was no standing water in November 2011, May, July and October 2012) and SFC, a 1-m PVC tube was inserted vertically into the soil under the chamber after all monthly gas sampling was complete, allowing two hours for the water level to equilibrate before measuring the level. The water table of site NF was calculated according to the elevation measured by a

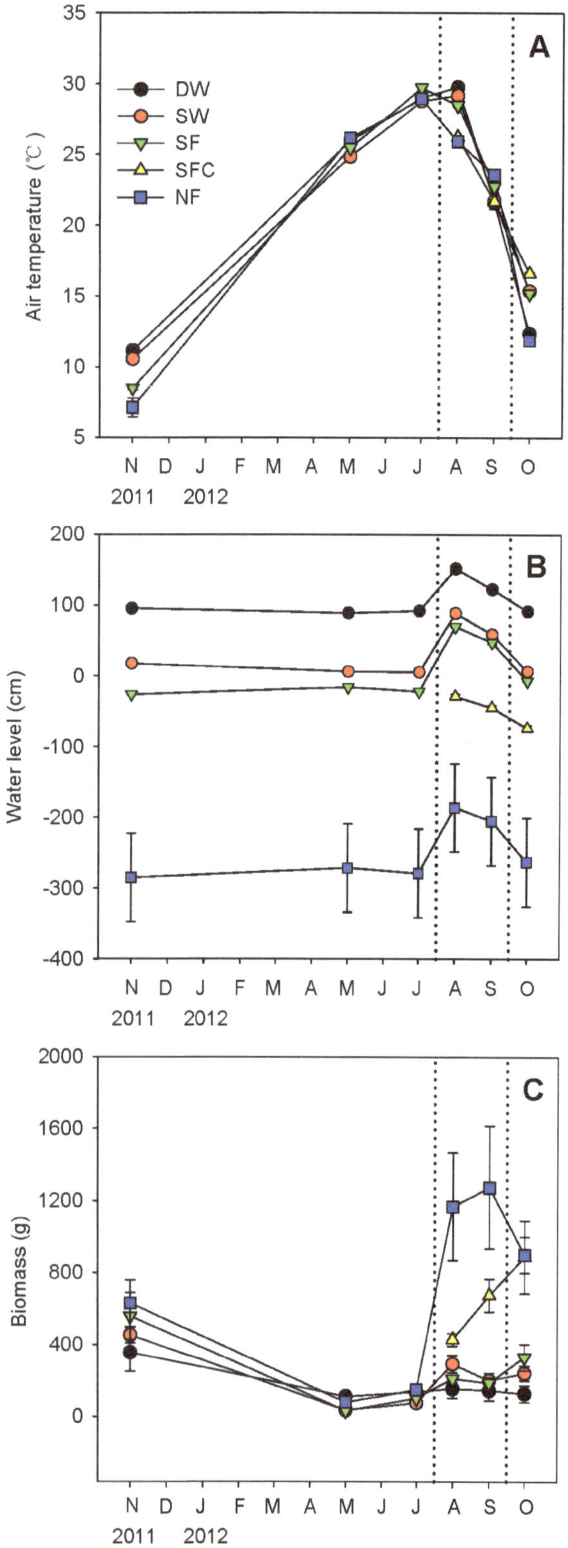

Figure 2. Monthly variation of air temperature, water level and biomass of each site. Days between dotted lines was the high water level period and thus the seasonally flooded site (SF) was under water. DW: deep water site; SW: shallow water site; SF: seasonally flooded site; SFC: 'control site' for the seasonally flooded site; NF: non-flooded site. A negative value of water level indicates the ground water depth. Error bars represent SE.

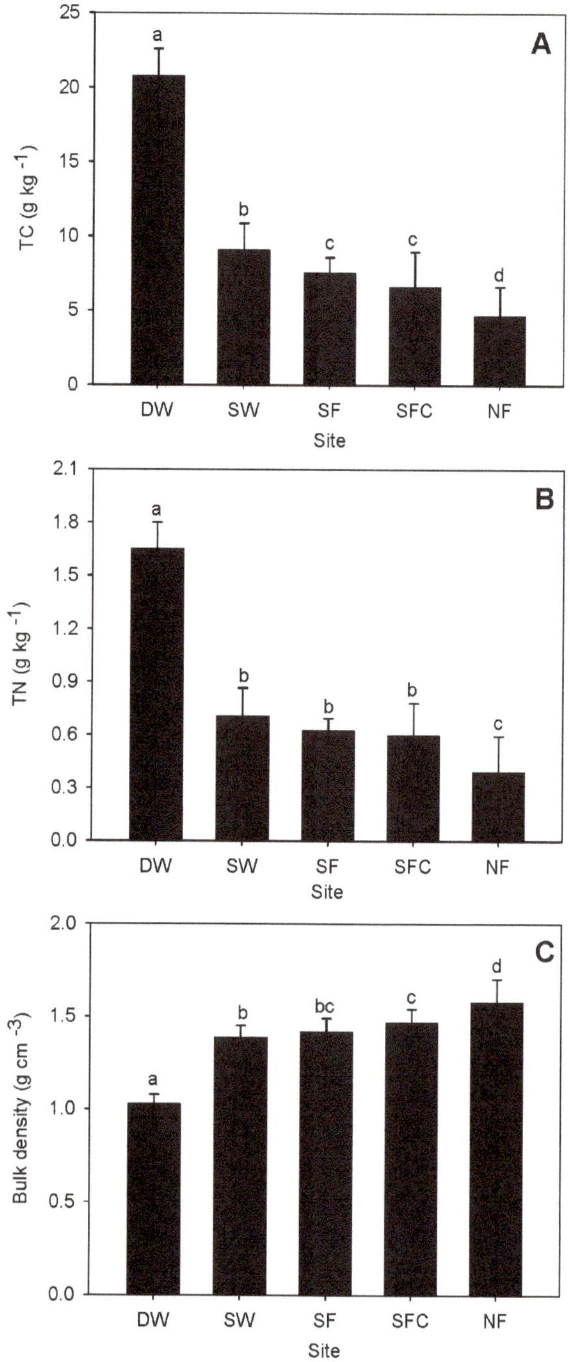

Figure 3. Soil properties (mean ± SE) of each site. Bars with different letters indicate a significant difference at $p<0.05$. TC: total carbon, TN: total nitrogen. DW: deep water site; SW: shallow water site; SF: seasonally flooded site; SFC: 'control site' for the seasonally flooded site; NF: non-flooded site.

Global Navigation Satellite System receiver (BLH-L90, Daheng International, China). DO in water was measured during the gas sampling by a handheld multi-parameter meter (Professional Plus, YSI, USA), after flooding.

Soil samples at site DW, SW, SF and NF were collected from three different layers (0–10 cm, 10–20 cm, and 20–30 cm below ground) at each replicate location in November 2011, except site SFC in October 2012. After air-drying and grinding (passing

Table 1. Dominant plant species at each plot in different months.

Site		Nov 2011	May 2012	Jul 2012	Aug 2012	Sep 2012	Oct 2012
DW	A	*Echinochloa olonum*^{AE}	*Myriophyllum* sp.		*Trapa*^{AE} sp.		
	B	no vegetation					
	C	*Typha angustifolia*^{AE}					
SW	A	*Xanthium sibiricum*^{E}	*Scirpus planiculmis*^{AE}		*Echinochloa colonum*^{AE}		
	B	*Setaria viridis*^{E}	*Bidens pilosa*^{E}		*Echinochloa colonum*^{AE}		
	C	*Zea mays*^{E}	*Polygonum lapathifolium*^{E}				*Typha angustifolia*^{AE}
SF	A	*Xanthium sibiricum*	*Cirsium setosum*		*Cirsium setosum*^{E}		*Cirsium setosum*
	B	*Setaria viridis*	*Hemarthria altissima*		*Hemarthria altissima*^{E}		*Hemarthria altissima*
	C	*Zea mays*	*Polygonum lapathifolium*		*Polygonum lapathifolium*^{E}		*Polygonum lapathifolium*
SFC	A	#			*Cirsium setosum*		
	B	#			*Hemarthria altissima*		
	C	#			*Zea mays*		
NF	A	*Xanthium sibiricum*					
	B	*Setaria viridis*	*Artemisia argyi*				
	C	*Zea mays*					

# indicates no data;
DW: deep water site, SW: shallow water site, SF: seasonally flooded site, SFC: 'control site' for seasonally flooded site, NF: non-flooded site;
A, B, C indicates sample plot with different vegetation;
Species with aerenchyma are denoted ^{A}, species that are emergent are denoted ^{E}.

through a 100 mesh sieve), TC and TN were analyzed using an elemental analyzer (vario MACRO cube, Elementar, Germany). Soil bulk density was measured following Chinese national standards NY/T 1121.4-2006 [26].

To determine which species had aerenchyma tissues, thin hand sections were made of roots and stems (photomicrographs were made). The aboveground biomass of every replicate in the chamber was weighed after drying at 80°C to constant mass.

Diurnal air temperature was measured by a digital thermometer (JM624, Jinming, China) at the start and end of each gas sampling at every plot.

Statistical analysis

Reported daily fluxes were obtained by averaging the means of the three-hourly values. To estimate how much CH_4 was transported by vegetation and to calculate the differences between species, plant-mediated flux was estimated on the basis of the difference between the flux before and after vegetation removal. We followed Kankaala et al. [27] to derive a mass-based plant transport index, PTI:

$$PTI = \frac{\Delta F}{B} = \frac{\text{flux with aboveground vegetation} - \text{flux without aboveground vegetation}}{\text{biomass}}$$

The Kruskal–Wallis H test was used to test for spatial difference of flux and the Mann–Whitney U test for further multiple comparisons. Spatial differences of soil properties were analyzed with one-way ANOVA, and then LSD for multiple comparisons. A $\log_{10}$ transformation was used to show the correlation between water depth and CH_4 flux. A value of 0.5 was added to the flux data before transformation to make sure that all of the data were positive. Spearman analyses were used for correlations between flux and environmental factors. All the analyses above were performed using IBM SPSS Statistics (19.0, IBM, USA). Charts were made using SigmaPlot (11.0, SYSTAT, USA).

Results

Spatial and seasonal variation of CH_4 flux

CH_4 fluxes at the five sites showed significant differences ($n = 324$, $p<0.05$; Figure 4). CH_4 fluxes at permanently flooded sites, both the deep and shallow water sites (DW: 3.1 ± 0.5 mg $m^{-2}\ h^{-1}$; SW: 1.3 ± 0.2 mg $m^{-2}\ h^{-1}$), and the seasonally flooded site (2.1 ± 0.4 mg $m^{-2}\ h^{-1}$) were significantly higher than non-flooded sites. The seasonally flooded site SF, which was flooded for two months, emitted higher (1.6 times) CH_4 than the permanent flooded site SW with shallow water. The seasonally flooded site SF was dramatically higher than its control site, SFC (the flux was 247 times higher). Although the never-flooded site NF presented as a weak sink ($-2.7\pm9.2\ 10^{-3}$ mg $m^{-2}\ h^{-1}$), while site SFC was a weak source ($8.7\pm8.9\ 10^{-3}$ mg $m^{-2}\ h^{-1}$), there was no significant difference between them ($n = 144$, $p>0.05$). The average CH_4 emission from all the flooded sites (DW, SW and SF) was about 1120 times higher than the non-flooded sites (SFC and NF).

There were different seasonal patterns among different sites (Figure 5). CH_4 flux at site DW and SW continuously increased from November 2011 to August 2012, and thereafter decreased gradually. CH_4 flux at site SF increased slightly from November 2011 to July 2012. After flooding it rose sharply (from 0.05 mg $m^{-2}\ h^{-1}$ to 6.4 mg $m^{-2}\ h^{-1}$) to a high peak, and then remained the highest emitter among all sites until the autumn. The seasonal pattern of site NF was totally contrary to sites DW, SW and SF; although the amplitude of flux was narrow. It decreased from autumn to the following summer, and then increased gently. CH_4 flux at site SFC was as low as site SF before it was flooded.

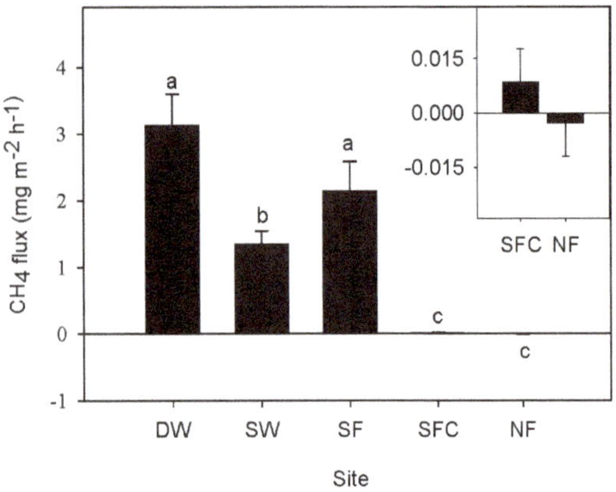

Figure 4. CH_4 flux (mean ± SE) at different sites. Bars with different letters indicate a significant difference at $p<0.05$. Positive values of flux indicate a CH_4 source. Inset in the top right-hand corner shows SFC and NF on a different scale. DW: deep water site; SW: shallow water site; SF: seasonally flooded site; SFC: 'control site' for the seasonally flooded site; NF: non-flooded site.

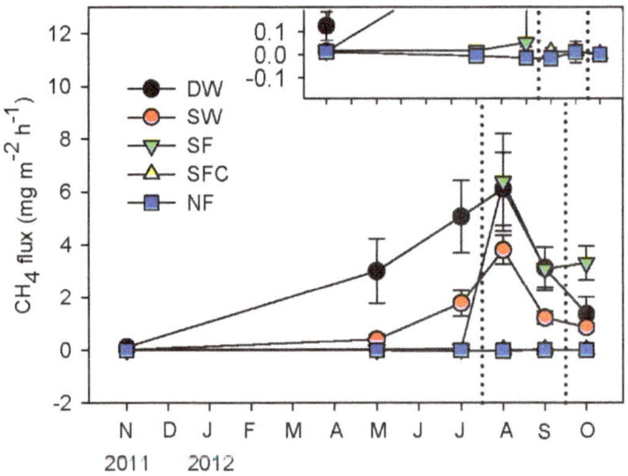

Figure 5. Monthly variation of CH_4 flux (mean ± SE) at different sites. A positive value of flux indicates a CH_4 source. Inset in the top right-hand corner shows SFC and NF on a different scale. Days between dotted lines was the high water level period and thus the seasonally flooded site (SF) was under water. DW: deep water site; SW: shallow water site; SF: seasonally flooded site; SFC: 'control site' for the seasonally flooded site; NF: non-flooded site.

Effects of environmental factors

Considering all the cases where the water depth was 2 m or less, the water depth was positively correlated with CH_4 flux ($n = 324$, $p<0.01$; Table 2). When there was standing water (water depth >0 cm; Figure 6), the habitat always emitted more CH_4 than when there was no standing water (water depth <0 cm).

Soil properties varied according to the zone: TC and TN were highest at site DW and lowest at site NF, while bulk density showed an opposite trend ($n = 60$, $p<0.05$; Figure 3). Soil TC and TN were positively correlated with flux at all sites ($n = 324$, $p<0.01$; Table 2), as well as individually at sites DW and SFC ($n = 72$, $p<0.01$). Bulk density was negatively correlated with flux at the whole-site scale ($n = 324$, $p<0.01$; Table 2), and no significant correlations were found at separate sites ($n = 72$, $p>0.05$).

There were large differences in CH_4 flux before and after the removal of aboveground vegetation at all sites (Figure 7). After vegetation removal, fluxes at sites DW and SW decreased by approximately 50% (47% and 53%, respectively), while the contribution of transportation at site SF was much smaller (6%). For the two emergent sites, SFC and NF, where fluxes were very small, opposite patterns were found. After the plants were removed, emissions at site SFC increased, while at site NF they decreased. The plant transportation index differed among the five sites and 14 plots (Figure 8). It showed seasonal variation, with the highest transportation efficiency found in spring or summer at most of the sites.

CH_4 emissions showed different patterns with different plant species after flooding at site SF (Figure 9). High CH_4 emissions (from 9.6 mg m^{-2} h^{-1} to 21.1 mg m^{-2} h^{-1}; average: 14.2±2.5 mg m^{-2} h^{-1}) were recorded at the plot with *Cirsium setosum* after the flooding (August; flooded for 20 days), which was 6.8 and 5.1 times higher than the site with *Polygonum lapathifolium* (from 1.6 mg m^{-2} h^{-1} to 2.6 mg m^{-2} h^{-1}; average: 2.1±0.2 mg m^{-2} h^{-1}) and *Hemarthria altissima* (from 2 mg m^{-2} h^{-1} to 4.1 mg m^{-2} h^{-1}; average: 2.8±0.5 mg m^{-2} h^{-1}), respectively. At the same time, the lowest water DO (30%–33%) was observed at the plot with *C. setosum* when the flux was the highest. Water DO was negatively correlated with flux at all flooded sites (DW: $n = 72$, $p<0.01$; SW: $n = 72$, $p<0.01$; SF: $n = 24$, $p<0.01$; Table 2).

The vegetation biomass began to increase in July and peaked in September to October (Figure 2C). There was a significantly negative correlation ($n = 324$, $p<0.01$; Table 2) between flux and biomass when all sites were taken as a whole. However significantly positive correlations were observed at sites DW and SF ($n = 72$, $p<0.01$; $n = 72$, $p<0.05$). No significant correlations were found at the other three sites ($n = 72$, $p>0.05$).

Air temperature positively correlated with CH_4 flux at sites DW and SW ($n = 72$, $p<0.01$; Table 2), but was negatively correlated at site NF ($n = 72$, $p<0.05$). No significant correlations were observed at sites SF and SFC ($n = 72$, $p>0.05$; $n = 36$, $p>0.05$).

Discussion

High CH_4 emissions from the littoral zone

The average CH_4 emission rate (1.3 mg m^{-2} h^{-1}) from the littoral zone of Miyun Reservoir was 6.5 times higher than the flux from the open water area (0.2 mg m^{-2} h^{-1}) [28]. The low emissions from open water are similar to those from reservoirs at higher or lower latitudes, but in those cases the littoral zone has not been investigated [29,30]. Reservoirs differ enormously in the percentage of their area that can be deemed 'littoral'. In the case

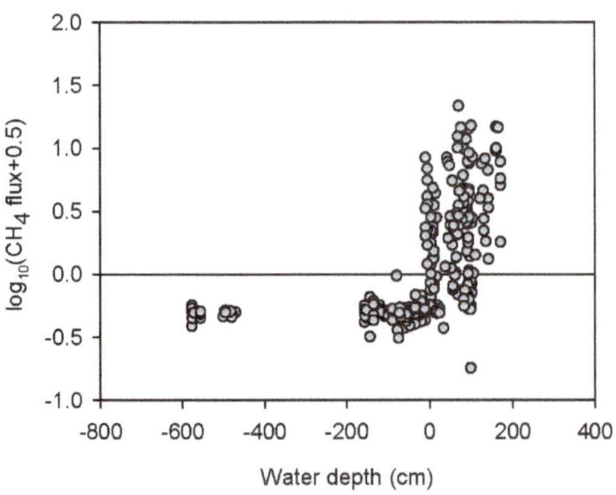

Figure 6. Relationship between water depth and CH_4 flux.

Table 2. Correlation coefficient *R* between CH_4 flux and environmental parameters.

Site	Water level	Soil TC	Soil TN	Soil bulk density	Biomass	Water DO	Air temperature
all sites	0.75**(324)	0.62**(324)	0.56**(324)	−0.53**(324)	−0.26**(324)	−0.28**(168)	0.25**(324)
DW	0.38**(72)	0.45**(72)	0.43**(72)	−0.02(72)	0.42**(72)	−0.31**(72)	0.63**(72)
SW	0.42**(72)	−0.06(72)	−0.05(72)	−0.03(72)	0.08(72)	−0.32**(72)	0.65**(72)
SF	0.74**(72)	0.19(72)	0.12(72)	0.07(72)	0.24*(72)	−0.51*(24)	0.08(72)
SFC	0.47**(36)	0.57**(36)	0.57**(36)	0.24(36)	−0.31(36)	#	0.10(36)
NF	−0.02(72)	−0.14(72)	−0.15(72)	0.05(72)	−0.03(72)	#	−0.28*(72)

# indicates no data;
Numbers in () indicate the sample size;
TC: total carbon, TN: total nitrogen, DO: dissolved oxygen;
** indicates significant correlation ($P<0.01$), * indicates significant correlation ($P<0.05$).

of Miyun, in the period from 1984 to 2005, the average area of the water surface was 103±15 km^2, while its annul difference was 9.5±2 km^2 [23]. Although the seasonally flooded area is as high as 8.5% of the total area of the reservoir, it explains how the littoral component of emissions is as much as 40% of the total. The high level of emissions from the littoral zone emphasizes its status as a methane hotspot in this case, but not all reservoirs will have a littoral zone comprising such a high percentage of the total. Although the present study involving 15 typical habitats, there are still uncertainties in the estimation because of high spatiotemporal flux variation and environmental heterogeneity, as demonstrated by the complex effects of plant, soil and temperature on fluxes found in this research. Furthermore, the dynamics of the littoral zone exacerbates the uncertainty. Nevertheless, emissions from the littoral zone are likely to form a substantial part of the reservoir's methane budget, and deserve further research [10,31].

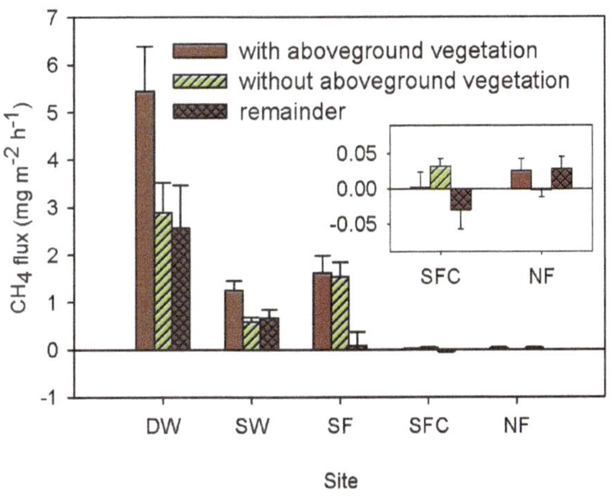

Figure 7. Contribution of aboveground plants to CH_4 flux (mean ± SE) of each site. A positive value of flux indicate a CH_4 source. A positive remainder means methane is transported from the vegetation to the atmosphere. Inset in the right-hand center shows SFC and NF on a different scale. DW: deep water site; SW: shallow water site; SF: seasonally flooded site; SFC: 'control site' for the seasonally flooded site; NF: non-flooded site.

Water level fluctuation and increased CH_4 emissions from the littoral zone

After 20 days of flooding, CH_4 emissions from site SF increased sharply to the highest level among all sites, and this high rate lasted to October (Figure 5). In contrast, a decreasing of CH_4 emissions was reported at a site in Finland and was explained by diminished plant-mediated transportation and decreased carbon supply from growing plants to the rhizosphere after plants were totally submerged [16]. In another study, different patterns were shown when flooding occurred in different seasons; CH_4 emissions were negligible after flooding in autumn or winter, but increased sharply after flooding in summer [32]. Based on the present work, we suggest that anaerobic plant decomposition, which apparently occurred at site SF after flooding, is the main reason for high emissions. Elsewhere, it has been shown to be the cause of CH_4 loss from tropical reservoirs [33]. Our situation was similar, but one in which the effect was more rapid due to the herbaceous environment.

Different fluxes among flood-tolerant and -intolerant plants were observed. In the case of the flood-intolerant *C. setosum*, flooding caused a sharp increase in emissions, but for the flood-tolerant *P. lapathifolium* and *H. altissima*, under the same water depth, no sharp increase was seen. This phenomenon might be the reason for the difference between the aforementioned research in Finland and that of our own; their research was carried out at places where *Carex* is the main vegetation, which does not rot immediately after flooding [16]. Different responses of plant species to flooding help to explain the spatial variation, and this could represent a potential approach to reducing methane emissions, i.e., through careful vegetation management.

The lowest water DO found in the plots with flood-intolerant species (Figure 9) may be the result of oxygen consumption during plant tissue decomposition [34]. Variations in CH_4 emissions have sometimes been attributed to variations in methane oxidation [35]. The observed negative relationship between DO and CH_4 flux (Table 2) asserts the importance of available DO and plant decomposition on CH_4 loss. A dramatic rise in CH_4 emissions appeared when the DO fell to approximately 30%, suggesting that 30% might be the lower bound of usable DO concentration for methanotrophs. However, more work on this aspect is needed, since CH_4 oxidation is a complex process affected by many factors, including microorganisms, temperature, water content, and redox potential [36,37]. It is therefore probable that the DO threshold will vary across different habitat types.

It is conceivable that wetlands featuring a littoral zone constituting a relatively large proportion of the total area, as well

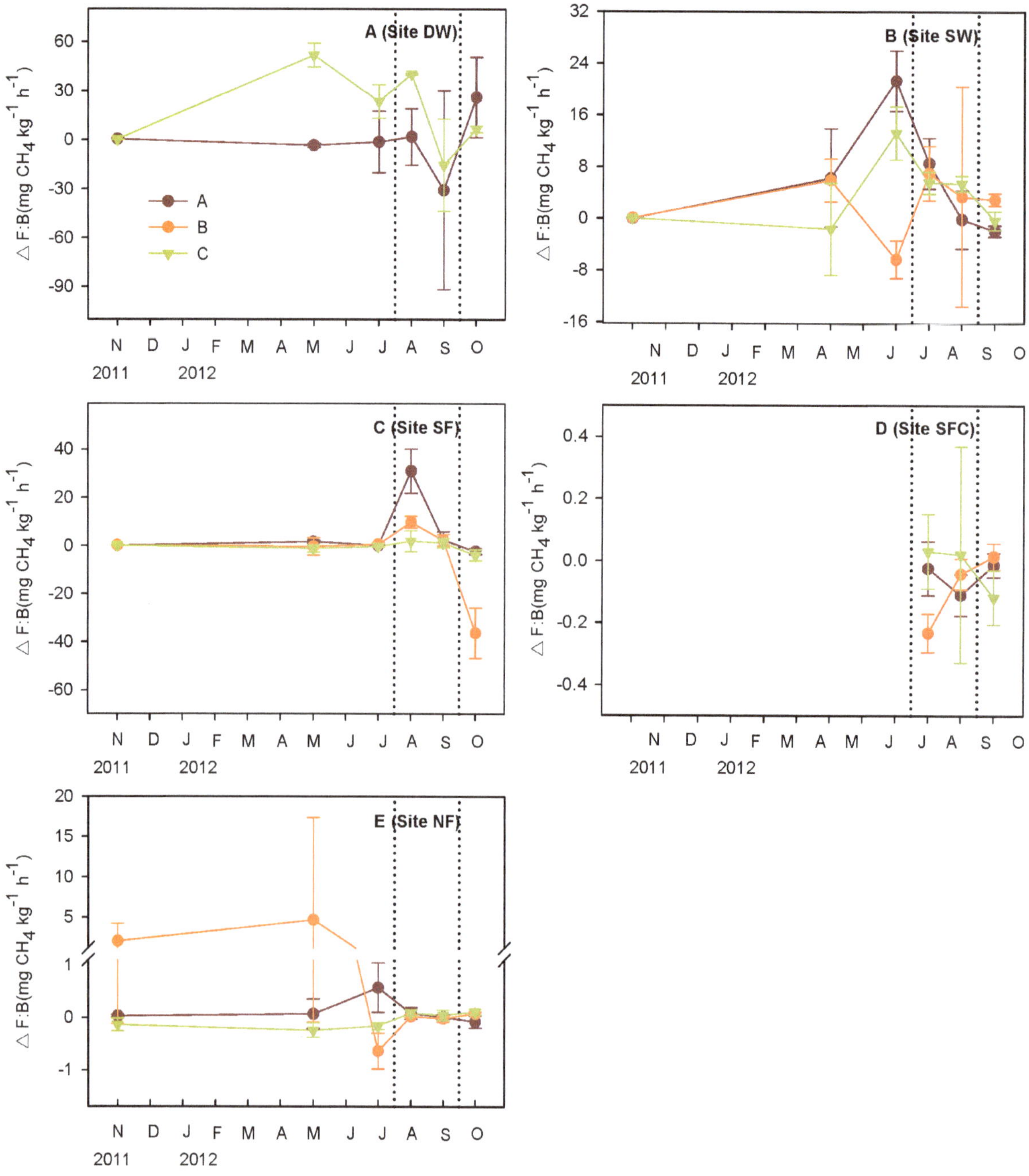

Figure 8. Above-ground plant transport index (mean ± SE). ΔF (mg CH_4 m^{-2} h^{-1}) is the flux with vegetation minus the flux without vegetation, i.e. the flux of methane that can be attributed to the presence of the vegetation. B (kg m^{-2}) is biomass. A positive flux means methane is transported from the vegetation to the atmosphere. Days between dotted lines was the high water level period and thus the seasonally flooded site (SF) was under water. DW: deep water site; SW: shallow water site; SF: seasonally flooded site; SFC: 'control site' for the seasonally flooded site; NF: non-flooded site.

as frequent water level fluctuations, might produce especially high emissions when the majority of the plant community comprises flood-intolerant species. It is further conceivable that climatic variability, i.e., an increased risk of flood and drought [38], will exacerbate methane production because of CH_4 emissions caused by unusual water level fluctuations, and those associated with re-flooding. For example, these conditions occurred in China after an extreme drought in 2010–2011, which caused the drying out of numerous lakes, giving them the appearance of grasslands, exemplified by a well-documented case at the largest freshwater lake in China, Poyang, which has an area of 3,150 km^2 [39].

Effects of water depth on the flux through soil and plants

Water depth determines soil and plant conditions, which in turn influence CH_4 flux through different mechanisms. However, we found these effects to be sometimes limited by the water depth itself.

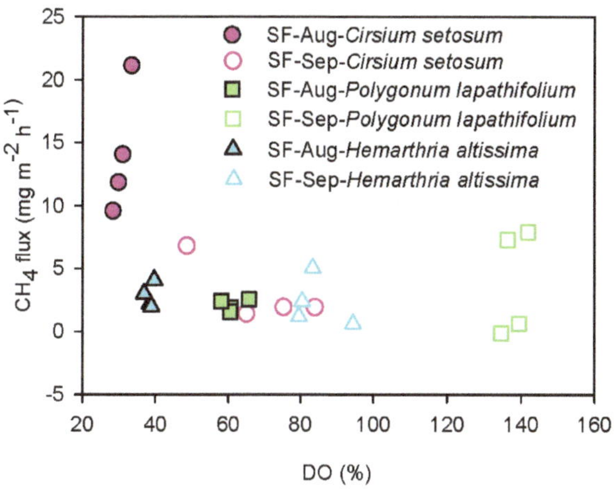

Figure 9. Relationship between dissolved oxygen (DO) of water and flux of the seasonally flooded site (SF).

Flooding decreases the soil redox potential [40–42], which affects both methanogenesis activity and gas transfer through plant tissues [43]. Besides, the anaerobic conditions inhibit aerobic respiration and thus allow more organic matter accumulation [44,45], which was the likely reason for the soil nutrients pattern found in our research (Figure 3). Carbon and nitrogen compounds form the substrate for CH_4 production [46,47]. Positive correlations are usually observed between soil nutrients and CH_4 emissions [48,49]. However, the insignificant correlations at SW, SF and NF suggested that the effect of soil nutrients was sometimes weak, and even dominated by other factors. A similar result has also been shown at a grassland site in Germany, where flux was found to be positively correlated with soil moisture but not nitrogen fertilizer application from 0 to 450 kg N ha^{-1} yr^{-1} [50]. High accumulation rates of organic matter forms a soil that has a low bulk density [51] and high porosity [52], which might accelerate CH_4 diffusion in soil pores, leading to methane release [53]. Our results agree with the literature, showing that the water level influenced CH_4 production through the soil redox potential and availability of organic substrates.

Plants are another important reason for the spatial pattern of flux. An apparent interspecific difference in the plant-mediated flux was found at site DW (Figure 8A), i.e., the flux transported by emergent plants was much higher than that of totally submerged plants. Higher emissions from emergent plants have also been observed at a Qinghai–Tibetan Plateau wetland, although they did not reduce the flux through the water surface [54]. Emergent macrophytes with a well-developed vascular system and with aerenchyma can be an important pathway for CH_4 transportation from sediment to atmosphere [55]. Although submerged plants have also developed aerenchyma, even a thin film of water over the leaves and stems might inhibit gas release from plant tissues since the diffusive velocity of gases in water is much slower than in air, by a factor of 10^4 [56]. In such cases, emissions will be suppressed until the time when ebullition begins. Further interpretation in this regard came from a study in which it was found that CH_4 emissions in a sedge-dominated zone decreased significantly, when the flood level was high enough to submerge the venting structures of the plants [16].

Much variation in plant-mediated CH_4 emissions among the five sites was observed (Figure 7), decreasing by 99% along the water depth gradient from water to dry land. Around 50% of the CH_4 emissions was transported by plant tissues at the permanently flooded sites (DW and SW), as was also found in a marsh with *Spartina alterniflora* [57]. Plant-mediated fluxes from the same species growing under different water depths were different, e.g., sites SF and SFC (Figures 9C and 9D). This shows that, although the transportation ability was different among plant species, it was nevertheless highly dependent on flooding.

Besides the spatial pattern, plant development is also considered as an explanation for the seasonal variation of CH_4 flux. There were positive correlations between flux and the biomass of sites DW and SF (Table 2). Gas transportation through aerenchyma is the most likely explanation [55,57]. Besides, we also calculated methane flux per kilogram of biomass as an index of transportation efficiency (Figure 8). The arched patterns of plant transportation at sites DW and SW suggest that vegetation activity might be correlated with plant transportation capacity [58–60]; and the high level of emissions in summer were caused not only by the high biomass, but were also influenced by the high transpiration rate of vegetation (methane dissolved in plant water can be transported from soil to the leaves, and hence reach the atmosphere). The interaction between biomass and transportation efficiency might be a reason for the lack of correlation between biomass and emissions at sites SW, SFC and NF (Table 2). Taking all sites as a whole, a negative correlation between flux and biomass was observed (Table 2), but this relationship is affected by the fact that the lowest CH_4 emissions and highest biomass were found at the driest site.

Acknowledgments

We thank Yifei Jia, Yunzhu Liu, Shengwu Jiao, Linlu Shi, Lei Guan, Yamian Zhang, Nana Li, Jialin Lei, Rui Li, Duoduo Feng, Hairui Duo, Lei Jing, Qing Zeng, Chu Lang, Xu Luo, Jiayuan Li, Yonghong Gao, Defeng Bai, Siyu Zhu, and Jianing Xu for their great support during the course of this study. We also thank Beijing North Miyun Reservoir Eco-agriculture Co. Ltd for granting us permission to conduct the study on its land.

Author Contributions

Conceived and designed the experiments: GL MY Y. Zhou CL. Performed the experiments: MY XG CL Y. Zhu. Analyzed the data: MY JG XG. Contributed reagents/materials/analysis tools: GL MY XG Y. Zhu CL. Wrote the paper: MY JG CL.

References

1. Solomon S, Qin D, Manning M, Chen Z, Marquis M, et al. (2007) Climate Change 2007: The Physical ScienceBasis. Contribution of Working Group I to the Fourth Assessment Report of the IntergovernmentalPanel on Climate Change. Cambridge University Press, Cambridge, United Kingdom and New York, NY, USA.
2. Denman KL, Brasseur G, Chidthaisong A, Ciais P, Cox PM, et al. (2007) Couplings Between Changes in the Climate System and Biogeochemistry. In: Climate Change 2007: The Physical Science Basis. Contribution of Working Group I to the Fourth Assessment Report of the Intergovernmental Panel on Climate Change [Solomon, S., D. Qin, M. Manning, Z. Chen, M. Marquis, K.B. Averyt, M.Tignor and H.L. Miller (eds.)]. Cambridge University Press, Cambridge, United Kingdom and New York, NY, USA.
3. Wildi W (2010) Environmental hazards of dams and reservoirs. Near Curriculum in Natural Environmental Science 88: 187–197.
4. Gunkel G (2009) Hydropower - A Green Energy? Tropical Reservoirs and Greenhouse Gas Emissions. Clean-Soil Air Water 37: 726–734.
5. Mäkinen K, Khan S (2010) Policy considerations for greenhouse gas emissions from freshwater reservoirs. Water Alternatives 3: 91–105.
6. Li S, Lu X (2012) Uncertainties of carbon emission from hydroelectric reservoirs. Natural Hazards 62: 1343-1345.
7. Bridgham SD, Cadillo-Quiroz H, Keller JK, Zhuang QL (2013) Methane emissions from wetlands: biogeochemical, microbial, and modeling perspectives from local to global scales. Global Change Biology 19: 1325–1346.

8. Chen X (2009) Gazette of the ministry of water resources of the People's Republic of China. Beijing: Ministry of Water Resources of the People's Republic of China.
9. Hu Y, Cheng H (2013) The urgency of assessing the greenhouse gas budgets of hydroelectric reservoirs in China. Nature Climate Change 3: 708–712.
10. Chen H, b NW, Yao S, Gao Y, Zhu D, et al. (2009) High methane emissions from a littoral zone on the Qinghai-Tibetan Plateau. Atmospheric Environment 43: 4995–5000.
11. Bergström I, Mäkelä S, Kankaala P, Kortelainen P (2007) Methane efflux from littoral vegetation stands of southern boreal lakes: An upscaled regional estimate. Atmospheric Environment 41: 339–351.
12. Juutinen S (2004) Methane fluxes and their environmental controls in the littoral zone of boreal lakes. Joensuu, Finland: University of Joensuu.
13. Light HM, Darst MR, MacLaughlin MT, Sprecher SW (1993) Hydrology, vegetation, and soils of four North Florida river flood plains with an evaluation of state and federal wetland determinations. Tallahassee, Florida: US Department of the Interior, US Geological Survey.
14. Laanbroek HJ (2010) Methane emission from natural wetlands: interplay between emergent macrophytes and soil microbial processes. A mini-review. Annals of Botany 105: 141–153.
15. Huang Y, Jiao Y, Zong L, Zheng X, Sass RL, et al. (2002) Quantitative dependence of methane emission on soil properties. Nutrient Cycling in Agroecosystems 64: 157–167.
16. Juutinen S, Alm J, Larmola T, Huttunen JT, Morero M, et al. (2003) Methane (CH4) release from littoral wetlands of Boreal lakes during an extended flooding period. Global Change Biology 9: 413–424.
17. Petrescu A, Van Beek L, Van Huissteden J, Prigent C, Sachs T, et al. (2010) Modeling regional to global CH4 emissions of boreal and arctic wetlands. Global Biogeochemical Cycles 24, GB4009, DOI: 10.1029/2009GB003610.
18. Petrescu AMR, van Huissteden J, Jackowicz-Korczynski M, Yurova A, Christensen TR, et al. (2008) Modelling CH4 emissions from arctic wetlands: effects of hydrological parameterization. Biogeosciences 5: 111–121.
19. Gorres CM, Conrad R, Petersen SO (2013) Effect of soil properties and hydrology on Archaeal community composition in three temperate grasslands on peat. FEMS Microbiology Ecology 85: 227–240.
20. Mander Ü, Maddison M, Soosaar K, Karabelnik K (2011) The Impact of Pulsing Hydrology and Fluctuating Water Table on Greenhouse Gas Emissions from Constructed Wetlands. Wetlands 31: 1023–1032.
21. Yamamoto A, Hirota M, Suzuki S, Oe Y, Zhang P, et al. (2009) Effects of tidal fluctuations on CO2 and CH4 fluxes in the littoral zone of a brackish-water lake. Limnology 10: 229–237.
22. Gao ZQ (1989) Beijing Chronicles of Water Conservancy,Volume III Beijing: Editing Committee of the Beijing Chronicles of Water Conservancy.
23. Cao RL, Li CJ, Liu LY, Wang JH, Yan GJ (2008) Extracting Miyun reservoirs water area and monitoring its change based on a revised normalized different water index. Science o f Surveying and Mapping 33: 158–160.
24. Moore TR, Roulet NT (1991) A comparison of dynamic and static chambers for methane emission measurements from subarctic fens. Atmosphere-Ocean 29: 102–109.
25. Chen H, Wu N, Wang Y, Zhu D, Yang G, et al. (2013) Inter-Annual Variations of Methane Emission from an Open Fen on the Qinghai-Tibetan Plateau: A Three-Year Study. PloS ONE 8: e53878.
26. MAPRC Ministry of agriculture of People's Republic of China (2007) Available: http://cx.spsp.gov.cn/index.aspx.Accessed 2013 Nov 10.
27. Kankaala P, Käki T, Mäkelä S, Ojala A, Pajunen H, et al. (2005) Methane efflux in relation to plant biomass and sediment characteristics in stands of three common emergent macrophytes in boreal mesoeutrophic lakes. Global Change Biology 11: 145–153.
28. Yang M, Li HL, Lei T, Zhou Y, Lu C (2011) Spatial-temporal variation of CH_4 flux and its environmental factors at Miyun Reservoir. Wetland Science 9: 191–197.
29. Demarty M, Bastien J, Tremblay A (2011) Annual follow-up of gross diffusive carbon dioxide and methane emissions from a boreal reservoir and two nearby lakes in Quebec, Canada. Biogeosciences 8: 41–53.
30. Zheng H, Zhao X, Zhao T, Chen F, Xu W, et al. (2010) Spatial–temporal variations of methane emissions from the Ertan hydroelectric reservoir in southwest China. Hydrological Processes 25: 1391–1396.
31. Bergier I, Novo EM, Ramos FM, Mazzi EA, Rasera MF (2011) Carbon dioxide and methane fluxes in the littoral zone of a tropical Savanna Reservoir (Corumba, Brazil). Oecologia Australis 15: 666–681.
32. Boon PI, Mitchell A, Lee K (1997) Effects of wetting and drying on methane emissions from ephemeral floodplain wetlands in south-eastern Australia. Hydrobiologia 357: 73–87.
33. Abril G, Parize M, Pérez MAP, Filizola N (2013) Wood decomposition in Amazonian hydropower reservoirs: An additional source of greenhouse gases. Journal of South American Earth Sciences 44: 104–107.
34. Cunha-Santino MBd, Pacobahyba LD, Bianchini I Jr. (2010) Decomposition of aquatic macrophytes from Cantá stream (Roraima, Brazil): kinetics approach. Acta Limnologica Brasiliensia 22: 237–246.
35. Le Mer J, Roger P (2001) Production, oxidation, emission and consumption of methane by soils: A review. European Journal of Soil Biology 37: 25–50.
36. Chowdhury TR, Dick RP (2013) Ecology of aerobic methanotrophs in controlling methane fluxes from wetlands. Applied Soil Ecology 65: 8–22.
37. Knittel K, Boetius A (2009) Anaerobic oxidation of methane: progress with an unknown process. Annual Review of Microbiology 63: 311–334.
38. Kundzewicz ZW, Mata LJ, Arnell NW, Döll P, Kabat P, et al. (2007) Freshwater resources and their management. Climate Change 2007: Impacts, Adaptation and Vulnerability. Contribution of Working Group II to the Fourth Assessment Report of the Intergovernmental Panel on Climate Change, M.L. . Parry, O.F. . Canziani, J.P. . Palutikof, P.J. . van der Linden and C.E. . Hanson, Eds., Cambridge University Press, Cambridge, UK.
39. Anonymous (2011) China's largest inland lake Poyang Lake turns into grassland in worst drought. Available: http://www.whatsonxiamen.com/news19375.html. Accessed 10 November 2013.
40. Koh H-S, Ochs CA, Yu K (2009) Hydrologic gradient and vegetation controls on CH4 and CO2 fluxes in a spring-fed forested wetland. Hydrobiologia 630: 271–206.
41. Wang ZP, DeLaune RD, Patrick WH, Masscheleyn PH (1993) Soil Redox and pH Effects on Methane Production in a Flooded Rice Soil. Soil Science Society of America Journal 57: 382–385.
42. Swamy YV, Nikhil GN, Venkanna R, Das SN, Chaudhury GR (2012) Emission of methane and nitrous oxide from Vigna mungo and Vigna radiata legumes in India during the dry cropping seasons. Atmosfera 25: 107–120.
43. Kludze HK, Delaune RD (1995) Gaseous exchange and wetland plant-response to soil redox intensity and capacity. Soil Science Society of America Journal 59: 939–945.
44. Kolka RK, Thompson JA (2007) Wetland Geomorphology, Soils, and Formative Processes. In: Batzer DP, Sharitz RR, editors. Ecology of Freshwater and Estuarine Wetlands. Berkeley, CA: University of California Press.
45. Bernal B, Mitsch W (2013) Carbon sequestration in freshwater wetlands in Costa Rica and Botswana. Biogeochemistry 115: 77–93.
46. Schimel J (2000) Global change: Rice, microbes and methane. Nature 403: 375–377.
47. Fazli P, Man HC, Shah UKM, Idris A (2013) Characteristics of Methanogens and Methanotrophs in Rice Fields: A Review. Asia-Pacific Journal of Molecular Biology and Biotechnology 21: 3–17.
48. Liu DY, Ding WX, Jia ZJ, Cai ZC (2011) Relation between methanogenic archaea and methane production potential in selected natural wetland ecosystems across China. Biogeosciences 8: 329–338.
49. Li CF, Zhou DN, Kou ZK, Zhi Sheng Z, Jin Ping W, et al. (2012) Effects of tillage and nitrogen fertilizers on CH4 and CO2 emissions and soil organic carbon in paddy fields of central China. PLoS ONE 7: e34642.
50. Kammann C, Grünhage L, Jäger HJ, Wachinger G (2001) Methane fluxes from differentially managed grassland study plots: the important role of CH4 oxidation in grassland with a high potential for CH4 production. Environmental Pollution 115: 261–273.
51. Sakin E (2012) Organic carbon organic matter and bulk density relationships in arid-semi arid soils in Southeast Anatolia region. African Journal of Biotechnology 11: 1373–1377.
52. Chaudhari PR, Ahire DV, Ahire VD, Chkravarty M, Maity S (2013) Soil Bulk Density as related to Soil Texture, Organic Matter Content and available total Nutrients of Coimbatore Soil. International Journal of Scientific and Research Publications 3: 1–8.
53. Allaire SE, Lafond JA, Cabral AR, Lange SF (2008) Measurement of gas diffusion through soils: Comparison of laboratory methods. Journal of Environmental Monitoring 10: 1326–1336.
54. Hirota M, Tang Y, Hu Q, Hirata S, Kato T, et al. (2004) Methane emissions from different vegetation zones in a Qinghai-Tibetan Plateau wetland. Soil Biology and Biochemistry 36: 737–748.
55. Bhullar GS, Edwards PJ, Olde Venterink H (2013) Variation in the plant-mediated methane transport and its importance for methane emission from intact wetland peat mesocosms. Journal of Plant Ecology 6: 298–304.
56. Haynes WM, Lide DR, Bruno TJ (2012) CRC Handbook of Chemistry and Physics 2012-2013. Boca Raton, Florida: CRC press.
57. Huang J, Tong C, Liu Z, Xiao H, Zhang L (2011) Plant-mediated Methane Transport and Emission from a Spartina alterniflora Marsh(in chinese). Chinese Bulletin of Botany 46: 534–543.
58. Enrica P, Rossano B, Marco B, Pierluigi V (2010) Net primary production and seasonal CO2 and CH4 fluxes in a Trapa natans L. meadow. Journal of Limnology 69: 225–234.
59. Duan X, Wang X, Zhiyun O (2006) Plant-mediated CH4 emission from a Phragmites-dominated wetland in an arid region, China. Journal of Freshwater Ecology 21: 139–145.
60. Gauci V, Gowing DJG, Hornibrook ERC, Davis JM, Dise NB (2010) Woody stem methane emission in mature wetland alder trees. Atmospheric Environment 44: 2157–2160.

13

Water Developments and Canids in Two North American Deserts: A Test of the Indirect Effect of Water Hypothesis

Lucas K. Hall[1]*, Randy T. Larsen[2], Robert N. Knight[3], Kevin D. Bunnell[4], Brock R. McMillan[1]

1 Department of Plant and Wildlife Sciences, Brigham Young University, Provo, Utah, United States of America, 2 Department of Plant and Wildlife Sciences and The Monte L. Bean Life Sciences Museum, Brigham Young University, Provo, Utah, United States of America, 3 Environmental Programs, United States Army Dugway Proving Ground, Dugway, Utah, United States of America, 4 Utah Division of Wildlife Resources, Salt Lake City, Utah, United States of America

Abstract

Anthropogenic modifications to landscapes intended to benefit wildlife may negatively influence wildlife communities. Anthropogenic provisioning of free water (water developments) to enhance abundance and distribution of wildlife is a common management practice in arid regions where water is limiting. Despite the long-term and widespread use of water developments, little is known about how they influence native species. Water developments may negatively influence arid-adapted species (e.g., kit fox, *Vulpes macrotis*) by enabling water-dependent competitors (e.g., coyote, *Canis latrans*) to expand distribution in arid landscapes (i.e., indirect effect of water hypothesis). We tested the two predictions of the indirect effect of water hypothesis (i.e., coyotes will visit areas with free water more frequently and kit foxes will spatially and temporally avoid coyotes) and evaluated relative use of free water by canids in the Great Basin and Mojave Deserts from 2010 to 2012. We established scent stations in areas with (wet) and without (dry) free water and monitored visitation by canids to these sites and visitation to water sources using infrared-triggered cameras. There was no difference in the proportions of visits to scent stations in wet or dry areas by coyotes or kit foxes at either study area. We did not detect spatial (no negative correlation between visits to scent stations) or temporal (no difference between times when stations were visited) segregation between coyotes and kit foxes. Visitation to water sources was not different for coyotes between study areas, but kit foxes visited water sources more in Mojave than Great Basin. Our results did not support the indirect effect of water hypothesis in the Great Basin or Mojave Deserts for these two canids.

Editor: Matt Hayward, Bangor University, United Kingdom

Funding: Funding for this project was provided by the United States Army Dugway Proving Ground and the Utah Division of Wildlife Resources. RN Knight (US Army Dugway Proving Ground) and KD Bunnell (Utah Division of Wildlife Resources) were involved in the preparation of the manuscript. The funders had no role in study design, data collection and analysis, or decision to publish.

Competing Interests: The authors have declared that no competing interests exist.

* E-mail: lucashall4@gmail.com

Introduction

Anthropogenic manipulations to landscapes or resources that are designed to benefit wildlife may have negative indirect effects on wildlife communities [1–3]. Indirect effects from anthropogenic manipulations are likely most pronounced when alteration influences resources that are limiting. In arid regions, water available for drinking (hereafter free water) is a limiting resource for some species that is often manipulated to increase distribution or density of animals [4–7]. Despite the widespread and long-term manipulation of free water (e.g., water developments for wildlife and livestock), there is little supporting information and much controversy concerning how this anthropogenic manipulation influences native species [8–10]. Water developments may be beneficial for some native species [11–13]. For example, water developments sustained suitable habitat for bighorn sheep (*Ovis canadensis*) where natural sources of free water were in decline [14]. However, manipulating a limiting resource, such as free water, may impose negative indirect effects on native species that have adapted to minimal availability of that resource.

The manipulation of free water in arid landscapes (i.e., addition of water developments) potentially weakens the advantages that arid-adapted species have accrued to minimize interspecific competition and predation from species that are water-dependent. In the Great Basin Desert, for example, it has been argued that water developments remove the limitation of arid systems to coyotes (*Canis latrans*) which compete with kit foxes (*Vulpes macrotis*) for habitat, space, and food [15–17]. This association is largely based on the differential physiological demand of free water by coyotes and kit foxes. To obtain enough preformed water (available in prey items) to survive in the absence of free water, both coyotes and kit foxes need to consume more prey than required to meet energetic demands [18]. However, coyotes need to consume twice the amount of prey per unit of mass relative to kit foxes to acquire sufficient preformed water to survive without free water creating an advantage for kit foxes in arid landscapes [18]. Furthermore, coyotes depend on evaporative cooling to expel heat and therefore have higher rates of water loss compared to kit foxes that rely on thermal conductance which reduces water loss [19]. Kit foxes also adhere more strictly than coyotes to behavioral adaptations that minimize water loss such as subterranean living and nocturnal activity [18,20]. Thus, it is less energetically feasible for coyotes to inhabit areas that lack free water relative to kit foxes. Water developments may, therefore, indirectly affect the arid-adapted kit fox by enabling the water-dependent coyote to occupy an otherwise inhospitable system and exert asymmetric interspe-

Table 1. Distances from scent stations to nearest known source of free water in wet and dry areas, 2011 to 2012.

Study area	Year	Stations in wet areas		Stations in dry areas	
		Mean distance km (±SE)	*N*	Mean distance km (±SE)	*N*
Great Basin	2011	1.49 (±0.14)	19	4.37 (±0.42)	13
	2012	1.60 (±0.12)	22	4.56 (±0.27)	17
Mojave	2011	1.07 (±0.16)	20	3.82 (±0.33)	12
	2012	0.93 (±0.17)	24	3.74 (±0.29)	15

cific competition on kit foxes (i.e., indirect effect of water hypothesis) [15,16].

The indirect effect of water hypothesis is comprised of two testable predictions: water-dependent competitors 1) will occur more frequently in areas near free water and 2) will spatially and/or temporally displace subordinate competitors. The predictions of the indirect effect of water hypothesis, however, have not been formally evaluated for canid communities and it is unclear whether this hypothesis is broadly applicable in arid systems. Our objective was to test the indirect effect of water hypothesis in the Great Basin and Mojave Deserts using coyotes and kit foxes as a model community. Specifically, we 1) evaluated support for the two predictions of the indirect effect of water hypothesis and 2) assessed relative use of free water by coyotes and kit foxes. This information will provide new insight into how anthropogenic modification of landscapes and resources may influence interspecific interactions and community dynamics.

Methods

Ethics Statement

Fieldwork was approved and sanctioned by the United States Department of Defense and the Utah Division of Wildlife Resources. Permission to access land in Great Basin was obtained from the United States Department of Defense (United States Army Dugway Proving Ground). Permission to access public land in Mojave was not required.

The kit fox is a protected species and we did not handle any individuals. Thus, our study was strictly observational in nature (using infrared camera traps) and no specific permits or regulations for animal care were needed according to the Institutional Animal Care and Use Committee of Brigham Young University.

Study Areas

This study was conducted at sites in both the Great Basin and Mojave Deserts. The Great Basin Desert study area consisted of 915 km^2 of land managed by the United States Department of Defense, United States Army Dugway Proving Ground in west-central Utah. The terrain was typical of Lake Bonneville lakebed characterized by dune systems and alkaline flats that were dominated by black greasewood (*Sarcobatus vermiculatus*). Where wildfires had occurred along the foothills, cheatgrass (*Bromus tectorum*) was common within communities of big sagebrush (*Artemisia tridentata*), rabbitbrush (*Chrysothamnus* spp.), and juniper (*Juniperus osteosperma*) [15]. Elevations across the study area ranged from approximately 1300 to 1800 m. Annual weather consisted of mean air temperatures of 12.69°C (range: −20.02 to 40.58°C) and

Figure 1. Photo captures of canids from infrared-triggered cameras. Clockwise from top left: kit fox (*Vulpes macrotis*) at a scent station, coyote (*Canis latrans*) at a scent station, coyote at a water development, and kit fox at a water development. Data were collected in the Great Basin and Mojave Deserts, USA, 2011 to 2012.

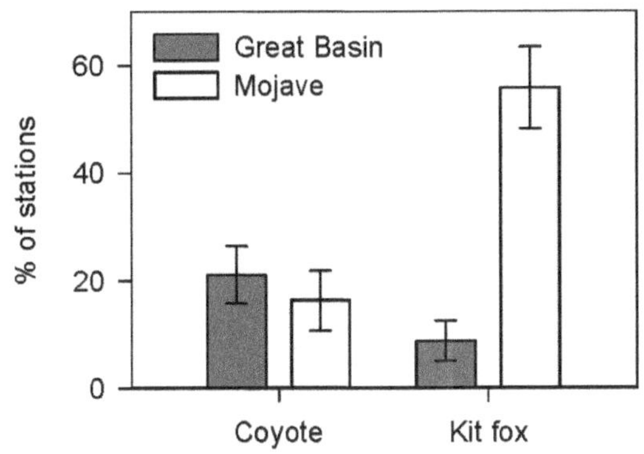

Figure 2. Overall proportions (±SE) of scent stations visited by canids. Data were collected on coyotes (*Canis latrans*) and kit foxes (*Vulpes macrotis*) in the Great Basin and Mojave Deserts, USA, 2011 to 2012.

mean precipitation of 150 mm (MesoWest, Bureau of Land Management & Boise Interagency Fire Center). The US Army Dugway Proving Ground has not been grazed by domestic livestock for the last 60 years [16]. In this study area, we identified 22 permanent water sources consisting of 11 water developments for wildlife, six natural springs, and five man-made ponds. The median distance of 10,000 random points to a natural water source (e.g., springs) and any water source was 7.10 km (range: 0.04 to 19.84 km) and 3.23 km (range: 0.06 to 8.54 km), respectively.

The Mojave Desert study area consisted of 1,064 km^2 of public land managed by the United States Department of the Interior, Bureau of Land Management. The Mojave study area was located in extreme southwestern Utah, northwestern Arizona, and southeastern Nevada and was approximately 360 km south of Great Basin. This study area was characterized by an alternating landscape of rolling hills/ridges and dry desert washes radiating from the Beaver Dam Mountains and emptying into the Beaver Dam Wash to the southwest near the intersection of the Utah-Nevada-Arizona state borders [21]. In areas that burned within the last decade, red brome (*B. rubens*) was well established among surviving creosote (*Larrea tridentata*), Joshua-tree (*Yucca brevifolia*), and black-brush (*Coleogyne ramosissima*) communities [22]. Along the foothills, the vegetation primarily consisted of sagebrush and juniper, transitioning to pinyon pine (*Pinus edulis*) at higher elevations. Elevations across the Mojave study area ranged from approximately 800 to 2000 m. Annual weather consisted of mean air temperatures of 19.18°C (range: −10.04 to 41.70°C) and mean precipitation of 113 mm (MesoWest, Bureau of Land Management & Boise Interagency Fire Center). The Mojave study area was grazed by livestock from October to May. We identified 66 permanent water sources in this study area consisting of 35 water developments for wildlife, 18 water troughs/tanks for livestock, 11 natural springs, and two man-made ponds. The median distance of 10,000 random points to a natural water source and any water source was 4.75 km (range: 0.03 to 14.01 km) and 2.10 km (range: 0.02 to 8.72 km), respectively.

Experimental Design and Sampling

To verify if presence of coyotes was greater in areas with free water (hereafter wet) compared to areas without (hereafter dry), we first established wet and dry areas in both study areas. Using ArcGIS (version 10.0, Environmental Systems Research Institute, Redlands, California), we created a uniform pattern of sample points with a distance of 4 km apart for both study areas. Each of these sample points was buffered with a 2.6 km radius based on the square root of a home range for coyotes inhabiting a semi-arid environment similar to our study areas [17]. The square root of the home range is a linear measure used to approximate daily movements of mammals and birds [23,24]. If free water was located within a buffer zone for a given sample area, we considered it a wet area. We identified water sources using databases with geospatial information for springs and water developments provided by the US Army Dugway Proving Ground and the Utah Division of Wildlife Resources. In addition, we consulted with local ranchers concerning water sources for livestock that were not in our databases. We were confident in our efforts to identify all known water sources in both study areas.

We established 32 scent stations in 2011 and 39 in 2012 and monitored stations for two-week periods during July to August (hottest part of the year; Table 1). Approximately 60% of the scent stations were located in wet areas and 40% in dry areas (Table 1). Scent stations in dry areas were 2.92 km farther from a known water source compared to scent stations in wet areas (Table 1). At each scent station, we placed a scent lure (2011: fatty acid scented disc [Pocatello Supply Depot, Pocatello, Idaho, USA]; 2012: liquid scent [Murray's Lures, Walker, West Virginia, USA]) on the ground and an infrared-triggered camera (PC 900, Reconyx©, Holmen, Wisconsin) approximately two meters from the scent either directly north or south to avoid false camera triggers by the sun.

To evaluate relative rates of visitation to water sources by canids, we monitored all known water sources at the Great Basin and Mojave study areas. From May to October, 2010 to 2012, we used infrared-triggered cameras to photograph canids visiting water sources. We randomly sampled water sources with cameras at both study areas for approximately two-week periods for a total of 78 weeks. To determine which water sources to sample for a given period, we generated random points within each study area using ArcMap. We then identified the nearest water source to a random point and camera-sampled as many water sources as possible (in 2010, we used six cameras in each study area for sampling compared to 15 cameras in 2011–2012). We attached cameras to metal posts and placed them approximately two meters from the edge of water where animals gained access to drink. At water sources with multiple locations of drinking access (e.g., paired tanks of water, ponds), we placed cameras at a minimum of two locations where animals could drink. We considered proximity to trails and recent sign to determine the location of cameras at ponds and large springs [25]. Our estimates of visitation by canids at large water sources were likely conservative due to the inability to monitor all potential locations where canids could access water. We assumed, however, that any potential bias was similar at large water sources from both study areas.

Statistical Analyses

To test prediction one associated with the indirect effect of water hypothesis, we used z-tests for proportions [26] to compare the proportion of scent stations in wet and dry areas visited by canids at both study areas. We tested the spatial segregation component of the second prediction by comparing visits to scent stations by each canid using Kendall's *Tau-b* correlation analyses. Kendall's *Tau-b* correlation accounted for ties that occurred because of zero visit data due to only one species of canid primarily visiting a given station [27]. We excluded all stations that were not visited by at least one species of canid from these correlation analyses. To test the temporal segregation component of prediction two, we used time stamps available on photographs

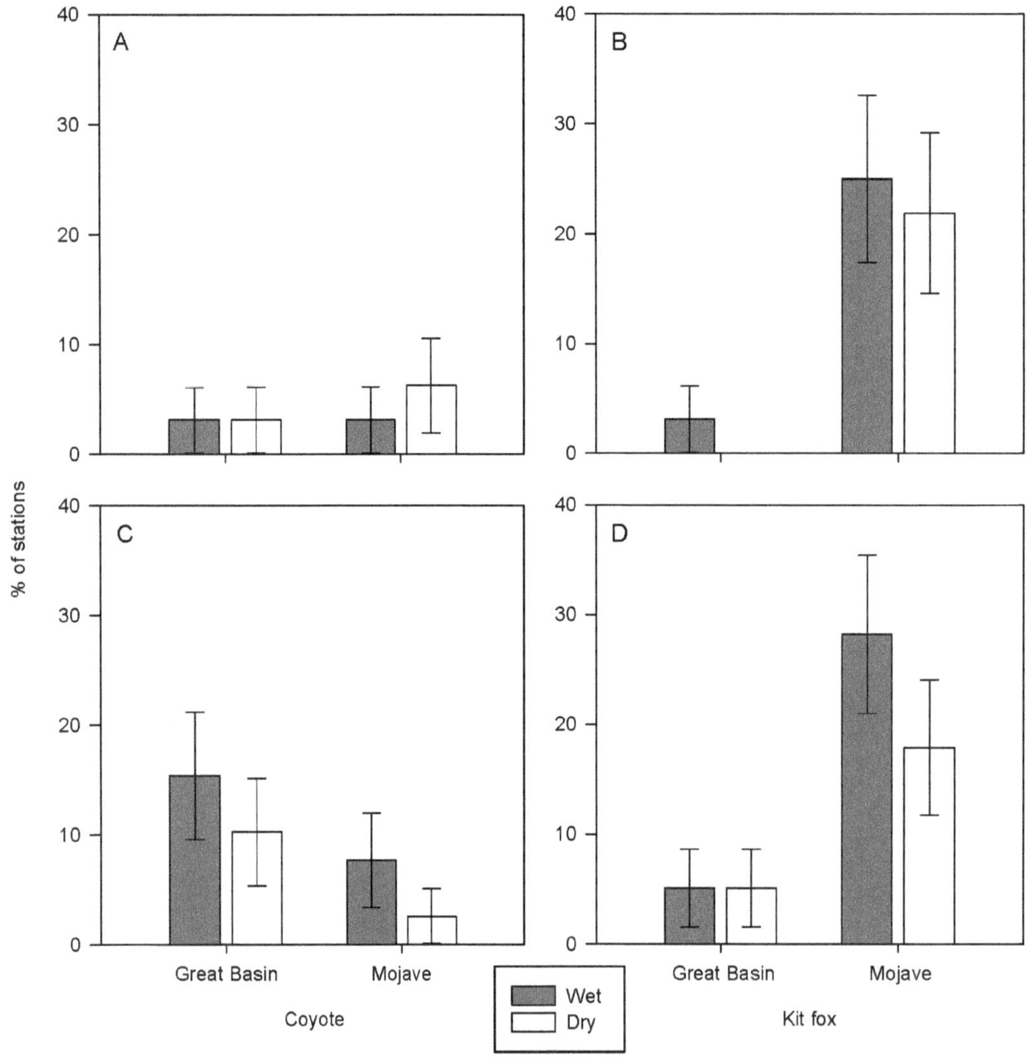

Figure 3. Proportions (±SE) of scent stations visited by canids in wet and dry areas in A) 2011 and B) 2012. Stations in wet areas were ≤2.6 km from free water (mean = 1.25 km) whereas stations in dry areas were ≥2.6 km from free water (mean = 4.17 km). Data were collected on coyotes (*Canis latrans*) and kit foxes (*Vulpes macrotis*) in the Great Basin and Mojave Deserts, USA.

of canids at scent stations and compared these times using Mann-Whitney *U*-tests. To account for the transition from 23:59 to 00:00 h (midnight) in our analyses and thus avoid the potential difference between photographs taken immediately prior to and after midnight we marked the change between days by setting 21:00 h (approximately sunset) as 00:00 h. This adjustment allowed us to compare times of visitation within a continuous scale of time marked by nocturnal hours which are likely more biologically relevant to both coyotes and kit foxes [28,29].

We used Mann-Whitney *U*-tests to compare the mean daily visitation rates (# of visits/# of operable camera trap days) of canids to water sources between study areas. We defined a visit as all photo captures of a species occurring within 30 min. Thus, photo captures occurring more than 30 min apart were considered independent [30]. We performed all analyses using Program R [31]. We conducted series of statistical analyses to test our hypotheses and we used a Bonferroni correction to avoid type I errors. We set the family-wise level of significance for all statistical tests at $\alpha = 0.05$ (Bonferroni adjusted $\alpha = 0.01$).

Results

We detected both coyotes and kit foxes at scent stations in Great Basin and Mojave Deserts (Fig. 1, Dataset S1). Other potential competitors and predators of kit foxes and coyotes that we observed at scent stations included badgers (*Taxidea taxus*), bobcats (*Lynx rufus*), and gray foxes (*Urocyon cinereoargenteus*); however, visits from these intraguild species were too few for meaningful comparison (Table 2). In Great Basin, we detected coyotes at more stations than kit foxes, but in Mojave we found the opposite relationship (Fig. 2). There was no difference between the proportion of stations visited by coyotes in Great Basin or Mojave (2011: $z = 0.47$, $P = 0.64$; 2012: $z = 1.77$, $P = 0.08$; Fig. 2). Conversely, the proportion of stations visited by kit foxes in Mojave was greater than in Great Basin (2011: $z = 4.04$, $P < 0.01$; 2012: $z = 3.52$, $P < 0.01$; Fig. 2). There was no difference in the proportions of visits to wet and dry stations by coyotes or kit foxes in 2011 (Great Basin coyotes: $z = 0.28$, $P = 0.78$; Great Basin kit foxes: $z = 0.84$, $P = 0.40$; Mojave coyotes: $z = 1.10$, $P = 0.27$; Mojave kit foxes: $z = 1.01$, $P = 0.31$; Fig. 3a) or 2012 (Great Basin

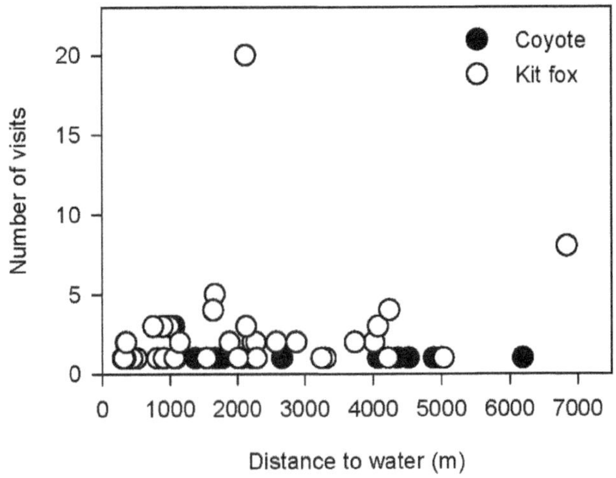

Figure 4. Visitation by canids to scent stations in relation to distance from free water. We defined a visit as all photo captures of a species occurring within 30 min. Data were collected on coyotes (*Canis latrans*) and kit foxes (*Vulpes macrotis*) in the Great Basin and Mojave Deserts, USA, 2011 to 2012.

coyotes: $z = 0.27$, $P = 0.79$; Great Basin kit foxes: $z = 0.27$, $P = 0.79$; Mojave coyotes: $z = 0.58$, $P = 0.56$; Mojave kit foxes: $z = 0.05$, $P = 0.96$; Fig. 3b). We detected coyotes and kit foxes at scent stations relatively close to and far from free water and did not observe a clear pattern between visitation and distance to free water up to 7 km for either canid (Fig. 4).

To determine spatial and temporal segregation there were too few visits by canids to scent stations in Great Basin during 2011 for statistical comparison. We did not observe a significant negative correlation between visits of coyotes and kit foxes to scent stations at Great Basin in 2012 ($Tau\text{-}b = -0.68$, $P = 0.02$, $df = 11$, $N = 13$). Similarly, in Mojave we did not observe a negative correlation between visits of both canid species during 2011 ($Tau\text{-}b = -0.27$, $P = 0.26$, $df = 15$, $N = 17$) or 2012 ($Tau\text{-}b = -0.23$, $P = 0.29$, $df = 17$, $N = 19$). There was no difference between the times of day when coyotes and kit foxes visited scent stations in 2011 at Mojave (Mann-Whitney $U = 63.00$, $P = 0.42$; Fig. 5a). In 2012, we did not detect a difference in times when canids visited scent stations in Great Basin (Mann-Whitney $U = 29.00$, $P = 0.12$; Fig. 5b) or Mojave (Mann-Whitney $U = 118.00$, $P = 0.25$; Fig. 5c).

We observed coyotes and kit foxes at water sources in both deserts (Fig. 1, Dataset S2). In 6,476 camera trap days at water sources in Great Basin, we observed 924 coyote visits and four kit fox visits. In 4,803 camera trap days at water sources in Mojave, we observed 353 coyote visits and 1,530 kit fox visits. In Great Basin, coyotes visited 19 of the 22 available water sources whereas kit foxes only visited two. In Mojave, coyotes visited 38 of the 66 available water sources and kit foxes visited 25. There was no difference between mean daily visitation rates for coyotes across study areas (Mann-Whitney $U = 490.00$, $P = 0.07$; Fig. 6). Alternatively, mean daily visitation rate was higher for kit foxes in Mojave than kit foxes in Great Basin (Mann-Whitney $U = 471.50$, $P < 0.01$; Fig. 6).

Table 2. Number of visits to scent stations by species of carnivores in 2011 and 2012.

Species	Great Basin	Mojave	Total
Coyote	13	12	25
Kit fox	11	82	93
Badger	2	1	3
Bobcat	1	1	2
Gray fox	0	2	2

Discussion

Our study was the first to evaluate the potential indirect effect that anthropogenic water developments may have on canid communities in two deserts. We tested the two predictions of the indirect effect of water hypothesis in the Great Basin and Mojave Deserts. We did not find any support for the first prediction as we detected coyotes equally in wet and dry areas in both deserts (and likewise for kit foxes). Also, we did not find support for the second prediction due to lack of spatial and temporal segregation between coyotes and kit foxes in both study areas. Our data indicated that factor(s) other than the presence or distribution of free water were associated with occurrence of coyotes.

Our spatial scale of inference was restricted to 7 km (the farthest distance between a water source and a scent station) and coyotes can travel farther than 7 km [32]. However, frequent travel beyond the distance of daily movements to visit a water source is energetically costly. If water sources served as a spatial "anchor" and coyotes were liberal in their daily movements, we would have expected to detect more coyotes in areas closer to water relative to areas farther from water. Furthermore, if coyotes were dependent on free water, we would have expected the number of visits by coyotes to scent stations to decrease as the distance of scent stations to the nearest water source increased. Moreover, given the availability of naturally occurring water sources in each area (median distance to water for 10,000 random points was 7.10 and 4.75 km for Great Basin and Mojave, respectively) and the change in availability due to human influence (median distances of 3.23 and 2.10 km for Great Basin and Mojave, respectively) this scale was relevant to both landscapes. Our inference, however, is limited to these scales and we cannot speak to distances beyond 7 km.

Although kit foxes have been observed to spatially and temporally avoid coyotes, results from our scent station experiment did not reveal avoidance [17,33,34]. Similar to our observations, other studies have documented kit foxes coexisting with coyotes without avoidance [35,36]. Observations with other carnivore communities have demonstrated that subordinate competitors can coexist with larger, dominant competitors. For example, coyotes (subordinate) did not spatially adjust their use of habitat to avoid wolves (*C. lupus*), rather coyotes altered behaviors near wolf-killed carcasses [37]. For kit foxes, the availability of resources and refuges (i.e., burrows) likely plays a role in how they partition space and time with coyotes [36,38].

Our data suggested that kit foxes were less abundant in Great Basin than Mojave, supporting previous reports of reduced populations in Great Basin [15]. However, abundance of coyotes appeared to be similar in both deserts, based on visits to scent stations and water sources. We suggest that coyotes may not solely regulate populations of kit foxes, though they can account for high rates of mortality [39,40]. Previous work has demonstrated that removal of coyotes did not influence survival of kit foxes, indicating that coyote-induced mortality may be compensatory and that other factors affect population dynamics of kit foxes, such as prey availability [41–43].

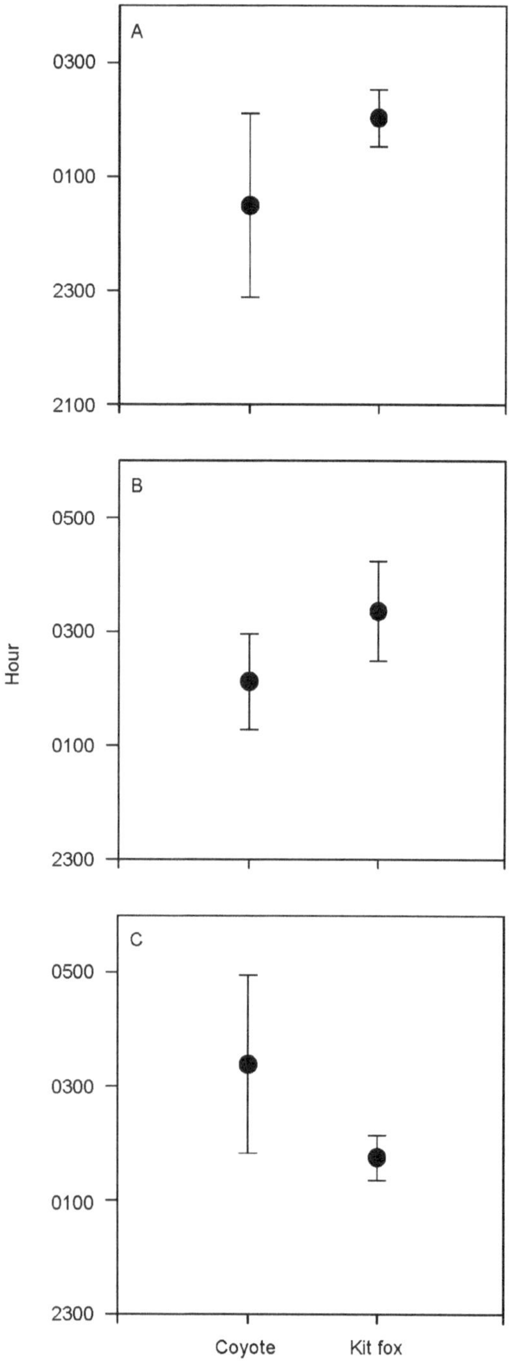

Figure 5. Mean times of day (±SE) when canids visited scent stations. Temporal visitation patterns were observed for coyotes (*Canis latrans*) and kit foxes (*Vulpes macrotis*) in A) Mojave during 2011, B) Great Basin during 2012, and C) Mojave during 2012.

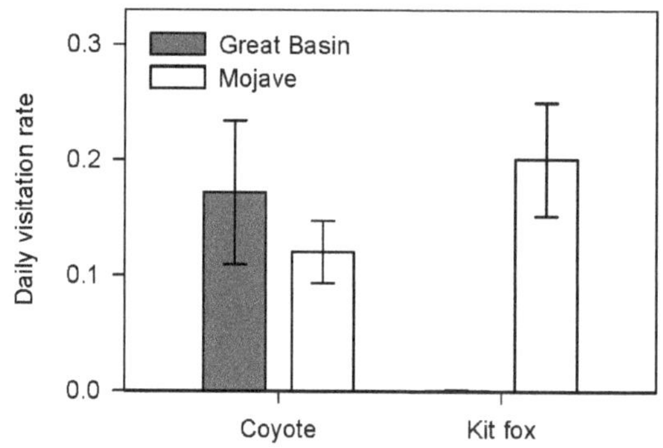

Figure 6. Mean daily visitation rates (±SE) of canids to sources of free water. We defined daily visitation rate as the number of species visits/the number of operable camera trap days per water source. Data were collected on coyotes (*Canis latrans*) and kit foxes (*Vulpes macrotis*) in the Great Basin and Mojave Deserts, USA, 2010 to 2012.

Historical variation in availability and distribution of free water in western North America may provide, in part, explanation for the lack of support for the indirect effect of water hypothesis in the Great Basin and Mojave Deserts. Western North America has experienced dramatic fluctuations in climate (and associated availability of water) over the last several thousand years. For example, 12,000 y BP much of Great Basin and Mojave was a wetland environment with large lakes [44,45]. Since that time, this region has alternated between levels of extreme drought and wet conditions [46]. The relatively recent addition of free water (i.e., water developments) in western North America, therefore, may not be novel to species inhabiting this region as both coyotes and kit foxes have experienced these conditions in their evolutionary histories. This natural variation in availability of free water over time has rarely been considered in controversies surrounding anthropogenic modification of water availability [47].

Kit foxes have been perceived to be independent of free water based on physiological and behavioral adaptations [18,19]. Moreover, historical distributions of kit foxes typically include areas located far from known sources of water, further supporting the notion that this species of canid can exist without free water [48]. Nonetheless, published accounts have reported sporadic use of free water by kit foxes [20,49]. Our study revealed an extreme rate of visitation to water developments by kit foxes in Mojave not previously reported in other areas of western North America. In Mojave, kit foxes were the most photographed carnivore at water developments and one of the most commonly photographed mammals [21]. The intensity of visitation to water developments by kit foxes in Mojave indicates that arid-adapted species may use water developments more frequently than previously believed [50].

Drinking free water may alleviate physiological stresses and improve survival even for species that are adapted to arid climates [51]. For kit foxes to persist without free water they need to consume nearly twice as much prey per day than what is solely required for energetic demands [18]. By drinking free water, kit foxes may reduce energy and time associated with securing additional prey items to satisfy water demand. Less time spent foraging and less distance traveled in search of prey also reduces the likelihood of encounters with other competitors and potential predators [38]. Furthermore, drinking free water may benefit female kit foxes during lactation due to additional loss of water via production of milk [7]. The frequent visitation to free water by kit foxes in Mojave suggests that water developments may be more beneficial to this species than what has been previously understood [52].

The difference between visitation rates of kit foxes to free water between deserts may be due to at least two factors. First, according to our scent station data, there are likely more kit foxes in Mojave than Great Basin, resulting in an increased probability of detection at free water. Second, nighttime temperatures at Great Basin (mean = 14.42°C, SE = 0.09) were on average 8.70°C cooler than Mojave (mean = 23.12°C, SE = 0.08) although maximum daytime temperatures at both study areas were similar (~41°C). Relatively warmer nighttime temperatures in Mojave likely resulted in a reduced thermal gradient for kit foxes to dissipate heat from nocturnal activities which may have resulted in higher rates of water loss [19] compared to kit foxes in Great Basin. Relatively warmer nights and subsequent higher rates of water loss likely created a greater physiological demand for free water by kit foxes in Mojave compared to Great Basin.

Our results did not provide support for the indirect effect of water hypothesis at Great Basin or Mojave Deserts. The apparent high abundance of kit foxes that we observed in Mojave, an arid landscape with many water developments, created a paradox based on the logic that water developments indirectly influence these canids via increased distribution of coyotes [15,16]. We did not find any support for the assertion that free water played a negative indirect role on kit foxes. Furthermore, our data did not indicate partitioning of activity in space or time. We reject the indirect effect of water hypothesis as operational in our study areas during our study years at the spatial scale evaluated. The ultimate factor(s) that influence the distribution of coyotes and kit foxes in these two deserts are unknown and warrant further study.

Supporting Information

Dataset S1 All visits to scent stations recorded by infrared-triggered cameras. We defined a visit as all photo captures of a species occurring within 30 min. Data were collected in the Great Basin and Mojave Deserts, USA, 2011 to 2012.

Dataset S2 All visits to water sources recorded by infrared-triggered cameras. We defined a visit as all photo captures of a species occurring within 30 min. Data were collected on coyotes (*Canis latrans*) and kit foxes (*Vulpes macrotis*) in the Great Basin and Mojave Deserts, USA, 2011 to 2012.

Acknowledgments

We thank R. Williams, J. Huber, D. Smedley, A. Raynor, and T. Bates for field assistance and picture analysis. We also thank B. Kluever, E. Gese, and three anonymous reviewers for comments that improved our study and manuscript. Special thanks to the Monte L. Bean Life Sciences Museum at Brigham Young University for lodging and use of facilities at the Lytle Ranch Preserve while we conducted this study.

Author Contributions

Conceived and designed the experiments: LKH RTL BRM. Performed the experiments: LKH RTL BRM. Analyzed the data: LKH RTL BRM. Contributed reagents/materials/analysis tools: RTL RNK KDB BRM. Wrote the paper: LKH RTL RNK KDB BRM.

References

1. Hamilton AM, Freedman AH, Franz R (2002) Effects of deer feeders, habitat and sensory cues on predation rates on artificial turtle nests. Am Midl Nat 147: 123–134.
2. Martinson TJ, Flaspohler DJ (2003) Winter bird feeding and localized predation on simulated bark-dwelling arthropods. Wildl Soc Bull 31: 510–516.
3. Cooper SM, Ginnett TF (2000) Potential effects of supplemental feeding of deer on nest predation. Wildl Soc Bull 28: 660–666.
4. Vallentine JF (1980) Range development and improvements. Provo: Brigham Young University Press. 545 p.
5. Bleich VC, Andrew NG, Martin MJ, Mulcahy GP, Pauli AM, et al. (2006) Quality of water available to wildlife in desert environments: comparisons among anthropogenic and natural sources. Wildl Soc Bull 34: 627–632.
6. Rautenstrauch KR, Krausman PR (1989) Influence of water availability and rainfall on movements of desert mule deer. J Mammal 70: 197–201.
7. Cain III JW, Krausman PR, Rosenstock SS, Turner JC (2006) Mechanisms of thermoregulation and water balance in desert ungulates. Wildl Soc Bull 34: 570–581.
8. Rosenstock SS, Ballard WB, Devos Jr JC (1999) Viewpoint: benefits and impacts of wildlife water developments. J Range Manag 52: 302–311.
9. Broyles B (1995) Desert wildlife water developments: questioning use in the southwest. Wildl Soc Bull 23: 663–675.
10. Krausman PR, Rosenstock SS, Cain III JW (2006) Developed waters for wildlife: science, perception, values, and controversy. Wildl Soc Bull 34: 563–569.
11. Cutler TL, Morrison ML (1998) Habitat use by small vertebrates at two water developments in southwestern Arizona. Southwest Nat 43: 155–162.
12. Hervert J, Krausman PR (1986) Desert mule deer use of water developments in Arizona. J Wildl Manag 50: 670–676.
13. Owen-Smith N (1996) Ecological guidelines for watering points in extensive protected areas. S Afr J Wildl Res 26: 107–112.
14. Longshore KM, Lowrey C, Thompson DB (2009) Compensating for diminishing natural water: predicting the impacts of water development on summer habitat of desert bighorn sheep. J Arid Environ 73: 280–286.
15. Arjo WM, Gese EM, Bennett TJ, Kozlowski AJ (2007) Changes in kit fox-coyote-prey relationships in the Great Basin Desert, Utah. West N Am Nat 67: 389–401.
16. Kozlowski AJ, Gese EM, Arjo WM (2008) Niche overlap and resource partitioning between sympatric kit foxes and coyotes in the Great Basin Desert of western Utah. Am Midl Nat 160: 191–208.
17. Nelson JL, Cypher BL, Bjurlin CD, Creel S (2007) Effects of habitat on competition between kit foxes and coyotes. J Wildl Manag 71: 1467–1475.
18. Golightly Jr RT, Ohmart RD (1984) Water economy of two desert canids: coyote and kit fox. J Mammal 65: 51–58.
19. Golightly Jr RT, Ohmart RD (1983) Metabolism and body temperature of two desert canids: coyotes and kit foxes. J Mammal 64: 624–635.
20. O'Brien CS, Waddell RB, Rosenstock SS, Rabe MJ (2006) Wildlife use of water catchments in southwestern Arizona. Wildl Soc Bull 34: 582–591.
21. Hall LK, Day CC, Westover MD, Edgel RJ, Larsen RT, et al. (2013) Vigilance of kit foxes at water sources: a test of competing hypotheses for a solitary carnivore subject to predation. Behav Process 94: 76–82.
22. Horn KJ, McMillan BR, St Clair SB (2012) Expansive fire in Mojave Desert shrubland reduces abundance and species diversity of small mammals. J Arid Environ 77: 54–58.
23. Bowman J, Jaeger JAG, Fahrig L (2002) Dispersal distance of mammals is proportional to home range size. Ecology 83: 2049–2055.
24. Bowman J (2003) Is dispersal distance of birds proportional to territory size? Can J Zool 81: 195.
25. Atwood TC, Fry TL, Leland BR (2011) Partitioning of anthropogenic watering sites by desert carnivores. J Wildl Manag 75: 1609–1615.
26. Zar JH (1999) Biostatistical analysis. New Jersey: Prentis-Hall. 663 p.
27. Kendall MG (1962) Rank correlation methods. London: Hodder Arnold. 210 p.
28. Neale JCC, Sacks BN (2001) Resource utilization and interspecific relations of sympatric bobcats and coyotes. Oikos 94: 236–249.
29. McGrew JC (1979) Vulpes macrotis. Mamm Species 123: 1–6.
30. Michalski F, Peres CA (2007) Disturbance-mediated mammal persistence and abundance-area relationships in Amazonian forest fragments. Conserv Biol 21: 1626–1640.
31. R Development Core Team (2011) R: a language and environment for statistical computing. Vienna: R Foundation for Statistical Computing, ISBN 3-900051-07-0. http://www.R-project.org.
32. Mills LS, Knowlton FF (1991) Coyote space use in relation to prey abundance. Can J Zool 69: 1516–1521.
33. Cypher BL, Spencer KA (1998) Competitive Interactions between coyotes and San Joaquin kit foxes. J Mammal 79: 204–214.
34. Warrick GD, Cypher BL (1998) Factors affecting the spatial distribution of San Joaquin kit foxes. J Wildl Manag 62: 707–717.
35. White PJ, Ralls K, Garrott RA (1994) Coyote - kit fox interactions as revealed by telemetry. Can J Zool 72: 1831–1836.
36. White PJ, Ralls K, White CAV (1995) Overlap in habitat and food use between coyotes and San Joaquin kit foxes. Southwest Nat 40: 342–349.
37. Atwood TC, Gese EM (2010) Importance of resource selection and social behavior to partitioning of hostile space by sympatric canids. J Mammal 91: 490–499.
38. Moehrenschlager A, List R, Macdonald DW (2007) Escaping intraguild predation: Mexican kit foxes survive while coyotes and golden eagles kill Canadian swift foxes. J Mammal 88: 1029–1039.
39. Cypher BL, Warrick GD, Otten MRM, O'Farrell TP, Berry WH, et al. (2000) Population dynamics of San Joaquin kit foxes at the Naval Petroleum Reserves in California. Wildl Monogr 145: 1–43.

40. Warrick GD, Scrivner JH, O'Farrell TP (1999) Demographic responses of kit foxes to supplemental feeding. Southwest Nat 44: 367–374.
41. Cypher BL, Scrivner JH (1992) Coyote control to protect endangered San Joaquin kit foxes at the Naval Petroleum Reserves, California. Proceedings of the Vertebrate Pest Conference 15: 42–47.
42. Dennis B, Otten MRM (2000) Joint effects of density dependence and rainfall on abundance of San Joaquin kit fox. J Wildl Manag 64: 388–400.
43. White PJ, White CAV, Ralls K (1996) Functional and numerical responses of kit foxes to a short-term decline in mammalian prey. J Mammal 77: 370–376.
44. Broecker WS, Kaufman A (1965) Radiocarbon chronology of Lake Lahontan and Lake Bonneville II, Great Basin. Geol Soc Am Bull 76: 537–566.
45. Currey DR (1990) Quaternary paleolakes in the evolution of semidesert basins, with special emphasis on Lake Bonneville and the Great Basin, U.S.A. Palaeogeogr Palaeoclimatol Palaeoecol 76: 189–214.
46. Cook ER, Woodhouse CA, Eakin CM, Meko DM, Stahle DW (2004) Long-term aridity changes in the western United States. Science 306: 1015–1018.
47. Larsen RT, Bissonette JA, Flinders JT, Whiting JC (2012) Framework for understanding the influences of wildlife water developments in the western United States. Calif Fish Game 98: 148–163.
48. Egoscue HJ (1956) Preliminary studies of the kit fox in Utah. J Mammal 37: 351–357.
49. O'Farrell TP (1999) Kit Fox. In: Novak M, Baker JA, Obbard ME, Malloch B, editors. Wild furbearer management and conservation in North America. Ontario: Queen's Printer for Ontario. 422–431.
50. Morgart JR, Hervert JJ, Krausman PR, Bright JL, Henry RS (2005) Sonoran pronghorn use of anthropogenic and natural water sources. Wildl Soc Bull 33: 51–60.
51. Brawata RL, Neeman T (2011) Is water the key? Dingo management, intraguild interactions and predator distribution around water points in arid Australia. Wildl Res 38: 426–436.
52. Simpson NS, Stewart KM, Bleich VC (2011) What have we learned about water developments for wildlife? Not enough! Calif Fish Game 97: 190–209.

14

Differential sRNA Regulation in Leaves and Roots of Sugarcane under Water Depletion

Flávia Thiebaut[1], Clícia Grativol[1], Milos Tanurdzic[2], Mariana Carnavale-Bottino[1], Tauan Vieira[1], Mariana Romeiro Motta[1], Cristian Rojas[3], Renato Vincentini[4], Sabrina Moutinho Chabregas[5], Adriana Silva Hemerly[1], Robert A. Martienssen[6], Paulo Cavalcanti Gomes Ferreira[1]*

1 Laboratório de Biologia Molecular de Plantas, Instituto de Bioquímica Médica, Universidade Federal do Rio de Janeiro, Rio de Janeiro, Rio de Janeiro, Brazil, **2** School of Biological Sciences, The University of Queensland, Brisbane St Lucia, Queensland, Australia, **3** Universidade Federal da INTEGRAÇÃO Latino-Americana, Foz do Iguaçu, Paraná, Brazil, **4** Centro de Biologia Molecular e Engenharia Genética, Universidade Estadual de Campinas, Campinas, São Paulo, Brazil, **5** Centro de Tecnologia Canavieira, Fazenda Santo Antônio – Laboratório de Biologia Molecular, Piracicaba, São Paulo, Brazil, **6** Howard Hughes Medical Institute and Gordon and Betty Moore Foundation, Cold Spring Harbor Laboratory, Cold Spring Harbor, New York, United States of America

Abstract

Plants have developed multiple regulatory mechanisms to respond and adapt to stress. Drought stress is one of the major constraints to agricultural productivity worldwide and recent reports have highlighted the importance of plant sRNA in the response and adaptation to water availability. In order to increase our understanding of the roles of sRNA in response to water depletion, cultivars of sugarcane were submitted to treatment of ceasing drip irrigation for 24 hours. Deep sequencing analysis was carried out to identify the sRNA regulated in leaves and roots of sugarcane cultivars with different drought sensitivities. The pool of sRNA selected allowed the analysis of different sRNA classes (miRNA and siRNA). Twenty-eight and 36 families of conserved miRNA were identified in leaf and root libraries, respectively. Dynamic regulation of miRNA was observed and the expression profiles of eight miRNA were verified in leaf samples from three biological replicates by stem-loop qRT-PCR assay using the cultivars: SP90–1638 - sensitive cultivar- and; SP83–2847 and SP83–5073 - tolerant cultivars. Altered miRNA regulation was correlated with changes in mRNA levels of specific targets. Two leaf libraries from individual sugarcane cultivars with contrasting drought-tolerance properties were also analyzed. An enrichment of 22-nt sRNA species was observed in leaf libraries. 22-nt miRNA triggered siRNA production by cleavage of their targets in response to water depletion. A number of genes of the sRNA biogenesis pathway were down-regulated in tolerant genotypes and up-regulated in sensitive in response to water depletion treatment. Our analysis contributes to increase the knowledge on the roles of sRNA in sugarcane submitted to water depletion.

Editor: Baohong Zhang, East Carolina University, United States of America

Funding: Research by the group is supported by the Conselho Nacional de Desenvolvimento Científico e Tecnológico (CNPq), Instituto Nacional de Ciência e Tecnologia em Fixação Biológica de Nitrogênio (INCT), Financiadora de Estudos e Projetos (FINEP), Fundação de Amparo à Pesquisa do Rio de Janeiro (FAPERJ) and Coordenação de Aperfeiçoamento de Pessoal de Nível Superior (CAPES). The funders had no role in study design, data collection and analysis, decision to publish, or preparation of the manuscript.

Competing Interests: The authors have declared that no competing interests exist.

* E-mail: paulof@bioqmed.ufrj.br

Introduction

Epigenetic regulation has gained increasing interest recently as a possible source of variation for plant breeding [1]. Several studies have shown that epigenetic regulation is essential for plant developmental processes [2], modifying our understanding of how plants can respond and adapt to different stress situations [3,4]. Epigenetic regulation is coupled with small RNA (sRNA) to bring about transcriptional gene silencing via DNA and histone methylation, or can act at post-transcriptional level by targeting mRNA for degradation via microRNA (miRNA)-directed cleavage [5]. sRNA are increasingly being investigated as an epigenetic mechanism involved in responses to stresses in plants [6].

The discovery of small non-coding RNA with regulatory functions has modified our notion of gene, both in terms of its definition and gene regulation. These small RNA provide regulatory plasticity, and are divided into two categories: microRNA (miRNA) and small interfering RNA (siRNA), which include *trans*-acting small interfering RNA (ta-siRNA) [7]. These sRNA classes are generated by the processing of a double-stranded RNA (dsRNA) precursor by RNase III-like Dicer proteins (DCL) [8]. However, the type of precursor, the enzymes of the biogenesis pathways, and mechanism of action can distinguish the sRNA classes.

The canonical miRNA are endogenous 21–24-nucleotide (nt) long sequences transcribed from MIR genes [9]. In plants, the primary transcripts of MIR genes are stem-loop partially double-stranded RNA structures, which are processed by DCL1 and two other proteins, HYPONASTIC LEAVES 1 (HYL1) and SERRATE (SE) to generate a miRNA/miRNA* duplex, which is methylated in both strands by HUA ENHANCER 1 (HEN1) [10] and exported by HASTY (HST) from the nucleus to the cytoplasm [11]. Next, one strand, the mature miRNA, is loaded into the RNA-Induced Silencing Complex (RISC), a ribonucleoprotein complex which includes the ARGONAUTE1 (AGO1) protein

[12]. AGO1 then binds the target mRNA transcript via complementarity to the miRNA sequence previously loaded in the active RISC. In plants, the regulation of expression occurs by target mRNA cleavage with near-perfect complementarity, or, less frequently, by translational inhibition or DNA methylation [13,14].

Precursors of siRNA, in contrast, are originated mainly from transcripts from repetitive DNA in the genome, but also by exogenous viral sources [15]. These transcripts are then made double stranded via the action of an RNA Dependent RNA polymerase activity (mainly RDR2) and further processed into siRNA by DCL3 [16] which, similarly to miRNA, are loaded in AGO4, and participate in directing DNA and histone methylation to the target locus, thereby causing its transcriptional silencing [17].

Trans-acting siRNA, on the other hand, are reminiscent of miRNA in terms of their role in post-transcriptional gene regulation; however, their biogenesis shares common processes with miRNA and siRNA. The primary ta-siRNA (pri-ta-siRNA) is a single-stranded RNA, which contains a target site for miRNA binding, necessary to trigger the ta-siRNA formation. After the cleavage of the pri-ta-siRNA, RNA-dependent RNA polymerase 6 (RDR6) generates a double-strand from 3′ fragment, the ta-siRNA precursor (pre-ta-siRNA). This precursor is processed by DCL4 into a duplex that will originate 21-nt ta-siRNA [18]. Remarkably, ta-siRNA mediated regulation is an ancient pathway, once they occur in mosses and all the way to flowering plants [19,20].

An important area of investigation is the role of sRNA regulation upon environmental stresses [21]. Given that RNAi mechanisms have been retained in all major plant clades, it is not surprising that miRNA and different classes of siRNA have been reported to participate in plant responses to abiotic stress [22]. A large fraction of the miRNA families, the most abundant in the majority of the plants, was first described in model species, such as *Arabidopsis thaliana*, where almost the entire genome sequence is available [5,23]. However, in polyploid plants with large genomes, where a complete genomic sequence is still missing, such as sugarcane (*Saccharum spp.*); miRNA discovery is significantly more complex. On the other hand, high-throughput sequencing technologies and bioinformatics analyzes enable the identification of conserved microRNA sequences by comparison with databases of small RNA sequences [24,25].

Sugarcane is a monocot plant from the Poaceae family, whose production is responsible for 60% of the raw sugar manufactured worldwide; it also represents an important source of biofuel. Despite of its economic importance, sugarcane gene regulatory pathways have not been studied extensively, in part due to its genomic complexity [26]. The description of gene regulatory pathways in sugarcane is an important step in a possible creation of tolerant varieties to drought, which is major goal in countries where sugarcane is cultivated.

As a first step to investigate the role sRNA regulation in the response of sugarcane to water depletion, we constructed and sequenced sRNA libraries from different cultivars of sugarcane in control and stress situation. Bioinformatics analyses identified 28 and 36 conserved miRNA families in leaf and root libraries, respectively. Two leaf libraries from individual sugarcane cultivars that have contrasting to drought-tolerance properties were also analyzed. MiRNA are differentially expressed in leaves and roots upon treatment with water depletion. The expression profiles of eight miRNA (miR156, miR159 (two isoforms), miR164, miR167, miR168, miR169 and miR397) differentially expressed in leaves were examined using stem–loop RT-PCR assay. The biological functional of miRNA was accessed by analyzing their putative targets. Furthermore, a large-scale investigation of sRNA in sugarcane using a computational approach has identified siRNA candidates from targets of miRNA with 22-nt in the length. The expression of one 21 nt siRNA, which is present in all libraries, was up-regulated in three different cultivars in response to water depletion. To investigate the distribution of the siRNA fraction was aligned in cluster of sugarcane repeats, and the T0h libraries have shown a distinct distribution of siRNAs. Furthermore, expression analyzes of genes from the machinery of sRNA biogenesis, including DCL1, DCL2, RDR6, HYL1, and three genes encoding of AGO1 and AGO4 have shown down-regulation in drought tolerant cultivars and up-regulation in sensitive cultivar in response to 24 h of water depletion. Our findings contribute to increase the knowledge of sRNA roles in response to water depletion in the complex, polyploidy sugarcane plant.

Materials and Methods

Plant Material and Experimental Procedure

The Centro de Tecnologia Canavieira (CTC) provided stalks of sugarcane cultivars, with different drought sensitivities. Based on chlorophyll and water content measurements, cultivars CTC15, CTC6, SP83–2847 and SP83–5073, and CTC9, CTC13, SP90–1638 and SP90–3414 were considered as drought tolerant and sensitive, respectively. Stalks were planted germinated and grown in 5 L pots with a mixture of soil and sand (2:1) in a greenhouse at 28°C, and irrigated by dripping. After three months, the plants were exposed to drought by withholding water for 24 hours. Treated and control roots and shoots were harvested at 0 and 24 hours of treatment, respectively. Preparation of small RNA libraries of root samples was described previously [27]. Similarly, four leaf sRNA libraries were constructed from RNA pools from sensitive and tolerant sugarcane cultivars submitted to water depletion or from control plants. Additionally, three experiments were carried out using the same experimental procedure to obtain plant material to validate the bioinformatics analysis of the libraries. However, for this experiment only three cultivars (SP83–2847 and SP83–5073 and SP90–1638) were used. Besides, only shoots were harvested.

RNA Extraction and Sequencing Small RNA Library Construction

Sensitive and tolerant cultivars in each experiment were grouped as a pool of treated plants (plants that received the treatment of water depletion) and control plants. Total RNA for each pool was isolated from fresh root and leaf plant material using Trizol (Invitrogen, CA, USA) as described by the manufacturer. The amount of RNA was measured using Thermo Scientific NanoDrop 2000c Spectrophotometer and then the quality was verified by electrophoresis on a 1% agarose gel. The RNA was used to construct the small RNA libraries, using Illumina-based small RNA cloning protocol [28]. Detailed protocol is available upon request. In addition, using the same strategy cited above, more two sRNAs libraries were constructed using leaves from individual cultivar one tolerant cultivar (CTC6) and one sensitive cultivar (SP903414) submitted to 24 h of water deficit. After this, sRNAs were sequenced on the Illumina GAII sequencer at Cold Spring Harbor Laboratory. The sequence data from this study have been submitted to NCBI - Gene Expression Omnibus (http://www.ncbi.nlm.nih.gov/geo/) under accession number GSE42483. The quality of the sequences was evaluated by measuring the quality of the reads according to the percentage of bases having a base quality greater or equal than 30 (Q30). On average, 80% of each flow cell lane had Q30 of quality. Next, adapters and "N" bases were trimmed of the reads, and reads in the 18–28 nt range were separated for further analysis. Using the UEA sRNA toolkit-Plant version filter pipeline (http://srna-tools.cmp.uea.ac.uk/) [29], and three different databases - all RNA from Rfam, all Arabidopsis tRNA from The Genomic tRNA Database and all plant t/rRNA sequences from EMBL release - reads with low-complexity (less than 3 different bases) and both sense and antisense matches with different types of RNA (e.g. sno/t/rRNA) were removed.

Bioinformatics Analysis

Identification and electronic northern of miRNA. Filtered sRNAs reads were submitted to the University of East Anglia sRNA toolkit-Plant version miRProf, which determines the expression levels of sRNAs matching them to known miRNAs at miRBase and at a specific genome. The expression level of a miRNA represents the number of occurrences of the sequence in the library. miRProf was run with sRNA minimum size of 18 nt, maximum size of 28 nt. The sRNA libraries were matched to known Viridiplantae mature miRNA deposited in the miRBase database release 17 (http://www.mirbase.org/ftp.shtml) and to *Sorghum bicolor* genome (JGI, v1.0) by using PatMaN (a miRProf inside program). The output of the miRProf showed sequences of miRNA that had a maximum of three mismatches with the miRBase and sorghum genome and their unique sequence counts. To allow comparison between libraries, unique counts were normalized in reads *per* 1 million (RPM) and the total number of filtered reads for each library was used for normalization. With the identified miRNA sequences and their normalized expression numbers, we represented the miRNA expression levels on each sample as electronic northern. The changes of miRNA expressions between libraries were evaluated by using the IDEG6 software available at (http://telethon.bio.unipd.it/bioinfo/IDEG6_form/). The statistical analysis (Fisher exact test) was performed with a p-value cutoff <0.05 and Bonferroni correction. The normalized miRNA expression data have been submitted to NCBI - Gene Expression Omnibus (http://www.ncbi.nlm.nih.gov/geo/) under accession number GSE42483.

MiRNAs from sRNAs libraries of CTC6 (tolerant cultivar) and SP903414 (sensitive cultivar) submitted to water depletion (24 h) were identified by a similar pipeline. The following comparisons of miRNA profile among pool libraries and individual cultivar libraries were performed: Comparison 1– influence of genetic variation on miRNA response to water depletion (T24h/S24h vs. CTC6/SP903414); Comparison 2– influence of water depletion treatment in miRNA response on each genotype (CTC6/T0h vs. T24h/T0h) and (SP903414/S0h vs. S24h/S0h).

miRNA target prediction. To identify putative miRNA targets, we used the standalone-based UEA sRNA toolkit-Plant target prediction pipeline. This tool allows choosing transcript databases for searching targets. In this investigation, we have used *Saccharum officinarum* ESTs – DFCI gene Index release 3. MiRNA/target duplexes must obey the following rules: i) no more than four mismatches between sRNA and target (G-U bases count as 0.5 mismatches); ii) no more than two adjacent mismatches in the miRNA/target duplex; iii) no adjacent mismatches in positions 2–12 of the miRNA/target duplex (5′ of miRNA); iii) no mismatches in positions 10–11 of miRNA/target duplex; no more than 2.5 mismatches in positions 1–12 of the of the miRNA/target duplex; iv) and the Minimum Free Energy (MFE) of the miRNA/target duplex should be> = 74% of the MFE of the miRNA bound to its perfect complement.

Analysis of siRNA candidates. Analysis of size distribution of reads unraveled a large fraction of sRNA of 22-nts, among the sRNA. Putative targets of these 22 nt sRNA were selected to identify putative ta-siRNA candidates. These targets were aligned to sRNA libraries without the previously identified miRNA sequences, using BLASTN tools with no gaps and minimum alignment of 18 bases.

To classify the fraction of sRNA present in all libraries, we aligned the sRNA present in all libraries, excluding the list of miRNA, in clusters of siRNA. The clusters were constructed with a dataset of small RNA libraries (approximately 20 libraries, with miRNA sequences excluded) and the annotation was based s in the classification of a sugarcane repeat database (Grativol *et* al., in preparation) and in the sugarcane EST database.

Expression Analysis by qRT-PCR

The expression profiles of eight sugarcane mature miRNA (miR156, miR159 (two isoforms), miR164, miR167, miR168, miR169 and miR397) and ta-siRNA candidates were assayed by stem–loop reverse transcription-PCR [30,31]. Total RNA extracted from leaves used in the small RNA libraries construction was treated with DNaseI (Promega). Total RNA was then reverse transcribed into cDNA using Super-Script III reverse transcriptase (Invitrogen). In the same reaction, RT primers specific of each sRNA and random primers were used to enable the amplification of the constitutively expressed control gene and the miRNA targets. Moreover, the expression profile of the genes involved in the synthesis of sRNA (DCL1, DCL2, RDR6, HYL1, AGO1 and AGO4) was investigated by qRT-PCR with samples obtained from sugarcane plants obtained using the same experimental conditions and the same cDNA described above.

Using this cDNA as template, qRT-PCR was done with SYBR Green PCR Master Mix (Applied Biosystems) using: 1 μL of first strand cDNA, 5 μL of SYBR Green solution, 2 μL of the forward primer (10 μM) and 2 μL of reverse primer (10 μM) [30]. Glyceraldehyde-3-phosphate dehydrogenase - GAPDH (CA2546 72.1 - F/5′CACGGCCACTGGAAGCA3′, R/5′TCCTCAGG-GTTCCTGATGCC3′) and 28S (F/5′GCGAAGCCAGAGGA-AACT3′, R5′GACGAACGATTTGCACGTC3′) were used as the internal control. GAPDH and 28S were validated as sugarcane housekeeping genes in different conditions [32–35]. qRT-PCR was performed using an Applied Biosystems 7500 Real-Time PCR Systems. To confirm the expression of miRNA and their targets three biological replicates were analyzed.

Results

Enrichment of 22-nt sRNA Species is Tissue Specific

Eight sRNA libraries were constructed and sequenced (four from leaf RNA and four from root RNA) to explore the role of sRNA in sugarcane plants under water depletion. In each case, we have used RNA isolated from pools of sugarcane cultivars submitted to water depletion for 24 hours and control plants. These cultivars were previously classified as tolerant or sensitive to water depletion according to sugarcane production, chlorophyll content and water content under drought stress (data not shown). Illumina-based sequencing yielded approximately 35 million and 24 million reads in leaf and root libraries, respectively (Table 1). To analyze the sRNA population, we followed the bioinformatics pipeline described in the Figure 1 and methods. Approximately 14 million filtered reads in leaves - corresponding to 164,422 unique sequences; and approximately 16 million in roots – were obtained, corresponding to 1,210,502 unique sequences (Table 1). After trimming and filtering adaptors and tRNA/rRNA sequences, the remaining sRNA sequences were analyzed for size distribution (Figure 2). In all root libraries, the most abundant sRNAs were the 21 and 24 nucleotide (nt) in length, respectively (Figure 2A). The more complex fraction was the one containing the 24 nt sRNA (Figure 2B). However, in leaf libraries an additional peak of 22 nt was observed in the distribution of redundant reads (Figure 2C, D). Accordingly, the distribution of leaf non-redundant reads also showed three peaks, 21 nt, 22 nt and 24 nt.

The miRNA set is more Complex in Root Libraries

In order to identify the sugarcane miRNAs, the set of sRNAs recovered from the libraries were aligned to the miRBase database, accepting up to three mismatches. A total of 205 and 312 unique miRNA sequences were found in leaf and root libraries, respectively. Out of these, 88 and 273 (leaf and root, respectively) are conserved miRNA or miRNA* (Table 1), previously identified in other plants. The remaining sequences were classified as new miRNA candidates and their characterization has been described before [27]. Sixty-four sequences are common among libraries from the two tissues (Figure 3A). Overall, the identified conserved-miRNA corresponded to 28 miRNA families in leaves, and 36 miRNA families in roots. Eleven different miRNA* were found in the leaf libraries and 54 in the root libraries. The distribution of miRNA according to length in root and leaf libraries is shown in the figure 3B. The majority of the conserved miRNA had 21-nt in both tissues and the majority of miRNA had Uracil (U) at the 5′-end (Figure 3C).

Dynamic Regulation of miRNA in Roots and Leaves under Water Depletion

The abundance of the different miRNAs can be calculated from their frequency in each library. To compare the abundance distribution of miRNAs in libraries from control samples and samples treated with water depletion, we normalized the miRNA reads and compared their detectable expression levels on each sample through electronic northern (Materials and Methods). Twenty known miRNA - from 88 identified in leaves - were expressed in all leaf libraries and 169 - from 273 were shared between root libraries (File S1 and S2, respectively). In the control, 94% of miRNA from leaf libraries have distinct expression patterns among cultivars (T0h and S0h). Among tolerant and sensitive cultivars submitted to water depletion (T24h and S24h), the percentage of miRNA differentially expressed ($p<0.05$) was reduced to 83% (File S3). A similar comparison was made in root samples; and 59% of miRNAs were differentially expressed in the cultivars, while 56% changed their expression in stressed plants (File S4). Independently of the condition, most miRNA were differentially expressed in libraries prepared from drought sensitive cultivars.

In order to evaluate if the miRNA profiles observed in pooled samples (tolerant and sensitive libraries) were replicated in individual genotypes, eight mature miRNA, identified in both tissues, were selected and their expression profile was analyzed by qRT-PCR using leaf samples from three biological replicates. In each experiment, leaves of the cultivars collected after 24 h of water depletion and control plants were used for this analysis. The selection of the cultivars was based on results of experiments performed in Centro de Tecnologia Canavieira – CTC, which evaluated of drought stress responses of 20 commercial sugarcane

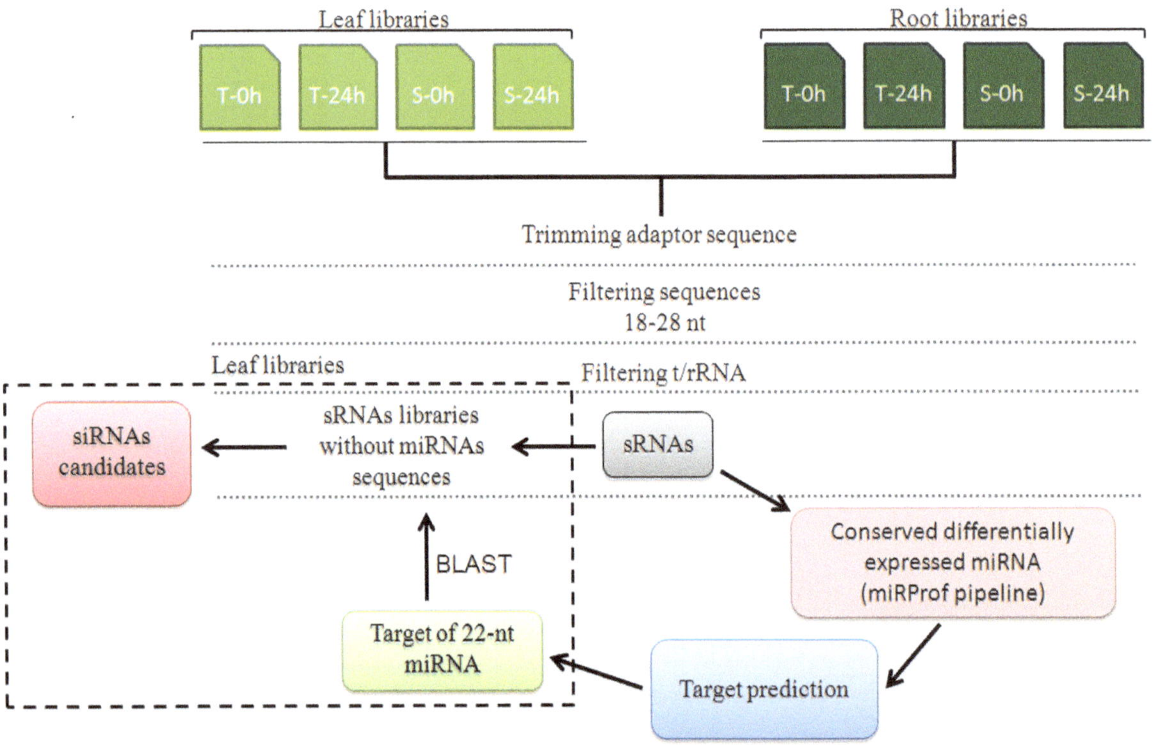

Figure 1. Pipeline for analysis of sRNA libraries. Pipeline used to identify sugarcane sRNA regulated by water depletion from leaf and root sRNA libraries. The identification of siRNA candidate (dotted box) was performed only in leaf libraries. T = tolerant sugarcane cultivar, S = sensitive sugarcane cultivar. 0 h = control of stress, 24 h = time of water depletion treatment.

Table 1. Summary of results obtained after computational data mining of small RNA libraries.

Description	Leaf libraries	Root libraries
Total		
All reads	35,755,991	24,243,352
Filtered reads[a]	14,193,780	16,956,062
Unique		
Filtered reads[a]	164,422	1,210,502
Conserved sRNA[b]	88	273

[a]Filtering for tRNA, rRNA, low-complexity sequence and trimming for "N" bases and 3′ adapters.
[b]Conserved miRNA deposited at miRBase database identified by miRProf pipeline.

cultivars (Sabrina M. Chabregas, personal communication). The measurement was carried out after cessation of irrigation by dripping. The physiological parameters analyzed - chlorophyll content, fluorescence and relative water content - established different levels of drought tolerance among sugarcane cultivars, although the mechanisms triggered by the stress can vary, including leaf curling, stomata closure and cell wall thickening. Based on these parameters, the cultivars SP901638 was considered the most sensitive cultivar, while SP832847 and SP835073 were considered as tolerant cultivars. These two tolerant cultivars have important characteristics: SP832847 is distinguished by its high productivity, while SP835073 has elevated sucrose and fiber content.

In four analyzed miRNAs (miR159X, miR168II, miR397II and miR164II), the differential expression profiles obtained by sequencing were similar with the results of the validation by RT-qPCR in at least one of biological replicates. Figure 4A shows the number of reads *per* million found in each leaf library for these miRNAs. These graphs allow the comparison of the abundance of these miRNA differentially expressed in response to the water depletion treatment. Similar profiles between the relative expressions obtained by pulsed RT-qPCR and bioinformatics-based expression analysis were observed for some, but not all, microRNAs. For the validation by RT-qPCR, samples from three experiments were used (Rep1, Rep2 and Rep3) to analyze the expression profile of the miRNAs in the cultivars SP901638,

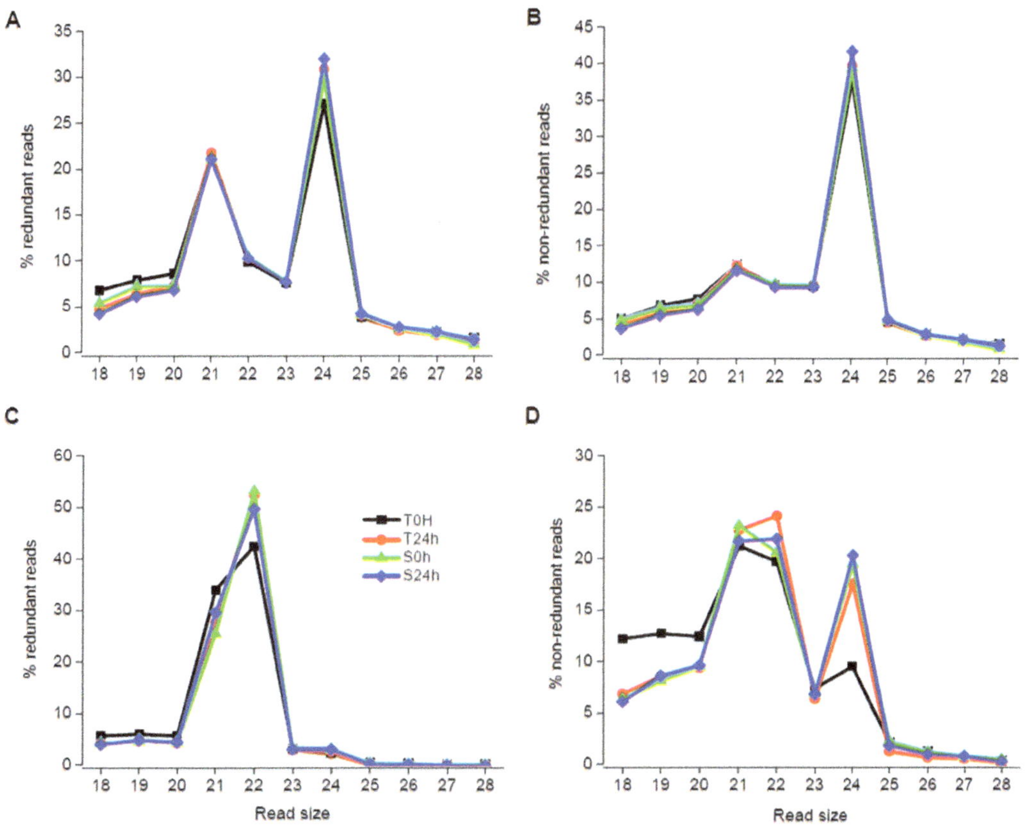

Figure 2. Small RNA reads sizes distribution. Graphs depict the length distribution of redundant and non-redundant small RNA dataset in root (A and B) and in leaf (C and D). T = tolerant sugarcane cultivar, S = sensitive sugarcane cultivar.

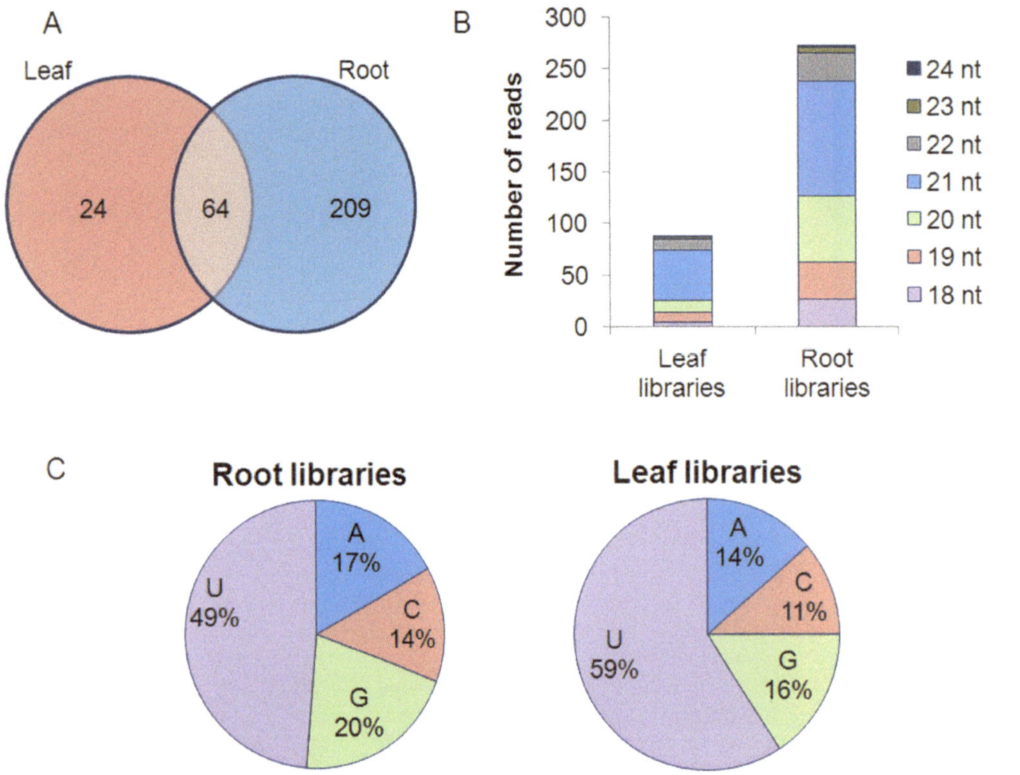

Figure 3. Distribution of conserved miRNA identified. (A) Diagram shows number of miRNA in common and unique in leaf and root libraries. (B) Distribution of miRNA length in leaf and root libraries. (C) Percentage of miRNA identified based on 5′ terminus nucleotide. A: adenine; U: uracile (thymine conversion); C: cytosine; G: guanine.

SP832847 and SP835073 (Figure 4B). In the sequencing analysis, miR159 was induced in tolerant cultivars submitted to water depletion (Figure 4A), and this expression profile was confirmed by RT-qPCR using the cultivar SP832847. However, the other tolerant cultivar showed different miR159 expression profiles between replicates samples, although in the Rep3, this miRNA was up-regulated (Figure 4B). In the sensitive cultivar its expression did not vary in bioinformatics analysis and in RT-qPCR analysis for Rep1 and Rep3 samples, while for Rep2 samples, miR159 was induced in response to water depletion (p-value <0.05). MiR164 II was significantly down-regulated in sRNA libraries of sensitive and tolerant cultivars. In two biological replicates this miRNAs was repressed in response to water depletion (Figure 4B), confirming the expression profile observed in sRNA libraries analyses (Figure 4A). This result could suggest that this miRNA acts in the early stage of drought stress response, as has been demonstrated in another sugarcane genotype [36]. MiR168 was also down-regulated in the sRNA libraries analysis (Figure 4A); nevertheless, this miRNA was down-regulated only in one biological replicates of each cultivar (Figure 4B). These differences of miRNAs regulation observed between biological replicates and the profiles obtained in the sRNA libraries can be reflecting the fact that each miRNA is regulating genes involved in quite diverse biological phenomena. For instance, while some miRNAs are regulating leaf morphogenesis, [39,40], plants can trigger different physiological mechanism in response to water stress [37], such as the leaf curling, leaf wilting and reduced leaf length [38]. The other miRNA analyzed - the miR397 - was repressed in response to water stress in sensitive cultivars and induced in tolerant cultivars (Figure 4A). The result of qRT-PCR (Figure 4B) showed a repression of this miRNA in sensitive cultivar from Rep1 and Rep2 (p-value <0.05), similar to the sequencing analysis. In contrast, miR397II shows a tendency to increase its expression in SP832847 and SP835073, confirming the profile observed in the sRNA libraries of tolerant cultivars.

In addition, we also verified the expression of four other miRNA (miR156V, miR167 VII, miR169 III and miR159 II) by qRT-PCR using samples from Rep1 (File S5). These four miRNA were induced in tolerant genotype SP832847 exposed to drought stress; however, only for miR156 V this up-regulation was statistically significant. The expression of miR159II was inverse to what was observed in the sRNA libraries. On the other hand, the profiles of miRNA expression found in pooled samples (tolerant and sensitive libraries) were validated for other miRNA (miR156 V, miR167 VII and miR169 III) in the qRT-PCR experiments.

To verify whether the regulation of miRNAs observed in the pool libraries is mostly due to water depletion or could be influenced by genetic variation, two leaf libraries were constructed and sequenced using isolated sugarcane cultivars submitted to treatment of water depletion. These two cultivars, CTC6 and SP90–3414, were also used in pool libraries analysis, and according to the CTC, they have contrasting drought-tolerance. The identification of miRNA from these libraries was performed using a similar pipeline previously described for the pooled samples libraries, and the normalized abundance distribution of miRNAs identified is available in File S1. A total of 54 miRNAs had their expression profile compared in two conditions: comparison 1- to verify if miRNA regulation was influenced by genetic variation in response to stress treatment (T24h/S24h vs. CTC6/SP903414); and comparison 2– to verify if miRNA

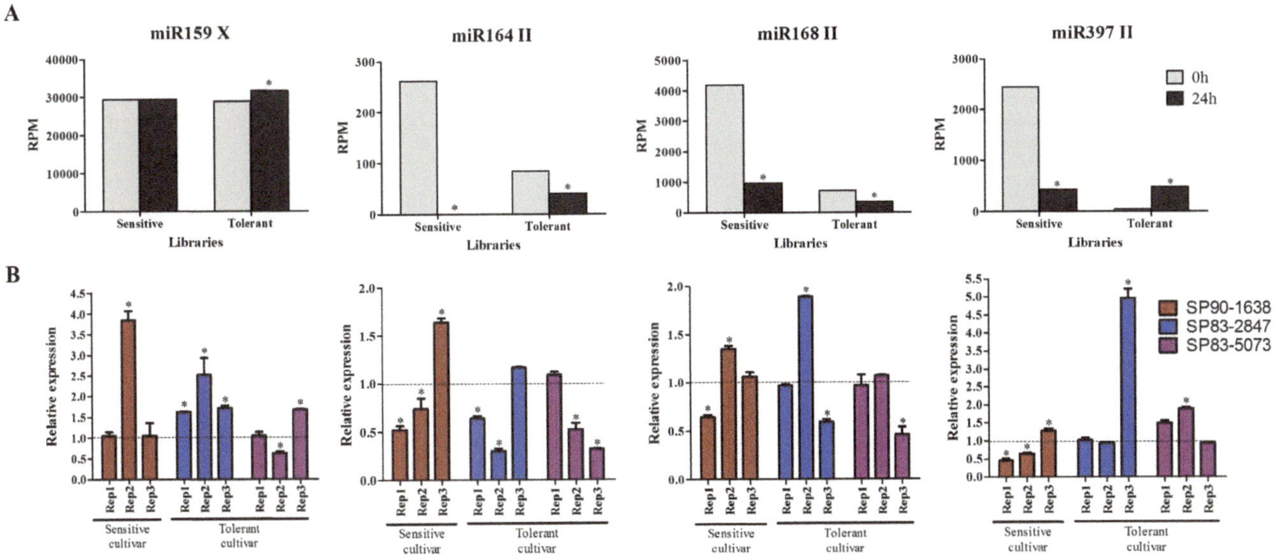

Figure 4. Comparison of miRNA expression profile. The expression profile of four miRNA were analyzed: miR159 X, miR164 II, miR168 II and miR397 II. (A) The graphs show the expression profile of miRNA based in bioinformatics analysis for each library (reads *per* million - RPM). *represent significantly changing of miRNA expression between control and treatment libraries (p-value <0.05). (B) The graphs show the relative expression of miRNA in each cultivar by qRT-PCR in three biological replicates. Three cultivars were used in these analyses (SP90–1638– sensitive cultivar; SP83–2847 and SP83–5073– tolerant cultivars). In each case, control condition had relative expression equal 1 (dotted line). *represent significantly changing of miRNA expression between control and treatment samples (p-value <0.05).

regulation was influenced by water depletion on tolerant and sensitive genotypes (CTC6/T0h vs. T24h/T0h) and (SP903414/S0h vs. S24h/S0h). These analyses resulted in a list of 39 miRNA that are positive for at least one of these comparisons (Table 2). Seventeen miRNAs had the same regulation profile in pool and individual cultivar libraries (Figure 5). Our results showed that 10 miRNAs share the similar expression profile only in comparison 1, suggesting an influence of genetic variation in miRNA response (Figure 5). In contrast, expression of 12 miRNAs was only influenced by water depletion (Figure 5). Among the miRNAs used for qRT-PCR, miR159 X and miR168 II had the same profile in both comparisons, suggesting that the of these miRNA regulation were influenced by both genotype and water depletion. These comparisons suggested that the regulation of miR164 II and miR397II was influenced by the treatment. Finally, for 15 miRNAs it was not possible to conclude if their regulation is mostly influenced by water depletion or by genetic variation (Figure 5).

Target Regulation

To achieve a better understanding of the biological function of the identified miRNA, we searched for putative miRNA targets using a bioinformatics approach. Based on perfect or near-perfect match between miRNA/target in plants [37], we used the plant target prediction tools available in a public website (http://srna-tools.cmp.uea.ac.uk) to identify targets for those miRNA identified. We searched for miRNA targets in the *Saccharum officinarum* ESTs – DFCI gene index v.3 sugarcane database. This analysis identified 677 putative targets for miRNA found in leaf libraries (File S6). The annotation of the miRNA targets shows that several of them are transcription factors. Interestingly, putative targets of some miRNA* were also identified (miR160* I, miR160* II, miR160* III, miR166* II, miR166* III, miR169* I, miR396* I and miR399* I).

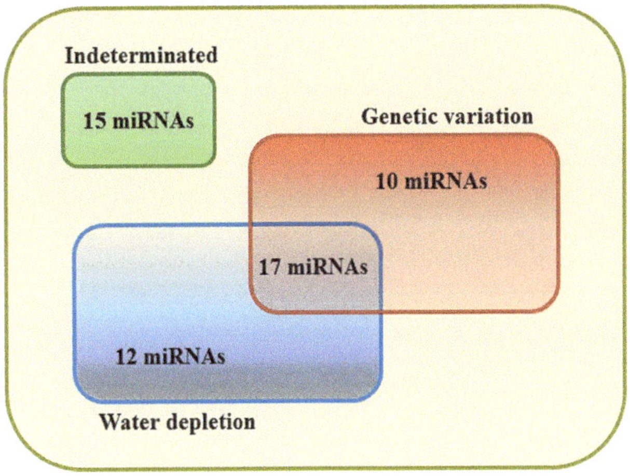

Figure 5. Regulation of miRNAs whether by genotype or/and water depletion. Comparison of miRNA regulation identified in individual cultivar libraries and pool libraries.

Targets from some of the miRNA whose expression was validated were selected to confirm their regulation. The expression of GAMyb (target of miR159 X), NAC1 (target of miR164 I), AGO1 (target of miR168 II) and Laccase (target of miR397) was verified by qRT-PCR, using the same cDNA samples used for the analysis of mature miRNA expression. The results showed a negative correlation among the expression patterns of some these possible target/miRNA pairs (Figure 6). For instance, GAMyb was significantly repressed, around 50% or more, in the SP832847 tolerant cultivar submitted to water depletion treatment in three biological replicates, while the miR159X was induced in these

Table 2. Comparison between miRNAs profile of pool plants and individual cultivar libraries.+= the miRNA profile is the same; - contrasting miRNA profile.

miRNA	Sequence	Comparison 1[a]	Comparison 2[b]
miR1439 I	ATTTGGAACGGAGGGAGTACT	+	+
miR156 I	GACAGAAGAGAGTGAGCACA	+	+
miR156 II	TGACAGAAGAGAGCGAGCAC	+	−
miR156 VI	TTGACAGAAGAGAGCGAGCAC	−	+
miR159 II	GGATTGAAGGGAGCTCTG	−	+
miR159 III	GTTGGATTGAAGGGAGCTCTG	+	−
miR159 V	TCTTTGGATTGAAGGGAGCTCTG	−	+
miR159 VII	TGGATTGAAGGGAGCTCTG	+	+
miR159 X	TTGGATTGAAGGGAGCTCTG	+	+
miR159 XI	TTGGATTGAAGGGAGCTCTGC	+	+
miR159 XIII	TTTGGATTGAAGGGAGCT	+	−
miR159 XVI	TTTGGATTGAAGGGAGCTCTG	+	+
miR159 XVII	TTTGGATTGAAGGGAGCTCTGC	+	−
miR162* I	GGGCGCAGTGGTTTATCGATC	−	+
miR164 I	TGGAGAAGCAGGGCACGTGCA	+	+
miR164 II	TGGAGAAGCAGGGCACGTGCT	−	+
miR166 I	TCTCGGACCAGGCTTCATTCC	−	+
miR166 II	TTCGGACCAGGCTTCATTCCC	+	+
miR166* I	AATGTTGTCTGGCTCGAGGTG	+	−
miR166* III	GAATGATGTCCGGTCCGAAGA	+	+
miR166* IV	GGAATGTTGTCTGGCTCGGGGG	+	+
miR167 III	TGAAGCTGCCAGCATGATCTA	+	+
miR167 IV	TGAAGCTGCCAGCATGATCTG	+	+
miR167 V	TGAAGCTGCCAGCATGATCTGA	+	−
miR167 VII	TGAAGCTGCCAGCATGATCTGG	−	+[c]
miR168 I	CCCGCCTTGCACCAAGTGAAT	−	+
miR168 II	TCGCTTGGTGCAGATCGGGAC	+	+
miR169 III	CAGCCAAGGATGACTTGCCGG	−	+
miR172 I	AGAATCTTGATGATGCTGCAT	+	−
miR1878 I	ATTTGTAGTGTTCAGATTGAGTTT	+	−
miR390 I	AAGCTCAGGAGGGATAGCGCC	+	+
miR390 II	AGCTCAGGAGGGATAGCGCC	−	+
miR395 III	TGAAGTGTTTGGGGGAACTC	+	+
miR396 II	TCCACAGGCTTTCTTGAACTG	+	+
miR397 II	TTGAGTGCAGCGTTGATGAGC	−	+[c]
miR398 II	TGTGTTCTCAGGTCGCCCCCG	+	+
miR444 II	TGCAGTTGTTGCCTCAAGCTT	+	−
miR528 I	TGGAAGGGGCATGCAGAGGAG	+	−
miR529 I	AGAAGAGAGAGAGTACAGCCT	−	+

[a]Comparison 1: influence of genetic variation on miRNA response to water depletion (T24h/S24h vs. CTC6/SP903414).
[b]Comparison 2: influence of water depletion treatment in miRNA response on each genotype (CTC6/T0h vs. T24h/T0h) and (SP903414/S0h vs. S24h/S0h).
[c]occur only in sensitive cultivars.

same samples. The other pair that showed an inverse expression pattern of the correspondent miRNA was the miR164II/NAC1 transcription factor. NAC1 was up-regulated and miR164II was down-regulated in a drought-stress sensitive cultivar (p-value < 0.05) in Rep1 and Rep2 samples. The NAC1 gene was induced (about 50%) while the miR164II was repressed (about 40%) in a tolerant cultivars, indicating a possible regulation of the NAC1 transcription factor by the miRNA. However, in the Rep3 of SP835073, we observed a repression of the target (p-value <0.05) although the expression of miRNA164II under water depletion did not change. In this case, the target was most likely regulated not only by miRNA164II. Alternatively, the miRNA could be induced earlier, reflecting in the late repression of target at 24 hs of treatment of water depletion. For miR168II and miR397II,

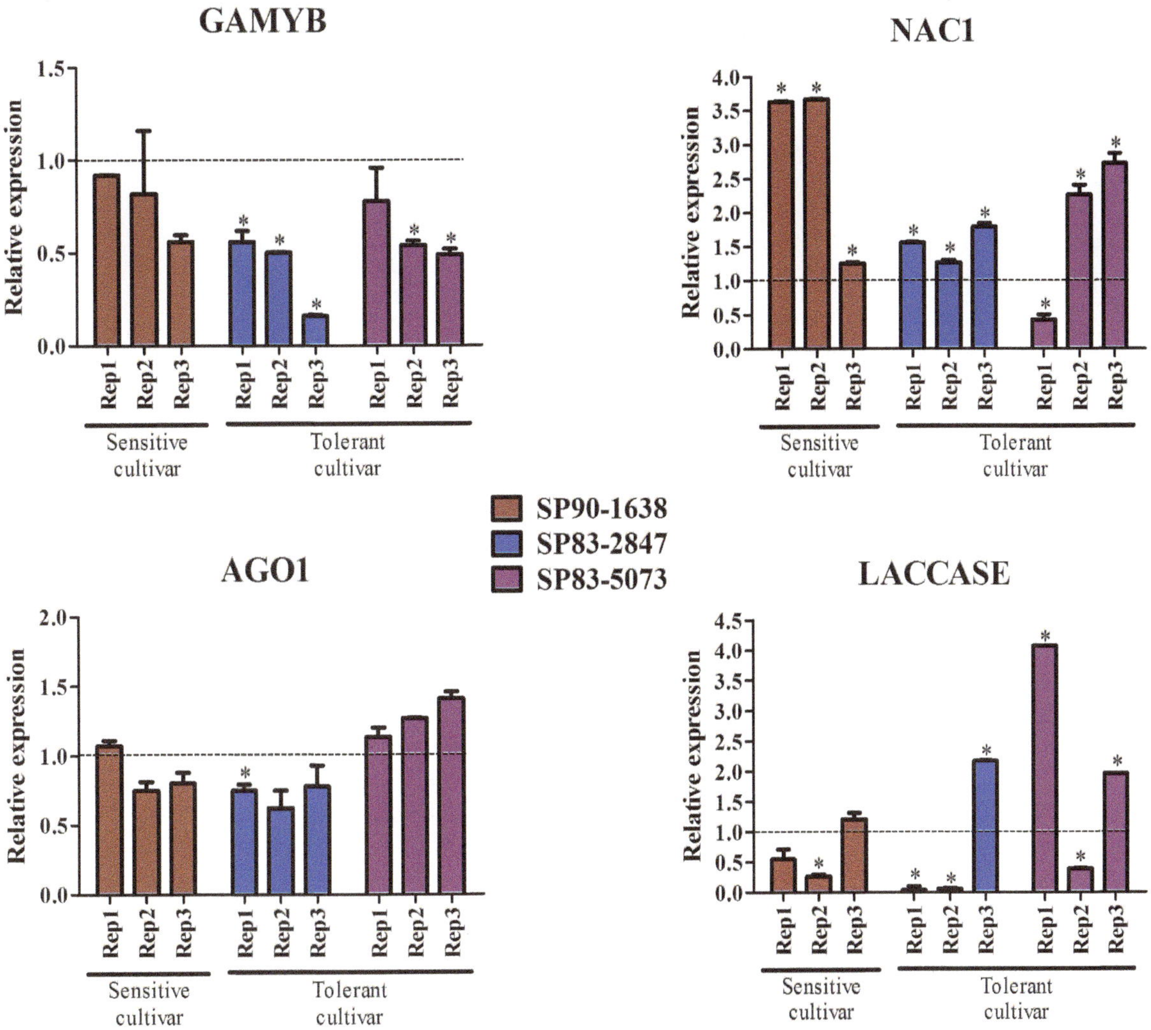

Figure 6. Relative expression profile of miRNA targets. Analysis of relative expression of miR159 target (GAMYB), miR164 target (NAC1), miR168 target (AGO1) and miR397 target (laccase) by qRT-PCR using samples of three sugarcane cultivar submitted to water depletion. In each case, control condition had relative expression equal 1 (dotted line). *represent significantly changing of miRNA expression between control and treatment samples (p-value <0.05).

their putative targets also did not show an inverse regulation (Figure 6).

22-nt miRNA Trigger siRNA Production in Leaves of Plants Under Stress

Recent studies revealed that 22-nt miRNA triggers the production of secondary siRNA [38,39]. In order to identify siRNA regulated in sugarcane submitted to water depletion, we first identified putative targets of miRNA of 22 nt (Table 3). Next, a BLASTN was performed using the selected targets against the fraction of sRNA present in all libraries, excluding the list of miRNA sequences. A list of siRNA candidate sequences that have more than 50 reads in each library with their abundance is available in the File S7. The putative targets of miR319I, miR167V and three putative targets of miR159I probably originated siRNAs (Table 4). Sequences of siRNA from these miRNA targets varied 1–2 nt in 5′- or 3′- end of the sequence, but the most representative sequence in each case is highlighted in Table 4. Two miR159I-targets (CA229394 and TC134732) give rise the same siRNA candidates sequence. Moreover, the position of miRNA cleavage site in the target is the same of siRNA candidate, and these siRNA sequences are present in all libraries.

The abundance of the siRNA candidates in each library was calculated counting the number of reads of each exclusive sequence. To compare libraries, we normalized these values in transcript (reads) *per* million, using total number of reads in each library. Because siRNAs generated by the CA229394 and TC134732 resulted in identical sequences of siRNA, we combined these two targets. Figure 7A shows the number of normalized reads of siRNA candidates in each library. Only the siRNA_CA229394/TC134732 was presented in the four libraries, being the most abundant siRNA candidate identified. The others candidates were expressed only in one library. The expression of siRNA_CA229394/TC134732 was verified by qRT-PCR using the Pulsed qRT-PCR. The expression profile is similar to the observed in the bioinformatics analysis (Figure 7). This siRNA was significantly up-regulated in all genotypes when plants were submitted to water depletion during 24 h (Figure 7B), indicating a response to drought stress independent of the genotype of the cultivar.

Table 3. 22-nt miRNA and their target prediction.

miRNA ID	Target gene accession	Start-end position of target[a]	Target description
miR159I	CA229394	139–157	similar to UniRef100_A3LXX5 Cluster: Predicted mutarotase
miR159I	TC134732	551–568	similar to UniRef100_Q96464 Cluster: GAMyb protein
miR159I	CA185455	320–339	similar to UniRef100_A8TZU8 Cluster: AsmA
miR159XII	TC152911	775–795	similar to UniRef100_Q0DUE9 Cluster: Os03g0189900 protein
miR159XVII	CA229394	137–157	similar to UniRef100_A3LXX5 Cluster: Predicted mutarotase
miR159XVII	TC126115	268–290	similar to UniRef100_A7NXD7 Cluster: Chromosome chr5 scaffold_2, whole genome shotgun sequence
miR160*I	CA181132	416–433	
miR166*IV	No target found		
miR167I	TC137034	433–454	similar to UniRef100_Q0DLC3 Cluster: Os05g0109600 protein
miR167V	CA232593	420–439	similar to UniRef100_Q653U3 Cluster: Auxin response factor 17
miR167V	CA201181	176–195	homologue to UniRef100_A2YG67 Cluster: Auxin response factor 17
miR167V	CA292372	75–94	homologue to UniRef100_A2YG67 Cluster: Auxin response factor 17
miR167V	TC132917	399–418	similar to UniRef100_Q653U3 Cluster: Auxin response factor 17
miR167V	TC116192	466–485	similar to UniRef100_Q653U3 Cluster: Auxin response factor 17
miR167V	TC134152	357–376	homologue to UniRef100_Q653U3 Cluster: Auxin response factor 17
miR167V	TC114828	576–595	similar to UniRef100_Q6L8U0 Cluster: Auxin response factor 4
miR167V	TC151621	670–689	similar to UniRef100_A7Q186 Cluster: Chromosome chr10 scaffold_43, whole genome shotgun sequence
miR167V	TC137034	433–454	similar to UniRef100_Q0DLC3 Cluster: Os05g0109600 protein
miR167VII	TC137034	432–454	similar to UniRef100_Q0DLC3 Cluster: Os05g0109600 protein
miR319I	TC153622	223–241	similar to UniRef100_Q0D8K4 Cluster: Os07g0152000 protein
miR319I	TC115735	883–902	homologue to UniRef100_Q9M7E5 Cluster: Elongation factor 1-alpha
miR398I	TC112854	935–956	homologue to UniRef100_Q94KS7 Cluster: Beta-expansin 7 precursor
miR398I	TC136123	131–150	similar to UniRef100_A7Q737 Cluster: Chromosome chr5 scaffold_58, whole genome shotgun sequence
miR398I	TC120003	173–192	similar to UniRef100_Q93WS1 Cluster: Selenium binding protein
miR398I	CA124904	137–156	similar to UniRef100_A7Q737 Cluster: Chromosome chr5 scaffold_58, whole genome shotgun sequence
miR398I	CA233608	219–238	similar to UniRef100_Q1M2Z3 Cluster: Selenium binding protein
miR398I	TC136146	180–199	similar to UniRef100_A7Q737 Cluster: Chromosome chr5 scaffold_58, whole genome shotgun sequence
miR398I	CA170115	320–339	similar to UniRef100_Q6ZXK6 Cluster: Kelch repeat containing protein
miR398I	TC131025	327–346	similar to UniRef100_Q6ZXK6 Cluster: Kelch repeat containing protein
miR398I	CA124259	329–348	similar to UniRef100_Q6ZXK6 Cluster: Kelch repeat containing protein
miR398I	CA196471	318–337	similar to UniRef100_Q6ZXK6 Cluster: Kelch repeat containing protein
miR398I	CA231219	160–179	similar to UniRef100_Q6ZXK6 Cluster: Kelch repeat containing protein
miR398I	CA155584	568–589	–

[a]start and end position of miRNA and target alignment.

Table 4. miRNA target that originate siRNA candidates.

miRNA ID	Target	Cleavage site[a]	Library with siRNA[b]	Alignment position of siRNA[c]
miR159I	CA229394	139–157	T0h, T24h, S0h e S24h	157–139/294–311
miR159I	TC134732	551–568	T0h, T24h, S0h e S24h	568–551
miR159I	CA185455	320–339	S24h	495–475
miR167V	TC116192	466–485	T24h	755–773
miR319I	TC115735	883–902	S0h	115–137

[a]start and end position of miRNA and target alignment.
[b]libraries where found siRNA candidates. T = tolerant cultivar; S = sensitive cultivar.
[c]position that originate siRNA.

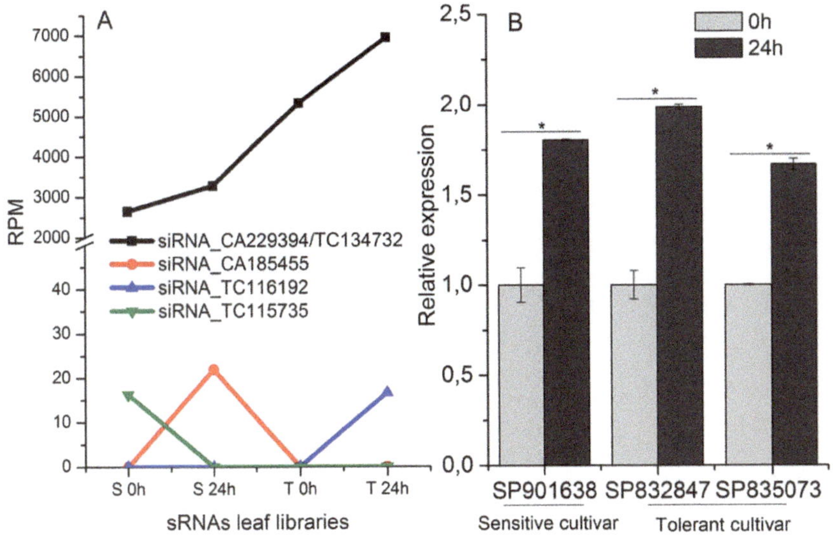

Figure 7. Analysis of siRNA candidates. Line graph is the abundance of siRNA candidates in each sRNA library normalized in reads *per* million. Bar graph is the relative expression of siRNA_CA229394/TC134732, which appeared in the four libraries, in different sugarcane cultivar submitted to water depletion by qRT-PCR. *represent significantly changing of miRNA expression between analyzed samples (p-value <0.05).

Water Depletion-affected siRNA Loci Regulation in Tolerant Cultivars

For further investigate the regulation of siRNA by water depletion in sugarcane, the fraction of sRNA present in all libraries, excluding the miRNA sequences, were aligned in clusters of siRNA. The clusters were constructed using a dataset of small RNA libraries consisting of 20 libraries (with miRNA sequences excluded). Next, the clusters of siRNA were annotated using information from a sugarcane repeat database (Grativol *et* al., in preparation) and in the sugarcane EST databases. The results showed a unique profile in the T0h leaf library (Figure 8) with more reads matching the Tentative Consensus (TC) EST (Expressed Sequence Tag) assembled and unclassified repeats than others libraries. Included are siRNA clusters that aligned with repeats that could be assigned to a repeat class (e.g. Retrotransposon, MITE) in Plant Repeat Database. Moreover, this library showed less reads, which aligned with retrotransposons, unannotated repeats and unannotated-siRNA-clusters. Interestingly, the profile of siRNAs aligned in other three leaf libraries is similar. Among the alterations in the retrotransposon families, the major change was in the LTR-gypsy. Stressed tolerant cultivars showed an increase of siRNA aligned with LTR-gypsy. Moreover, libraries of tolerant cultivar under water depletion (T24h) and sensitive cultivar control (S0h) are more similar. In all libraries, the fractions with less aligned reads were transposons, MITEs (miniature inverted-repeat transposable elements), centromere, telomere and ribosomal sequences (Figure 8). The classification of unannotated repeat refers to siRNA clusters that matched to unannotated repeats in the sugarcane repeats database. Unannotated-siRNA-clusters did not have matches with sugarcane repeats databases, but based on its abundance it is a putative siRNA cluster. We analyzed this distribution in root libraries and similar profiles were observed in the four libraries (File S8).

The size distribution of siRNA candidates in leaf libraries showed that the T0h sample showed a profile unlike the other libraries (Figure 9). It does not contain 21 nt reads and it has an additional peak of 23 nt. In the other three libraries, we observed two peaks, 21 nt (highest peak) and 24 nt. Next, we investigated the distribution of siRNA candidate sequences in the sRNA libraries (Figure 9 B). The result showed that 28 siRNA candidates were shared between leaf libraries. The S0h library showed the highest number of expressed siRNA candidates. Sensitive cultivars showed a higher number of exclusive siRNA candidates, while the tolerant cultivars showed a lower number of siRNA candidates. In addition, we observed a high number (45) of siRNA candidates in common between S0h, T24h and S24h.

Core Components of sRNA Biogenesis are Regulated in Response to Water Depletion

To investigate whether the mechanisms of sRNA biogenesis could be altered in the libraries, we selected enzymes that participate in the pathways responsible for sRNA production. The biosynthesis of plant sRNA starts with the process of double strand RNA precursor. In plants, the precursor interacts with DICER-LIKE (DCL) endonucleases, which catalyzes excision of mature miRNA [16]. Four DCL have been identified in Arabidopsis [40], and we selected two sugarcane DCLs homologues to verify whether their expression profiles in sugarcane leaf submitted to water depletion were altered: DCL1 and DCL2 (Figure 10). Both DCLs are significantly induced in sensitive cultivar submitted to water depletion and this profile was not observed in the tolerant cultivars. These results suggest that the biogenesis of sRNA is increased in the sensitive cultivar. Accordingly, we have identified more miRNA in sensitive cultivars than in the tolerant cultivars. In addition, we verified the expression profile of HYPONASTIC LEAVES (HYL) 1, an enzyme in the miRNA biogenesis pathway [41]. The result showed a down-regulation only in one tolerant cultivar: SP835073 (p-value <0.05). In the sensitive cultivar, HYL1 was significantly up-regulated by water depletion. Plants encode RNA-dependent RNA polymerases (RDRs) to produce dsRNA which will give rise to siRNA [42]. Therefore, we analyze the expression profile of RDR6, and the results showed a small reduction in RDR6 expression in the tolerant cultivars after the stress. Finally, we analyze the expression of sugarcane genes AGO proteins. Arabidopsis has 10 genes encoding AGO proteins [43,44]. In sugarcane, we identified four genes that are

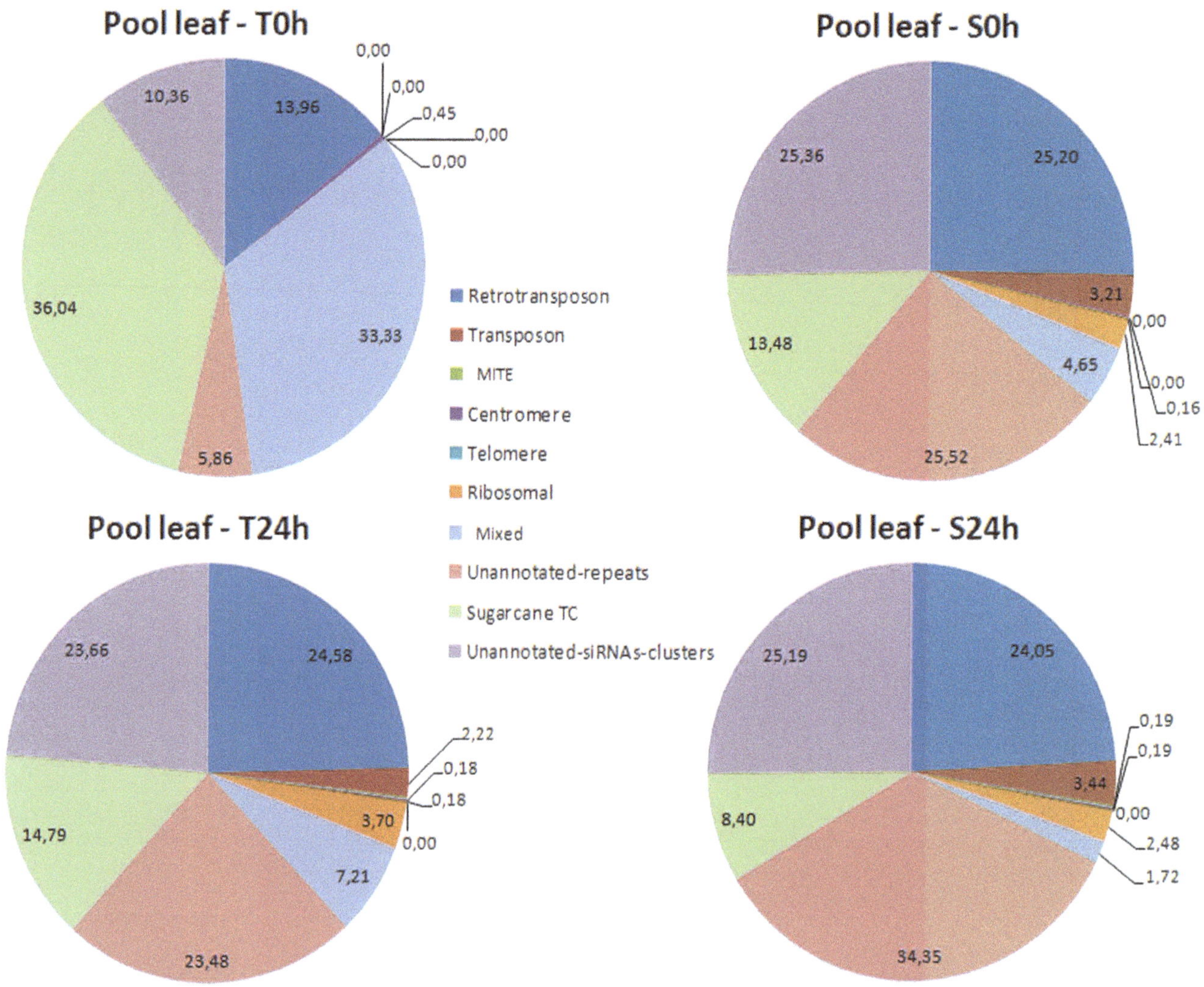

Figure 8. Distribution of leaf siRNA candidates aligned in cluster of siRNA. Each graph represents one leaf sRNA library. T = tolerant sugarcane cultivar, S = sensitive sugarcane cultivar. 0 h = control of stress, 24 h = time of water depletion treatment.

homologous to AtAGO1 and one homologous to AtAGO4. We investigated the expression of two of the AGO1 genes (AGO1_2 and AGO1_3) and AGO4 (Figure 10). The results showed that AGO1_ 2 was repressed in tolerant cultivars (about 40%); however, it was slightly induced in the sensitive cultivar (p-value <0.05). AGO1_3 was significantly down-regulated in the sensitive

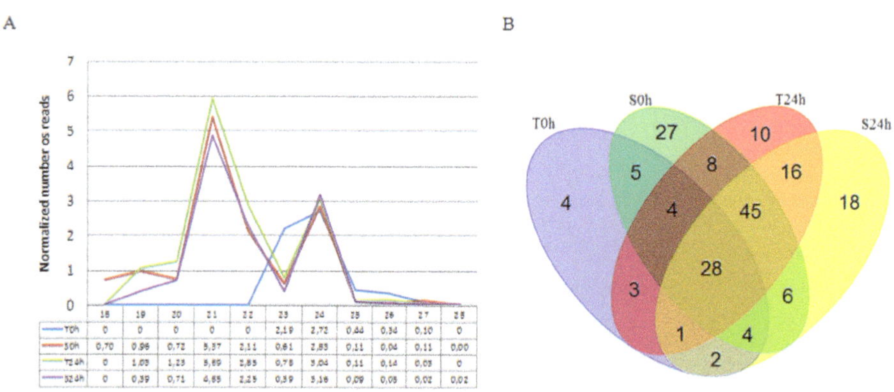

	18	19	20	21	22	23	24	25	26	27	28
T0h	0	0	0	0	0	2,19	2,72	0,44	0,34	0,10	0
S0h	0,70	0,96	0,72	5,37	2,11	0,61	2,83	0,11	0,04	0,11	0,00
T24h	0	1,03	1,23	5,89	2,83	0,78	3,04	0,11	0,14	0,03	0
S24h	0	0,39	0,71	4,85	2,25	0,39	3,16	0,09	0,05	0,02	0,02

Figure 9. Analysis of siRNA candidates that aligned in siRNA cluster. (A) Length distribution of siRNA in each leaf library. (B) Venn diagram comprising the siRNA distribution in each leaf library. T = tolerant sugarcane cultivar, S = sensitive sugarcane cultivar. 0 h = control of stress, 24 h = time of water depletion treatment.

cultivar, but it showed different regulation patterns in the tolerant cultivars (SP832847– repressed; SP835073– unalterable). Likewise, AGO4 expression showed a tendency of repression in the cultivar SP832847, while the expression of AGO4 in the cultivar SP835073 did not vary. These results suggest that AGO4 expression increases in the sugarcane sensitive cultivar upon water depletion.

Discussion

Plants are sessile organism and under natural conditions, they are often exposed to stresses caused by the environment. To respond and adapt to stress, plant have developed multiples regulatory mechanisms [45]. Recent studies have shown that many genes are epigenetically regulated in plants [46,47]. Among the epigenetic modifications, DNA methylation is the most well-studied; and sRNA production is a novel epigenetic system that recently has been under intense investigation [6]. High-throughput sRNA sequencing has helped to accelerate the discovery of sRNA. In the present study, we have used deep sequencing to identify sRNA regulated in sugarcane plants submitted to water depletion. Drought stress is one of the major constraints to agricultural productivity worldwide and recent reports have highlighted the importance of sRNA in the response and adaptation to water availability to plants [48]. This study investigated the role of two classes of sRNA, the miRNA and the siRNA in the early responses of sugarcane to drought stress. First, we constructed and sequenced sRNA libraries from roots and leaves of sugarcane genotypes with different drought sensitivities. Consistent with most recent reports [49,50], roots libraries had two major peaks of sRNA of 21 and 24 nt (Figure 2). Unexpectedly, a peak of sRNA of 22 nt in length was observed in leaf libraries. In a recent studies a peak of 22 nt was also observed in the size distribution of *Cucumis sativus* sRNA [51,52].

Recent reports have revealed that plant miRNA are responsive to abiotic stress such as drought, cold, heat and salinity [48]. The goals of this work were to identify miRNAs present in sugarcane, and investigate their regulation in response to water depletion in different genotypes. Based on characteristics of plant miRNA conservation [24,25], all plant miRNA deposited on the miRBase were used to identify sugarcane miRNA. Bioinformatics analysis identified 88 conserved miRNA or miRNA* in leaves and 273 in roots, being 64 sequences in common between the two sugarcane tissues. MiR159 is the more representative miRNA family in all cultivars, in both tissues. However, analysis of miRNA expression revealed that many of them are expressed only in certain tissues and/or cell specific, and at stage of development [53–55]. This could explain the difference in miRNA abundances found in roots and leaves of sugarcane. DCL1, the enzyme that cleaves the miRNA precursor in plants, releases preferentially 21-nt shorts RNA from dsRNA [56]; and immunoprecipitation with anti-AGO1 antibodies revealed the preferential association of AGO1 (the canonical enzyme of miRNA pathway) with sRNA containing 5′-terminal U [12]. Accordingly, the majority of the sugarcane miRNA identified were 21 nt in length and have uracil (U) at the

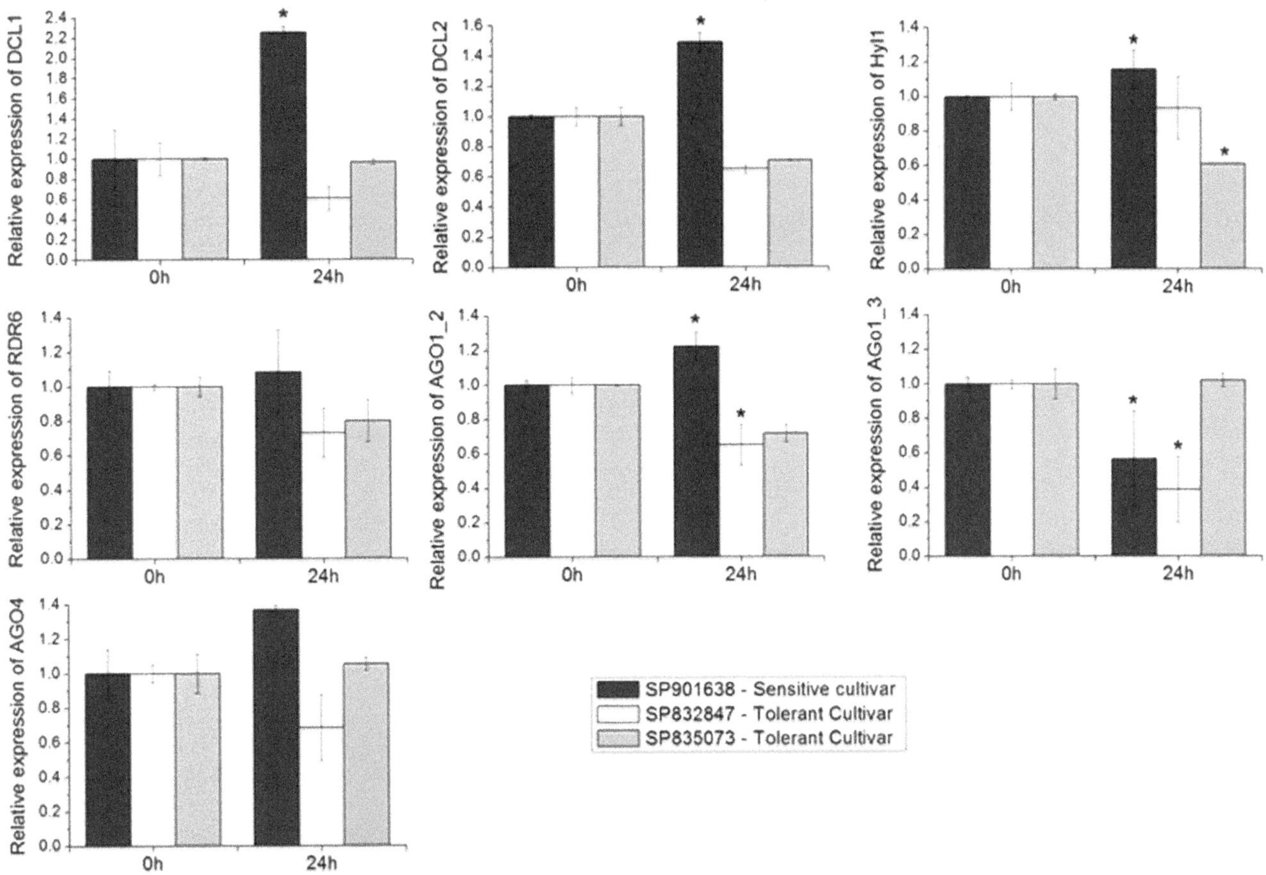

Figure 10. Relative expression of sRNA pathway genes. Analysis by qRT-PCR using samples of three sugarcane cultivar submitted to water depletion. *represent significantly changing of miRNA expression between analyzed samples (p-value <0.05).

5′-end in both tissues. Furthermore, although miRNA* are normally less abundant [48], some studies have indicated that plant miRNA* could play a role in plant stress tolerance [57,58]. In this study, 11 and 54 miRNA* were identified in leaf and in root libraries, respectively, and putative targets for leaf miRNA* were found. MiR169* had a larger number of targets, and most of them encode Elongation factor 1-alpha (EF 1-α). Accordingly, the role of EF 1-α has been analyzed in response to abiotic stress, including in response to water deficit [59]. Others miRNA* also have putative targets related to abiotic stresses, such as Kelch repeat-containing F-box like protein (target of miR396*) and cold induced protein like (target of miR399*).

Recent studies have shown that some miRNA are responsive to drought stress. In sugarcane, it has also been shown that miRNA are differently expressed in sRNA libraries of cultivars with contrasting tolerance to drought [27,36]. MiR156, miR159, miR164, miR167, miR168, miR168 and miR397 are examples of miRNA that have been described to be regulated under drought stress conditions [60–62]. In this study, we investigated whether these miRNA were also regulated by water depletion in sugarcane. The qRT-PCR results confirmed the majority of the expression profiles observed in the bioinformatics analysis; however, there were a few that were not confirmed and this lack of correlation could be explained because a pool of genotypes was used in the construction of sRNA libraries, while in the qRT-PCR analyzes we examined miRNA expression in individual genotypes. In order to verify whether these microRNAs are differentially regulated by water depletion or by the genotype, two sRNA libraries of individual cultivars submitted to water depletion (one tolerant and one sensitive) were sequenced. In both rice [63] and sugarcane [33] cultivars, differential miRNA in response to water deficiency be at least partially explained by genetic variation. Our results have shown that at least a fraction of the microRNAs regulated in the pool libraries have similar regulation in cultivars with contrasting tolerance to drought, suggesting an influence of the environment in miRNA regulation.

The biological function of miRNA can be inferred by the knowledge of their targets. Targets can be regulated by cleavage, repression of translation, or DNA and histone methylation [14,64]. In plants, bioinformatics/*in silico* analysis of miRNA and their targets genes is possible due to the perfect or near-perfect complementarity between miRNA/target in the binding sites [37]. Several sugarcane miRNA targets are transcription factors, in agreement with previous reports on conserved miRNA targets [34,65,66] and some of these are involved in a signaling pathway of plant adaptation to drought stress [67]. For instance, the transcription factor GAMyb was identified as a miR159 X target. GAMyb is a plant growth regulator and its regulation has been described during abiotic stress [4,34]. The target of miR164 I is NAC1 (No Apical Meristem -NAM, Arabidopsis transcription activation factor- ATAF, Cup-shaped cotyledon- CUC), a transcription factor that has been reported as involved in tolerance to abiotic stress [68]. In rice, the overexpression of NAC1 produced plants more tolerant to drought and salt stress [69]. The modulation of miRNA (up- or down-regulation) can result in a modification of their target abundance under stress condition [70,71]. An inverse expression correlation of GAMyb and NAC1 and their respective miRNAs was confirmed for one tolerant and the sensitive cultivar. The differences in miRNA regulation observed between the two tolerant cultivars could be associated of their different-physiological response to the stress. Accordingly, the reduction in chlorophyll content in response to drought stress is higher in SP835073 than SP835073, indicating an increased sensitive to this stress [72]. Expression of others targets, Laccase (target of miR397 II) and AGO1 (target of miR168 II) was analyzed. In both cases, there is not an inverse correlation between target and miRNA, although, the predicted targets are canonical targets for their respectively miRNA. In this case, it is possible that the post-transcriptional regulation of the canonical target by miRNA-directed cleavage is not limiting the accumulation of mRNA under the stress condition, and another layer of regulation is acting, similarly to what has been described in other plants [73]. In addition, different cultivars develop contrasting morphological characteristics in response to water depletion that could result in differences in the miRNAs regulation. For example, although miR397 is similarly regulated in tolerant cultivars, its target, the enzyme laccase, which is involved in lignin biosynthesis [74], is highly induced in SP835073. This cultivar is classified as high fiber content, and upon water depletion, could increase cell wall biosynthesis to prevent water loss.

Another important role of miRNA is in the production of siRNA. The class of siRNA that is produced after cleavage of the target by miRNA is the *trans*-acting (ta-) siRNA [75,76]. Recently, it has been described that miRNA of 22 nt in length also triggers secondary siRNA biogenesis [77]. Based on this observation, and in their abundance in leaf libraries, 22-nt miRNA species and their targets were selected to identify putative ta-siRNA. Several *TAS* genes processed by miRNA to generate ta-siRNA are well characterized in plants [75]. Examples of these genes are *TAS1*, *TAS2*, *TAS3* and *TAS4* that are transcripts recognized by miR173 (*TAS1* and *TAS2*), miR390 (*TAS3*) and miR828 (*TAS4*). However, a recent study has showed the cleavage of NB-LRR defense genes by miRNA produces ta-siRNA [39]. In addition, *MYB*-derived siRNAs were described, and the miRNA that regulated this gene was miR828 in apple [78]. In our analysis, the target of miR159I (encoding a Myb family protein) originated ta-siRNA candidates present in all libraries. Interestingly, this sugarcane siRNA was up-regulated in samples from water depletion treatment, independent of the tolerance of the cultivar to stress, suggesting the involvement of this siRNA in the response to water depletion. The production of the discrete type of ta-siRNA (not in phase) described here could be a novel mechanism of sRNA-based regulation. Contrary to the canonical ta-siRNAs whose production silence additional targets, the sugarcane ta-siRNA matches the same gene as the original microRNA and could be a tool to reinforce the silencing. Nevertheless, this hypothesis needs experimental confirmation, once the failure to detect low abundance in-phase siRNA could be due to the lack of sugarcane genome information.

The population of sRNA is complex and in addition of the ta-siRNA class described above, a large portion of plants sRNA is originated from repeats [6]. This study revealed altered distributions of siRNA- derived from repeats and gene- among sugarcane cultivars with different tolerance of drought stress. The profile of siRNA clusters derived from TC EST observed in the sensitive cultivars without water depletion is similar to tolerant cultivars under treatment. Interestingly, tolerant cultivars without treatment showed more siRNA candidates from siRNA clusters, classified as TC EST, than the others libraries. Analyzing these TC ESTs, we observed that many are related to genes of photosystem I and II (data not shown). It is known that water depletion decreases the activities of photosynthetic carbon reduction cycle enzymes [79], and possibly this reduction could be the result in an increase in siRNA derived from these TC EST in sugarcane submitted to water depletion.

Different classes of sRNA are involved in changes in DNA methylation (24-nt species) and in the regulation after transcription (21-nt species) of TEs (transposable elements). In wheat, under normal conditions, most TEs families were regulated by sRNA with 24-nt [80]. In contrast, both sensitive and tolerant sugarcane genotypes under water depletion showed an increase of 21-nt species regulating retroelements. A recent study showed that drought stress induced changes in DNA methylation. MSAP analysis in rice revealed that the retrotransposon like LTR-gypsy was a polymorphic DNA methylated fragments in response to drought stress [81]. The wheat tolerant cultivar without stress showed a similar profile of 24-nt-based regulation (DNA methylation) of LTR-gypsy. Interestingly, in tolerant sugarcane cultivars, water depletion induced a shift in sized of LTR-gypsy derived siRNA from 24-nt to 21-nt species. These results suggest water deficiency may change methylated sites of TEs starting a sRNA-directed posttranscriptional silencing suppression of TE activity in sugarcane.

An important characteristic that differentiates sRNA classes (miRNA and siRNA) is their origin and biogenesis, even though some classes share common enzymes in their pathways [82]. DCL, HYL, RDR and AGO are examples of enzyme involved in the plant sRNA pathways. DCL1 is involved in the first step of canonical miRNA biogenesis and can be involved in production of miRNA-like siRNA and DCL1-dependent 22-nt siRNA; while, DCL2 is involved in cleavage of long inverted repeat precursor of miRNA-like siRNA [82]. The expression profile of these DCLs in sugarcane suggested a more effective activity of sRNA biogenesis in the sensitive than in tolerant sugarcane cultivars. The result is in agreement with the observation that more miRNA have been identified in sensitive cultivars than in tolerant cultivars. RDR6 is required for siRNA biogenesis, including ta-siRNA biosynthesis which convert the cleaved target transcript into double-strand RNA to be further processed by DCL [42,83]. Differential expression of RDR6 was shown in sugarcane cultivars with contrasting tolerance to drought stress, with RDR6 expression being down-regulated in the tolerant cultivars. The first steps in miRNA processing requires the DCL1-HYL1 complex [84]. However, HYL1 was down-regulated only in the SP835073 (tolerant cultivar); therefore, it was not possible to conclude whether regulation of HYL1 expression is involved in the plant responses to water depletion. The mature miRNA is incorporated in a ribonucleoprotein complex (RISC) containing the AGO protein [85]. Following the mechanism of sRNA action, the AGO1 can trigger the cleavage of targeted RNA or inhibit mRNA translation [12,13]. The expression profile of three AGO1 genes (AGO1-miR168II target, AGO1_2 and AGO1_3) was analyzed. Still, AGO1_2 and AGO1_3 were down-regulated in tolerant cultivar. The expression of the AGO4 gene was not significantly different in sugarcane cultivars submitted to water depletion. AGO4 has affinity for miRNA or siRNA and trigger the cleavage of target or mediate cytosine DNA methylation [86]. Moreover, generally, AGO4 binds sRNA that has 5′ adenosine 24-nt [87]. Altogether, these results suggest that some components of sRNA biogenesis are involved in the plants responses to water depletion.

Our results characterized the regulation of sRNA and explored the expression profile of sRNA pathway genes in sugarcane cultivars submitted to water depletion. We confirm the findings of recent studies that correlate the response of miRNA and their targets genes to availability of water. Furthermore, we showed that sRNA responses to water depletion could differ between sugarcane cultivars, contributing to increase the knowledge of sRNA roles in response to drought stress in the complex, polyploidy sugarcane plant.

Supporting Information

File S1 Electronic Northern of leaf miRNA. The number of conserved miRNA reads in each library were normalized by transcripts per million. The miRNA sequences and their length were also shown.

File S2 Electronic Northern of root miRNA. The number of conserved miRNA reads in each library were normalized by transcripts per million. The miRNA sequences and their length were also shown.

File S3 Pairwise statistical analysis of leaf miRNAs. Statistical analysis of bioinformatics data using pairwise Fisher exact test, p-value <0.05.

File S4 Pairwise statistical analysis of root miRNAs. Statistical analysis of bioinformatics data using pairwise Fisher exact test, p-value <0.05.

File S5 Relative expression of miRNA in different sugarcane cultivars by qRT-PCR. *represent significantly changing of miRNA expression between analyzed samples (p-value <0.05).

File S6 Putative targets of all leaf miRNA in sugarcane. EST sugarcane data from Gene Index version 3.0 was used to search for potential new miRNA targets by UEA sRNA toolkit-Plant target prediction pipeline.

File S7 siRNA candidates from targets of miRNA 22-nt in each library.

File S8 Distribution of root siRNA candidates aligned in cluster of siRNA. Each graph represents one-leaf sRNA library. T = tolerant sugarcane cultivar, S = sensitive sugarcane cultivar. 0 h = control of stress, 24 h = time of water depletion treatment.

Acknowledgments

We are grateful to Centro de Tecnologia Canavieira (CTC) for providing the drought experiment. We thank Andreia Cordeiro for technical assistance. We thank André H Ferreira for language revision of the manuscript.

Author Contributions

Conceived and designed the experiments: FT PCGF ASH. Performed the experiments: FT TV MRM MCB CAR SMC. Analyzed the data: FT CG. Contributed reagents/materials/analysis tools: PCGF ASH MT RAM. Wrote the paper: FT CG. Helped with Informatic tools: RV.

References

1. Mirouze M, Paszkowski J (2011) Epigenetic contribution to stress adaptation in plants. Curr Opin Plant Biol 14: 267–274. Available: http://www.ncbi.nlm.nih.gov/pubmed/21450514. Accessed 2012 August 14.
2. Kidner CA, Martienssen RA (2005) The developmental role of microRNA in plants. Curr Opin Plant Biol 8: 38–44. Available: http://www.ncbi.nlm.nih.gov/pubmed/15653398. Accessed 2011 June 11.

3. Madlung A, Comai L (2004) The effect of stress on genome regulation and structure. Ann Bot 94: 481–495. Available: http://www.ncbi.nlm.nih.gov/pubmed/15319229. Accessed 2012 August 4.
4. Khraiwesh B, Zhu JJ-K (2012) Role of miRNAs and siRNAs in biotic and abiotic stress responses of plants. Biochim Biophys Acta 1819: 137–148. Available: http://www.pubmedcentral.nih.gov/articlerender.fcgi?artid=3175014&tool=pmcentrez&rendertype=abstract. Accessed 2012 July 15.
5. Bartel DP, Lee R, Feinbaum R (2004) MicroRNAs: Genomics, Biogenesis, Mechanism, and Function Genomics: The miRNA Genes. 116: 281–297.
6. Simon SA, Meyers BC (2011) Small RNA-mediated epigenetic modifications in plants. Curr Opin Plant Biol 14: 148–155. Available: http://www.ncbi.nlm.nih.gov/pubmed/21159545. Accessed 2012 March 20.
7. Guleria P, Mahajan M, Bhardwaj J, Yadav SK (2011) Plant small RNAs: biogenesis, mode of action and their roles in abiotic stresses. Genomics, proteomics Bioinforma/Beijing Genomics Inst 9: 183–199. Available: http://www.ncbi.nlm.nih.gov/pubmed/22289475. Accessed 2012 July 16.
8. Bernstein E, Caudy AA, Hammond SM, Hannon GJ (2001) Role for a bidentate ribonuclease in the initiation step of RNA interference. Nature 409: 363–366. Available: http://www.ncbi.nlm.nih.gov/pubmed/11201747.
9. Jamalkandi SA, Masoudi-Nejad A (2009) Reconstruction of Arabidopsis thaliana fully integrated small RNA pathway. Funct Integr Genomics 9: 419–432. Available: http://www.ncbi.nlm.nih.gov/pubmed/19802639. Accessed 2012 August 6.
10. Yu B, Yang Z, Li J, Minakhina S, Yang M, et al. (2005) Methylation as a crucial step in plant microRNA biogenesis. Science (80-) 307: 932–935. Available: http://www.ncbi.nlm.nih.gov/pubmed/15705854. Accessed 2011 August 1.
11. Park MY, Wu G, Gonzalez-Sulser A, Vaucheret H, Poethig RS (2005) Nuclear processing and export of microRNAs in Arabidopsis. Proc Natl Acad Sci U S A 102: 3691–3696. Available: http://www.pubmedcentral.nih.gov/articlerender.fcgi?artid=553294&tool=pmcentrez&rendertype=abstract.
12. Baumberger N, Baulcombe DC (2005) Arabidopsis ARGONAUTE1 is an RNA Slicer that selectively recruits microRNAs and short interfering RNAs. Proc Natl Acad Sci U S A 102: 11928–11933. Available: http://www.pubmedcentral.nih.gov/articlerender.fcgi?artid=1182554&tool=pmcentrez&rendertype=abstract.
13. Mallory AC, Vaucheret H (2006) Functions of microRNAs and related small RNAs in plants. Nat Genet 38 Suppl: S31–6. Available: http://www.ncbi.nlm.nih.gov/pubmed/16736022. Accessed 2011 July 17.
14. Wu L, Zhou H, Zhang Q, Zhang J, Ni F, et al. (2010) DNA methylation mediated by a microRNA pathway. Mol Cell 38: 465–475. Available: http://www.ncbi.nlm.nih.gov/pubmed/20381393. Accessed 2011 July 17.
15. Ruiz-Ferrer V, Voinnet O (2009) Roles of plant small RNAs in biotic stress responses. Annu Rev Plant Biol 60: 485–510. Available: http://www.ncbi.nlm.nih.gov/pubmed/19519217. Accessed 2011 July 17.
16. Xie Z, Johansen LK, Gustafson AM, Kasschau KD, Lellis AD, et al. (2004) Genetic and functional diversification of small RNA pathways in plants. PLoS Biol 2: E104. Available: http://www.pubmedcentral.nih.gov/articlerender.fcgi?artid=350667&tool=pmcentrez&rendertype=abstract. Accessed 2011 July 5.
17. Zilberman D, Cao X, Jacobsen SE (2003) ARGONAUTE4 control of locus-specific siRNA accumulation and DNA and histone methylation. Science 299: 716–719. Available: http://www.ncbi.nlm.nih.gov/pubmed/12522258.
18. Allen E, Howell MD (2010) miRNAs in the biogenesis of trans-acting siRNAs in higher plants. Semin Cell Dev Biol 21: 798–804. Available: http://www.ncbi.nlm.nih.gov/pubmed/20359543. Accessed 2012 August 14.
19. Cho SH, Addo-Quaye C, Coruh C, Arif MA, Ma Z, et al. (2008) Physcomitrella patens DCL3 is required for 22–24 nt siRNA accumulation, suppression of retrotransposon-derived transcripts, and normal development. PLoS Genet 4: e1000314. Available: http://www.pubmedcentral.nih.gov/articlerender.fcgi?artid=2600652&tool=pmcentrez&rendertype=abstract. Accessed 2012 November 21.
20. Krasnikova MS, Milyutina IA, Bobrova VK, Ozerova LV, Troitsky AV, et al. (2009) Novel miR390-dependent transacting siRNA precursors in plants revealed by a PCR-based experimental approach and database analysis. J Biomed Biotechnol 2009: 952304. Available: http://www.pubmedcentral.nih.gov/articlerender.fcgi?artid=2762245&tool=pmcentrez&rendertype=abstract. Accessed 2012 November 21.
21. Chinnusamy V, Zhu J-K (2009) Epigenetic regulation of stress responses in plants. Curr Opin Plant Biol 12: 133–139. Available: http://www.pubmedcentral.nih.gov/articlerender.fcgi?artid=3139470&tool=pmcentrez&rendertype=abstract. Accessed 2012 July 16.
22. Sunkar R, Chinnusamy V, Zhu J, Zhu J-K (2007) Small RNAs as big players in plant abiotic stress responses and nutrient deprivation. Trends Plant Sci 12: 301–309. Available: http://www.ncbi.nlm.nih.gov/pubmed/17573231. Accessed 2012 July 18.
23. Zhang B, Pan X, Cobb GP, Anderson TA (2006) Plant microRNA: a small regulatory molecule with big impact. Dev Biol 289: 3–16. Available: http://www.ncbi.nlm.nih.gov/pubmed/16325172. Accessed 2012 July 17.
24. Zhang B, Pan X, Cannon CH, Cobb GP, Anderson TA (2006) Conservation and divergence of plant microRNA genes. Plant J 46: 243–259. Available: http://www.ncbi.nlm.nih.gov/pubmed/16623887. Accessed 2012 July 16.
25. Sunkar R, Jagadeeswaran G (2008) In silico identification of conserved microRNAs in large number of diverse plant species. BMC Plant Biol 8: 37. Available: http://www.pubmedcentral.nih.gov/articlerender.fcgi?artid=2358906&tool=pmcentrez&rendertype=abstract. Accessed 2011 July 20.
26. Grivet L, Arruda P (2002) Sugarcane genomics: depicting the complex genome of an important tropical crop. Curr Opin Plant Biol 5: 122–127. Available: http://www.ncbi.nlm.nih.gov/pubmed/11856607.
27. Thiebaut F, Grativol C, Carnavale-bottino M, Rojas CA, Farinelli L, et al. (2012) Computational identification and analysis of novel sugarcane microRNAs. BMC Genomics 13. doi:10.1186/1471-2164-13-290.
28. Brennecke J, Aravin AA, Stark A, Dus M, Kellis M, et al. (2007) Discrete small RNA-generating loci as master regulators of transposon activity in Drosophila. Cell 128: 1089–1103. Available: http://www.ncbi.nlm.nih.gov/pubmed/17346786. Accessed 2011 July 17.
29. Moxon S, Schwach F, Dalmay T, Maclean D, Studholme DJ, et al. (2008) A toolkit for analysing large-scale plant small RNA datasets. Bioinformatics 24: 2252–2253. Available: http://www.ncbi.nlm.nih.gov/pubmed/18713789. Accessed 2011 July 5.
30. Varkonyi-Gasic E, Wu R, Wood M, Walton EF, Hellens RP (2007) Protocol: a highly sensitive RT-PCR method for detection and quantification of microRNAs. Plant Methods 3: 12. Available: http://www.pubmedcentral.nih.gov/articlerender.fcgi?artid=2225395&tool=pmcentrez&rendertype=abstract. Accessed 2011 March 2.
31. Chen C, Ridzon DA, Broomer AJ, Zhou Z, Lee DH, et al. (2005) Real-time quantification of microRNAs by stem-loop RT-PCR. Nucleic Acids Res 33: e179. Available: http://www.pubmedcentral.nih.gov/articlerender.fcgi?artid=1292995&tool=pmcentrez&rendertype=abstract. Accessed 2011 June 10.
32. Vargas L, Carvalho TLG, Ferreira PCG, Baldani VLD, Baldani JI, et al. (2012) Early responses of rice (Oryza sativa L.) seedlings to inoculation with beneficial diazotrophic bacteria are dependent on plant and bacterial genotypes. Plant Soil 356: 127–137. doi:10.1007/s11104-012-1274-8.
33. Zanca AS, Vicentini R, Ortiz-Morea FA, Del Bem LEV, da Silva MJ, et al. (2010) Identification and expression analysis of microRNAs and targets in the biofuel crop sugarcane. BMC Plant Biol 10: 260. Available: http://www.pubmedcentral.nih.gov/articlerender.fcgi?artid=3017846&tool=pmcentrez&rendertype=abstract. Accessed 2011 July 21.
34. Thiebaut F, Rojas CA, Almeida KL, Grativol C, Domiciano GC, et al. (2011) Regulation of miR319 during cold stress in sugarcane. Plant Cell Environ: no–no. Available: http://doi.wiley.com/10.1111/j.1365-3040.2011.02430.x. Accessed 2011 October 10.
35. Carnavale Bottino M, Rosario S, Grativol C, Thiebaut F, Rojas CA, et al. (2013) High-throughput sequencing of small RNA transcriptome reveals salt stress regulated microRNAs in sugarcane. PLoS One 8: e59423. Available: http://www.pubmedcentral.nih.gov/articlerender.fcgi?artid=3609749&tool=pmcentrez&rendertype=abstract. Accessed 2013 November 13.
36. Ferreira H, Gentile A, Vilela RD, Gilson G, Costa L (2012) microRNAs Associated with Drought Response in the Bioenergy Crop Sugarcane (Saccharum spp.). PLoS One 7: e46703. doi:10.1371/journal.pone.0046703.
37. Rhoades MW, Reinhart BJ, Lim LP, Burge CB, Bartel B, et al. (2002) Prediction of plant microRNA targets. Cell 110: 513–520. Available: http://www.ncbi.nlm.nih.gov/pubmed/20430753.
38. Chen H-M, Chen L-T, Patel K, Li Y-H, Baulcombe DC, et al. (2010) 22-Nucleotide RNAs trigger secondary siRNA biogenesis in plants. Proc Natl Acad Sci U S A 107: 15269–15274. Available: http://www.pubmedcentral.nih.gov/articlerender.fcgi?artid=2930544&tool=pmcentrez&rendertype=abstract. Accessed 2010 July 20.
39. Zhai J, Jeong D-H, De Paoli E, Park S, Rosen BD, et al. (2011) MicroRNAs as master regulators of the plant NB-LRR defense gene family via the production of phased, trans-acting siRNAs. Genes Dev 25: 2540–2553. Available: http://www.pubmedcentral.nih.gov/articlerender.fcgi?artid=3243063&tool=pmcentrez&rendertype=abstract. Accessed 2012 March 5.
40. Liu Q, Feng Y, Zhu Z (2009) Dicer-like (DCL) proteins in plants. Funct Integr Genomics 9: 277–286. Available: http://www.ncbi.nlm.nih.gov/pubmed/19221817. Accessed 2012 April 10.
41. Vazquez F, Gasciolli V, Crété P, Vaucheret H (2004) The nuclear dsRNA binding protein HYL1 is required for microRNA accumulation and plant development, but not posttranscriptional transgene silencing. Curr Biol 14: 346–351. Available: http://www.ncbi.nlm.nih.gov/pubmed/14972688. Accessed 2012 March 21.
42. Willmann MR, Endres MW, Cook RT, Gregory BD (2011) The Functions of RNA-Dependent RNA Polymerases in Arabidopsis. Arabidopsis Book 9: e0146. Available: http://www.pubmedcentral.nih.gov/articlerender.fcgi?artid=3268507&tool=pmcentrez&rendertype=abstract. Accessed 2012 July 20.
43. Morel J, Godon C, Mourrain P, Béclin C, Boutet S, et al. (2002) Fertile Hypomorphic ARGONAUTE (ago1) Mutants Impaired in Post-Transcriptional Gene Silencing and Virus Resistance. 14: 629–639. doi:10.1105/tpc.010358.are.
44. Zheng X, Zhu J, Kapoor A, Zhu J-K (2007) Role of Arabidopsis AGO6 in siRNA accumulation, DNA methylation and transcriptional gene silencing. EMBO J 26: 1691–1701. Available: http://www.pubmedcentral.nih.gov/articlerender.fcgi?artid=1829372&tool=pmcentrez&rendertype=abstract. Accessed 2012 March 8.
45. Grativol C, Hemerly AS, Ferreira PCG (2012) Genetic and epigenetic regulation of stress responses in natural plant populations. Biochim Biophys Acta 1819:

176–185. Available: http://www.ncbi.nlm.nih.gov/pubmed/21914492. Accessed 2012 March 22.

46. Zhang X, Germann S, Blus BJ, Khorasanizadeh S, Gaudin V, et al. (2007) The Arabidopsis LHP1 protein colocalizes with histone H3 Lys27 trimethylation. Nat Struct Mol Biol 14: 869–871. Available: http://www.ncbi.nlm.nih.gov/pubmed/17676062. Accessed 2012 March 29.
47. Turck F, Roudier F, Farrona S, Martin-Magniette M-L, Guillaume E, et al. (2007) Arabidopsis TFL2/LHP1 specifically associates with genes marked by trimethylation of histone H3 lysine 27. PLoS Genet 3: e86. Available: http://www.pubmedcentral.nih.gov/articlerender.fcgi?artid = 1885283&tool = pmcentrez&rendertype = abstract. Accessed 2012 March 12.
48. Sunkar R, Li Y-F, Jagadeeswaran G (2012) Functions of microRNAs in plant stress responses. Trends Plant Sci 17: 1–8. Available: http://www.ncbi.nlm.nih.gov/pubmed/22365280. Accessed 2012 March 13.
49. Wang T, Chen L, Zhao M, Tian Q, Zhang W (2011) Identification of drought-responsive microRNAs in Medicago truncatula by genome-wide high-throughput sequencing. BMC Genomics 12: 367. Available: http://www.biomedcentral.com/1471-2164/12/367. Accessed 2011 July 17.
50. Szittya G, Moxon S, Santos DM, Jing R, Fevereiro MPS, et al. (2008) High-throughput sequencing of Medicago truncatula short RNAs identifies eight new miRNA families. BMC Genomics 9: 593. Available: http://www.pubmedcentral.nih.gov/articlerender.fcgi?artid = 2621214&tool = pmcentrez&rendertype = abstract. Accessed 2012 July 16.
51. Martínez G, Forment J, Llave C, Pallás V, Gómez G (2011) High-throughput sequencing, characterization and detection of new and conserved cucumber miRNAs. PLoS One 6: e19523. Available: http://www.pubmedcentral.nih.gov/articlerender.fcgi?artid = 3095615&tool = pmcentrez&rendertype = abstract. Accessed 2011 July 18.
52. Nobuta K, Lu C, Shrivastava R, Pillay M, De Paoli E, et al. (2008) Distinct size distribution of endogenous siRNAs in maize: Evidence from deep sequencing in the mop1-1 mutant. Proc Natl Acad Sci U S A 105: 14958–14963. Available: http://www.ncbi.nlm.nih.gov/pmc/articles/PMC2567475/.
53. Mica E, Piccolo V, Delledonne M, Ferrarini A, Pezzotti M, et al. (2010) Correction: High throughput approaches reveal splicing of primary microRNA transcripts and tissue specific expression of mature microRNAs in Vitis vinifera. BMC Genomics 11: 109. Available: http://www.pubmedcentral.nih.gov/articlerender.fcgi?artid = 2831844&tool = pmcentrez&rendertype = abstract.
54. Wei LQ, Yan LF, Wang T (2011) Deep sequencing on genome-wide scale reveals the unique composition and expression patterns of microRNAs in developing pollen of Oryza sativa. Genome Biol 12: R53. Available: http://www.ncbi.nlm.nih.gov/pubmed/21679406. Accessed 2011 July 20.
55. Breakfield NW, Corcoran DL, Petricka JJ, Shen J, Sae-Seaw J, et al. (2011) High-resolution experimental and computational profiling of tissue-specific known and novel miRNAs in Arabidopsis. Genome Res. Available: http://www.ncbi.nlm.nih.gov/pubmed/21940835. Accessed 2011 September 29.
56. Dong Z, Han M-H, Fedoroff N (2008) The RNA-binding proteins HYL1 and SE promote accurate in vitro processing of pri-miRNA by DCL1. Proc Natl Acad Sci U S A 105: 9970–9975. Available: http://www.pubmedcentral.nih.gov/articlerender.fcgi?artid = 2481344&tool = pmcentrez&rendertype = abstract.
57. Zhang X, Zhao H, Gao S, Wang W-C, Katiar-Agarwal S, et al. (2011) Arabidopsis Argonaute 2 regulates innate immunity via miRNA393*-mediated silencing of a Golgi-localized SNARE gene MEMB12. Mol Cell 42: 356–366. doi:10.1016/j.molcel.2011.04.010.Arabidopsis.
58. Hsieh L-C, Lin S-I, Shih AC-C, Chen J-W, Lin W-Y, et al. (2009) Uncovering small RNA-mediated responses to phosphate deficiency in Arabidopsis by deep sequencing. Plant Physiol 151: 2120–2132. Available: http://www.pubmedcentral.nih.gov/articlerender.fcgi?artid = 2785986&tool = pmcentrez&rendertype = abstract. Accessed 2012 March 10.
59. Fu J, Momčilović I, Prasad PVV (2012) Roles of Protein Synthesis Elongation Factor EF-Tu in Heat Tolerance in Plants. J Bot 2012: 1–8. Available: http://www.hindawi.com/journals/jb/2012/835836/. Accessed 2012 July 20.
60. Kong Y (2010) Differential Expression of microRNAs in Maize Inbred and Hybrid Lines during Salt and Drought Stress. Am J Plant Sci 01: 69–76. Available: http://www.scirp.org/journal/PaperDownload.aspx?DOI = 10.4236/ajps.2010.12009. Accessed 2012 July 18.
61. Zhou L, Liu Y, Liu Z, Kong D, Duan M, et al. (2010) Genome-wide identification and analysis of drought-responsive microRNAs in Oryza sativa. J Exp Bot 61: 4157–4168. Available: http://www.ncbi.nlm.nih.gov/pubmed/20729483. Accessed 2012 July 18.
62. Li B, Qin Y, Duan H, Yin W, Xia X (2011) Genome-wide characterization of new and drought stress responsive microRNAs in Populus euphratica. J Exp Bot 62: 3765–3779. Available: http://www.pubmedcentral.nih.gov/articlerender.fcgi?artid = 3134338&tool = pmcentrez&rendertype = abstract. Accessed 2011 August 15.
63. He G, Zhu X, Elling AA, Chen L, Wang X, et al. (2010) Global epigenetic and transcriptional trends among two rice subspecies and their reciprocal hybrids. Plant Cell 22: 17–33. Available: http://www.pubmedcentral.nih.gov/articlerender.fcgi?artid = 2828707&tool = pmcentrez&rendertype = abstract. Accessed 2013 August 7.
64. Khraiwesh B, Arif MA, Seumel GI, Ossowski S, Weigel D, et al. (2010) Transcriptional control of gene expression by microRNAs. Cell 140: 111–122. Available: http://www.ncbi.nlm.nih.gov/pubmed/20085706. Accessed 2012 July 13.
65. Guo H, Xie Q, Fei J, Chua N (2005) MicroRNA Directs mRNA Cleavage of the Transcription Factor NAC1 to Downregulate Auxin Signals for Arabidopsis Lateral Root Development. Development 17: 1376–1386. doi:10.1105/tpc.105.030841.illustrated.
66. Yang T, Xue L, An L (2007) Functional diversity of miRNA in plants. Plant Sci 172: 423–432. Available: http://linkinghub.elsevier.com/retrieve/pii/S0168945206002986. Accessed 2012 March 15.
67. Golldack D, Lüking I, Yang O (2011) Plant tolerance to drought and salinity: stress regulating transcription factors and their functional significance in the cellular transcriptional network. Plant Cell Rep 30: 1383–1391. Available: http://www.ncbi.nlm.nih.gov/pubmed/21476089. Accessed 2012 March 7.
68. Puranik S, Sahu PP, Srivastava PS, Prasad M (2012) NAC proteins: regulation and role in stress tolerance. Trends Plant Sci: 1–13. Available: http://www.ncbi.nlm.nih.gov/pubmed/22445067. Accessed 2012 March 29.
69. Zheng X, Chen B, Lu G, Han B (2009) Overexpression of a NAC transcription factor enhances rice drought and salt tolerance. Biochem Biophys Res Commun 379: 985–989. Available: http://www.ncbi.nlm.nih.gov/pubmed/19135985. Accessed 2012 July 21.
70. Jagadeeswaran G, Saini A, Sunkar R (2009) Biotic and abiotic stress down-regulate miR398 expression in Arabidopsis. Planta 229: 1009–1014. Available: http://www.ncbi.nlm.nih.gov/pubmed/19148671. Accessed 2011 July 17.
71. Lv D-K, Bai X, Li Y, Ding X-D, Ge Y, et al. (2010) Profiling of cold-stress-responsive miRNAs in rice by microarrays. Gene 459: 39–47. Available: http://www.ncbi.nlm.nih.gov/pubmed/20350593. Accessed 2012 July 21.
72. Silva M, Jifon J, Da Silva J, Sharma V (1997) Use of physiological parameters as fast tools to screen for drought tolerance in sugarcane. Brazilian J Plant Physiol 19: 193–201.
73. Jeong D-H, German MA, Rymarquis LA, Thatcher SR, Green PJ (2009) Abiotic stress-associated miRNAs: detection and functional analysis. In: Meyers BC, Green PJ, editors. Plant MicroRNAs, Methods in Molecular Biology. 203–230.
74. O'Malley DM, Whetten R, Bao W, Chen C-L, Sederoff RR (1993) The role of of laccase in lignification. Plant J 4: 751–757.
75. Allen E, Xie Z, Gustafson AM, Carrington JC (2005) MicroRNA-directed phasing during trans-acting siRNA biogenesis in plants. Cell 121: 207–221. Available: http://www.ncbi.nlm.nih.gov/pubmed/15851028. Accessed 2011 July 17.
76. Axtell MJ, Jan C, Rajagopalan R, Bartel DP (2006) A two-hit trigger for siRNA biogenesis in plants. Cell 127: 565–577. Available: http://www.ncbi.nlm.nih.gov/pubmed/17081978. Accessed 2012 July 26.
77. Chen H-M, Chen L-T, Patel K, Li Y-H, Baulcombe DC, et al. (2010) 22-nucleotide RNAs trigger secondary siRNA biogenesis in plants. Proc Natl Acad Sci: 1–6. Available: http://www.pnas.org/cgi/doi/10.1073/pnas.1001738107. Accessed 2010 July 20.
78. Xia R, Zhu H, An Y, Beers EP, Liu Z (2012) Apple miRNAs and tasiRNAs with novel regulatory networks. Genome Biol 13: R47. Available: http://www.ncbi.nlm.nih.gov/pubmed/22704043. Accessed 2012 July 16.
79. Reddy AR, Chaitanya KV, Vivekanandan M (2004) Drought-induced responses of photosynthesis and antioxidant metabolism in higher plants. J Plant Physiol 161: 1189–1202. Available: http://linkinghub.elsevier.com/retrieve/pii/S0176161704000422. Accessed 2012 July 30.
80. Cantu D, Vanzetti LS, Sumner A, Dubcovsky M, Matvienko M, et al. (2010) Small RNAs, DNA methylation and transposable elements in wheat. BMC Genomics 11: 408. Available: http://www.pubmedcentral.nih.gov/articlerender.fcgi?artid = 2996936&tool = pmcentrez&rendertype = abstract.
81. Wang W, Pan Y, Zhao X, Dwivedi D, Zhu L, et al. (2010) Drought-induced site-specific DNA methylation and its association with drought tolerance in rice (Oryza sativa L.): 1–10. doi:10.1093/jxb/erq391.
82. Vazquez F, Legrand S, Windels D (2010) The biosynthetic pathways and biological scopes of plant small RNAs. Trends Plant Sci 15: 337–345. Available: http://www.ncbi.nlm.nih.gov/pubmed/20427224. Accessed 2012 March 9.
83. Cuperus JT, Carbonell A, Fahlgren N, Garcia-ruiz H, Burke RT, et al. (2010) Unique functionality of 22-nt miRNAs in triggering RDR6-dependent siRNA biogenesis from target transcripts in Arabidopsis. Online 17: 997–1004. Available: http://www.ncbi.nlm.nih.gov/pubmed/20562854. Accessed 2010 July 26.
84. Kurihara Y, Takashi Y, Watanabe Y (2006) The interaction between DCL1 and HYL1 is important for efficient and precise processing of pri-miRNA in plant microRNA biogenesis. RNA 12: 206–212. Available: http://www.pubmedcentral.nih.gov/articlerender.fcgi?artid = 1370900&tool = pmcentrez&rendertype = abstract. Accessed 2012 July 20.
85. Vaucheret H (2008) Plant ARGONAUTES. Trends Plant Sci 13(7): 350–358. doi:10.1016/j.tplants.2008.04.007.
86. Zilberman D, Cao X, Johansen LK, Xie Z, Carrington JC, et al. (2004) Role of Arabidopsis ARGONAUTE4 in RNA-Directed DNA Methylation Triggered by Inverted Repeats. Curr Biol 14: 1214–1220. doi:10.1016/j.
87. Havecker ER, Wallbridge LM, Hardcastle TJ, Bush MS, Kelly KA, et al. (2010) The Arabidopsis RNA-directed DNA methylation argonautes functionally diverge based on their expression and interaction with target loci. Plant Cell 22: 321–334. Available: http://www.pubmedcentral.nih.gov/articlerender.fcgi?artid = 2845420&tool = pmcentrez&rendertype = abstract. Accessed 2012 July 13.

15

Long-Term Monitoring of Rainfed Wheat Yield and Soil Water at the Loess Plateau Reveals Low Water Use Efficiency

Wei Qin[1,2], Baoliang Chi[1]*, Oene Oenema[2]*

1 Dryland Agriculture Research Centre, Shanxi Academy of Agricultural Sciences, Taiyuan, Shanxi, China, **2** Environmental Sciences, Wageningen University, Wageningen, The Netherlands

Abstract

Increasing crop yield and water use efficiency (WUE) in dryland farming requires a quantitative understanding of relationships between crop yield and the water balance over many years. Here, we report on a long-term dryland monitoring site at the Loess Plateau, Shanxi, China, where winter wheat was grown for 30 consecutive years and soil water content (0–200 cm) was measured every 10 days. The monitoring data were used to calibrate the AquaCrop model and then to analyse the components of the water balance. There was a strong positive relationship between total available water and mean cereal yield. However, only one-third of the available water was actually used by the winter wheat for crop transpiration. The remaining two-thirds were lost by soil evaporation, of which 40 and 60% was lost during the growing and fallow seasons, respectively. Wheat yields ranged from 0.6 to 3.9 ton/ha and WUE from 0.3 to 0.9 kg/m^3. Results of model experiments suggest that minimizing soil evaporation via straw mulch or plastic film covers could potentially double wheat yields and WUE. We conclude that the relatively low wheat yields and low WUE were mainly related to (i) limited rainfall, (ii) low soil water storage during fallow season due to large soil evaporation, and (iii) poor synchronisation of the wheat growing season to the rain season. The model experiments suggest significant potential for increased yields and WUE.

Editor: Eric Gordon Lamb, University of Saskatchewan, Canada

Funding: The authors have no support or funding to report.

Competing Interests: The authors have declared that no competing interests exist.

* E-mail: Baoliangchi@hotmail.com (BC); Oene.Oenema@wur.nl (OO)

Introduction

Water scarcity is a growing global concern [1–4]. For rainfed agriculture, this pressure may become more severe under climate change due to the expected more erratic rainfall and longer dry spells [5–8]. Currently, rainfed agriculture covers 80% of the world's cultivated land and accounts for 60% of crop production [9]. However, crop yield and water use efficiency (WUE) are often low in rainfed agriculture, especially in arid and semi-arid areas due to, for example, degraded soils, erratic rainfall and poor water management [10]. Many of these areas in Africa and Asia face also rapid population growth. Hence, there is a pressing need to increase crop yields and WUE in rainfed agriculture [11–15].

Water use efficiency is commonly defined as crop yield over evapotranspiration (ET), where ET is the sum of soil evaporation (E) and crop transpiration (T) [16–19]. The latter (T) is a direct consequence of crop production, while E is 'unproductive' water loss. The central question for rainfed agriculture in arid and semi-arid regions is 'how to transform unproductive water loss (E) into productive water use (T)'. Unfortunately, the partitioning between E and T is often not well-known, due to difficulties and high cost in distinguishing E and T in the field. As a result, water use of crops is commonly reported as evapotranspiration (ET) [16,17,20]. This lack of information makes it difficult to assess how much of the evaporative water loss can be used for increasing yields by appropriate measures.

Crop growth simulation models have the potential of providing more comprehensive insights into the functioning of soil-crop systems, and can be helpful to explore options for increasing yield and WUE [21–23]. These models though are simplified representations of parts of reality and therefore require testing in the real world. Fortunately, numerous field studies have examined the effect of water availability, with or without irrigation treatments, on crop yield and WUE [16–20]. In principle, these results can be used to calibrate and test the crop growth simulation models. However, most of these field studies are short-term (2–5 years) and focus on the growing season only, largely ignoring the water balance during the fallow season. Furthermore, crop yields and WUE show large variations due to differences in soils, climate conditions and crop husbandry practices. For example, in the North China Plain, annual precipitation ranges from 400 to 650 mm and wheat grain yields roughly range from 1 to 3 ton/ha/year under rainfed conditions. With 200 to 300 mm of irrigation, grain yield can be increased by 60 to 100% and WUE can be increased by 20 to 40% [16]. Globally, WUE of wheat shows even larger variation ranging from approximately 0.2 to 1.8 kg/m^3 [24]. Evidently, this wide range suggests considerable scope for improvement, but the underlying causes of WUE is not always well-known and irrigation water is not available on most places.

The study reported here has the objectives (i) to calibrate a water-driven crop growth model on the basis of monitoring data from a long-term field experiment with rainfed winter wheat, (ii) to analyse the components of the water balance of this field, and (iii) to explore the potential for increasing crop yield and WUE through model experiments. The field experiment was situated in the dryland of the Loess Plateau, Shanxi, China. Winter wheat was grown for 30 consecutive years and soil water content (0–200 cm) was measured every 10 days. The FAO AquaCrop model was chosen as simulation model because it is a water-driven crop growth model that can separately calculate E and T, and simulates the final crop yield as function of water use [23,25]. The AquaCrop model uses canopy ground cover as the basis to calculate T and to separate E and T. Crop yield is then calculated as the product of biomass and harvest index (HI). The principles and modules of the AquaCrop model are well-documented in a series of AquaCrop publications [23,25–28]. Compared to some other crop growth models, AquaCrop requires relatively few input parameters [23,28]. Such a limited number of input parameters facilitates model calibration and utilization for different crops and under different management strategies [23,26,27,29–36].

Materials and Methods

Site Information

The long-term monitoring site is located in Beizhang, Linyi county in Shanxi province on the Loess Plateau (35° 9′ 3.83″ N, 110° 34′ 25.40″ E, Altitude: 491 m) in China. The authority is Dryland Agriculture Research Centre, Shanxi Academy of Agricultural Sciences. We have the permission to conduct the study on this site. We confirm that the field studies did not involve endangered or protected species.

The site has a semi-arid climate with extensive monsoonal influence, which is dry and cold in winter, rainy and hot in summer. Rainfall in June to September accounts for more than 70% of annual rainfall. Average annual rainfall was 517 mm in the period of 1980–2010 with large annual variations, from a minimum of 331 mm to a maximum of 832 mm. Mean annual sunshine duration is 2270 h, annual average temperature 13.5°C and mean annual potential evaporation is 1340 mm. The soil is a typical Loessial soils (Calcic Luvisols) of the Loess Plateau [37–40]. Soil slope is <1% and soil texture is silt loam, with a small proportion of clay. The soil was rather homogeneous in texture and key physical properties (Table 1). Maximum soil water holding capacity was a significant fraction of the total annual rainfall (Table 2).

The top soil (30 cm) contained 5.8 g/kg of organic C and 0.55 g/kg of total N. Mean available N, P and K were 63 mg/kg, 14 mg/kg and 142 mg/kg, respectively, in 2007. Since 1983, mean available P and K have been increased by 7 mg/kg and 10 mg/kg, respectively [41]. Total soil organic C was measured by dry combustion combined with elemental C analysis. Total N was measured by Kjeldahl method. Soil mineral N was extracted with 1 M KCl and analysed by the cadmium reduction method. Available P was extracted with 0.5 M $NaHCO_3$. Available K was extracted with 1.0 M NH4OAc.

Experimental Design and Measurements

The long-term crop yield and soil moisture monitoring experiment started in 1980. There were no experimental treatments. The size of the field is 0.5 ha. Winter wheat was planted each year between 25 September and 5 October, depending on the actual climate and soil water conditions. Sowing rate was 150 kg seed/ha. The growing period of the winter wheat is approximately 245 days (from 1 October to 1 June of the next year). The fallow season is about 120 days (from 2nd June to 30th September) (Figure 1). Local bred cultivars (Jinmai) were used for all years. Information about crop parameters is listed in Table 3. Fertilizers were applied at planting at a rate of 127.5 kg/ha of N as urea and 90 kg/ha of P_2O_5 as superphosphate, for all years. At harvest, grain yields were measured in 5 random plots (2 m^2 per plot) selected from two diagonal lines of the field. Dry matter content of grain and straw was measured at the laboratory after drying at 70°C. Soil water contents were measured in 16 different layers up to 2 m depth every 10 days by using the gravity method. The top 10 cm was sampled in 5 cm intervals. From 0.1 to 1 m, samples were taken at 10 cm intervals, and from 1 to 2 m at 20 cm intervals (Figure 2). Each sample consisted of 5 subsamples, taken randomly on two diagonal lines across the field. Observations during soil sampling over years revealed that spatial variations in soil profile were small.

Water Balance and WUE Estimations

Mean monthly rainfall data were collected at a near-by meteorological station (Linyi station, 35° 10′ 7.02″ N, 110° 46′ 44.59″ E, Altitude: 441 m), which is 25 km away from the experimental field. The water balance for both fallow and crop growing seasons reads as follows:

$$R + I \pm \Delta S = E + T + R_r + P_e \quad (1)$$

where R is rainfall, I is irrigation, ΔS is the change of soil water content, E is soil evaporation, T is crop transpiration, R_r is runoff and P_e is percolation. All units are presented in mm or in m^3/ha.

Irrigation was not applied in this study. Furthermore, runoff and percolation (leaching) were small and disregarded. Hence, the water balance of the fallow seasons was simplified to:

$$R \pm \Delta S = E \quad (2)$$

The water balance of the growing season was simplified to:

$$R \pm \Delta S = E + T \quad (3)$$

We used formula (2) and the recorded rainfall and soil water content data to calculate evaporation during the fallow season and formula (3) to calculate ET during the growing season. The partitioning of E and T was done by the AquaCrop model.

Water use efficiency (WUE, kg/m^3) was calculated as:

$$WUE = grain\,yield \,/\, ET \quad (4)$$

where ET is evapotranspiration (mm), the sum of E and T during the growing season.

Calibration and Validation of the AquaCrop Model

The FAO AquaCrop model is a water-driven crop growth model for the simulation of crop biomass and yield as function of water availability [23,25]. AquaCrop requires 4 main sets of input data, i.e. climate data (rainfall, minimum and maximum temperature and reference evapotranspiration (ETo)), crop parameters, soil data and field management data. Climate data were collected from a near-by meteorological station (Linyi station, 35° 10′ 7.02″ N, 110° 46′ 44.59″ E, Altitude: 441 m), which is 25 km away from the experimental field. ETo was calculated by FAO Penman-Monteith equation as described in Allen et al. [42].

Table 1. Soil physical properties for different layers up to 200 cm.

Soil layer	Bulk density	Field Capacity	Wilting point	Texture, (particle size, µm in %)			
cm	g/cm^3	v/v in %	v/v in %	>63 µm	63-20 µm	20-2 µm	<2 µm
5	1.34	31.4	6.8	7.5	41.0	32.4	19.2
10	1.34	31.4	6.8	7.5	41.0	32.4	19.2
20	1.39	32.9	8.5	8.3	41.8	31.2	18.8
30	1.43	34.5	8.4	6.9	44.3	30.1	18.7
40	1.39	33.6	7.0	6.9	44.3	30.1	18.7
50	1.43	33.6	7.3	6.9	44.3	30.1	18.7
60	1.39	32.8	7.0	4.5	40.5	31.4	23.6
70	1.36	30.9	7.2	4.5	40.5	31.4	23.6
80	1.25	27.6	6.9	4.5	40.5	31.4	23.6
90	1.28	27.9	8.3	4.5	40.5	31.4	23.6
100	1.32	28.3	8.2	4.5	40.5	31.4	23.6
120	1.30	29.3	7.8	3.0	43.4	34.4	19.2
140	1.32	29.7	7.9	3.0	43.4	34.4	19.2
160	1.31	29.5	7.9	2.4	40.4	38.7	18.5
180	1.32	29.7	7.9	2.4	40.4	38.7	18.5
200	1.32	29.7	7.9	2.4	40.4	38.7	18.5

Soil data (Tables 1 and 2) and field management data were derived from measurements and from recordings at the monitoring site. Crop data and parameters were derived from measurements, literature [23,25–27] and by calibration (Table 3).

The objective of calibration and validation was to achieve the best match between simulated outputs and monitoring data (soil moisture and crop yield) for all 30 years, using a common procedure [27,29–36]. We randomly selected 15 years' data from the monitoring field for calibration and used the other 15 years' data for validation of the calibrated model. In the calibration step, we used the year-specific climate, soil and initial soil water data as fixed input data. Then we adjusted some of the crop parameters (see Table 3), based on our understanding of crop growth, development and crop responses to water deficits, until the differences between simulated output and monitoring data were minimal. We repeated this process for all selected 15 years and ultimately obtained a satisfactorily index of agreement (d = 0.92). Then, the calibrated model was validated with the other 15 years' data, and again obtained an acceptable index of agreement (d = 0.93).

Data Analysis

The root mean square error (RMSE) [43] has been widely used to evaluate the performance of a model [30,44–47]. However, Willmott and Matsuura [44] pointed out that, compared to RMSE, the mean absolute error (MAE) is a better indicator and therefore the evaluation of model performance should be based on the MAE. In this study, we used the MAE, the mean bias error (MBE) and the Willmott index of agreement (d) to evaluate the model performance.

The MAE measures the weighted average magnitude of the absolute errors and was calculated as follows:

$$MAE = \frac{1}{n}\sum_{i=1}^{n} |\mathrm{Mi} - \mathrm{Oi}| \qquad (5)$$

where n is the number of observations, Mi is the modelled yield or soil water content and Oi is the observed yield or soil water content.

The MBE indicates whether the model is under or over predicting the observed values and also indicates the uniformity of error distribution. Positive MBE values indicate over prediction, negative values indicate under prediction and a value of zero

Table 2. Soil moisture holding capacity of the soil profile (0–200 cm).

	Total water, mm(mm)	Available water, mm(mm)	In volume, % (v/v)
Saturation	860	711	43
Field capacity	596	447	30
70% FC	462	312	23
50% FC	373	223	19
Wilting point	149	0	8

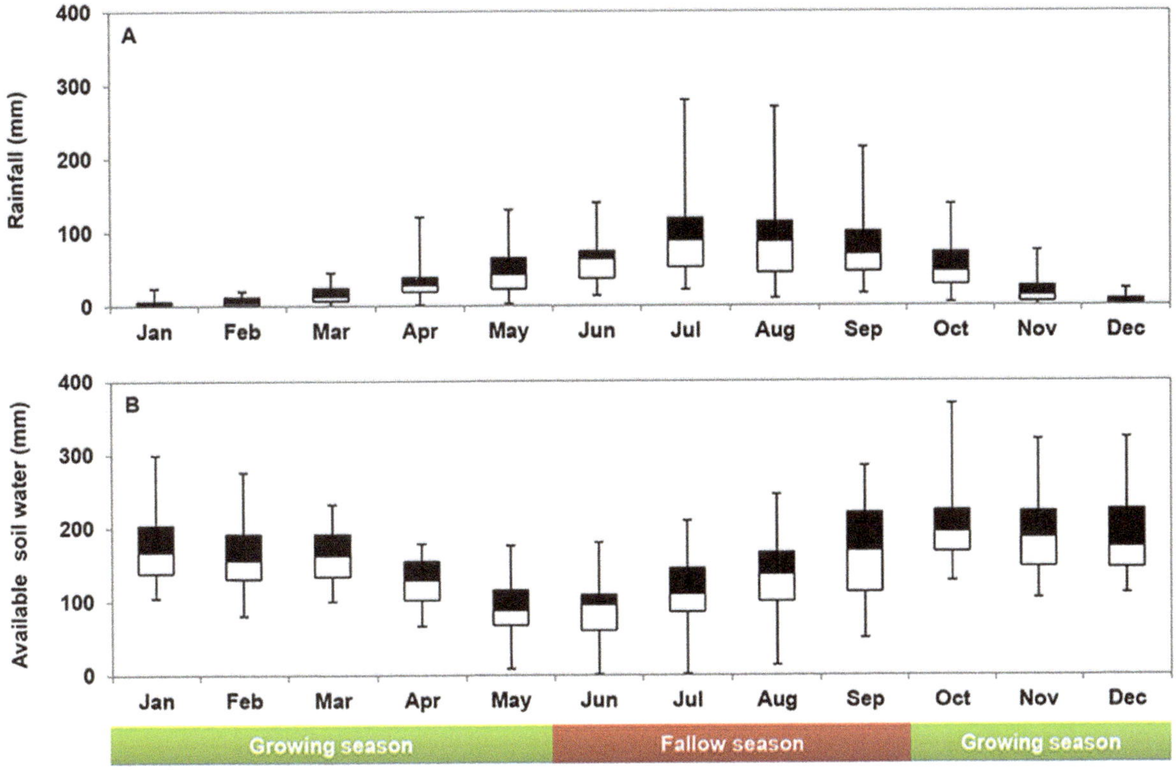

Figure 1. Monthly rainfall and available soil water during the period 1980–2010. A: monthly rainfall distribution; B: monthly available water content. Black and white boxes show 75 and 25% percentile values. Whiskers show maximum and minimum values (the same applies to other figures). The growing season of winter wheat was 245 days from 1 October to 1 June. of the next year, highlighted in green bar. Fallow season was 120 days from 2nd Jun. to 30th Sep., highlighted in red bar.

indicates equal distribution between negative and positive values. The MBE was calculated as follows:

$$MBE = \frac{1}{n}\sum_{i=1}^{n}(\mathrm{Mi} - \mathrm{Oi}) \quad (6)$$

The Willmott index of agreement (d) [43] has values ranging from 0 to 1. A value close to 1 suggests a good model performance. The agreement index d was calculated as follows:

$$d = 1 - \frac{\sum_{i=1}^{n}(\mathrm{Mi} - \mathrm{Oi})^2}{\sum_{i=1}^{n}\left(|\mathrm{Mi} - \bar{\mathrm{O}}| + |\mathrm{Oi} - \bar{\mathrm{O}}|\right)^2} \quad (7)$$

where $\bar{\mathrm{O}}$ is the average value of the observed yield or soil water content.

Model Experiments

To investigate the potentials of minimizing soil evaporation so as to increase crop yield and WUE, we set up six model experiments as follows:

Experiment 1 (E1): Reference A; winter wheat was planted at 50% of field capacity (FC) (low soil water content)

Experiment 2 (E2): E1+organic mulch (straw)

Experiment 3 (E3): E1+plastic film cover

Experiment 4 (E4): Reference B; winter wheat was planted at 70% of FC (high soil water content)

Experiment 5 (E5): E4+organic mulch (straw)

Experiment 6 (E6): E4+plastic film cover

In the model experiments, we used E1 and E4 as references to simulate crop growth under relatively low (with 50% FC) and high (with 70% FC) soil water content at winter wheat seeding (Table 2). The range from 50 to 70% FC largely represented the initial soil water content for the planting period. The low value is representative for no water harvesting, and the high values is representative for water harvesting during the fallow period via mulching and/or covers. Model experiments E2 and E5 aimed at testing the effects of organic mulch during the growing season, and experiments E3 and E6 aimed at testing the effects of plastic film during the growing season on crop yield and water balance. The effectiveness of the soil evaporation reduction by organic mulch and plastic film cover during the growing season were estimated at 50 and 90%, respectively, which are default values of the AquaCrop model [25]. However, since plastic film will cover only about 80% of the wheat planted field, the overall soil evaporation reduction by plastic film was set at 72%. In practice, the plastic film covers are used for the growing season only and destroyed or removed after harvesting the crops. We ran these six experiments for the period of 1980–2010.

Table 3. Full set of crop parameters used in this study.

Parameters	Contents and values	Source
Crop development	in calendar days	
From sowing to emergence	7 days	Field observation
From sowing to max. canopy	200 days	Field observation
From sowing to flowering	200 days	Field observation
From sowing to senescence	210 days	Field observation
From sowing to maturity	245 days	Field observation
Length for building up Harvest Index (HI)	44 days	Field observation
Duration of flowering	5 days	Field observation
Plant density	300 plants/m^2	Field observation
Sowing rate	150 kg seed/ha	Field observation
1000 seed mass	40 g	Field observation
Germination rate	80%	Field observation
Max. root depth	2 m	Field observation
Reference HI	42%	Field observation
Max HI	48%	Field observation
Crop water productivity	15 g/m^2	Calibrated
Canopy development		
Initial canopy cover	4.5%	Field observation
Canopy expansion	2.9%/day	Calculated by AquaCrop
Max. canopy cover	90%	Field observation
Canopy decline	7.2%/day (39 days)	Calculated by AquaCrop
Thresholds temperatures		
Base temperature for biomass production	0°C	Calibrated
Upper temperature for biomass production	26°C	Calibrated
Range of cold stress for biomass production	0–14°C	Calibrated
Range of cold stress for pollination	4–9°C	Calibrated
Range of heat stress for pollination	32–37°C	Calibrated
Water extraction pattern throughout the root zone		
Upper 1/4 (0–0.5 m)	40%	Calibrated
Second 1/4 (0.5–1 m)	30%	Calibrated
Third 1/4 (1–1.5 m)	20%	Calibrated
bottom 1/4 (1.5–2 m)	10%	Calibrated
Water stresses		
Canopy expansion	Moderately tolerant (upper = 0.25, lower = 0.65, shape factor = 5)	Calibrated
Stomatal closure	Extremely sensitive (upper = 0.25, shape factor = 2.5)	Calibrated
Early canopy senescence	Tolerant (upper = 0.75, shape factor = 2.5)	Calibrated
Aeration stress	Moderately tolerant (5 vol%)	Calibrated
Evapotranspiration		
Soil evaporation coefficient	Effect of canopy shelter in late season = 50%	AquaCrop default value
Crop transpiration coefficient	1.1 (reduction with age = 0.15%/day)	AquaCrop default value
Fertilities stresses	Not considered	AquaCrop default value

Results

Patterns of Monthly Rainfall and Soil Water Content

Most of the rainfall occurs in the summer from June to September, accounting for more than 70% of the total annual rainfall while winter wheat was seeded in the end of September or beginning of October, and harvested in the end of May or beginning of June. Hence, most of the rain fell when wheat had matured already, i.e., during the fallow period. Therefore, the growing season of winter wheat was poorly synchronized to the rain season. On the other hand, due to concentrated rainfall in June to September, mean soil water content in the upper two metre was highest in October when winter wheat was seeded. Thereafter, the soil water content gradually decreased, due to soil

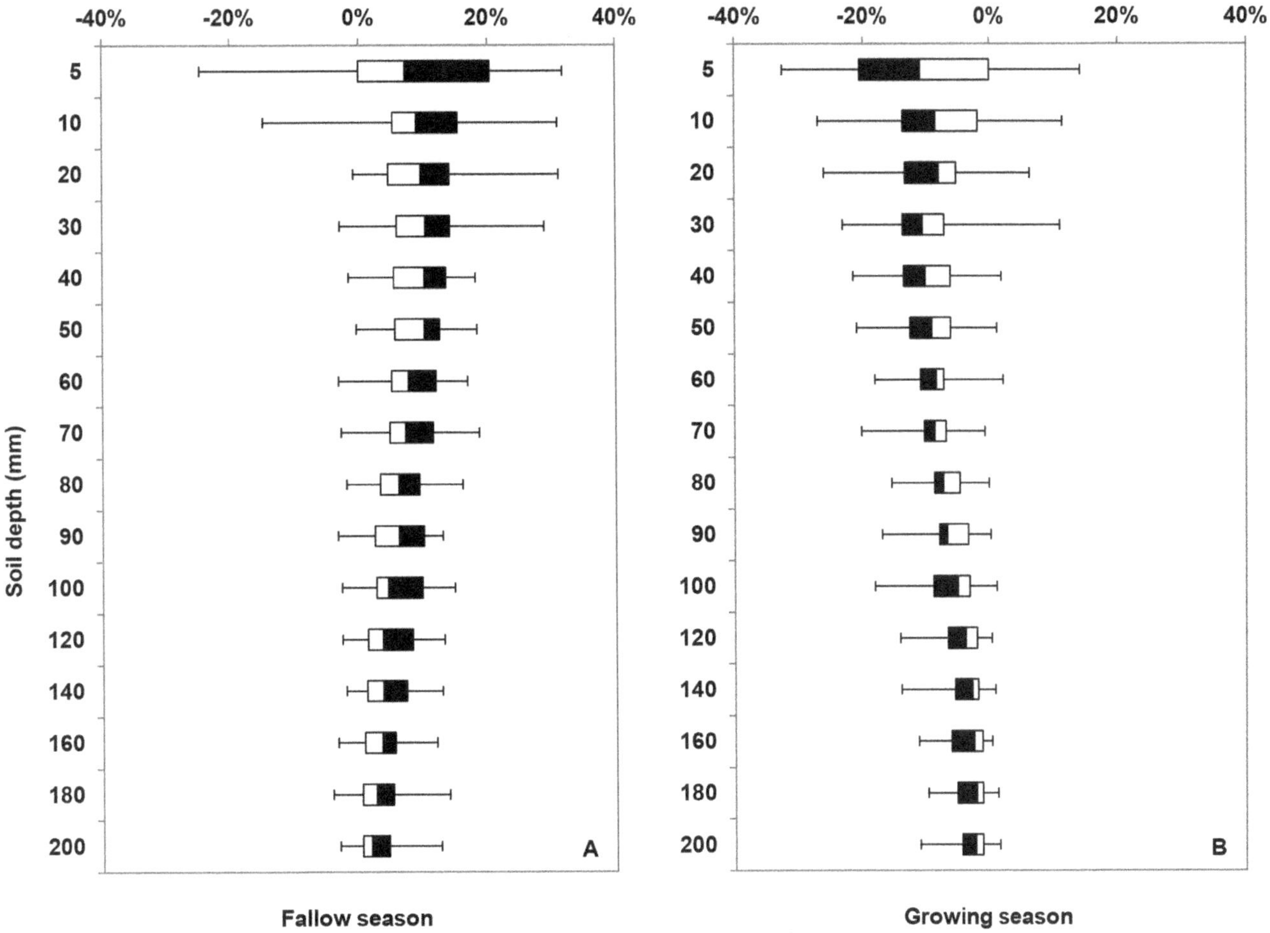

Figure 2. Changes in soil water content in the upper 2 meter of the soil during the period of 1980–2010. A: water stored in the upper 2 meter during the fallow season, presented in positive percentage (in volume, v/v). B: water depletion in the upper 2 m during the growing season, presented in negative percentage (in volume, v/v). The soil water changes for fallow season were calculated as the soil water content of October deducted by that of June. Similarly, the soil water changes for growing season were calculated as the soil water content of June deducted by that of October in the previous year.

evaporation and crop transpiration. Mean soil water content was lowest in June, at the beginning of the rain season when the winter wheat was harvested (Figure 1).

The changes of the soil water content (for 16 layers in the upper 2 meter of the soil) during the fallow and growing seasons are shown in Figure 2. The changes in water content in the fallow season mirror the changes during the growing season, i.e. water stored in the soil during the fallow season largely equalled to soil water depletion during the growing season. The soil water changes for fallow season were calculated as the soil water content of October deducted by that of June. Similarly, the soil water changes for growing season were calculated as the soil water content of June deducted by that of October in the previous year. Soil water content mostly changed in the top 1 m, accounting for 70% of total soil water change (±120 mm) (Figure 2). We cannot exclude that some of the seasonal variations are caused by slight spatial variations in soil characteristics.

Model Calibration and Performance

The AquaCrop model was calibrated using the climate data, soil data, field management data and monitoring data. The full set of crop parameters is listed in Table 3. The performance of the model on the simulated crop yield and soil water balance is shown in Figure 3. The relationship between observed and modelled grain yield of the calibrated model for the second set of 15-years data (validation step) was almost as good as for the whole set of data (not shown). For the 30 years' data, the relationship between observed and modelled grain yield had a correlation coefficient of (R^2) of 0.77, and the index of agreement (d) was 0.93 (Figure 3A). Similarly, the relationship between observed and modelled soil water content showed a R^2 of 0.78 and an index of agreement (d) of 0.93 (Figure 3B). Mean absolute errors (MAE) were 311 kg/ha and 25 mm, respectively. Mean bias errors (MBE) were −168 kg/ha and −12 mm in simulating yields and soil water balance, respectively, suggesting that the model slightly underestimated grain yields and soil water contents by 8 and 9%, respectively. We conclude that the performance of the AquaCrop model was acceptable for doing further simulations.

Water Balance and Partitioning of E and T

AquaCrop was used to estimate soil evaporation and crop transpiration, and the water balance for both fallow and growing seasons during the period 1980–2010 (Figure 4). There was only a very marginal change in soil water content when considering the water balance of the total season (fallow+growing season) over the 30 years' period; total rainfall was nearly equal to the sum of soil

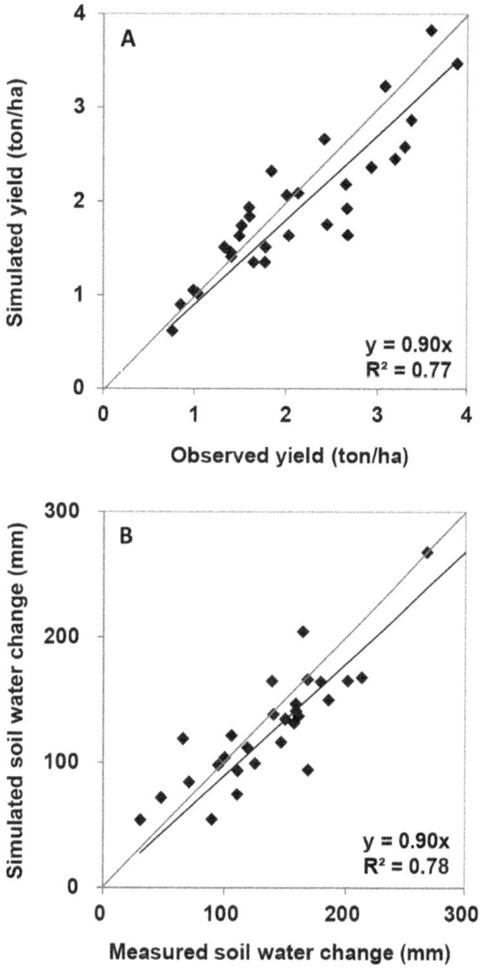

Figure 3. The AquaCrop Model simulations on yield (A) and soil water change (B). Diagonal lines represent 1:1 lines.

evaporation and crop transpiration. Crop transpiration accounted for approximately one-third of total seasonal rainfall. The remaining two-thirds were assumed to be lost by soil evaporation (Figure 4A), but we cannot exclude that a small fraction of this evaporative loss was actually lost by leaching and/or runoff. Mean rainfall in the fallow season was 323 mm, of which 207 mm (64%) was lost by soil evaporation and 116 mm (36%) was stored in the soil (Figure 4B). During the growing season, mean crop transpiration was 185 mm (57%) and soil evaporation was 137 mm (43%), of which rainfall and soil moisture contributed 194 (60%) and 128 mm (40%), respectively (Figure 4C). Mean rainfall of the growing season was slightly larger than mean crop transpiration.

Wheat Yield and WUE

Due to limited amounts of available water and the irregular rainfall pattern, wheat yield and WUE were rather low, ranging from 0.6 to 3.9 ton/ha/year and from 0.3 to 0.9 kg/m^3, respectively. Relationships between yield and annual rainfall ($p = 0.028$), and between yield and growing season rainfall ($p = 0.027$) were highly significant (Figures 5). The relationship between yield and rainfall during the fallow season was not significant ($p = 0.167$). Furthermore, there were significant linear relationships between wheat grain yield and ET, T, and WUE, with coefficients of determination (R^2) of 0.61, 0.68 and 0.72, respectively (Figure 5). Given the slope (0.01) between yield and T, an increase of 1 mm available water can produce 10 kg grain per hectare.

Model Experiments

In model experiment E1 (winter wheat planted at 50% of FC), mean wheat yield and WUE were 2 ton/ha and 0.6 kg/m^3, respectively (Figure 6). In model experiment E2, with organic mulch during the growing season, mean yield increased to 2.3 ton/ha and WUE increased to 0.8 kg/m^3. In model experiment E3, with plastic film during the growing season, mean yield increased further to 2.5 ton/ha and WUE to 1.0 kg/m^3.

Similarly, in model experiment E4 (winter wheat planted at 70% of FC), mean yield and WUE were 2.9 ton/ha and 0.8 kg/m^3, respectively. With organic mulch (E5), mean yield and WUE increased to 3.2 ton/ha and 0.9 kg/m^3, and with plastic film cover (E6), mean yield and WUE increased further to 3.5 ton/ha and 1.1 kg/m^3. Our model experiments show that both organic mulch and plastic film cover could significantly improve yield and WUE, but the impact of plastic cover was bigger than the organic mulch mainly due to high effectiveness in reducing soil evaporation. Moreover, increasing water storage in the soil during the fallow season, so that available soil water content increases from 50 to 70% of FC in autumn at the time of winter wheat seeding, is at least as effective as a plastic cover during the growing season.

Discussion

We successfully calibrated and validated the AquaCrop model on the basis of the long-term monitoring data (30 years) of rainfed winter wheat on the Loess Plateau of northern China. The full set of crop parameters (Table 3) may provide also guidance to future studies and further model calibrations and validations. We also quantified four key components of the water balance, i.e., rainfall, changes in soil water, soil evaporation and crop transpiration, for both fallow and growing seasons by combining empirical data from a long-term wheat monitoring site with calculated results using the AquaCrop model. In the end, we also explored the potential for increasing wheat yield and WUE through model experiments.

The relationship between observed and modelled wheat yield had a R^2 of 0.77, slope of 0.9 and an index of agreement (d) of 0.93 (Figure 3). Mkhabela and Bullock [47] reported a R^2 of 0.66, slope of 0.96, index of agreement (d) of 0.99 between observed and modelled wheat yields. Araya et al. [48] reported a R^2>0.80 when simulating barley biomass and grain yield. Stricevic et al. [49] reported a R^2>0.84 when simulating yields of maize, sunflower and sugar beet. Similarly for simulating soil water content, we found a R^2 of 0.78, slope of 0.9 and an index of agreement (d) of 0.93 (Figure 3). Mkhabela and Bullock [47] reported a R^2 of 0.9 and a slope of 0.9 for simulating soil water content. Hence, the performance of AquaCrop for our dryland wheat field is largely comparable with that of other modelling studies.

Water stress limited the crop yield at this site. According to our model simulations, water stress has led to suboptimal yields in essentially all years, for both low and relatively high wheat yields. For example, a very low grain yield (0.6 ton/ha) was recorded in the year 2000, when water stress for leaf expansion and stomatal closure started already at the 54th day after planting. In contrast, water stress for leaf expansion and stomatal closure started to occur only from day 156 after planting in 2003, when grain yield was 3.8 ton/ha. In both cases, water stress occurred before flowering stage (~200 days after planting). Water stress leads to

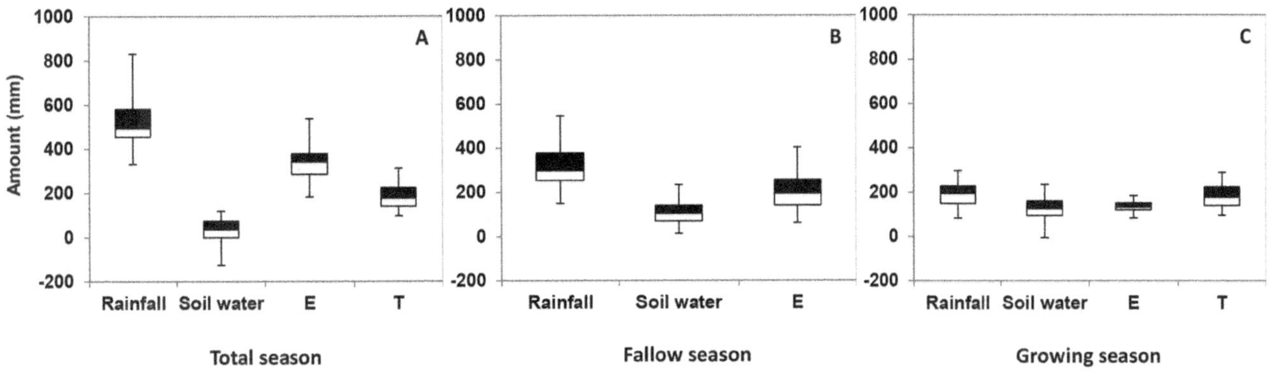

Figure 4. Water balance (rainfall, change in soil water, E, T) of the total season (A), fallow season (B) and growing season (C) during the period of 1980–2010. Total season means the sum of fallow season and growing season.

low grain yield and low WUE, but depending on the stage and duration of the water stress [17,50].

We found significant linear relationships between wheat yield and ET, and between wheat yield and T (Figure 5), in line with some other studies [17,51,52]. Mean T/ET ratio in our study was only 58% during the growing season, which was 8–12% lower than that reported by Liu et al. [53] and Kang et al. [54] but highly in line with that reported by Wang et al. [55]. The higher T/ET ratio in the studies of Liu et al. [53] and Kang et al. [54] was probably due to the irrigation treatments where the crop had more water for transpiration. It is well-known that crop yield and WUE are often lower in rainfed agriculture than in irrigated agriculture [17,20,50], but depending also on possible nutrient, weed, and disease stresses and irrigation management.

Advanced technologies, such as precision irrigation, are for a long time available but unfortunately not affordable and applicable to the farmers of the Loess Plateau, mainly because of the high cost relative to the low value of cereals [56]. Therefore, we focused on low-cost options, such as straw mulch and plastic film cover because those are the most accessible and low cost materials for farmers to implement in the field. Minimizing soil evaporation could save water for crop transpiration, and thereby increase wheat yield and WUE. Our model experiments suggest that wheat yield can be improved significantly by minimizing soil evaporation via organic mulch and plastic film cover, especially also during the fallow period. Mulching with crop residues can decrease soil evaporation and increase soil water retention. Plastic film cover can significantly increase crop yield and WUE, and promote crop growth during early growth when temperature is

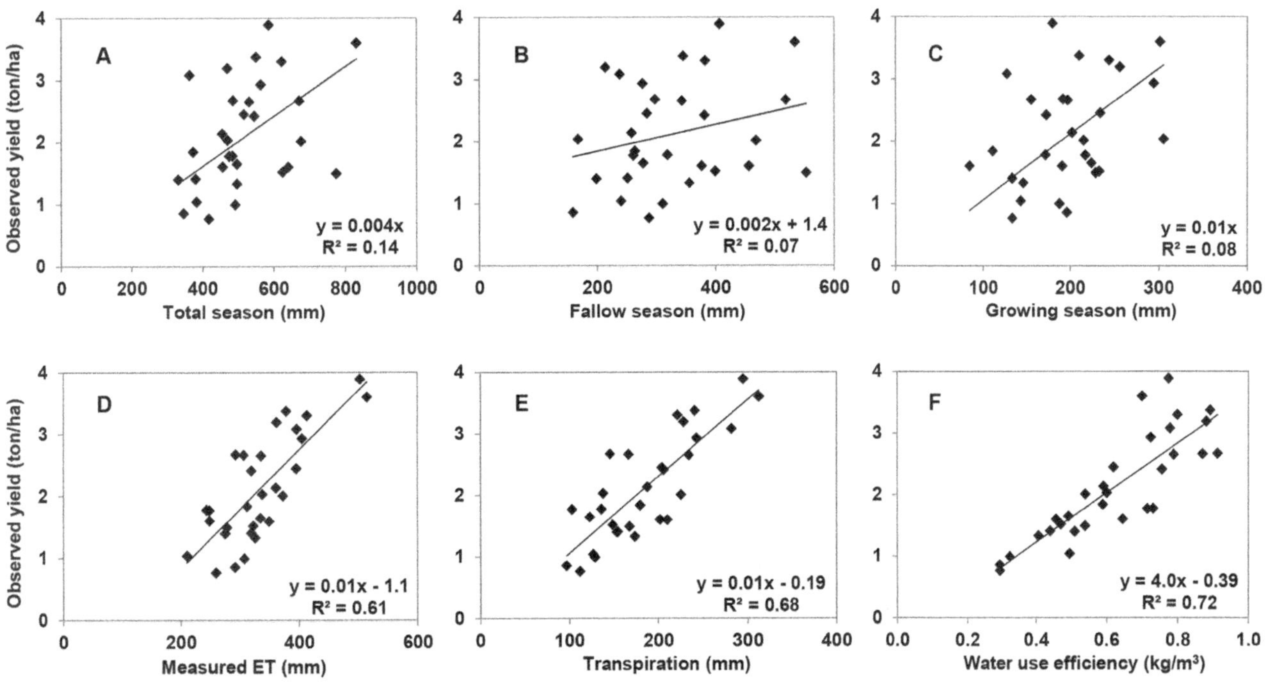

Figure 5. Relationships between observed yield and total rainfall (A), rainfall in fallow season (B), rainfall in growing season (C), measured ET (D), Transpiration (E) and WUE (F). The significant level is 0.028 for (A), 0.167 for (B), 0.027 for (C). For D, E and F, the significant level is all smaller than 0.01.

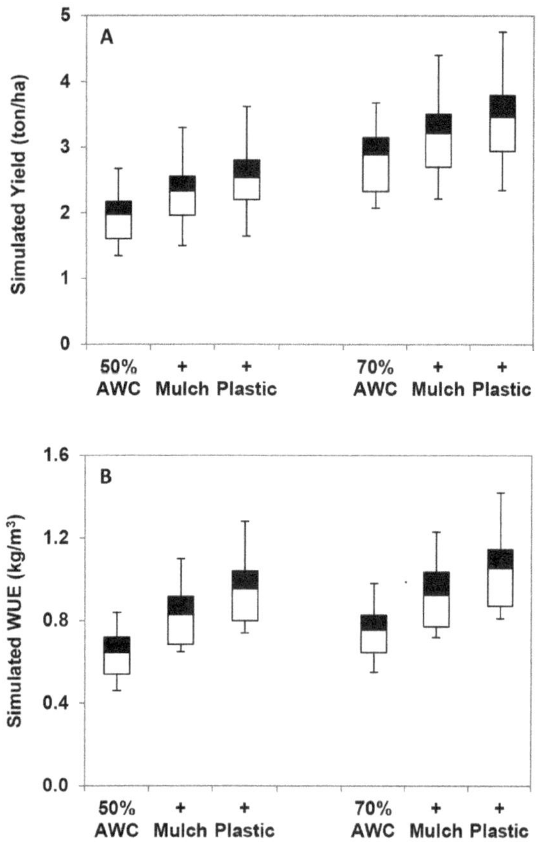

Figure 6. Results of six model experiments; effects of straw mulch and plastic film on wheat yield (A) and water use efficiency (B). Experiment E1: Planting at 50% of FC; Experiment E2: E1+organic mulch; Experiment E3: E1+plastic cover; Experiment E4: Planting at 70% of FC; Experiment E5: E4+organic mulch; Experiment E6: E4+plastic cover.

low. Our results show that crop yields can be increased by ~0.9 ton/ha through increasing soil water storage during the fallow period. Crop yields can be increased further by on average ~0.3 ton/ha through straw mulch and by ~0.5 ton/ha on through plastic film covers during the growing season. At the same time, WUE increases on average by 0.2 to 0.6 kg/m^3. These results are in line with results reported by Deng et al. [57] and others [58–60].

Evidently, increased rainwater harvesting during the fallow season is an effective option. Straw mulch significantly reduces the evaporative water losses during the fallow season and is conducive to the infiltration of rain water in the soil. Plastic film covers are less applicable during the fallow season, because they may limit the infiltration of rain water and thereby increase runoff. Reduced tillage can also improve soil water storage. According to a recent study of Hou et al. [61], rotational tillage (rotation of no-tillage and subsoiling) could significantly increase soil water storage during the summer fallow and wheat growing season compared with conventional tillage. They found that rotational tillage increased wheat yields by 10%, and WUE by 7.5%, respectively.

Conclusions

Low wheat yield at the monitoring site was largely due to (i) limited rainfall, (ii) low soil water storage during fallow season because of high water loss via soil evaporation, and (iii) the poor synchronisation of the wheat growing season to the rainfall distribution season. Although water was limited, on average only one-third of the total available water was actually used by the crop for transpiration. The remaining two-thirds was lost by soil evaporation, 60% during the fallow season and 40% during the growing season. Our model experiments suggest that minimizing soil evaporation via organic mulch or plastic film covers can significantly increase wheat yield and WUE. More importantly, these increases can be realized by the application of relatively low cost measures. Further studies are needed to test the effectiveness of these measures in the field.

Acknowledgments

We thank Xiaoguang Yang, from College of Resources and Environmental Sciences, China Agricultural University and Junqi Zhu, from Wageningen University and Research Centre, for providing the climate data for this study. We also thank Xinxin Wang, from Wageningen University and Research Centre, for helping us with statistics.

Author Contributions

Conceived and designed the experiments: WQ BC OO. Performed the experiments: WQ BC. Analyzed the data: WQ BC. Contributed reagents/materials/analysis tools: WQ BC OO. Wrote the paper: WQ BC OO.

References

1. de Wit CT (1958) Transpiration and crop yields. Verslag Landbouwkundig Onderzoek, Wageningen, Netherlands: Wageningen University.
2. Marschner H (1995) Mineral Nutrition of Higher Plants. Academic Press, San Diego: 889 p.
3. Piao SL, Ciais P, Huang Y, Shen ZH, Peng SS, et al. (2010) The impacts of climate change on water resources and agriculture in China. Nature 467: 43–51.
4. Hoekstra AY, Mekonnen MM (2012) The water footprint of humanity. Proceedings of the National Academy of Sciences of the United States of America 109: 3232–3237.
5. IPCC (2007) Climate Change 2007: Impacts, Adaptation and Vulnerability. Contribution of Working Group II to the Fourth Assessment Report of the International Panel of Climate Change. Cambridge Univ Press, UK.
6. Marris E (2008) Water: More crop per drop. Nature 452: 273–277.
7. Milly PCD, Betancourt J, Falkenmark M, Hirsch RM, Kundzewicz ZW, et al. (2008) Climate change - Stationarity is dead: Whither water management? Science 319: 573–574.
8. Hanjra MA, Qureshi ME (2010) Global water crisis and future food security in an era of climate change. Food Policy 35: 365–377.
9. World Water Assesment Programme (2009) The 3rd United Nations World Water Development Report: Water in a Changing World. Paris: UNESCO, and London: Earthscan.
10. Rockstrom J, Karlberg L, Wani SP, Barron J, Hatibu N, et al. (2010) Managing water in rainfed agriculture-The need for a paradigm shift. Agricultural Water Management 97: 543–550.
11. Vitousek PM, Mooney HA, Lubchenco J, Melillo JM (1997) Human domination of Earth's ecosystems. Science 277: 494–499.
12. Smil V (2000) Feeding the world. A challenge for the twenty-first century. MIT Press, Cambridge, Massachusetts, USA.
13. Tilman D, Cassman KG, Matson PA, Naylor R, Polasky S (2002) Agricultural sustainability and intensive production practices. Nature 418: 671–677.
14. Sachs JD (2008) Common Wealth. Economics for a Crowded Planet. The Penguin Press, New York: 386 p.
15. Godfray HC, Beddington JR, Crute IR, Haddad L, Lawrence D, et al. (2010) Food security: the challenge of feeding 9 billion people. Science 327: 812–818.
16. Zhang H, Wang X, You M, Liu C (1999) Water-yield relations and water-use efficiency of winter wheat in the North China Plain. Irrigation Science 19: 37–45.
17. Kang SZ, Zhang L, Liang YL, Hu XT, Cai HJ, et al. (2002) Effects of limited irrigation on yield and water use efficiency of winter wheat in the Loess Plateau of China. Agricultural Water Management 55: 203–216.
18. Zhang XY, Pei D, Chen SY, Sun HY, Yang YH (2006) Performance of double-cropped winter wheat-summer maize under minimum irrigation in the North China Plain. Agronomy Journal 98: 1620–1626.

19. Siahpoosh MR, Dehghanian E (2012) Water Use Efficiency, Transpiration Efficiency, and Uptake Efficiency of Wheat during Drought. Agronomy Journal 104: 1238–1243.
20. Zhang XY, Chen SY, Sun HY, Wang YM, Shao LW (2010) Water use efficiency and associated traits in winter wheat cultivars in the North China Plain. Agricultural Water Management 97: 1117–1125.
21. Whisler FD, Acock B, Baker DN, Fye RE, Hodges HF, et al. (1986) Crop Simulation-Models in Agronomic Systems. Advances in Agronomy 40: 141–208.
22. de Wit CT, Vankeulen H (1987) Modeling Production of Field Crops and Its Requirements. Geoderma 40: 253–265.
23. Steduto P, Hsiao TC, Raes D, Fereres E (2009) AquaCrop-The FAO Crop Model to Simulate Yield Response to Water: I. Concepts and Underlying Principles. Agronomy Journal 101: 426–437.
24. Zwart SJ, Bastiaanssen WGM, de Fraiture C, Molden DJ (2010) A global benchmark map of water productivity for rainfed and irrigated wheat. Agricultural Water Management 97: 1617–1627.
25. Steduto P, Hsiao TC, Fereres E, Raes D (2012) FAO Irrigation and drainage paper 66, Crop yield response to water, Food and Agriculture Organization of the United Nations.
26. Raes D, Steduto P, Hsiao TC, Fereres E (2009) AquaCrop-The FAO Crop Model to Simulate Yield Response to Water: II. Main Algorithms and Software Description. Agronomy Journal 101: 438–447.
27. Hsiao TC, Heng L, Steduto P, Rojas-Lara B, Raes D, et al. (2009) AquaCrop-The FAO Crop Model to Simulate Yield Response to Water: III. Parameterization and Testing for Maize. Agronomy Journal 101: 448–459.
28. Todorovic M, Albrizio R, Zivotic L, Saab MTA, Stockle C, et al. (2009) Assessment of AquaCrop, CropSyst, and WOFOST Models in the Simulation of Sunflower Growth under Different Water Regimes. Agronomy Journal 101: 509–521.
29. Farahani HJ, Izzi G, Oweis TY (2009) Parameterization and Evaluation of the AquaCrop Model for Full and Deficit Irrigated Cotton. Agronomy Journal 101: 469–476.
30. Dominguez A, Tarjuelo JM, de Juan IA, Lopez-Mata E, Breidy J, et al. (2011) Deficit irrigation under water stress and salinity conditions: The MOPECO-Salt Model. Agricultural Water Management 98: 1451–1461.
31. Geerts S, Raes D, Garcia M, Taboada C, Miranda R, et al. (2009) Modeling the potential for closing quinoa yield gaps under varying water availability in the Bolivian Altiplano. Agricultural Water Management 96: 1652–1658.
32. Geerts S, Raes D, Garcia M, Miranda R, Cusicanqui JA, et al. (2009) Simulating Yield Response of Quinoa to Water Availability with AquaCrop. Agronomy Journal 101: 499–508.
33. Andarzian B, Bannayan M, Steduto P, Mazraeh H, Barati ME, et al. (2011) Validation and testing of the AquaCrop model under full and deficit irrigated wheat production in Iran. Agricultural Water Management 100: 1–8.
34. Salemi H, Soom MAM, Lee TS, Mousavi SF, Ganji A, et al. (2011) Application of AquaCrop model in deficit irrigation management of Winter wheat in arid region. African Journal of Agricultural Research 6: 2204–2215.
35. Araya A, Keesstra SD, Stroosnijder L (2010) Simulating yield response to water of Teff (Eragrostis tef) with FAO's AquaCrop model. Field Crops Research 116: 196–204.
36. Abedinpour M, Sarangi A, Rajput TBS, Singh M, Pathak H, et al. (2012) Performance evaluation of AquaCrop model for maize crop in a semi-arid environment. Agricultural Water Management 110: 55–66.
37. ISS-CAS (2012) China Soil Scientific Database Website. Institute of Soil Science, Chinese Academy of Science (ISS-CAS), China. Available: http://www.soil.csdb.cn/. Accessed 2012 Oct 1.
38. IUSS (2006) World reference base for soil resources 2006. A framework for international classification, correlation and communication, 2nd edition. World Soil Resources Reports No 103 FAO, Rome.
39. Wang XB, Cai DX, Hoogmoed WB, Oenema O, Perdok UD (2007) Developments in conservation tillage in rainfed regions of North China. Soil & Tillage Research 93: 239–250.
40. Wang XB, Wu HJ, Dai K, Zhang DC, Feng ZH, et al. (2012) Tillage and crop residue effects on rainfed wheat and maize production in northern China. Field Crops Research 132: 106–116.
41. Xie WY, Zhou HP, Guan CL, Zhang JJ, Yan XY, et al. (2011) Spatial Distribution of Soil Active Nutrients in Farmland in Shanxi Province. Journal of Shanxi Agricultural Sciences.
42. Allen RG, Pereira LS, Raes D, Smith M (1998) Crop evapotranspiration. Guidelines for computing crop water requirements. FAO irrigation and drainage paper no. 56, 300 p.
43. Willmott CJ (1982) Some Comments on the Evaluation of Model Performance. Bulletin of the American Meteorological Society 63: 1309–1313.
44. Willmott CJ, Matsuura K (2005) Advantages of the mean absolute error (MAE) over the root mean square error (RMSE) in assessing average model performance. Climate Research 30: 79–82.
45. Hu K, Li B, Chen D, Zhang Y, Edis R (2008) Simulation of nitrate leaching under irrigated maize on sandy soil in desert oasis in Inner Mongolia, China. Agricultural Water Management 95: 1180–1188.
46. Hu KL, Li Y, Chen WP, Chen DL, Wei YP, et al. (2010) Modeling Nitrate Leaching and Optimizing Water and Nitrogen Management under Irrigated Maize in Desert Oases in Northwestern China. Journal of Environmental Quality 39: 667–677.
47. Mkhabela MS, Bullock PR (2012) Performance of the FAO AquaCrop model for wheat grain yield and soil moisture simulation in Western Canada. Agricultural Water Management 110: 16–24.
48. Araya A, Habtu S, Hadgu KM, Kebede A, Dejene T (2010) Test of AquaCrop model in simulating biomass and yield of water deficient and irrigated barley (Hordeum vulgare). Agricultural Water Management. 1838–1846.
49. Stricevic R, Cosic M, Djurovic N, Pejic B, Maksimovic L (2011) Assessment of the FAO AquaCrop model in the simulation of rainfed and supplementally irrigated maize, sugar beet and sunflower. Agricultural Water Management. 1615–1621.
50. Zhang HP, Oweis T (1999) Water-yield relations and optimal irrigation scheduling of wheat in the Mediterranean region. Agricultural Water Management 38: 195–211.
51. Huang MB, Gallichand J, Zhong LP (2004) Water-yield relationships and optimal water management for winter wheat in the Loess Plateau of China. Irrigation Science 23: 47–54.
52. Kang SZ, Zhang L, Liang YL, Dawes W (2003) Simulation of winter wheat yield and water use efficiency in the Loess Plateau of China using WAVES. Agricultural Systems 78: 355–367.
53. Liu CM, Zhang XY, Zhang YQ (2002) Determination of daily evaporation and evapotranspiration of winter wheat and maize by large-scale weighing lysimeter and micro-lysimeter. Agricultural and Forest Meteorology 111: 109–120.
54. Kang SZ, Gu BJ, Du TS, Zhang JH (2003) Crop coefficient and ratio of transpiration to evapotranspiration of winter wheat and maize in a semi-humid region. Agricultural Water Management 59: 239–254.
55. Wang P, Song XF, Han DM, Zhang YH, Zhang B (2012) Determination of evaporation, transpiration and deep percolation of summer corn and winter wheat after irrigation. Agricultural Water Management 105: 32–37.
56. Robert PC (2002) Precision agriculture: a challenge for crop nutrition management. Plant and Soil 247: 143–149.
57. Deng XP, Shan L, Zhang HP, Turner NC (2006) Improving agricultural water use efficiency in and and semiarid areas of China. Agricultural Water Management 80: 23–40.
58. Chakraborty D, Garg RN, Tomar RK, Singh R, Sharma SK, et al. (2010) Synthetic and organic mulching and nitrogen effect on winter wheat (Triticum aestivum L.) in a semi-arid environment. Agricultural Water Management 97: 738–748.
59. Gao YJ, Li Y, Zhang JC, Liu WG, Dang ZP, et al. (2009) Effects of mulch, N fertilizer, and plant density on wheat yield, wheat nitrogen uptake, and residual soil nitrate in a dryland area of China. Nutrient Cycling in Agroecosystems 85: 109–121.
60. Chakraboyty D, Nagarajan S, Aggarwal P, Gupta VK, Tomar RK, et al. (2008) Effect of mulching on soil and plant water status, and the growth and yield of wheat (Triticum aestivum L.) in a semi-arid environment. Agricultural Water Management 95: 1323–1334.
61. Hou XQ, Li R, Jia ZK, Han QF, Wang W, et al. (2012) Effects of rotational tillage practices on soil properties, winter wheat yields and water-use efficiency in semi-arid areas of north-west China. Field Crops Research 129: 7–13.

16

Performance and Reliability Analysis of Water Distribution Systems under Cascading Failures and the Identification of Crucial Pipes

Qing Shuang, Mingyuan Zhang*, Yongbo Yuan

Department of Construction Management, Dalian University of Technology, Dalian, Liaoning, China

Abstract

As a mean of supplying water, Water distribution system (WDS) is one of the most important complex infrastructures. The stability and reliability are critical for urban activities. WDSs can be characterized by networks of multiple nodes (e.g. reservoirs and junctions) and interconnected by physical links (e.g. pipes). Instead of analyzing highest failure rate or highest betweenness, reliability of WDS is evaluated by introducing hydraulic analysis and cascading failures (conductive failure pattern) from complex network. The crucial pipes are identified eventually. The proposed methodology is illustrated by an example. The results show that the demand multiplier has a great influence on the peak of reliability and the persistent time of the cascading failures in its propagation in WDS. The time period when the system has the highest reliability is when the demand multiplier is less than 1. There is a threshold of tolerance parameter exists. When the tolerance parameter is less than the threshold, the time period with the highest system reliability does not meet minimum value of demand multiplier. The results indicate that the system reliability should be evaluated with the properties of WDS and the characteristics of cascading failures, so as to improve its ability of resisting disasters.

Editor: Vanesa Magar, Plymouth University, United Kingdom

Funding: This work is supported by the National Natural Science Foundation of China under Grant No. 51208081. The funders had no role in study design, data collection and analysis, decision to publish, or preparation of the manuscript.

Competing Interests: The authors have declared that no competing interests exist.

* E-mail: myzhang@dlut.edu.cn

Introduction

The stability and reliability of Water distribution systems (WDSs) is one of the important factors in ensuring public safety and the continuous operation of urban functions. Such functions include water supply, infrastructure construction and industrial development, etc. It is also the key field for infrastructure construction. The WDS is a large scale network system with complex topological structure [1]. Its functions are designed to convey volumes of water to customers under adequate pressure. Nowadays, along with the increased population and population density, WDS is developing into wide-range supply which carries fluid under high or less pressure. A WDS can be represented as a spatially networks of multiple interconnected components. Pipes can be represented as links. Junctions, reservoirs and consumers can be represented as a collection of nodes. With the link-node representation of physical components in WDS, complex network analysis can be applied to evaluate the system reliability.

Complex networks are an essential part in the understanding of many natural systems [2]. A complex network is a network with non-trivial topological features, which often occur in real life. Complex networks analysis provides a way to understand the meaning and functions of the system [3]. It focuses on predict the networked system behavior on the basis of measured structure. Albert et al. [4] have found that the scale-free networks have strong robustness under random disturbance, but it is very vulnerable under intentional attacks. These important discoveries have made the network security under abnormal conditions become a hot issue in this field. Cascading failures is a conductive failure process in the field of network security [5]. When the network encounters natural or man-made disasters, i.e. network attacks and random failures, the minor anomalous event of a point may spread to the whole system through cascade reaction, leading to large-scale consequences and secondary failures. Many models have been provided to investigate the cascading failures. The present studies mainly focus on: (1) the network reliability and topology structure after remove some nodes or links [6–8]; (2) the formation conditions and reasons of cascading phenomena and the network dynamics in networks or weighted networks [9–10]; and (3) the metrics of network robustness and the sequent network optimization and design [11].

For the real-world networks, the cascading failures of infrastructure systems have been proposed as well. The power grids of North America and the Western United States are two key studies in this field [12–15]. Besides, the Internet network [16], the power grids of European [17–18] and Italian [19], and other kinds of power systems [20] and traffic networks [21] are also the focus of studies. In these studies, most cascading failures use the virtual network simulation method which measures the network load by the topological property, such as the betweenness and degree. The betweenness is defined as the total number of the shortest paths that pass through the vertex [22]. The degree is defined as the number of edges connected to the vertex [23]. Therefore betweenness and degree basically measure the topological structure of a network. This method fits for the disaster simulation under uncertainties and the rapid assessment on disasters. However, it ignores the properties of city lifeline system as an entity network. Using the betweenness or degree to represent the actual network flow cannot guarantee the accuracy of the calculation results. Therefore, the results cannot be directly applied to the decision-making.

System reliability is defined as the ability of the system to complete the scheduled functions in a certain period under the given working state [24–25]. There are two failure mechanisms of a WDS [26]: mechanical and hydraulic failures. Mechanical reliability is defined as the probability that the WDS and its components are operational. Mechanical reliability focuses on a topological perspective. Hydraulic reliability is defined as the probability that the WDS meet flow and pressure requirements. Hydraulic reliability considers failures in meeting consumer demand. The reliability in this paper combined these two types of reliability. With the topological dynamics changing, the scheduled functions refer to ensuring the water demand and hydraulic pressure required in the daily life of customs in normal conditions; under failure conditions, the water supply and hydraulic pressure would not be lower or higher than the specified limit.

The purpose of this paper is to study the propagation characteristics of cascading failures in the WDS and put forward the methods to identify the crucial pipes. The methods from complex networks and engineering are adopted. The simulation of the cascading failures in WDS is required to meet the equilibrium of water supply and demand. The main task is adopting the numerical simulation technology to depict the damage process of WDS in cascading failures. The network dynamics are used with failure propagation. The uncertain factors, i.e., the nodal head bounds, daily demand multipliers and the water demand have been taken into consideration. The crucial pipe is identified by its vulnerability and sensitivity to cascading failures. By identifying the most crucial pipes, one can effectively protect the network to avoid cascading failures and build attack-robust networks.

Methods

1. Reliability Assessment

The reliability is defined as the probability that the WDS meet flow and pressure requirements under the possible mechanical failure scenarios (e.g., pipe breaks). The definition of system reliability given by Zhuang et al. [27] is adopted. Mathematically, the reliability can be expressed as the ratio of the available flow to the require demand. The condition includes normal condition and failure condition. The reliability of node and system is expressed as:

$$R_j = \frac{Q_j^{avl}}{Q_j^{req}} \tag{5}$$

$$R_{sys} = \frac{\sum_{j=1}^{N_{Node}} Q_j^{avl}}{\sum_{j=1}^{N_{Node}} Q_j^{req}} \tag{6}$$

where R_j is the jth node reliability; R_{sys} is the reliability of WDS; N_{Node} is the number of nodes; Q_j^{avl} is the available flow at jth node (L/s); Q_j^{req} is the require demand when jth node is under normal condition.

2. WDS Topological Structure Expression

Before analogue simulation and calculation, the first step is to input the graphic information of the system into the computer and set up the model. A WDS can be analyzed by the methods of graph theory based on its topology structure. Graph theory is the study of graphs. A graph consists of a set of nodes and links, representing the interactions among them. A graph is customarily depicted the nature of the links between nodes. A directed graph is one in which links have orientation [28]. A WDN is a directed graph due to the operational flow and pressure requirements. The reservoirs, junctions and customers are described as nodes; while the pipes, pumps and valves are represented as links. The adjacent nodes in the graph are connected by the links in most of the cases.

The matrix is used as an effective tool to depict the properties of the graph in network modeling. The adjacency matrix and incidence matrix are the most common ones. The adjacency matrix A is used to represent the relationship between the nodes in the network. The values for the element a_{ij} are: 0 and 1. When the node i and node j is connected by a pipe, the value is 1; when the node i and node j is unconnected, the value is 0.

$$a_{ij} = \begin{cases} 1, & \textit{Node i and node j connected} \\ 0, & \textit{Node i and node j unconnected} \end{cases} \tag{1}$$

Incidence matrix N is for describing the relationship between nodes and pipes. The row represents the nodes and the column represents the pipes, respectively. In the network graph, each node and pipe is numbered by consecutive number from 1. The information of node i is recorded in the ith row and that of pipe j is recorded in the jth column. The element n_{ij} is expressed as:

$$n_{ij} = \begin{cases} 1, & \textit{Node i is the initial point of pipe j} \\ -1, & \textit{Node i is the terminal point of pipe j} \\ 0, & \textit{Node i is unconnected with pipe j} \end{cases} \tag{2}$$

3. Modeling of WDS Based on Cascading Failures

3.1 The Load of WDS Nodes. In view of the actual physical meaning of WDS, we adopt the nodal pressure head as the load. The load of nodes is a relevant quantity, which can be material, information and energy [5], and can be concrete or abstract. With the passage of time, the load of the nodes exchanges along the connected edge between each of the node pair. The WDS is a kind of material loading network. It distributes water from the reservoirs to the customers. In order to meet the customers' daily needs, each component of the WDS must be able to provide required water demand and pressure head under both normal and failure conditions. Therefore, we define the initial load of the node as the service head H_s. The service head ensures that all the imposed demands can be satisfied.

3.2 The Relationship between Nodal Capacity and Initial Load. The nodal capacity is the maximum load the node can bear. In the place of residence or business, the nodal pressure should not be too high or too low. This because the low pressure leads to flow reductions or blanking; high pressure causes the pipe leakage and even the burst of ageing pipes so that losing its service functions. In general, for each node, there are three kinds of pressure head:

(1) H_{min}, the acceptable minimum level of pressure. If the nodal pressure head is lower than H_{min}, the node loses its service function. Therefore, the minimum capacity is defined as the nodal minimum head.

(2) H_s, the service level of pressure to meet all the imposed demands. Only when each nodal pressure head is higher than service head, the WDS can be performing normally;

(3) H_{max}, the acceptable maximum level of pressure that a node can bear. Then, define the maximum capacity of the node. In a man-made system, the nodal capacity is severely limited by cost. For WDS, the pressure head, H, at each demand node is always within a specified range of a minimum head H_{min} and a maximum H_{max} [29] when it's in normal operating conditions. Besides, there is a direct relationship between pipe leakage and service pressure. The water leakage increases with pressure [30]. In order to avoid the leakage of ageing pipes caused by over-high pressure, the maximum capacity of the node should be defined. Suppose α is the tolerance parameter. It is possible to assume that the maximum capacity is proportional to the initial load H_s. The maximum capacity can be expressed as:

$$H_j^{\max} = (1+\alpha)H_j^s \tag{3}$$

where H_j^{max} is the jth nodal maximum head capacity; H_j^s is the jth nodal service head. α allows a systematic evaluation of the aggregated performance of water distribution network element during cascading propagation [31]. The bigger α is, the higher the capacity constraint will be, which means less likely the node would fail.

To represent actual flows supplied to customers under abnormal condition, the available nodal demand is expressed as a function of nodal pressure head. Considering the formulation proposed by Wagner et al. [32], and take the maximum head capacity into account, the function can be expressed as follows:

$$Q_j^{avl} = \begin{cases} 0 & H_j < H_j^{\min} \\ Q_j^{req}\sqrt{\dfrac{H_j - H_j^{\min}}{H_j^s - H_j^{\min}}} & H_j^{\min} \le H_j \le H_j^s \\ Q_j^{req} & H_j^s < H_j \le H_j^{\max} \\ 0 & H_j > H_j^{\max} \end{cases} \tag{4}$$

where Q_j^{avl} is the flow delivered to the jth node (L/s); Q_j^{req} is the require demand when jth node is under normal condition. H_j, H_j^{min}, H_j^{max} and H_j^s represent the calculated head, the minimum head, the maximum head and the service head at jth node, respectively.

For the above three kinds of pressure heads, there are four states of water supply: (1) nodes are completely shut off when the pressure head is lower than the minimum head; (2) the customers' demands is supplied at a reduction level when the pressure head is higher than the minimum head and lower than the service head; (3) nodes meet the customers' demand when the pressure is higher than the service head; and (4) nodes are closed when the pressure head is higher than the maximum head.

3.3 The Cascading Dynamics. The load on the network is in dynamic changes, especially when the network structure transforms. For example, the load is redistributed due to some nodes or pipes shut off. In general, the network node and edge have a limited bearing capacity. If the maximum load (capacity) is exceeded, the network equilibrium will be broken and the load will be redistributed.

When a pipe of the WDS is closed for failure condition, it is equivalent to removing an edge of the network. Then it triggers the network flow to be redistributed among all nodes. The artificial time step t ($t = 1, 2, \ldots$) is introduced to monitor the process of cascading failures. At time t, if the nodal pressure head H_j exceeds its capacity (above the maximum head or below the minimum head), this node fails to provide required water. The failure triggers the reduction of its downstream pipes. The WDS pressure is a spatial vector and the pressure on each node is interdependent. The pressure of one node changes leads to other node pressure changes to varying degrees. In this situation, a new round of load redistribution occurs and leads to cascading failures. The iterative process continues until there are no failure nodes or pipes produced, which implies the cascading can be considered stopped. The iterative process is described in Figure 1.

In the operation of WDS, two or more components come to failure together rarely happens [33]. Therefore, this paper only considers the situation of single-pipe failure. Note that this model can be easily extended to multiple-pipe failure.

4. Hydraulic Simulation

The EPANET [34] simulation engine has been used for the WDS hydraulic simulation. EPANET is a water distribution system modeling software developed by the United States Environmental Protection Agency's (EPA). EPANET is available as an open-source toolkit. Its Programmer's Toolkit is a dynamic link library (DLL) of functions that allow developers to customize EPANET to their own needs.

In actual operation, the network distributes water along the pipelines. The water demands of the customers are time-varying and uncertain. Therefore, the water demand multipliers need to be considered as an uncertain and dynamic variable. To evaluate the system reliability with the time-varying demands, a water demand pattern is established by the extended period hydraulic simulations module of EPANET. The time step is set as one hour. Use MATLAB 2010a to implement modeling of different stages and the EPANET Toolkits' functions.

5. Assumptions and Algorithm Flowchart

Assumptions:

1. The pipes of the WDS have just two states: operation or failure. Demand nodes have three states: operation, failure or intermediate state. The intermediate state means the water demand at the node is partly met. The amount of water flow is available but less than the require demand.
2. A pipe is operational if the water can flow smoothly from its initial node to its terminal node.

3a. A node is operational if its service function is effective, i.e., the pressure head of the node is no less than the minimum head and no higher than the maximum capacity.

3b. When a node is in failure, its downstream connected edge is in failure also.

3c. If the upstream pipes of a node are all in failure, the node is failed as an unintended isolated node due to it has no source of water supply.

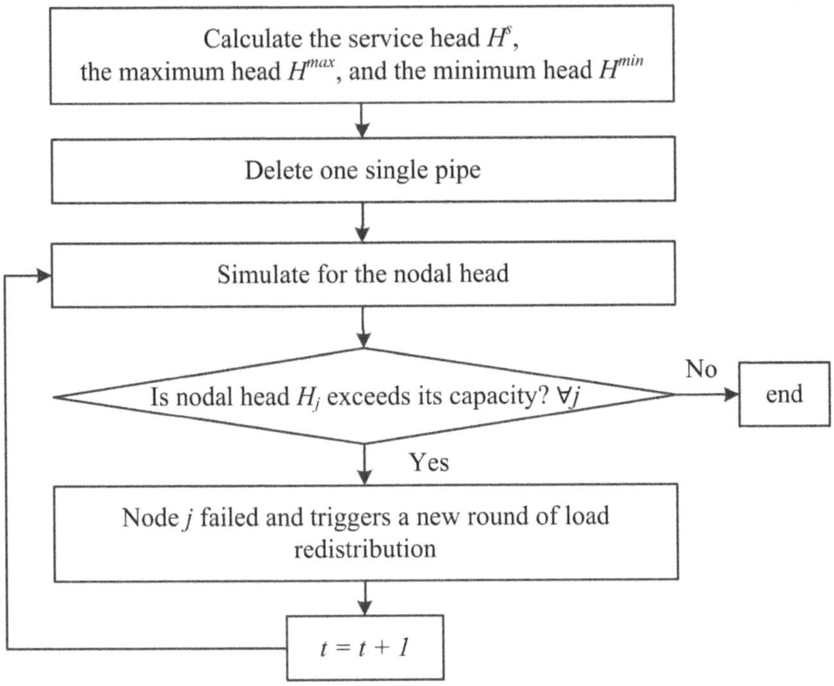

Figure 1. Iterative process of the cascading dynamics of WDS.

For a WDS, three kinds of failure states are existed based on the assumptions. (1) The pressure head at a node exceeds its maximum capacity; (2) The pressure head at a node is lower than its minimum capacity; and (3) A node is unintended isolated as a result of all its upstream pipes are in failure state.

According to the simulation flow of cascading failures, the algorithm flowchart is described as follows (Fig. 2):

1. Input the basic information of a WDS, including diameters, lengths and roughness coefficients of pipes, nodal elevations, base demands, and demand multipliers. Load the network topologic and construct the adjacency matrices A and the incidence matrices N.
2. Calculate the initial load (service pressure head) and its maximum and minimum capacity.
3. Adjust the require water demand of customers according to the demand multipliers. Set the initial time step as $t = 1$. $t = 1$ indicates that each node of a WDS meet its pressure head. $t = 2$ describes the hydraulic state after a certain pipe failed. $t \geq 3$ represents each stage of the cascading failures where new nodes or pipes fail as a result of flow redistribution
4. Assume each of the pipes is in failure successively. Define FailureNodeList and FailureLinkList as the set of failed nodes and failed pipes, respectively. Set FailureNodeList = {} and FailureLinkList = {}. If there is a subsequent failure, the node or pipe index is appended to the FailureNodeList and FailureLinkList, respectively. These two sets are used to update the system topological structure.
5. Run hydraulic analysis in failure conditions. Simulate the running state of the water distribution network by EPANET and obtain the pressure head on each node. A node is recognized as a subsequent failure node if its load exceeds the range of capacity. Update the available flow according to Eq. (4).
6. Update the topological structure under failure conditions. Close the downstream pipes of failure nodes. Judge whether there is unintended isolated node; if so, this node is in failure. Store the calculated failure nodes and pipes in FailureNodeList and FailureLinkList respectively and update the topological structure of the WDS.
7. Calculate the system and nodal reliability. Repeat steps (3)–(6) until the failure of all the pipes have been simulated. Use Eq. (5) and (6) to calculate the reliability of each node and the whole network, respectively.

Results and Discussion

The proposed methodology is applied to a WDS from Islam et al. [35–36]. The network includes two reservoirs, twenty-five water demand nodes and forty pipes. The topological structure, nodal elevation, base demand, pipe diameter and length as well as other basic information of the network has been shown in Figure 3. The total length of all pipes is 19.5 km. The water is supplied by gravity from the elevated reservoirs (reservoir 26 and reservoir 27) with the total heads of 90 m and 85 m, respectively. The pipe lengths range from 100 m to 680 m and the diameters vary from 200 mm to 700 mm. The basic demand is in the range of 33.33 l/s to 133.33 l/s. The demand multipliers (DMs) are considered as ranging from 0.38 at 2am to 1.53 at 7am. The Hazen-Williams formula is used to calculate the head loss.

As the original literature has not given the minimum head, we use the EPANET to solve hydraulic calculation on the network under normal condition (no pipe failure occurred). The pressure head at each node is shown in Table 1. We can find from the results that the pressure heads of node 24 and 25 are lower than 65 m for their long distance away from the reservoirs. Considering the minimum head have a certain tolerance, here assume $H_{min} = 50$ m. The maximum head can be calculated by Eq. (1)

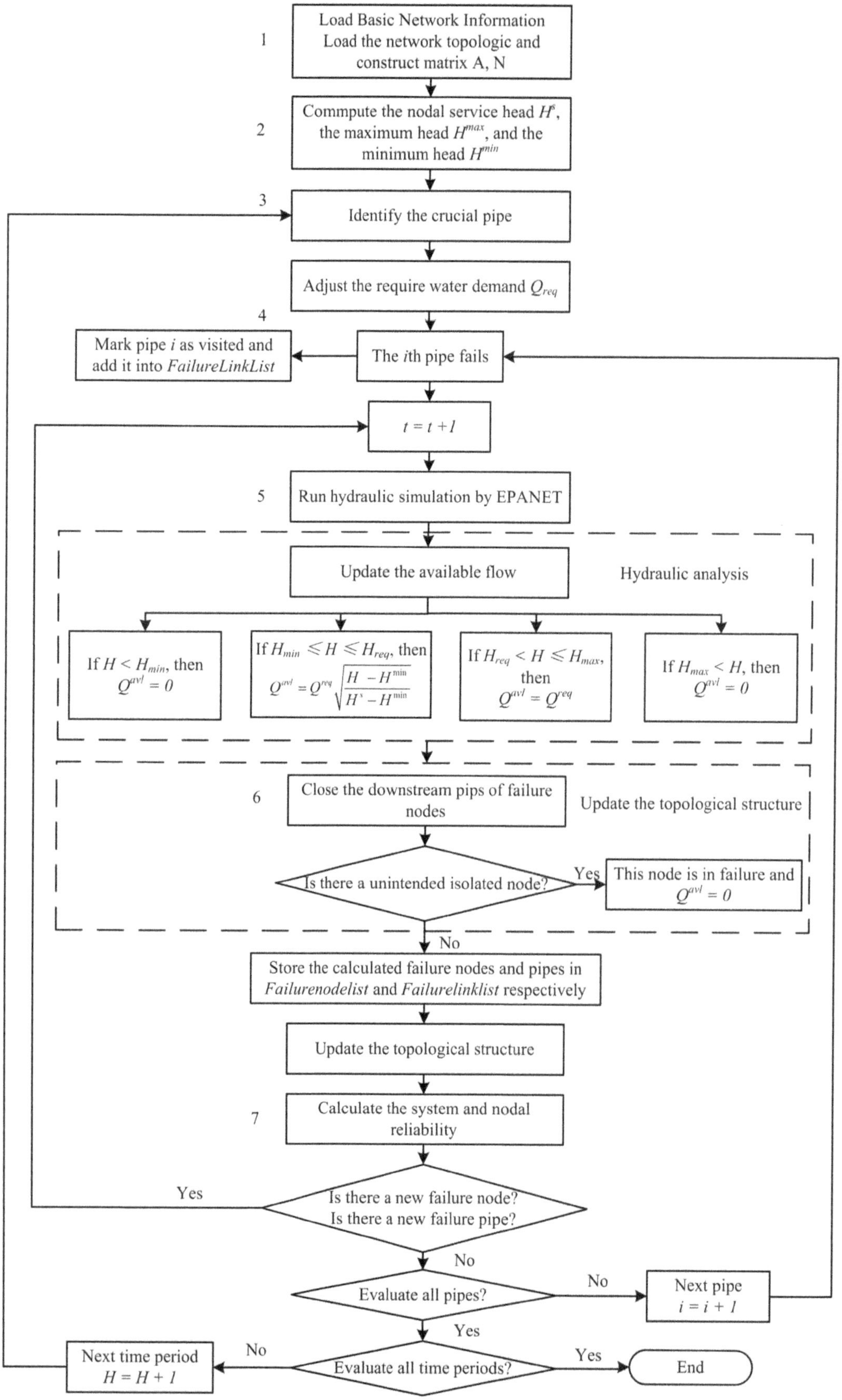

Figure 2. Simulation flowchart of pipe failure under WDS cascading failures.

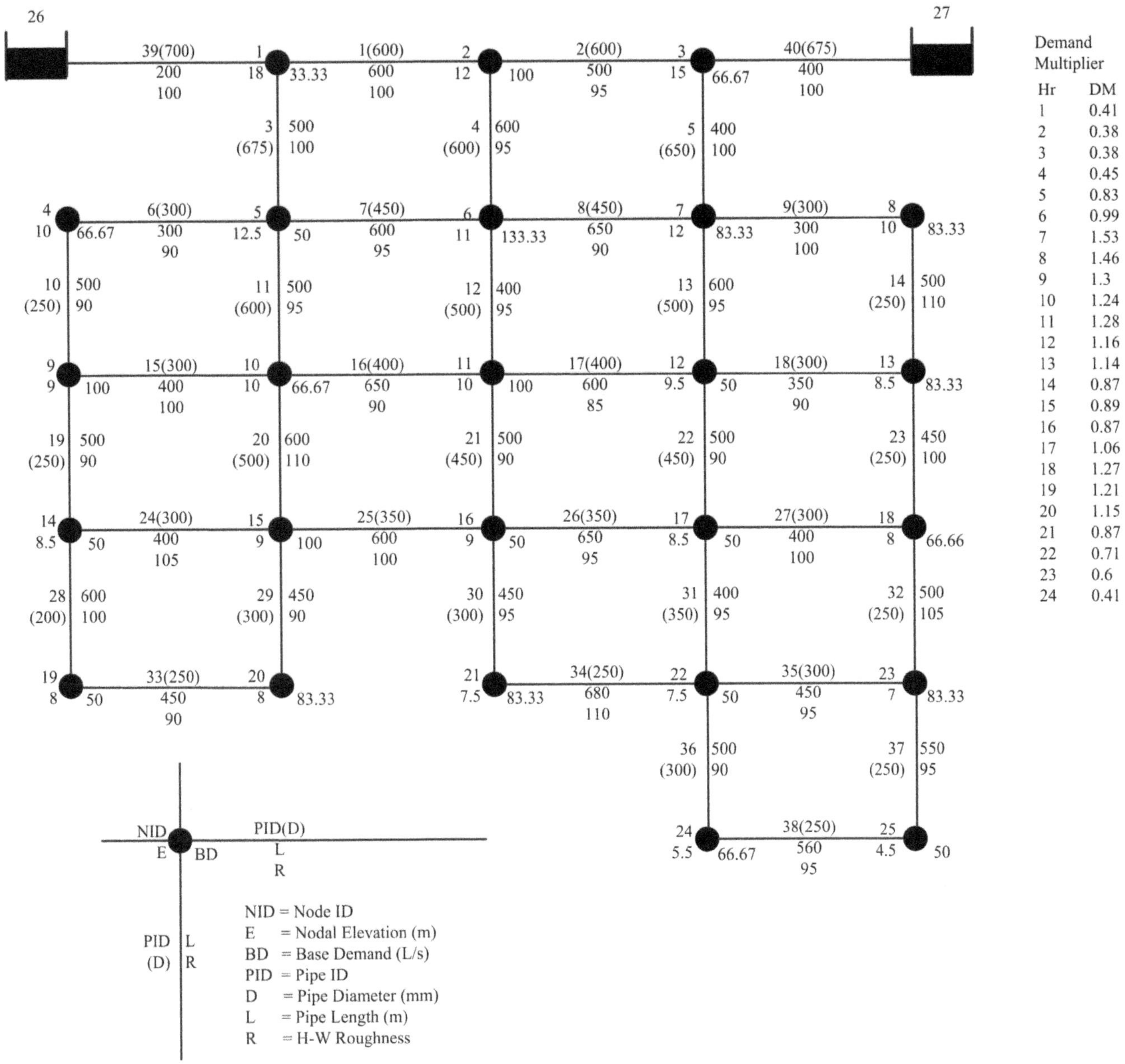

Figure 3. The layout of the example WDS.

Table 1. Pressure heads of each node under the normal condition.

Node ID	Pressure Head (m)	Node ID	Pressure Head (m)	Node ID	Pressure Head (m)
1	86.91	10	79.30	19	69.65
2	84.30	11	78.88	20	70.27
3	84.35	12	77.89	21	69.91
4	78.66	13	74.54	22	68.82
5	82.92	14	74.91	23	66.85
6	82.14	15	76.20	24	64.70
7	82.16	16	75.88	25	64.36
8	77.09	17	74.24		
9	76.14	18	71.53		

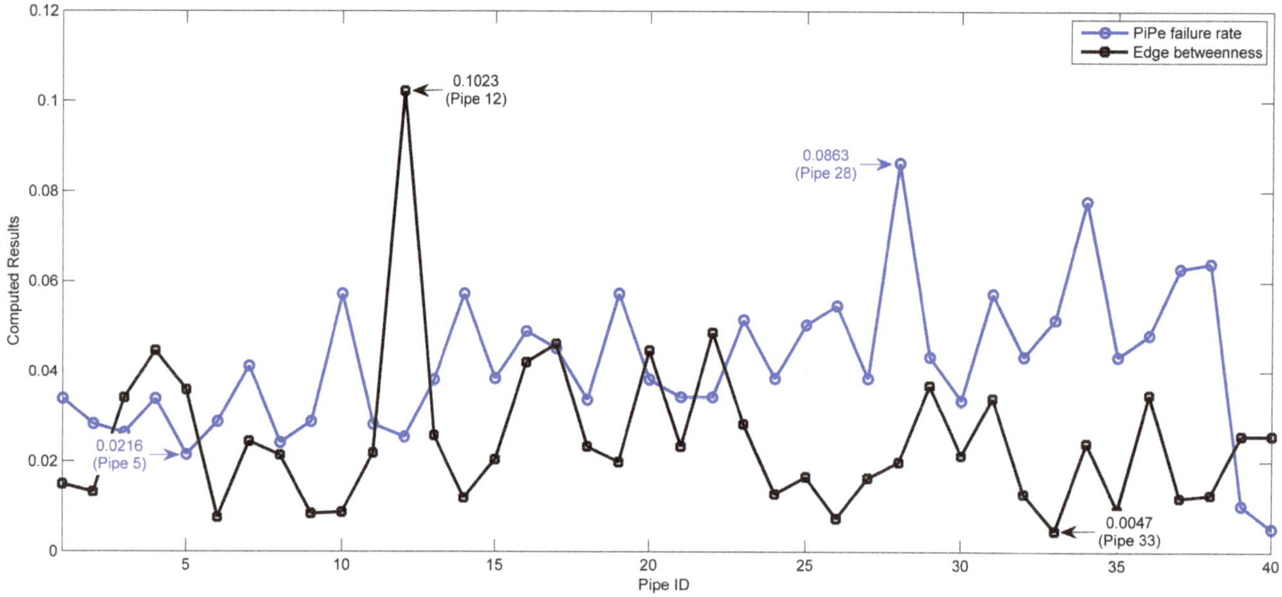

Figure 4. Failure rate and edge betweenness of WDS.

1. Failure rate and edge betweenness

Figure 4 shows the failure rate of each pipe and the edge betweenness of pipe network. The failure rate per year per unit length of pipe is calculated by the function proposed by Su and Mays [37].

Calculate the edge betweenness of the network by Matlab BGL toolbox, and then normalize the results. It is obvious that the pipe 39 and 40 which are directly connected with the reservoirs are the most crucial pipes. The failures of these two pipes directly lead to the failure of the downstream water demand nodes. Therefore, in the study of pipes, it is necessary to research the reliability of pipes except pipe 39 and 40. Figure 4 shows pipe failure rate and the edge betweenness. It is visible that except pipe 39 and 40, the minimum pipe failure rate is 0.0216 (pipe 5) and the maximum one is 0.0863 (pipe 28). So, it is difficult to distinguish the crucial pipe. From the edge betweenness, we can see that the top three maximums are 0.1023 (pipe 12), 0.0486 (pipe 22) and 0.0461 (pipe17), and the minimum is 0.0047 (pipe 33).

2. Peak and Period of Cascading Failures in in WDS

The value of α measures the limit of capacity caused by cost factors in the initial stage of WDS construction, or that caused by ageing and corrosion in the operation stage of WDS. Despite the simple meaning of α, it provides the method to evaluate the overall performance of the system in the cascading failures. Figure 5 shows the reliability of the system when *DM* value is in the range of 0.3~1.5 and α is ranging from 0 to 0.3. Figure 5 (a) presents the 3D view, (b) is an overhead view. Suppose a certain pipe fails, the simulation is carried out. The system reliability is calculated for the whole system if no subsequent failures are found. Simulate the 38 pipes successively. The system reliability in Figure 5(a) is the average reliability of these 38 pipes. The simulation result shows that, (1) the larger α is, the better the invulnerability of WDS against cascading failures will be. This is because under the given load redistribution strategies, the higher capacity of network design means a stronger potential ability of the system in assimilating and accommodating local failure and a better ability of the network to deal with of cascading failure. (2) as the computational condition of H_{max}, α enlarges the node pressure from maximum 86.91 m to 112.98 m, the relative amplification is 30%. With the increase of α, the system reliability increases in the whole with a remarkable change from 0 to 1.0 and the relative growth is 100%, 70% higher than the maximum pressure head. In the limited capacity of the node, the small disturbance of WDS has a significant effect on water supply.

Analysis from the perspective of *DM* shows that *DM* has a great effect on the propagation of cascading failures in WDS. The simulation results shows, (1) when $DM>1$, system reliability is low after coming to the stable state and the stable value becomes smaller with the increase of *DM*. For example, when $DM = 1.3$, the stable value of system reliability is 0.7179; $DM = 1.5$, the stable value of system reliability is 0.5557; when $DM<1$, the stable value of system reliability is much higher. For example, when $DM = 0.5$, the stable value of system reliability is 0.9993. (2) when α is given, there exists a threshold DM_c of *DM* making the system reliability always reach a peak over a period of time. When *DM* is at its threshold DM_c, the system reliability levels off to DM_c and rises rapidly. After reaching the peak, it decreases rapidly. Moreover, DM_c decreases with the increase of α. For example, when $\alpha = 0$ and $DM = 1.08$, the peak is $Rsys = 0.7161$; when $\alpha = 0.3$ and $DM = 0.33$, $Rsys = 1.0$. (3) When $DM<1$, as DM decreases, the system reliability needs more time to be stable and this time gets longer with the decrease of *DM*. For example, when $DM = 0.7$, the system reliability gets stable at $\alpha = 0.18$; when $DM = 0.3$, the system reliability has not yet reached a stable value at $\alpha = 0.3$. Therefore, in the simulation of the pipe network, the *DM* should be considered to make the evaluation result more accurate.

3. Identification of Crucial Pipe

The WDS simulation is carried out based on the flowchart in Figure 2. α is set in six types from 0.05 to 0.3. Steps 3–7 in Figure 2 are carried out after setting α and *DM* to a certain value. The subsequent failures are considered. If no new failures are found, the cascading failure stops and the system remain stable. The system reliability can be obtained when the system come to stable state again. The 38 pipes are simulated successively. The lowest

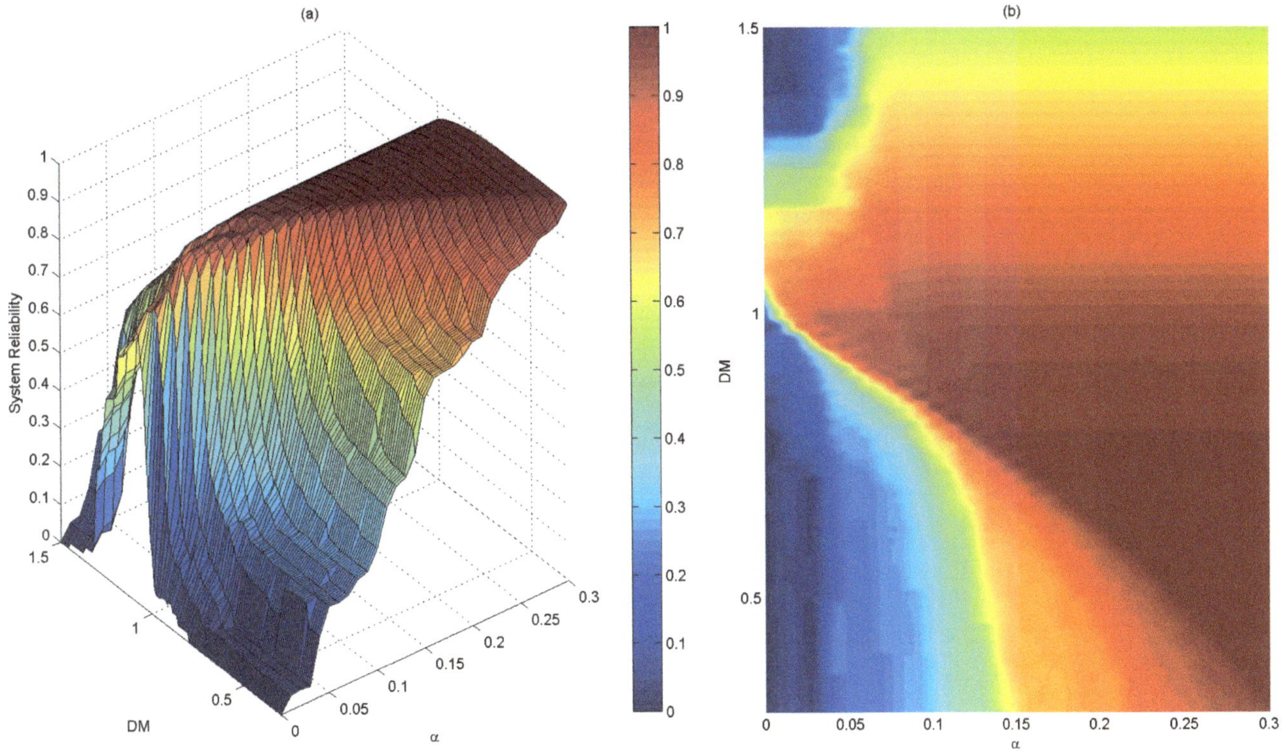

Figure 5. Variation diagram of WDS reliability with *DM* and *α*.

system reliability of this period and its related pipe ID are recorded in Table 2.

The simulation results are shown in Table 2. The first column is the *DM* per hour; the second column is hours; the column 3, 5, 7, 9, 11, and 13 list the ID of crucial pipes under different tolerance parameters α and their corresponding system reliability R_{sys}, respectively.

It can be found in Table 2 that the vulnerable pipes ID are 1, 2, 3, 5, 8, 9, 11, 22, 30, 32, and 36. Among them, in order to find which one is the most crucial pipe, we calculate the frequency of each pipe ID which appears in Table 2. It is done by calculating the sum of each pipe ID over sum of all pipes ID. The comparison of the frequency of crucial pipes, the failure rate and the edge betweenness is shown in Figure 6. We can see that pipe 3 is the most crucial pipes because its frequency is significantly higher than that of other pipes. The distribution period of pipe 3 covers 5–21, the corresponding $DM \geq 0.83$ and the maximum value reaches 1.53. In addition to pipe 3, the other crucial pipes are pipe 32, 5, 2, 1, (9, 30, 36) (figures in brackets denote the equally important). Besides, from the geographical distribution of pipes, pipe 1, 2, 3, 5 are relatively closer to the reservoirs while pipe 30, 32, 36 are relatively far away from the reservoirs.

Compare the pipe failure rate and edge betweenness shown in Figure 4, we can see that the crucial pipes 3, 32, 5, 2, and 1 got from the simulation of cascading failure are not the maximum values in the failure rate or the edge betweenness. On the contrary, pipe 2, 3, 5 have a relative small failure rate which lead to serious network avalanche. On the other hand, betweenness used as a way to measure network characteristics has been widely seen as the index of the importance. The existing studies on the robustness of the lifeline system have taken the edge or node betweenness as network initial load [12,14,38], based on what the further analysis on network characteristics and emergency strategies have been proposed. The simulation results show that the failure of pipe 12 with the largest betweenness does not cause a wide range of network avalanche. Using betweenness or degree instead of lifeline network entity flow is a method that analyses different kinds of lifeline networks (WDS, transportation, communication, power grid, etc.) in the same way. It ignores the characteristics of the network flow, service function and constraints. The results cannot be directly applied to adjusting the lifeline network relationship or adjusting the network flow.

4. Discussion of the Most Reliable Time Period

Except the multiple identical minimum values occurred at $\alpha = 0.05$, the minimum value under other conditions are within the time period of $H = 7$ period ($R_{sys} = 0.1961$). The $DM = 1.53$ at 7am is the largest water demand during the day. It can be verified that the peak of water usage is the most vulnerable period of a WDS. System reliability decreases with the increase of *DM*.

Table 3 presents the maximum value (relative maximum) of the minimum values of system reliability in one day. The value is selected by the maximum value of system reliability from each column in Table 2. The time period 2 and 3 has the minimum *DM* ($DM = 0.38$). Except for the situation of $\alpha = 0.3$, the relative maximum values of system reliability do not meet the minimum *DM*, but 0.41, 0.6, 0.71, 0.87, 0.99 and 1.21, respectively.

In order to make further analysis on the most reliable period in a day, Figure 7 censuses the frequency of $R_{sys} = 1$ in each period. In order to avoid the deviation of results, the frequency of $R_{sys} = 1$ when $\alpha > 0.3$ is calculated and analyzed. The statistical results show that the frequency of $R_{sys} = 1$ does not change with α increased when $\alpha \geq 0.3$. Hence, in this example, there is a threshold of tolerance parameter $\alpha_c = 0.3$. When $\alpha > \alpha_c$, the period with the highest system reliability meets the time period with

Table 2. Crucial pipes ID (CPID) under six types of α and their corresponding system reliability R_{sys}.

		$\alpha=0.3$		$\alpha=0.25$		$\alpha=0.2$		$\alpha=0.15$		$\alpha=0.1$		$\alpha=0.05$	
DM	Hour	CPID	R_{sys}	CPID	R_{sys}	CPID	R_{sys}	CPID	R_{sys}	CPID	R_{sys}	CPID	R_{sys}
0.41	1	32	0.9956	32	0.8861	1	0.7407	2	0.6482	2	0.2593	5	0.1111
0.38	2	32	0.9993	32	0.8885	1	0.7407	5	0.6019	2	0.2593	5	0.1111
0.38	3	32	0.9993	32	0.8885	1	0.7407	5	0.6019	2	0.2593	5	0.1111
0.45	4	32	0.9881	32	0.9233	1	0.7407	5	0.6759	2	0.2593	5	0.1111
0.83	5	3	0.9151	3	0.9151	3	0.9151	3	0.9151	32	0.463	2	0.1574
0.99	6	3	0.7651	3	0.7651	3	0.7651	3	0.7651	30	0.75	30	0.2315
1.53	7	3	0.1961	3	0.1961	3	0.1961	3	0.1961	3	0.1961	3	0.1961
1.46	8	3	0.2285	3	0.2285	3	0.2285	3	0.2285	3	0.2285	3	0.2285
1.3	9	3	0.3181	3	0.3181	3	0.3181	3	0.3181	3	0.3181	5	0.2315
1.24	10	3	0.3793	3	0.3793	3	0.3793	3	0.3793	3	0.3793	9	0.2315
1.28	11	3	0.3336	3	0.3336	3	0.3336	3	0.3336	3	0.3336	5	0.2315
1.16	12	3	0.4921	3	0.4921	3	0.4921	3	0.4921	3	0.4921	9	0.2315
1.14	13	3	0.5214	3	0.5214	3	0.5214	3	0.5214	3	0.5214	11	0.2315
0.87	14	3	0.8833	3	0.8833	3	0.8833	3	0.8833	36	0.463	2	0.1574
0.89	15	3	0.8662	3	0.8662	3	0.8662	3	0.8662	30	0.75	8	0.1574
0.87	16	3	0.8833	3	0.8833	3	0.8833	3	0.8833	36	0.463	2	0.1574
1.06	17	3	0.6649	3	0.6649	3	0.6649	3	0.6649	3	0.6649	22	0.2315
1.27	18	3	0.3465	3	0.3465	3	0.3465	3	0.3465	3	0.3465	5	0.2315
1.21	19	3	0.4299	3	0.4299	3	0.4299	3	0.4299	3	0.4299	5	0.2315
1.15	20	3	0.508	3	0.508	3	0.508	3	0.508	3	0.508	9	0.2315
0.87	21	3	0.8833	3	0.8833	3	0.8833	3	0.8833	36	0.463	2	0.1574
0.71	22	32	0.9259	32	0.9259	32	0.9259	2	0.7037	5	0.3148	5	0.1111
0.6	23	32	0.9305	32	0.9305	32	0.787	1	0.7037	5	0.2593	5	0.1111
0.41	24	32	0.9956	32	0.8861	1	0.7407	2	0.6482	2	0.2593	5	0.1111

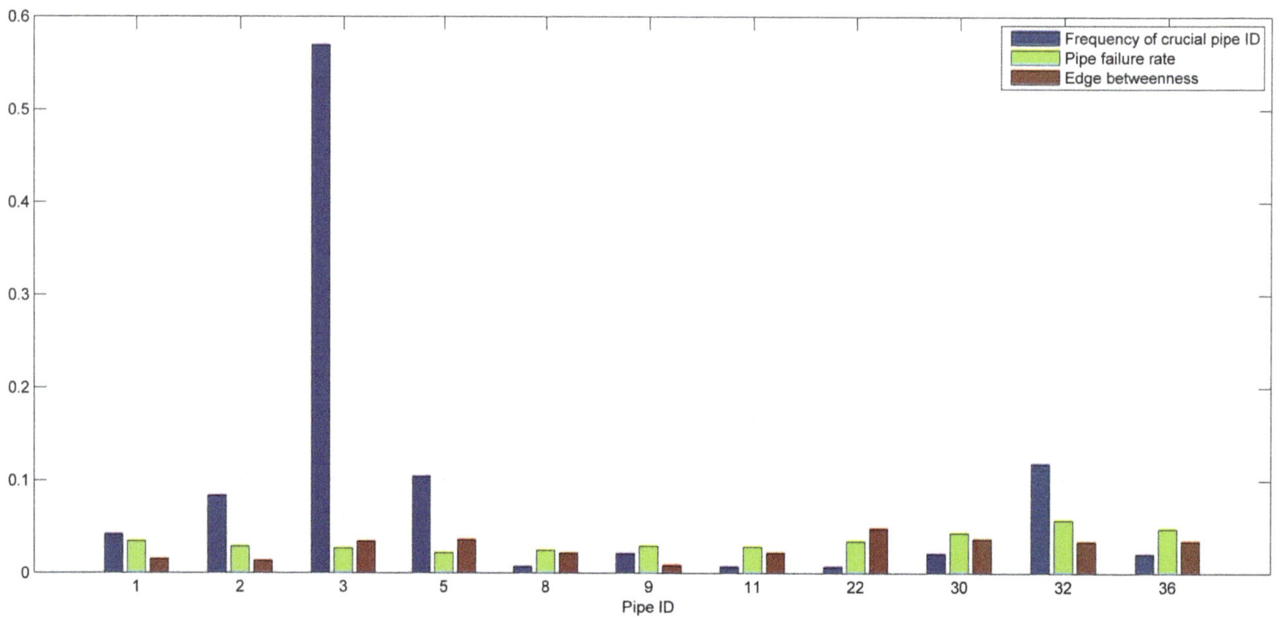

Figure 6. Frequency diagram of crucial pipes.

Table 3. The relative maximum of the minimum values of system reliability under six states of α.

α=0.3		α=0.25		α=0.2	
Hour (DM)	R_{sys}	**Hour (DM)**	R_{sys}	**Hour (DM)**	R_{sys}
2 (0.38), 3 (0.38)	0.9993	23 (0.6)	0.9305	22 (0.71)	0.9259
α=0.15		α=0.1		α=0.05	
Hour (DM)	R_{sys}	**Hour (DM)**	R_{sys}	**Hour (DM)**	R_{sys}
5 (0.83)	0.9151	6 (0.99)	0.75	6 (0.99)	0.2315

$DM<1$; when $\alpha<\alpha_c$, the period with the highest system reliability does not meet the minimum *DM*. Therefore, when select the repair period of a WDS, one cannot simply choose the period with the minimum *DM*, but choose the period with the highest system reliability based on the operation status. What's more, the crucial pipes apt to causing large-scale cascading failure should be avoided also.

Conclusions

The cascading dynamics of WDS in failure condition and the identification of crucial pipes have been discussed in this paper. The propagation of cascading failures in WDS is measured by the damage of certain pipe. The identify factor of crucial pipes is the system reliability after the network restores to stable state. The cascading failure simulation of WDS has taken the nodal pressure head, available water flow, daily demand multipliers and the topological structure in to account. Based on this method, using MATLAB to call EPANET source program is realized. The case study has demonstrated the applicability of this method. The results verified that this method is suitable for WDS and can effectively identify the crucial pipes.

In the network cascading dynamics modeling, it is generally assumed that once the load of a node or edge in the network exceeds its maximum capacity, the corresponding node or edge is avalanched and out of function, triggering the redistribution of network load and cascading failures. However, in the real-world network, there is always some kind of emergency mechanism and emergency response. When a failure occurs, the external emergency power can be involved in to exert its ability of emergency processing so as to repair the network and ensure its normal operation. Therefore, starting from the protection of critical infrastructure network, further research should focus on how to improve the utilization of limited emergency resources and resist the propagation of cascading failure in the whole network.

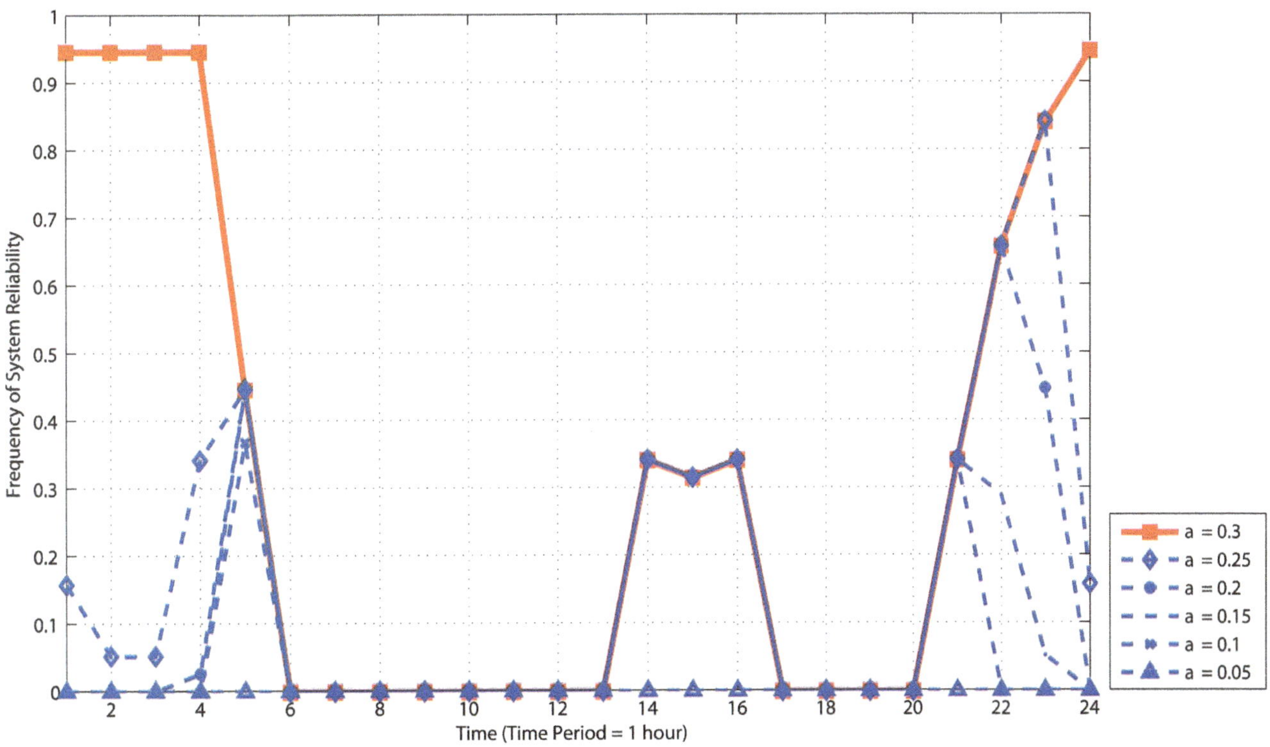

Figure 7. Frequency diagram of the maximum value (R_{sys}= 1) of system reliability within 24 hours.

Acknowledgments

The authors wish to thank the anonymous reviewer whose constructive comments were very helpful for strengthening the presentation of this paper.

Author Contributions

Conceived and designed the experiments: QS MZ YY. Performed the experiments: QS. Analyzed the data: QS MZ YY. Contributed reagents/materials/analysis tools: MZ YY. Wrote the paper: QS.

References

1. Yazdani A, Jeffrey P (2011) Complex network analysis of water distribution systems. Chaos: Interdisciplinary J Nonlinear Sci 21: 016111.
2. Pastor-Satorras R, Vazquez A, Vespignani A (2001) Dynamical and correlation properties of the Internet. Phys Rev E 87: 258701.
3. Newman MEJ (2003) The structure and function of complex networks. SIAM Rev 45: 167–256.
4. Albert R, Jeong H, Barabasi AL (2000) Error and attack tolerance of complex networks. Nature 409: 378–382.
5. Motter AE, Lai YC (2002) Cascade-based attacks on complex networks. Phys Rev E 66: 065102.
6. Holme P, Kim BJ, Yoon CN, Han SK (2002) Attack vulnerability of complex networks. Phys Rev E 65: 056109.
7. Buldyrev SV, Shere N, Cwilich GA (2011) Interdependent networks with correlated degrees of mutually dependent nodes. Phys Rev E 83: 016112
8. Buldyrev SV, Parshani R, Paul G, Stanley HE, Havlin S, et al. (2010) Catastrophic cascade of failures in interdependent networks. Nature 464: 1025–1028.
9. Simonsen I, Buzna L, Peters K, Bornholdt S, Helbing D (2008) Transient dynamics increasing network vulnerability to cascading failures. Phys Rev Lett 100: 218701.
10. Wang WX, Chen G (2008) Universal robustness characteristic of weighted networks against cascading failure. Phys Rev E 77: 026101.
11. Mishkovski I, Biey M, Kocarev L (2011) Vulnerability of complex networks. Commun Nonlinear Sci Numer Simul 16: 341–349.
12. Wang JW, Rong LL (2009) Cascade-based attack vulnerability on the US power grid. Saf Sci 47: 1332–1336.
13. Bompard E, Masera M, Napoli R, Xue F (2009) Assessment of structural vulnerability for power grids by network performance based on complex networks. Crit Inform Infrastruc Security 5508: 144–154.
14. Wang JW, Rong LL (2011) Robustness of the western United States power grid under edge attack strategies due to cascading failures. Saf Sci 49: 807–812.
15. Kinney R, Crucitti P, Albert R (2005) Modeling cascading failures in the North American power grid. Eur Phys J B 46: 101–107.
16. Xia YX, Hill DJ (2008) Attack vulnerability of complex communication networks. IEEE Trans Circuits Syst II Express Briefs 55: 65–69.
17. Rosas-Casals M, Valverde S, Sole RV (2007) Topological vulnerability of the European power grid under errors and attacks. Int J Bifurc Chaos 17: 2465–2475.
18. Sole RV, Rosas-Casals M, Corominas-Murtra B, Valverde S (2008) Robustness of the European power grids under intentional attack. Phys Rev E 77: 026102.
19. Cracitti P, Latora V, Marchiori M (2004) A topological analysis of the Italian electric power grid. Phys A 338: 92–97.
20. Ren H, Dobson I (2008) Using Transmission Line Outage Data to Estimate Cascading failure propagation in an electric power system. IEEE T Circuits-II 55: 927–931.
21. Wu JJ, Sun HJ, Gao ZY (2007) Cascading failures on weighted urban traffic equilibrium networks. Phys A 386: 407–413.
22. Freeman LC (1977) A Set of Measures of Centrality Based on Betweenness. Sociometry 40: 35–41.
23. Borgatti SP (2005) Centrality and network flow. Soc Net 27: 55–71.
24. Tolson BA, Maier HC, Simpson AR, Lence BJ (2004) Genetic Algorithms for Reliability-Based Optimization of Water Distribution Systems. J Water Resour Plann Manage 130:63–72.
25. Mays LW (2000) Water distribution systems handbook. New York: McGraw-Hill. 912p.
26. Ostfeld A, Shamir U (1993) Incorporating reliability in optimal design of water distribution networks-review and new concepts. Reliab Eng Syst Safe 42: 5–11.
27. Zhuang B, Lansey K, Kang D (2013) Resilience/availability analysis of municipal water distribution system incorporating adaptive pump operation. J Hydraul Eng 139: 527–537.
28. Bondy JA, Murty USR (1976) Graph theory with applications. New York: American Elsevier Pub. Co. 171p.
29. Su YC, Mays LW, Duan N, Lansey KE (1987) Reliability-based optimization model for water distribution systems. J Hydraul Eng 113:1539–1556.
30. Puust R, Kapelan Z, Savic DA (2000) A review of methods for leakage management in pipe networks. Urban Water J 7: 25–45.
31. Dueñas-Osorio L, Vemuru SM (2009) Cascading failures in complex infrastructure systems. Struct Saf 31: 157–167.
32. Wagner JM, Shamir U, Marks DH (1988) Water distribution reliability - Simulation methods. J Water Resour Plann Manage 114: 276–294.
33. Bao Y, Mays LW (1990) Model for water distribution system reliability. J Hydraul Eng 116: 1119–1137.
34. Rossman LA (2000) EPANET 2 user's manual, National Risk Management Research Laboratory, U.S. Environmental Protection Agency, Cincinnati, OH.
35. Islam MS, Sadiq R, Rodriguez MJ, Francisque A, Najjaran H, et al. (2011) Leakage detection and location in water distribution systems using a fuzzy-based methodology. Urban Water J 8: 351–365.
36. Islam MS, Sadiq R, Rodriguez M, Najjaran H, Hoorfar M (2013). Reliability Assessment for Water Supply Systems under Uncertainties. J Water Resour Plann Manage In press. doi: 10.1061/(ASCE)WR.1943-5452.0000349
37. Su Y, Mays LW, Duan N, Lansey KE (1987) Reliability-based optimization model for water distribution systems. J Hydraul Eng 113: 1539–1556.
38. Chai CL, Liu X, Zhang WJ (2011) Application of social network theory to prioritizing Oil & Gas industries protection in a networked critical infrastructure system. J Loss Prevent Proc 24: 688–694.

17

The Standard, Intervention Measures and Health Risk for High Water Iodine Areas

Peng Liu[1], Lixiang Liu[1], Hongmei Shen[1]*, Qingzhen Jia[2], Jinbiao Wang[3], Heming Zheng[4], Jing Ma[5], Dan Zhou[1], Shoujun Liu[1], Xiaohui Su[1]

1 Centre for Endemic Disease Control, Chinese Centre for Disease Control and Prevention, Harbin Medical University, Harbin, Heilongjiang, China, **2** Institute for Endemic Disease Control of Shanxi Province, Linfen, Shanxi, China, **3** Institute for Endemic Disease Control of Shandong Province, Jinan, Shandong, China, **4** Centre for Disease Control of Henan Province, Zhengzhou, Henan, China, **5** Centre for Disease Control of Hebei Province, Shijiazhuang, Hebei, China

Abstract

Our study aims to clarify the population nutrient status in locations with different levels of iodine in the water in China; to choose effective measurements of water improvement(finding other drinking water source of iodine not excess) or non-iodised salt supply or combinations thereof; to classify the areas of elevated water iodine levels and the areas with endemic goiter; and to evaluate the risk factors of water iodine excess on pregnant women, lactating women and the overall population of women. From Henan, Hebei, Shandong and Shanxi province of China, for each of 50~99 μg/L, 100~149 μg/L, 150~299 μg/L, and ≥300 μg/L water iodine level, three villages were selected respectively. Students of 6–12 years old and pregnant were sampled from villages of each water-iodine level of each province, excluded iodized salt consumer. Then the children's goiter volume, the children and pregnant's urinary iodine and water iodine were tested. In addition, blood samples were collected from pregnant women, lactating women and other women of reproductive age for each water iodine level in the Shanxi Province for thyroid function tests. These indicators should be matched for each person. When the water iodine exceeds 100 μg/L; the iodine nutrient of children are iodine excessive, and are adequate or more than adequate for the pregnant women. It is reasonable to define elevated water iodine areas as locations where the water iodine levels exceed 100 μg/L. The supply of non-iodised salt alone cannot ensure adequate iodine nutrition of the residents, and water improvement must be adopted, as well. Iodine excess increases the risk of certain thyroid diseases in women from one- to eightfold.

Editor: Qinghua Sun, The Ohio State University, United States of America

Funding: The world organisations of UNICEF and GAIN funded the project. The grant numbers of UNICEF is 2011 H&N IR1 Act 17 and that of GAIN is 104641. The URL of UNICEF' website is http://www.unicef.cn, that of GAIN is http://www.gainhealth.org/. No financial relationships with any organizations that might have an interest in the submitted work in the previous three years; and no other relationships or activities that could appear to have influenced the submitted work. The UNICEF and GAIN offer valuable technique support during design and found process of the project, they also reviewed and offered valuable comments for the research report. This does not alter the authors' adherence to all PLOS ONE policies on sharing data and materials.

Competing Interests: The authors have declared that no competing interests exist.

* E-mail: shenhm119@126.com

Introduction

Iodine excess can be caused by various sources. In 1960, Suzuki reported cases of goiter in Hokkaido caused by eating kelp[1],[2], and Mecullagh observed that the goiter prevalence increase after iodised oil injection[3]. In 1978[4] and 1987[5], China reported the high iodine goitres caused by water iodine. In 2013, there were ten countries with excessive iodine intake[6] secondary to numerous factors. Currently, although iodine excess caused by elevated water iodine has been reported mainly in China, some other countries also suffer this problem, such as Columbia[7], Sudan[8], Germany[9], Denmark[10], Niger[11], Saharawi[12] and Sri Lanka[13]. In China, water iodine levels vary greatly amongst the elevated water iodine areas spread across 12 provinces and municipalities[14]. However, the supply of non-iodised salt has been adopted for these areas as the only intervention measure, with the exception of Beijing and Fujian, where water improvement has also been used. Hence, several questions remain: 1) For different areas with elevated water iodine levels, what are the iodine levels in the population? 2) Should water improvement, the supply of non-iodised salt, or a combination of both interventions be adopted based on the specific context of relevant areas? 3) How should the areas with elevated water iodine and the areas with endemic iodine excess goitres be determined and classified? Therefore, the Study Project on the Standard and Intervention Measures for Areas of Elevated Water Iodine was implemented by the Endemic Disease Centre of China in the elevated water iodine provinces, including Shandong, Shanxi, Henan and Hebei Provinces, from July 2011 to March 2012. The project results are now shared as follows.

Materials and Methods

Survey scope

The number of surveyed villages was determined according to the historic IDD surveillance results in recent years and in reference to the various local water iodine levels of the four provinces. For each range of iodine levels (50–99 μg/L, 100–149 μg/L, 150–299 μg/L, and ≥300 μg/L), three villages were selected in each of the four provinces.

Figure 1. Distribution of the survey sites.

Survey subjects

Children aged six to 12 years-old were sampled in the primary schools of the survey villages (should no primary schools exist in the surveyed villages, students from the village studying at the township central primary school were sampled). For each pre-set water iodine group, at least 200 children taking non-iodised salt were sampled. Pregnant women of the survey villages were sampled from the maternal and child care service centres or health centres of the townships. For each pre-set water iodine group, at least 50 pregnant women taking non-iodised salt were sampled.

Survey indicators

Water iodine levels were verified at the surveyed villages with five water samples collected from the east, west, south, north and centre of each village with scattered water supplies, and two water samples were collected from villages with centralised water supplies, before the following survey was conducted. If the water iodine of the village fit the pre-set level, the drinking water of the children's and pregnant women's households was sampled to test the actual water iodine level. The table salt of the children's and pregnant women's households was sampled for testing through a semi-quantitative method (<5 mg/kg was defined non-iodised salt). Children and pregnant women from the households found to use iodised salt were excluded, and the remaining individuals continued with the following tests and examinations. One spot urine samples were collected from the above-mentioned children and pregnant women to test their urinary iodine levels. The urinary iodine levels were stratified according to the WHO's recommendation [15]. Goiter size was detected by ultrasound among the children aged six to 12 years-old, and goiter was defined by the Chinese diagnostic criteria for endemic goiter (WS276-2007)[16]. Approximately blood samples were collected from pregnant women, lactating women and other women of reproductive age for each water iodine level in Shanxi Province for FT3, FT4, TSH, anti-TPO and anti-Tg tests. The reference values were FT3: 2.8–7.1 pmol/L, FT4: 12–22 pmol/L, TSH: 0.27–4.20 IU/L, TPOAb: >34 IU/ml, TgAb:

Table 1. Drinking water iodine and urinary iodine grouped by the actual water iodine levels.

Province	Actual water iodine group (μg/L)	Children (median μg/L) Size	Water iodine	Urinary iodine	Pregnant (median μg/L) Size	Water iodine	Urinary iodine
Shandong	50–99	683	71.6	252.0	142	68.3	166.2
	100–149	256	113.8	336.9	85	120.2	202.9
	150–299	302	239.3	460.4	62	201.6	278.8
	≥300	215	350.3	620.2	79	405.7	452.0
Shanxi	50–99	196	73.8	274.2	21	82.0	238.6
	100–149	189	144.7	312.8	20	121.5	204.9
	150–299	159	258.5	445.6	21	258.5	373.9
	≥300	165	501.0	793.5	19	485.9	607.4
Henan	50–99	131	64.1	270.4	26	74.9	280.9
	100–149	338	139.6	336.0	147	138.1	326.0
	150–299	383	242.8	379.0	205	224.1	308.5
	≥300	310	475.7	345.0	150	475.7	349.5
Hebei	100–149	195	111.7	396.4	50	110.7	309.8
	150–299	226	236.5	316.8	65	235.9	270.4
	≥300	211	392.3	382.7	39	335.8	385.0
Total	50–99	1010	72.2	260.2	187	69.6	183.0
	100–149	978	118.2	338.2	299	121.5	284.2
	150–299	1070	242.8	401.0	353	230.2	304.0
	≥300	901	392.3	473.0	287	411.9	386.0

Note: P<0.001 when different groups of the lumped actual household water iodine are compared with a median test. This result suggests that a significant difference exists among the water iodine levels of various groups and that such difference also exists when any two groups are compared.

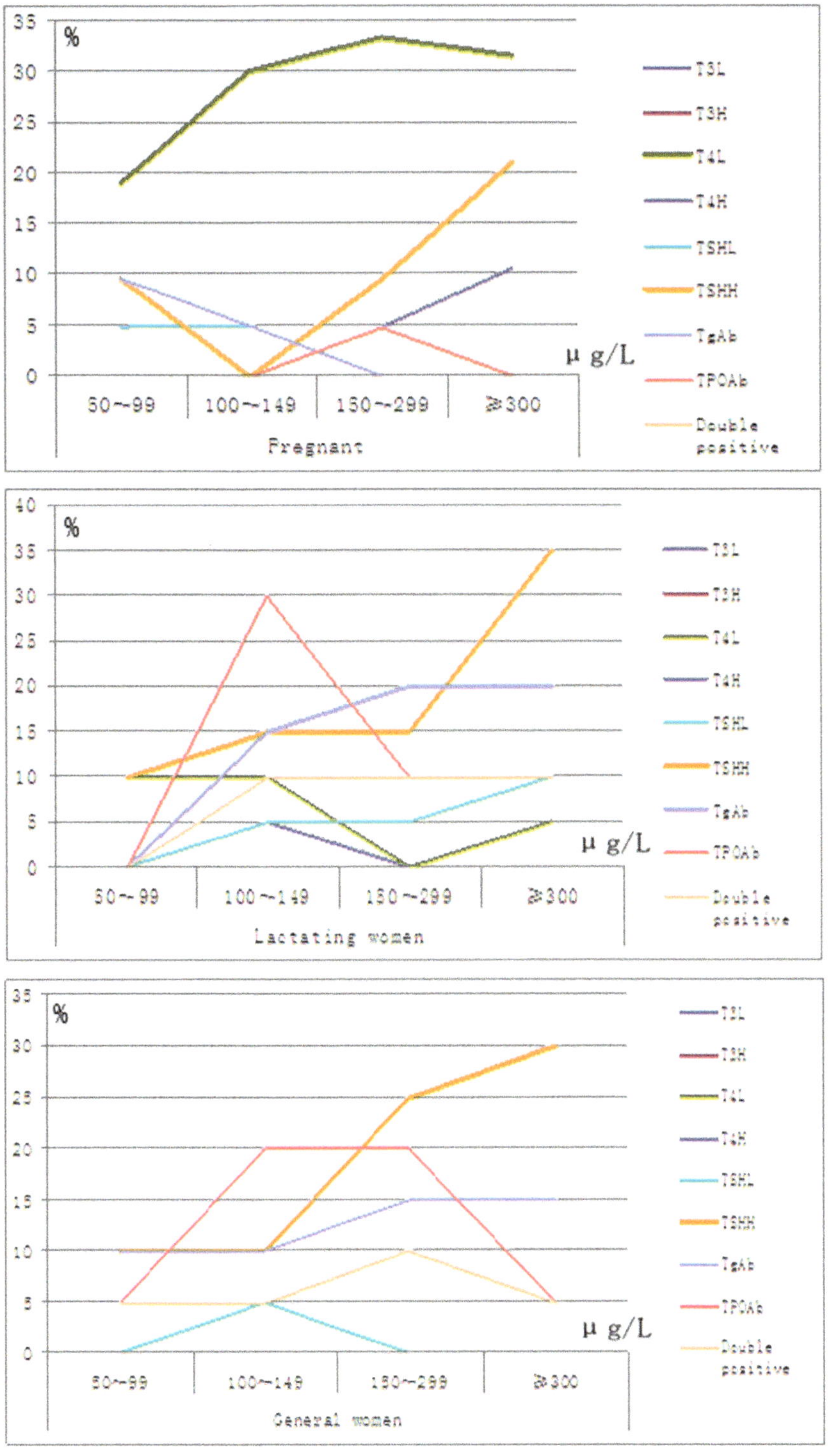

Figure 2. Blood tests results among pregnant (a), lactating (b) and child-bearing age woman (c).

>115 IU/ml. The urinary iodine levels, table salt iodine, drinking water iodine (and children's goiter size or pregnant women's thyroid function) was matched for each person.

Quality control

A detailed project implementation plan was developed and validated by experts to control the project quality. A project launching/training meeting and a review meeting were conducted by the Endemic Disease Centre, which the project managers from the project provinces, prefectures and counties attended. A centralised database was developed, and the submitted data were cleaned by the Endemic Disease Centre to ensure data quality. Project managers and relevant staff from the Endemic Disease Centre participated in the onsite work in the Shandong and Hebei Provinces. All laboratories responsible for the testing passed the external quality check by the National IDD Reference Lab.

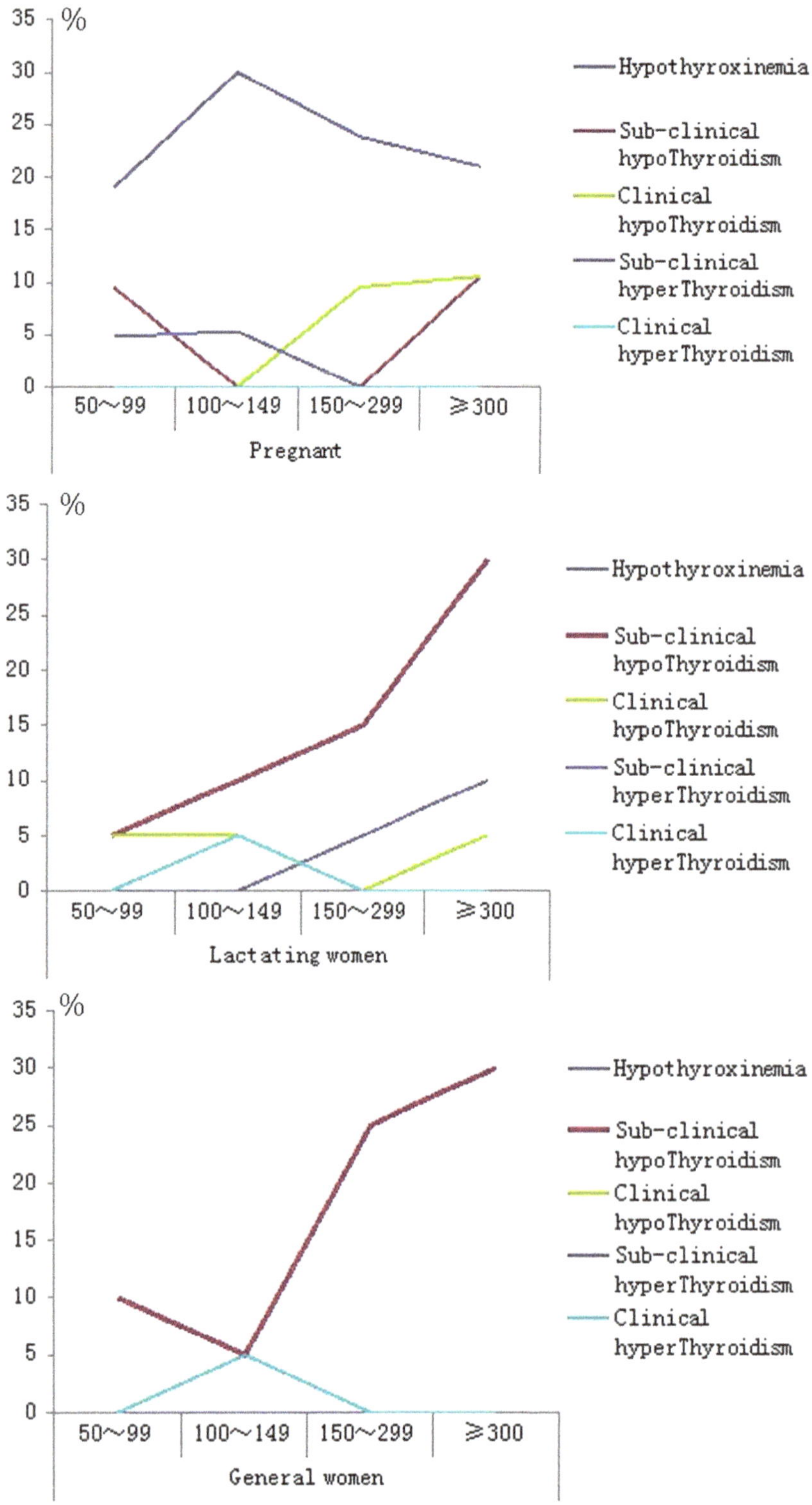

Figure 3. Detection rate among the pregnant (a), lactating (b) and child-bearing age woman (c).

Ethical statement

The research was approved by the review board of Harbin Medical University. The informed consent form was prepared before the research began. We obtained informed consent from the guardians on the behalf of the children and the women themselves, and the consent was written.

Table 2. ORs for different pre-set water iodine levels 50-99 µg/L as the control group.

Category	Pre-set water iodine groups (µg/L)	Hypo-hyroxinemia	Sub-clinical hypo-thyroidism	Clinical hypo-thyroidism	Sub-clinical hyper-thyroidism
Pregnant	100–149	1.82	0.00		1.05
	150–299	1.33	0.00		0.00
	≥300	1.13	1.12		0.00
Lactating	100–149	1.00	2.11	1.00	
women	150–299	0.00	3.35	0.00	
	≥300	0.47	8.14	1.00	
Other	100–149		0.47		
women	150–299		3.00		
	≥300		3.86		
Total	100~149	1.48	0.59	2.07	1.01
	150~299	1.00	1.69	2.03	1.00
	≥300	0.81	3.48	3.21	2.10

Results

Overview of project sites

The survey counties for Shandong were Xinxian, Guanxian and Chiping in Liaocheng City; those for Shanxi were Xiaodian and Qingxu in Taiyuan City; those for Henan were Xiayi and Yucheng in Shangqiu City and Taiqian in Puyang City; and those for Hebei were Dacheng in Langfang City, Huanghua in Cangzhou City, and Weixian in Xingtai City. Please see Figure 1, Table S1 in File S1 for the survey sites.

Profile of the survey subjects

Onsite work was then developed according to the implementation plan at the project sites. The total sample size met the size required by the implementation plan. Please see Table S2–S4 in File S1. Survey subjects were then further grouped based on the actual drinking water iodine levels. The sample size grouped by the actual water iodine levels differed only slightly from the required one and met the size requirement. The survey indicated that seven- to 11-year-old children accounted for a larger proportion of the children surveyed and that those 6 and 12 years old accounted for a smaller number. Please see Table S5 in File S1. The pregnant women surveyed were predominantly 20–30 years old, and women in this age group accounted for 80% of the total survey subjects. Please see Table S6 in File S1.

Results of the water iodine survey

The median water iodine was calculated according to the actual water iodine levels. The actual water iodine levels of the children and pregnant are consistent with the pre-set water group; hence, we only supply results from the actual water iodine levels. The water iodine levels of the children were consistent with those of the pregnant women for the same actual water iodine group. Please see Table S7 in File S1, Figure S1,S2 in File S1.

MUI (Median of Urinary Iodine)

The MUI of the children and pregnant women in the actual water iodine groups demonstrates that with water iodine levels of 50–99 µg/L, children exhibited more than adequate iodine levels with their MUI between 200 and 299 µg/L. When the water iodine levels exceeded 100 µg/L, the children in various groups exhibited excessive iodine levels with MUI exceeding 300 µg/L. This result is consistent across the four provinces. With water iodine levels of 50–99 µg/L, the urinary iodine of the pregnant women in the Shandong, Shanxi and Henan provinces fell within 150–249 µg/L. With the water iodine levels of 100–149 µg/L, the urinary iodine levels of the pregnant women from all provinces either fell within the adequate range of 150–249 µg/L or exceeded adequate levels (250–499 µg/L). Thus, it can be concluded that with the supply of non-iodised salt and water iodine levels of 50–99 µg/L, children exhibit more than adequate iodine levels and pregnant women exhibit adequate iodine levels; with water iodine levels between 100 and 149 µg/L, children exhibit excessive iodine levels, and pregnant women either exhibit adequate or excessive iodine levels. See Table 1, S8–S12 in File S1, Figure S3–S5 in File S1 for details.

Thyroid size

Our analysis of goiter size indicated that the goiter rate is not typically high, despite the fact that most of the groups surveyed were within areas with elevated water iodine levels. The goiter rate typically exceeded 5% when the water iodine exceeded 300 µg/L. The goiter rate for all water iodine levels in Shanxi was higher than that of other provinces, although the underlying cause remains to be further. Please see Figure S6–S7 in File S1.

Results of thyroid function testing among pregnant women (percentage tested positive)

Figure 2, Table S14 in File S1 indicates that the pregnant women with low T4 levels account for a high percentage across the groups. The proportion increases from 19.0% to 31.6% with the rising water iodine, which suggests decreasing storage of thyroid hormones. The rising water iodine levels are coupled with the rising percentage of lactating women with both low and high TSH. However, the hormone levels of the lactating women are quite stable, suggesting active physiological adjustment. In the groups of women of child-bearing age, the percentage of those with high TSH levels rises with increasing water iodine. However,

Table 3. MUI of the children under different water iodine levels.

Actual water iodine group(μg/L)	Urinary iodine of children		Urinary iodine of pregnant	
	Size	Median (μg/L)	Size	Median (μg/L)
290–299	18	551.2	2	350.5
280–289	18	576.7	14	259.9
270–279	28	510.1	7	299.4
260–269	127	468.0	11	329.9
250–259	270	411.8	52	338.7
240–249	135	482.0	70	367.5
230–239	122	360.6	54	289.6
220–229	26	443.6	43	318.2
210–219	23	523.3	13	370.7
200–209	14	339.0	16	338.8
190–199	15	514.0	15	335.4
180–189	14	300.0	14	235.0
170–179	114	277.8	13	172.0
160–169	30	397.8	19	298.4
150–159	27	254.3	29	254.3
140–149	347	344.1	124	301.4
130–139	37	363.9	21	209.0
120–129	80	346.3	46	216.4
110–119	372	339.0	93	306.4
100–109	108	332.7	30	268.7
90–99	120	309.0	16	241.8
80–89	195	285.0	39	235.6
70–79	237	259.9	33	153.2
60–69	289	256.4	57	165.9
50–59	169	241.6	30	183.9

their hormone levels do not register a remarkable increase, contrary to the normal situation. A possible explanation can be that the rising TSH level not only causes elevated hormone levels but also promotes the degradation of the hormones. A large number of lactating women and women of child-bearing age tested positive with either anti-TPO or anti-Tg, with the percentage reaching 30% in some groups. Furthermore, a high percentage of such women tested positive with two indicators (see Figure 2 a, b and c, Table S13, S14 in File S1).

The disease was diagnosed based on the lab test results of the blood samples. The diagnosed results demonstrate that accompanying the rising of water iodine, the incidence of hypothyroxinemia and sub-clinical hypothyroidism among pregnant women, lactating women and women of child-bearing age were all increased (see Figure 3 a, b and c). When the odds ratio (OR) of morbidity was calculated for all pre-set water iodine levels using 50–99 μg/L as the control group, the result demonstrate that the OR of hypothyroxinemia among the pregnant women decreased across all groups but remained higher than the control group. The OR of sub-clinical hypothyroidism among lactating women gradually rose to eight times that of the control group, whereas the OR of sub-clinical hypothyroidism among the women of child-bearing age gradually rose to four times of that of the control group. The OR of clinical hyperthyroidism and hypothyroidism failed to be obtained due to too few cases (see Table 2).

Discussion

The goiter rate of Shanxi is excessively higher compared with other provinces with similar median water iodine levels. A possible contributor could be a different type of iodine within the water, such as iodate, iodine ion or other substances causing goiter in the environment. However, the specific contributor is subject to further study.

The study results can be fed into the refined definition elevated water iodine areas and endemic goiter areas. Children of all of the four provinces exhibited MUIs greater than 300 μg/L in the group with water iodine levels of 100–149 μg/L. Most of the children in the group with water iodine between 50 and 99 μg/L exhibited urinary iodine between 200 and 299 μg/L or more than adequate levels, whereas the MUI of pregnant women was 150–249 μg/L or adequate levels. When the water iodine samples were further broken down, the urinary iodine levels of the children demonstrate that the MUI of the children was 309.0 μg/L when the water iodine level was 90 to 100 μg/L and that the children's MUIs drop when the water iodine decreases. However, a substantial MUI decrease has not been registered among the children, with the average MUI of 241.6 μg/L being observed when the water iodine decreases to 50 to 60 μg/L. When the water iodine continues to drop, the MUI of the pregnant women also decreases to the lower bound of the adequate level at 150 μg/

L. Therefore, it is safe and conservative to set the cut-off water iodine level at 100 μg/L in the new definition (see Table 3).

The prevalence of sub-clinical hypothyroidism among women was 8.1% among the population above 18 years-old in Jiangsu. We determined that 8.2% of the survey subjects exhibited sub-clinical hypothyroidism in the group with water iodine levels of 50–99 μg/L, which is close to the survey result. The other research in Jiangsu was 18.9% and 11.1% for sub-clinical hypothyroidism and hypothyroidism in child-bearing age women[17]. In the screening of 4,800 pregnant women at eight-week gestation by the research team led by Prof. Teng WP, 6.15% of the subjects were determined to have sub-clinical hypothyroidism (9.5% for this project), 1.11% exhibited hypothyroxinemia (19.0% for this project), and 9.6% were TPO Ab positive (1.6% for this project). Sang ZN et al. found excessive iodine intake during late pregnancy may lead to maternal thyroid dysfunction, particularly subclinical hypothyroidism when their urinary above 250 μg/L[18]. The data suggest that a significant number of pregnant women exhibit "mild" thyroid dysfunction.

Conclusions

The standard committee is revising the *Determination and classification of the areas of elevated water iodine and the endemic areas of iodine excess goiter* (GB/T19380-2003) according to the survey results. The new standard will upgrade the definition of areas of elevated water iodine (areas with water iodine exceeding 150 μg/L) to the areas with water iodine levels exceeding 100 μg/L. Furthermore, areas in which more than 5% of the local children between eight and 10 years-old exhibit goiter will be regarded as endemic areas. The new standard will be approved and released in the near future.

The *Surveillance Plan for the Areas with Elevated Water Iodine Levels in China (Trial)* has been developed, which determines that the surveillance will sample half of the elevated water iodine counties in Tianjin, Hebei, Shanxi, Jiangsu, Anhui, Shandong, Henan, and Shaanxi and that surveillance sites will be set up in the villages to survey water iodine levels, children's thyroid sizes, and children's urinary iodine. The plan will be revised and finalised within several years of trial implementation.

The survey results have indicated that the supply of non-iodised salt alone cannot ensure that the residents will have adequate iodine nutrition and that water improvement must be adopted as well. The need for water improvement is especially pressing for areas with water iodine levels exceeding 100 μg/L.

The goiter rate of the children in Shanxi is slightly higher than other provinces despite no significant differences in the water iodine levels. This observation suggests that other factors may be in play, such as a different type of iodine existing in the water or factors other than iodine.

With the increasing water iodine levels, the percentage of pregnant women suffering hypothyroxinemia rises before dropping, and the number of lactating women and women of child-bearing age with sub-clinical hypothyroidism increased eight-fold and four-fold, respectively. Additionally, the number of women who tested positive for TPO Ab increased, and the number of women who tested positive for TG Ab first increased then decreased, and those women positive for both TPO and TG Ab accounted for a high percentage.

It is reasonable to define areas with elevated water iodine as those with water iodine levels exceeding than 100 μg/L.

Supporting Information

File S1 This file contains Tables S1–S14 and Figures S1–S7. Table S1, Project sites selected. Table S2, The sample size for the pre-set water iodine levels. Note: Please note that all the samples reflected in this and following tables are non-iodized salt users. Table S3, The sample size for the actual water iodine levels. Table S4, The number of blood samples collected according to the pre-set water iodine and the actual water iodine levels. Table S5, Age group distribution of the surveyed children. Table S6, Age group distribution of the surveyed women. Table S7, Actual household drinking water iodine levels of the children and pregnant women grouped as per the pre-set water iodine levels. Table S8, Urinary iodine levels of the children and pregnant women in the pre-set water iodine groups. Table S9, Urinary iodine frequency distribution of the children in the pre-set water iodine groups. Table S10, Urinary iodine frequency distribution of the children in the actual water iodine groups. Table S11, Urinary iodine frequency distribution of the pregnant women in the pre-set water iodine groups. Table S12, Urinary iodine frequency distribution of the pregnant women grouped in the actual water iodine groups. Table S13, Goiter rate of the children in the pre-set water iodine groups. Table S14, Results of the blood tests among the women in the pre-set water iodine groups. Note: The reference range is the standard range provided by the testing method and applies to the pregnant, lactating and general women. No specific reference range has been provided for different groups of subjects. Figure S1, Water iodine levels of the children and women grouped by the pre-set water iodine levels. Figure S2, Water iodine levels of the children and pregnant women in the actual water iodine groups. Figure S3, MUI of the children in the pre-set and actual water iodine groups. Figure S4, MUI of the pregnant women in the pre-set and actual water iodine groups. Figure S5, Relations between median water iodine and MUI of the subjects in the pre-set water iodine groups. Figure S6, Goiter rate of the children in the pre-set water iodine groups Correlation between children's water iodine and urinary iodine. Figure S7, Goiter rate of the children in the actual water iodine groups.

Acknowledgments

The authors would like to thanks the organisations participated in this survey: The National Centre for Endemic Disease Control (CEDC), 4 provincial (municipality) Centres for Disease Control (CDC) or CEDCs, including Shandong, Shanxi, Henan and Hebei. The authors of this article thank those colleagues from the above organisations who took part in the survey for their hard work. We also thank the UNICEF and GAIN for their valuable technique supporting during design and found process of the project, they also helped review and offer valuable comments for the research report, We are particular grateful for the support of two international organisations.

Author Contributions

Conceived and designed the experiments: PL HMS SJL LXL XHS. Performed the experiments: QZJ JBW HMZ JM. Analyzed the data: PL DZ. Wrote the paper: PL HMS.

References

1. Suzuki H, Higuchi T, Hashimoto H, Otaki S (1962) A case of endemic goiter along the seashores of Hidaka District, Hokkaido. Jpn J Med Sci Biol 51: 781–786.
2. Suzuki H, Higuchi T, Sawa K, Ohtaki S, Horiuchi Y (1965) "Endemic coast goitre" in Hokkaido, Japan. Acta Endocrinol (Copenh) 50: 161–176.
3. Clarke KH, Mc CS, Winikoff D (1960) The use of an intramuscular depot of iodized oil as a long-lasting source of iodine. Med J Aust 47: 89–91.

4. Ma T, Yu ZH, Lu TZ, Wang SY, Dong CF, et al. (1982) High-iodide endemic goiter. Chin Med J (Engl) 95: 692–696.
5. Li M, Qu CY, Qian QD, Jia QZ, Creswell JE, et al. (1987) Endemic goitre in central China caused by excessive iodine intake. Lancet 2: 257–259.
6. Pearce EN, Andersson M, Zimmermann MB (2013) Global iodine nutrition: where do we stand in 2013? Thyroid 23: 523–528.
7. Gaitan E (1983) Endemic goiter in western Colombia. Ecol Dis 2: 295–308.
8. Medani AM, Elnour AA, Saeed AM (2012) Excessive iodine intake, water chemicals and endemic goitre in a Sudanese coastal area. Public Health Nutr 16:1–7.
9. Kirchner S, Stelz A, Muskat E (1996) Contribution of natural mineral water to the iodine supply of the population. Z Lebensm Unters Forsch 203:311–315.
10. Pedersen KM, Laurberg P, Nohr S, Jorgensen A, Andersen S (1999) Iodine in drinking water varies by more than 100-fold in Denmark. Importance for iodine content of infant formulas. Eur J Endocrinol 140: 400–403.
11. Pearce EN, Gerber AR, Gootnick DB, Khan LK, Li RW, et al. (2002) Effects of chronic iodine excess in a cohort of long-term American workers in West Africa. J Clin Endocrinol Metab 87: 5499–5502.
12. Henjum S, Barikmo I, Gjerlaug AK, Abderraháman ML, Arne O, et al. (2010) Endemic goitre and excessive iodine in urine and drinking water among Saharawi refugee children. Public Health Nutr 13: 1472–1477.
13. Fordyce FM, Johnson CC, Udaya RB, Appleton JD, Chandra B, et al. (2000) Selenium and iodine in soil, rice and drinking water in relation to endemic goitre in Sri Lanka. Sci Total Environ 263: 127–141.
14. Shen HM, Liu SJ, Sun DJ, Zhang SB, Su XH, et al. (2011) Geographical distribution of drinking-water with high iodine level and association between high iodine level in drinking-water and goitre: a Chinese national investigation. Br J Nutr 106: 243–247.
15. WHO UNICEF, ICCIDD (2007)Guide Manual: Assessment of iodine deficiency disorders and monitoring their elimination-A GUIDE FOR PROGRAMME MANAGERS.
16. Ministry of Health (2007) Diagnostic criteria for endemic goiter, People's Medical Publishing House: Beijing.
17. He YX, Wang PH, Zhou YL, Zhang QL, Chen ZG, et al. (2011) Investigation on the Thyroid Function of Childbearing Aged Women Exposed to Excessive Iodine. J Environ Hyg 1: 1–3.
18. Sang ZN, Wei W, Zhao N, Zhang GQ, Chen W, et al. (2012) Thyroid dysfunction during late gestation is associated with excessive iodine intake in pregnant women. J Clin Endocrinol Metab 97: 1363–1369.

18

Water Filtration Using Plant Xylem

Michael S. H. Boutilier[¶], Jongho Lee[¶], Valerie Chambers, Varsha Venkatesh, Rohit Karnik*

Department of Mechanical Engineering, Massachusetts Institute of Technology, Cambridge, Massachusetts, United States of America

Abstract

Effective point-of-use devices for providing safe drinking water are urgently needed to reduce the global burden of waterborne disease. Here we show that plant xylem from the sapwood of coniferous trees – a readily available, inexpensive, biodegradable, and disposable material – can remove bacteria from water by simple pressure-driven filtration. Approximately 3 cm^3 of sapwood can filter water at the rate of several liters per day, sufficient to meet the clean drinking water needs of one person. The results demonstrate the potential of plant xylem to address the need for pathogen-free drinking water in developing countries and resource-limited settings.

Editor: Zhi Zhou, National University of Singapore, Singapore

Funding: This work was supported by the James H. Ferry, Jr. Fund for Innovation in Research Education award to R.K. administered by the Massachusetts Institute of Technology. SEM imaging was performed at the Harvard Center for Nanoscale Systems, a member of the National Nanotechnology Infrastructure Network (NNIN), which is supported by the National Science Foundation under NSF award no. ECS-0335765. The funders had no role in study design, data collection and analysis, decision to publish, or preparation of the manuscript.

Competing Interests: The authors have declared that no competing interests exist.

* E-mail: karnik@mit.edu

¶ These authors contributed equally to this work.

Introduction

The scarcity of clean and safe drinking water is one of the major causes of human mortality in the developing world. Potable or drinking water is defined as having acceptable quality in terms of its physical, chemical, and bacteriological parameters so that it can be safely used for drinking and cooking [1]. Among the water pollutants, the most deadly ones are of biological origin: infectious diseases caused by pathogenic bacteria, viruses, protozoa, or parasites are the most common and widespread health risk associated with drinking water [1,2]. The most common water-borne pathogens are bacteria (e.g. *Escherichia coli*, *Salmonella typhi*, *Vibrio cholerae*), viruses (e.g. adenoviruses, enteroviruses, hepatitis, rotavirus), and protozoa (e.g. giardia) [1]. These pathogens cause child mortality and also contribute to malnutrition and stunted growth of children. The World Health Organization reports [3] that 1.6 million people die every year from diarrheal diseases attributable to lack of access to safe drinking water and basic sanitation. 90% of these are children under the age of 5, mostly in developing countries. Multiple barriers including prevention of contamination, sanitation, and disinfection are necessary to effectively prevent the spread of waterborne diseases [1]. However, if only one barrier is possible, it has to be disinfection unless evidence exists that chemical contaminants are more harmful than the risk from ingestion of microbial pathogens [1]. Furthermore, controlling water quality at the point-of-use is often most effective due to the issues of microbial regrowth, byproducts of disinfectants, pipeline corrosion, and contamination in the distribution system [2,4].

Common technologies for water disinfection include chlorination, filtration, UV-disinfection, pasteurization or boiling, and ozone treatment [1,2,5]. Chlorine treatment is effective on a large scale, but becomes expensive for smaller towns and villages. Boiling is an effective method to disinfect water; however, the amount of fuel required to disinfect water by boiling is several times more than what a typical family will use for cooking [1]. UV-disinfection is a promising point-of-use technology available [1], yet it does require access to electricity and some maintenance of the UV lamp, or sufficient sunlight. While small and inexpensive filtration devices can potentially address the issue of point-of-use disinfection, an ideal technology does not currently exist. Inexpensive household carbon-based filters are not effective at removing pathogens and can be used only when the water is already biologically safe [1]. Sand filters that can remove pathogens require large area and knowledge of how to maintain them [1], while membrane filters capable of removing pathogens [2,4] suffer from high costs, fouling, and often require pumping power due to low flow rates [6] that prevents their wide implementation in developing countries. In this context, new approaches that can improve upon current technologies are urgently needed. Specifically, membrane materials that are inexpensive, readily available, disposable, and effective at pathogen removal could greatly impact our ability to provide safe drinking water to the global population.

If we look to nature for inspiration, we find that a potential solution exists in the form of plant xylem – a porous material that conducts fluid in plants [7]. Plants have evolved specialized xylem tissues to conduct sap from their roots to their shoots. Xylem has evolved under the competing pressures of offering minimal resistance to the ascent of sap while maintaining small nanoscale pores to prevent cavitation. The size distribution of these pores – typically a few nanometers to a maximum of around 500 nm, depending on the plant species [8] – also happens to be ideal for filtering out pathogens, which raises the interesting question of whether plant xylem can be used to make inexpensive water filtration devices. Although scientists have extensively studied plant xylem and the ascent of sap, use of plant xylem in the context of water filtration remains to be explored. Measurements of sap flow in plants suggest that flow rates in the range of several liters per hour may be feasible with less than 10 cm-sized filters, using only gravitational pressure to drive the flow [7].

We therefore investigated whether plant xylem could be used to create water filtration devices. First, we reason which type of plant xylem tissue is most suitable for filtration. We then construct a simple water filter from plant xylem and study the resulting flow rates and filtration characteristics. Finally, we show that the xylem filter can effectively remove bacteria from water and discuss directions for further development of these filters.

Materials and Methods

Materials

Branches were excised from white pine growing on private property in Massachusetts, USA, with permission of the owner and placed in water. The pine was identified as *pinus strobus* based on the 5-fold grouping of its needles, the average needle length of 4.5 inches, and the cone shape. Deionized water (Millipore) was used throughout the experiments unless specified otherwise. Red pigment (pigment-based carmine drawing ink, Higgins Inks) was dissolved in deionized water. Nile-red coated 20 nm fluorescent polystyrene nanospheres were obtained from Life Technologies. Inactivated, Alexa 488 fluorescent dye labeled *Escherichia coli* were obtained from Life Technologies. Wood sections were inserted into the end of 3/8 inch internal diameter PVC tubing, sealed with 5 Minute Epoxy, secured with hose clamps, and allowed to cure for ten minutes before conducting flow rate experiments.

Construction of the Xylem Filter

1 inch-long sections were cut from a branch with approximately 1 cm diameter. The bark and cambium were peeled off, and the piece was mounted at the end of a tube and sealed with epoxy. The filters were flushed with 10 mL of deionized water before experiments. Care was taken to avoid drying of the filter.

Filtration and Flow Rate Experiments

Approximately 5 mL of deionized water or solution was placed in the tube. Pressure was supplied using a nitrogen tank with a pressure regulator. For filtration experiments, 5 psi (34.5 kPa) pressure was used. The filtrate was collected in glass vials. For dye filtration, size distribution of the pigments was measured for the input solution and the filtrate. Higgins pigment-based carmine drawing ink, diluted $\sim 1000\times$ in deionized water, was used as the input dye solution. For bacteria filtration, the feed solution was prepared by mixing 0.08 mg of inactivated *Escherichia coli* in 20 mL of deionized water ($\sim 1.6\times10^7$ mL^{-1}) after which the solution was sonicated for 1 min. The concentration of bacteria was measured in the feed solution and filtrate by enumeration with a hemacytometer (inCyto C-chip) mounted on a Nikon TE2000-U inverted epifluorescence microscope. Before measurement of concentration and filtration experiments, the feed solution was sonicated for 1 min and vigorously mixed.

Imaging

Xylem structure was visualized in a scanning electron microscope (SEM, Zeiss Supra55VP). Samples were coated with gold of 5 nm thickness before imaging. To visualize bacteria filtration, 5 mL of solution at a bacteria concentration of $\sim 1.6\times10^7$ mL^{-1} was flowed into the filter. The filter was then cut longitudinally with a sharp blade. One side of the sample was imaged using a Nikon TE2000-U inverted epifluorescence microscope and the other was coated with gold and imaged with the SEM. Optical images were acquired of the cross section of a filter following filtration of 5 mL of the dye at a dilution of $\sim 100\times$.

Particle Sizing

Dynamic light scattering measurements of particle size distributions were performed using a Malvern Zetasizer Nano-ZS.

Results

Xylem Structure and Rationale for use of Conifer Xylem

The flow of sap in plants is driven primarily by transpiration from the leaves to the atmosphere, which creates negative pressure in the xylem. Therefore, xylem evolution has occurred under competing pressures of providing minimal resistance to the flow of sap, while protecting against cavitation (i.e. nucleation) and growth of bubbles that could stop the flow of sap and kill the plant, and to do this while maintaining mechanical strength [7]. The xylem structure comprises many small conduits that work in parallel and operate in a manner that is robust to cavitation [7,8] (Figure 1). In woody plants, the xylem tissue is called the sapwood, which often surrounds the heartwood (i.e. inactive, non-conducting lignified tissue found in older branches and trunks) and is in turn surrounded by the bark (Figure 1b,c). The xylem conduits in gymnosperms (conifers) are formed from single dead cells and are called tracheids (Figure 1c), with the largest tracheids reaching diameters up to 80 μm and lengths up to 10 mm [7]. Angiosperms (flowering plants) have xylem conduits called vessels that are derived from several cells arranged in a single file, having diameters up to 0.5 mm and lengths ranging from a few millimeters to several meters [7]. These parallel conduits have closed ends and are connected to adjacent conduits via "pits" [8] (Figure 1d,e). The pits have membranes with nanoscale pores that perform the critical function of preventing bubbles from crossing over from one conduit to another. Pits occur in a variety of configurations; Figure 1d,e shows torus-margo pit membranes that consist of a highly porous part shaped like a donut (margo) and an impermeable part in the center called torus, occurring in conifers [8]. The porosity of the pit membranes ranges in size from a few nanometers to a few hundred nanometers, with pore sizes in the case of angiosperms tending to be smaller than those in gymnosperms [8,9]. Pit membrane pore sizes have been estimated by examining whether gold colloids or particles of different sizes can flow through [8,10]. Remarkably, it was observed that 20 nm gold colloids could not pass through inter-vessel pit membranes of some deciduous tree species [10], indicating an adequate size rejection to remove viruses from water. Furthermore, inter-tracheid pit membranes were found to exclude particles in the 200 nm range [8], as required for removal of bacteria and protozoa.

Since angiosperms (flowering plants, including hardwood trees) have larger xylem vessels that are more effective at conducting sap, xylem tissue constitutes a smaller fraction of the cross-section area of their trunks or branches, which is not ideal in the context of filtration. The long length of their xylem vessels also implies that a large thickness (centimeters to meters) of xylem tissue will be required to achieve any filtration effect at all – filters that are thinner than the average vessel length will just allow water to flow through the vessels without filtering it through pit membranes (Figure 1a). In contrast, gymnosperms (conifers, including softwood trees) have short tracheids that would force water to flow through pit membranes even for small thicknesses (<1 cm) of xylem tissue (Figure 1a). Since tracheids have smaller diameters and are shorter, they offer higher resistance to flow, but typically a greater fraction of the stem cross-section area is devoted to

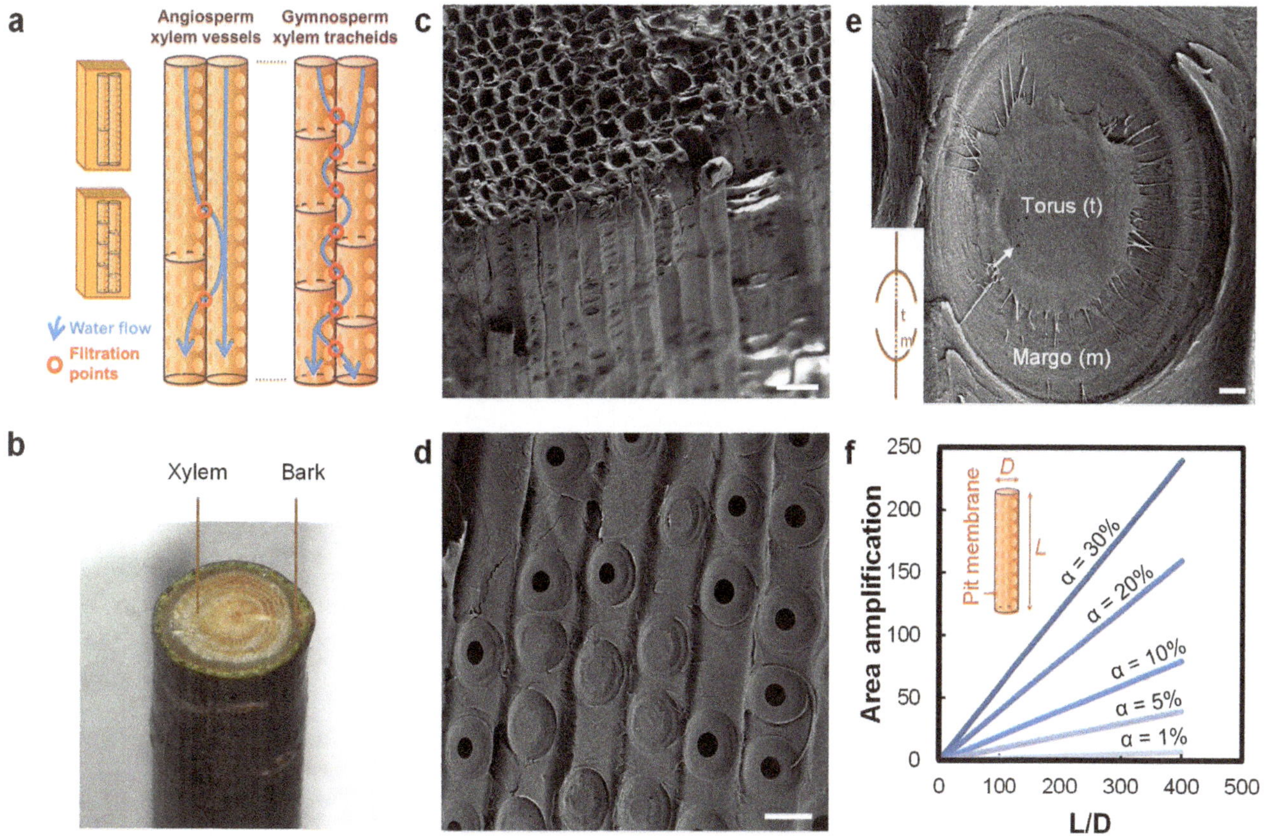

Figure 1. Xylem structure. a) Structure of xylem vessels in flowering plants and tracheids in conifers. Longer length of the vessels can provide pathways that can bypass filtration through pit membranes that decorate their circumference. b) Photograph of ~1 cm diameter pine (*pinus strobus*) branch used in the present study. c) Scanning electron microscope (SEM) image of cut section showing tracheid cross section and lengthwise profile. Scale bar is 40 μm. d) SEM image showing pits and pit membranes. Scale bar is 20 μm. e) Pit membrane with inset showing a cartoon of the pit cross-section. The pit cover has been sliced away to reveal the permeable margo surrounding the impermeable torus. Arrow indicates observed hole-like structures that may be defects. The margo comprises radial spoke-like structures that suspend the torus, which are only barely visible overlaying the cell wall in the background. Scale bar is 1 μm. f) Dependence of area amplification, defined as the pit membrane area divided by the nominal filter area, on the tracheid aspect ratio *L/D* and fractional area α occupied by pit membranes.

conducting xylem tissue. For example, in the pine branch shown in Figure 1b used in this study, fluid-conducting xylem constitutes the majority of the cross-section. This reasoning leads us to the conclusion that in general the xylem tissue of coniferous trees – i.e. the sapwood – is likely to be the most suitable xylem tissue for construction of a water filtration device, at least for filtration of bacteria, protozoa, and other pathogens on the micron or larger scale.

The resistance to fluid flow is an important consideration for filtration. Pits can contribute a significant fraction (as much as 30–80%) [7,8] of the resistance to sap flow, but this is remarkably small considering that pit membrane pore sizes are several orders of magnitude smaller than the tracheid or vessel diameter. The pits and pit membranes form a hierarchical structure where the thin, highly-permeable pit membranes are supported across the microscale pits that are arranged around the circumference of the tracheids (Figure 1a). This arrangement permits the pit membranes to be thin, offering low resistance to fluid flow. Furthermore, the parallel arrangement of tracheids with pits around their circumference provides a high packing density for the pit membranes. For a given tracheid with diameter D and length L, where pit membranes occupy a fraction α of the tracheid wall area, each tracheid effectively contributes a pit membrane area of $\pi DL\alpha/2$, where the factor of 2 arises as each membrane is shared by two tracheids. However, the nominal area of the tracheid is only $\pi D^2/4$, and therefore, the structure effectively amplifies the nominal filter area by a factor of $2\alpha(L/D)$ (Figure 1f). The images in Figure 1c indicate that typical $D \sim 10$–15 μm and $\alpha \sim 0.2$ yield an effective area amplification of ~20 for tracheid lengths of 1–2 mm. Therefore, for a filter made by cutting a slice of thickness $\sim L$ of the xylem, the actual membrane area is greater by a large factor due to vertical packing of the pit membranes. Larger filter thicknesses further increase the total membrane area, but the additional area of the membrane is positioned in series rather than in parallel and therefore reduces the flow rate, but potentially improves the rejection performance of the filter due to multiple filtration steps as shown in Figure 1a.

Construction of the Xylem Filter and Measurement of Flow Rate

The xylem filter device was constructed by simply peeling off the bark and cambium from a section of the pine branch and inserting it into a tube (Figure 2a). Although a simple tube fastener could provide a leak-tight seal between the tube and the xylem, we used epoxy to ensure that there was no inadvertent leakage. When deionized water was loaded into the tube above the xylem and subjected to pressure in the 0.5–5 psi (3.45 to 34.5 kPa) range, we found that water readily flowed through the xylem. The flow rate was proportional to applied pressure (Figure 2b), which allowed for the extraction of the hydrodynamic conductivity K (m^2 Pa^{-1} s^{-1}) of the filter, defined by

$$Q = KA\frac{\Delta P}{l} \quad (1)$$

where Q is the volumetric flow rate (in m^3 s^{-1}) under pressure difference ΔP across the filter, while l and A are the thickness and the cross-section area of the filter, respectively. The observed conductivities for three different filters were in the range of ~5–6×10^{-10} m^2 Pa^{-1} s^{-1} (Figure 2c), or equivalently, ~0.5–0.6 kg s^{-1} m^{-1} MPa^{-1} when defined with respect to mass flow rate of water.

Biologists have performed similar flow rate measurements by cutting a section of a plant stem under water, flushing to remove any bubbles, and applying a pressure difference to measure the flow rate [11,12]. Xylem conductivities of conifers [7] typically range from 1–4 kg s^{-1} m^{-1} MPa^{-1}, which compares very well with the conductivities measured in our experiments. Lower conductivities can easily result from introduction of bubbles [11] or the presence of some non-conducting heartwood. We can therefore conclude that the flow rate measurements in our devices are consistent with those expected from prior literature on conductivity of conifer xylem.

Filtration of Pigment Dye

After construction of the filter, we tested its ability to filter a pigment dye with a broad particle size distribution. The red color of the feed solution disappeared upon filtration (Figure 3a) indicating that the xylem filter could effectively filter out the dye.

Since the dye had a broad pigment size distribution, we investigated the size-dependence of filtration by quantifying the pigment size distribution before and after filtration using dynamic light scattering. We found that the feed solution comprised particles ranging in size from ~70 nm to ~500 nm, with some larger aggregates (Figure 3b). In contrast, the filtrate particle size distribution peaked at ~80 nm, indicating that larger particles were filtered out. In a separate experiment, we observed that 20 nm fluorescent polystyrene nanoparticles could not be filtered by the xylem filter, confirming this size dependence of filtration.

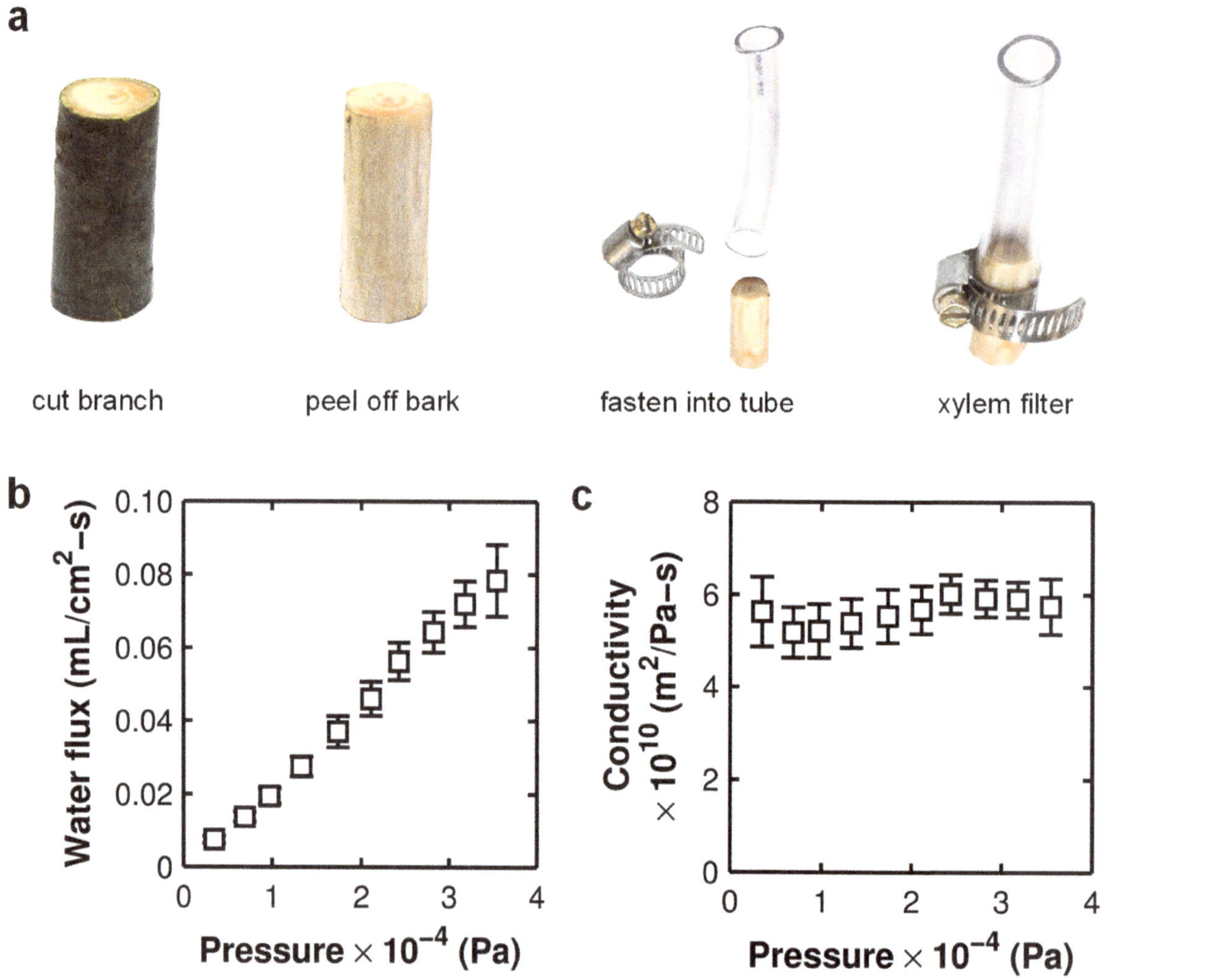

Figure 2. Xylem filter. a) Construction of xylem filter. b) Effect of applied pressure on the water flux through the xylem filter. c) Hydrodynamic conductivity of the filter extracted at each measured pressure using the total filter cross-section area and thickness as defined by Equation 1. Error bars indicate ±S.D. for measurements on three different xylem filters.

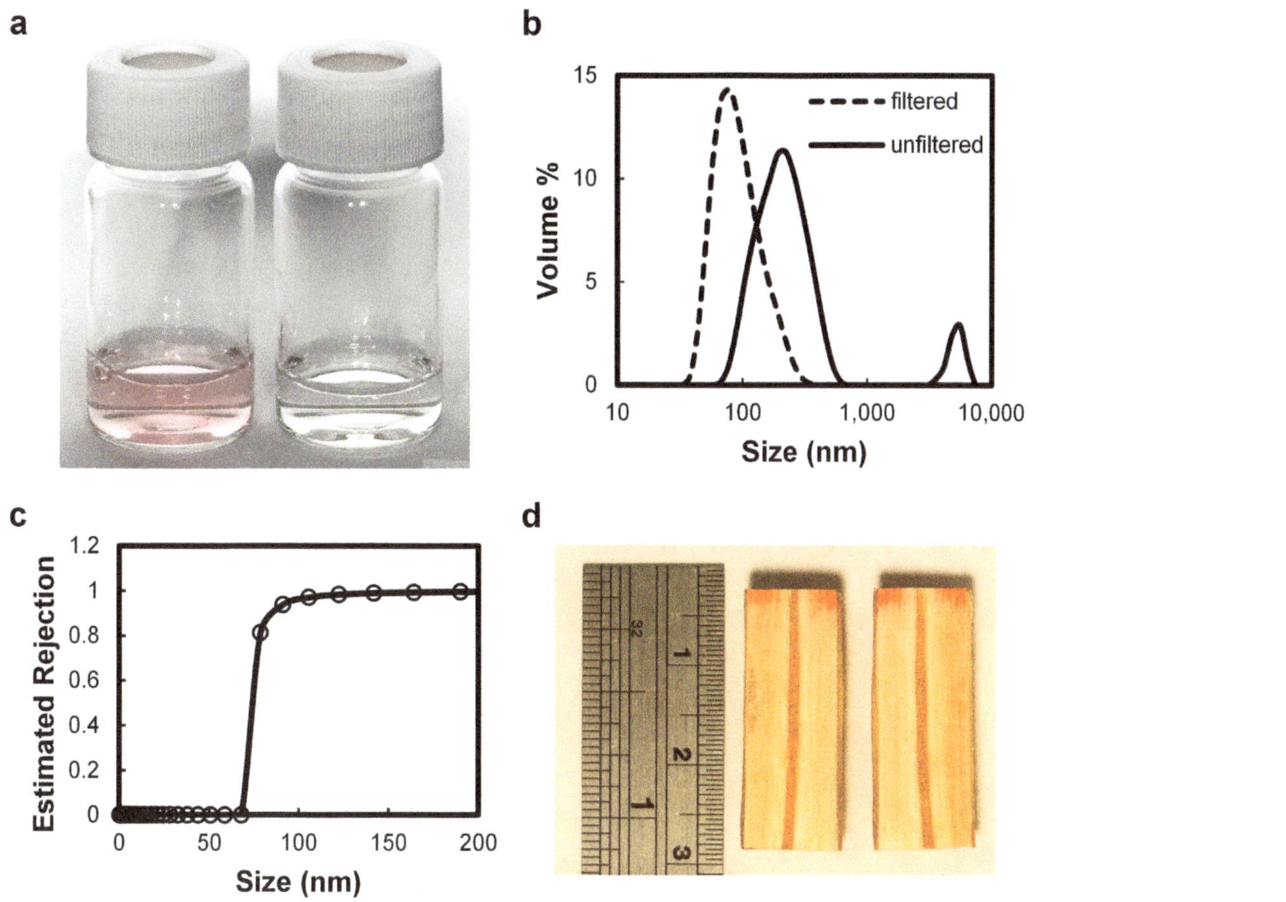

Figure 3. Filtration performance of the xylem filter. a) Feed solution of a pigment dye before filtration (left), compared to the filtrate (right). b) Size distribution of the pigment particles in the feed and filtrate solutions measured by dynamic light scattering. c) Dependence of the rejection on the particle size estimated from the data in (b). d) Cross-section of the xylem filter after filtration. Scale is in centimeters and inches.

Assuming that pigment particles 70 nm or less in size were not rejected, the size distributions before and after filtration enable calculation of the rejection performance of the xylem filter as a function of particle size (Figure 3c). We find that the xylem filter exhibits excellent rejection for particles with diameters exceeding 100 nm, with the estimated rejection exceeding 99% for particles over 150 nm. Smaller particles are expected to pass through the larger porosity of the pit membrane: SEM images in Figure 1e indicate sub-micron spacing between the radial spoke-like margo membrane through which the pigment particles can pass, although the porosity is difficult to resolve in the SEM image.

After filtration, we cut the xylem filter parallel to the direction of flow to investigate the distribution of dye in the filter. We observed that the dye was confined to the top 2–3 millimeters of the xylem filter (Figure 3d), which compares well with the tracheid lengths on the millimeter scale expected for coniferous trees [7]. These results show that the majority of the filtration occurred within this length scale, and suggests that the thickness of the xylem filter may be reduced to below 1 cm while still rejecting the majority of the dye.

Filtration of Bacteria from Water

Finally, we investigated the ability of the xylem filter to remove bacteria from water. As a model bacterium, we used fluorescently labeled and inactivated *Escherichia coli* bacteria that are cylindrical in shape with a diameter of ~1 µm. Use of fluorescently labeled bacteria enabled easy enumeration of their concentrations, and also allowed us to track the location in the xylem filter where they were trapped. Since filtration is dominated by size-exclusion at this length scale, we do not expect modification with the dye to significantly affect filtration characteristics. Filtration using three different xylem filters showed nearly complete rejection of the bacteria (Figure 4a). Using a hemacytometer to count the bacteria, we estimate that the rejection was at least 99.9%.

To investigate the mechanism of filtration, the xylem filter was cut parallel to the direction of flow after filtration. When examined under a fluorescence microscope, we observed that the bacteria accumulated over the donut-shaped margo pit membranes (Figure 4b). This distribution is consistent with the expectation that the bacteria are filtered by the porous margo of the pit membranes. The distribution of trapped bacteria was not uniform across the cross section of the filter. Similar to the case of the dye, bacteria were observed only within the first few millimeters from the end through which the solution was infused (indicated by the white arrow in Figure 4c). In addition, the low-magnification fluorescence image shows that the bacteria had accumulated primarily over pit membranes at the bottom of the tracheids, which is again not unexpected. Further investigation by SEM clearly showed individual bacterial cells accumulated on the pit membranes over the porous margo (Figure 4d,e). These results confirm the pit membranes as the functional units that provide the filtration effect in the xylem filter.

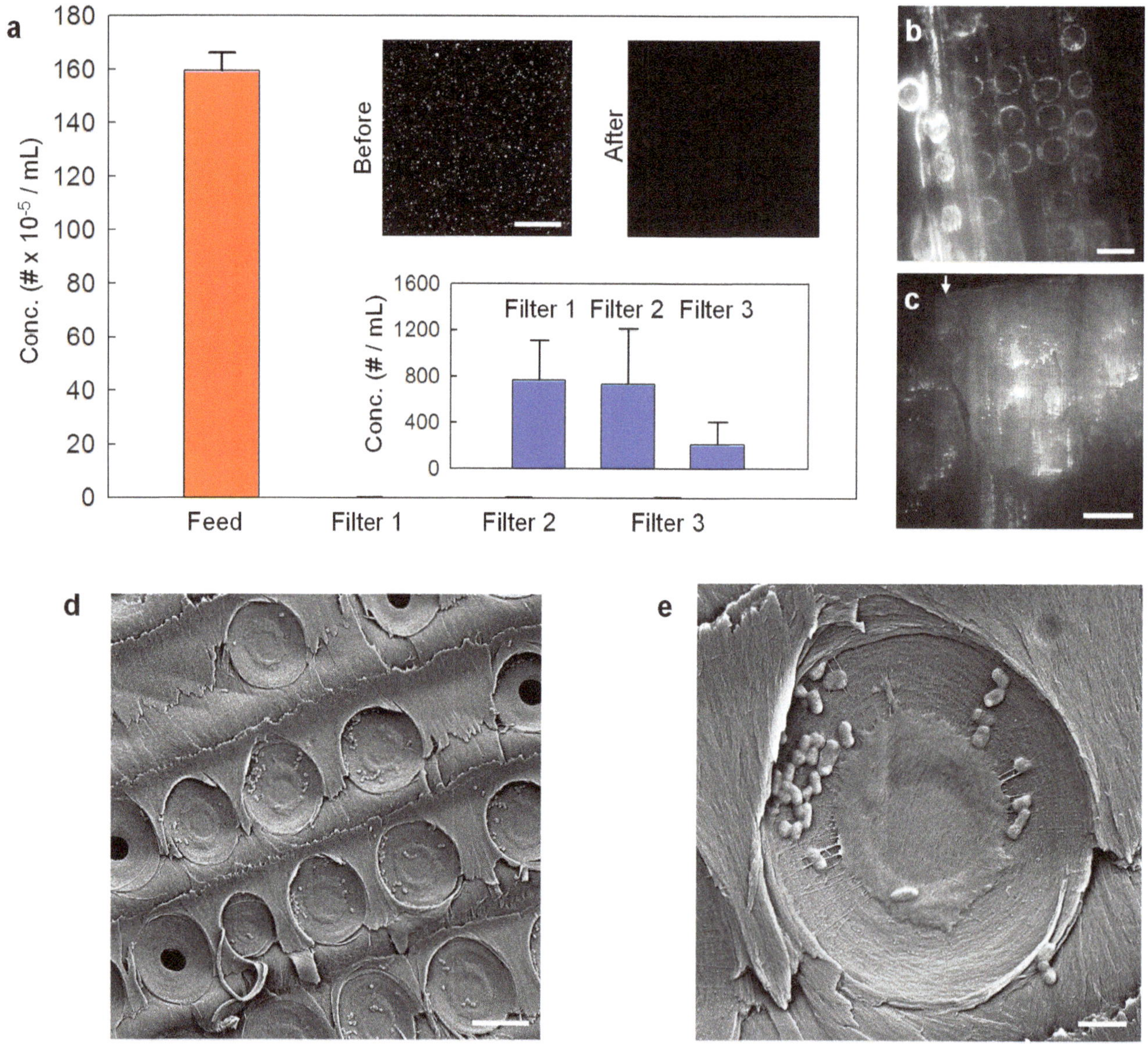

Figure 4. Filtration of model bacteria by the xylem filter. a) Concentrations of bacteria in the feed and filtrate solutions. Inset shows fluorescence images of the two solutions. Scale bar is 200 μm. Error bars indicate ±S.D. for experiments performed on three different xylem filters. b) Fluorescence image of xylem filter cross-section showing accumulation of bacteria over the margo pit membranes. Scale bar is 20 μm. c) Low-magnification fluorescence image shows that bacteria are trapped at the bottoms of tracheids within the first few millimeters of the top surface. Scale bar is 400 μm. Arrow indicates top surface of the xylem filter and also the direction of flow during filtration. Autofluorescence of the xylem tissue also contributes to the fluorescence signal in (b) and (c). d), e) SEM images showing bacteria accumulated on the margo pit membranes after filtration. Scale bars are 10 μm and 2 μm, respectively.

Discussion

Wood has been investigated recently as a potential filtration material [13], showing moderate improvement of turbidity. While we showed filtration using freshly cut xylem, we found that the flow rate dropped irreversibly by over a factor of 100 if the xylem was dried, even when the xylem was flushed with water before drying. We also examined flow through commercially available kiln-dried wood samples cut to similar dimensions. Wood samples that exhibited filtration showed two orders of magnitude smaller flow rates than in the fresh xylem filter, while those that had high flow rates did not exhibit much filtration effect and seemed to have ruptured tracheids and membranes when observed under SEM. Wetting with ethanol or vacuuming to remove air did not significantly increase the flow rate in the wood samples that exhibited the filtration effect, suggesting that the pit membranes may have a tendency to become clogged during drying. These results are consistent with literature showing that the pit membranes can become irreversibly aspirated against the cell wall, blocking the flow [14]. In fact, the pit membranes in the SEM images (Figure 1d,e and Figure 4d,e), which were acquired after drying the samples, appear to be stuck to the walls. Regardless, our results demonstrate that excellent rejection (> 99.9%) of bacteria is possible using the pit membranes of fresh plant xylem, and also provide insight into the mechanism of filtration as well as guidelines for selection of the xylem material.

Peter-Varbanets et al. [2] have outlined the key requirements for point-of-use devices for water disinfection: a) performance (ability to effectively remove pathogens), b) ease of use (no time-consuming maintenance or operation steps), c) sustainability (produced locally with limited use of chemicals and non-renewable energy), and d) social acceptability. Meeting all of these requirements has proved to be challenging, but point-of-use methods that have been successfully used for low-cost water treatment in developing countries include free-chlorine/solar disinfections, combined coagulant-chlorine disinfection, and biosand/ceramic filtrations [5]. While chlorine is a very effective biocide, its reaction with organic matter can produce carcinogenic by-products [15] and some waterborne pathogens such as *Cryptosporidium parvum* and *Mycobacterium avium* are resistant to the chlorine [16]. Solar disinfection based on ultraviolet irradiation can effectively inactivate *C. parvum*, but this requires low turbidity of source water [17] and is not effective for control of viruses [16]. Filtration based on biosand and ceramic filters is also effective at removing pathogens, but the effectiveness against viruses is low or unknown [18]. Coagulation combined with chlorine disinfection removes or inactivates viruses and pathogens effectively. However, necessity of an additional filtration step and relatively high cost are potential barriers for practical use [18]. Among these methods, a review on field studies by Sobsey et al. [5] suggested that biosand and ceramic filtration are the most effective methods in practice, because once the apparatus is installed, the effort for use and dosage is significantly reduced and therefore promotes persistent use compared to disinfection approaches. Although membrane-based filtration is the most widely used for water treatment in industrialized nations and the cost of membranes has significantly decreased, membranes are still unaffordable to poor communities in the developing world [2]. Ultrafiltration systems run by hydrostatic pressure [19] and some recently invented point-of-use devices using ultrafiltration membranes may provide water to developing regions at reasonable cost [2]. However, membranes still require specialized chemicals and processes for manufacture, and need cleaning or replacement.

Xylem filter technology could be an attractive option for low-cost and highly efficient point-of-use water treatment by filtration, overcoming some of the challenges associated with conventional membranes. Xylem filters could provide the advantage of reduced human effort compared to existing point-of-use water treatment options, requiring only simple periodic filter replacement. In addition, the pressures of 1–5 psi used here are easily achievable using a gravitational pressure head of 0.7–3.5 m, implying that no pumps are necessary for filtration. The measured flow rates of about 0.05 mL/s using only $\sim$1 cm^2 filter area correspond to a flow rate of over 4 L/d, sufficient to meet the drinking water requirements of one person [20]. This is comparable to chlorination and biosand filtration, which have the highest production rates of prevalent point-of-use water treatment methods, and far exceeds typical production rates for solar disinfection. Xylem filters could potentially be produced locally and inexpensively, and disposed of easily owing to their biodegradability. The high flow rates and low cost would certainly help address the issues of maintenance and replacement. For example, 200 filters of 10 cm^2 area and 0.5 cm thickness could be packaged into a volume of about 1 L, which will be inexpensive and last a few years even with weekly replacement. Furthermore, as suggested by the dye filtration experiment, xylem filters should be able to significantly reduce water turbidity, enhancing the aesthetic qualities of the drinking water, which is hardly achieved by chlorination and solar disinfection.

Wood is an easily available material. While use of fresh xylem does not preclude its use as a filter material, dried membranes have definite practical advantages. Therefore, the process of wood drying and its influence on xylem conductivity needs further study. In particular, processes that yield intact yet permeable xylem tissues that can be stored long-term will be helpful for improving the supply chain if these filters are to be widely adopted. In addition, flow through xylem of different plants needs to be studied to identify locally available sources of xylem, which will truly enable construction of filters from locally available materials. In the present study, we report results using xylem derived from only one species. These xylem filters could not filter out 20 nm nanoparticles, which is a size comparable to that of the smallest viruses. It will be interesting to explore whether there exist any coniferous species that have pit membranes with smaller pore sizes that can filter out viruses, or whether conifer xylem can be impregnated with particles such as carbon black to improve rejection of viruses. In their absence, angiosperms with short vessels that yield practical filter lengths may be the best alternative due to their smaller pit membrane pore sizes [8]. Further exploration of xylem tissues from different plants with an engineering perspective is needed to construct xylem filters that can effectively reject viruses, exhibit improved flow rates, or that are amenable to easy storage. It is also conceivable that plants could be selected or developed for enhanced filtration characteristics, as has been the norm in agriculture for enhancement of many desirable characteristics including resistance to pests, flavor, or productivity.

Conclusions

Plant xylem is a porous material with membranes comprising nanoscale pores. We have reasoned that xylem from the sapwood of coniferous trees is suitable for disinfection by filtration of water. The hierarchical arrangement of the membranes in the xylem tissue effectively amplifies the available membrane area for filtration, providing high flow rates. Xylem filters were prepared by simply removing the bark of pine tree branches and inserting the xylem tissue into a tube. Pigment filtration experiments revealed a size cutoff of about 100 nm, with most of the filtration occurring within the first 2–3 mm of the xylem filter. The xylem filter could effectively filter out bacteria from water with rejection exceeding 99.9%. Pit membranes were identified as the functional unit where actual filtration of the bacteria occurred. Flow rates of about 4 L/d were obtained through $\sim$1 cm^2 filter areas at applied pressures of about 5 psi, which is sufficient to meet the drinking water needs of one person. The simple construction of xylem filters, combined with their fabrication from an inexpensive, biodegradable, and disposable material suggests that further research and development of xylem filters could potentially lead to their widespread use and greatly reduce the incidence of waterborne infectious disease in the world.

Acknowledgments

The authors thank Yukiko Oka for assistance with preparation of illustrations and Sunandini Chopra for help with dynamic light scattering measurements.

Author Contributions

Conceived and designed the experiments: MSHB JL VV VC RK. Performed the experiments: MSHB JL VV VC. Analyzed the data: MSHB JL RK. Contributed reagents/materials/analysis tools: VC. Wrote the paper: MSHB JL RK.

References

1. Gadgil A (1998) Drinking water in developing countries. Annual Review of Energy and the Environment 23: 253–286.
2. Peter-Varbanets M, Zurbrügg C, Swartz C, Wouter Pronk W (2009) Decentralized systems for potable water and the potential of membrane technology. Water Research 43: 245–265.
3. World Health Organization. Available: http://www.who.int/water_sanitation_health/mdg1/en/. Accessed 2014 Jan 14.
4. Madaeni SS (1999) The application of membrane technology for water disinfection. Water Research 33: 301–308.
5. Sobsey MD, Stauber CE, Casanova LM, Brown JM, Elliott MA (2008) Point of use household drinking water filtration: A practical, effective solution for providing sustained access to safe drinking water in the developing world. Environmental Science and Technology 42: 4261–4267.
6. Loo SL, Fane AG, Krantz WB, Lim TT (2012) Emergency water supply: A review of potential technologies and selection criteria. Water Research 46: 3125–3151.
7. Sperry JS (2003) Evolution of water transport and xylem structure. International Journal of Plant Sciences 164: S115–S127.
8. Choat B, Cobb AR, Jansen S (2008) Structure and function of bordered pits: new discoveries and impacts on whole-plant hydraulic function. New Phytologist 177: 608–625.
9. Jansen S, Choat B, Pletsers A (2009) Morphological variation of intervessel pit membranes and implications to xylem function in angiosperms. American Journal of Botany 96: 409–419.
10. Choat B, Ball M, Luly J, Holtum J (2003) Pit membrane porosity and water stress-induced cavitation in four co-existing dry rainforest tree species. Plant Physiology 131: 41–48.
11. Espino S, Schenk HJ (2011) Mind the bubbles: achieving stable measurements of maximum hydraulic conductivity through woody plant samples. Journal of Experimental Botany 62: 1119–1132.
12. van Ieperen W, van Meeteren U, van Gelder H (2000) Fluid ionic composition influences hydraulic conductance of xylem conduits. Journal of Experimental Botany 51: 769–776.
13. Sens ML, Emmendoerfer ML, Muller LC (2013) Water filtration through wood with helical cross-flow. Desalination and Water Treatment.
14. Petty JA (1972) Aspiration of bordered pits in conifer wood. Proceedings of the Royal Society Series B-Biological Sciences 181: 395-+.
15. Dunnick J, Melnick R (1993) Assessment of the carcinogenic potential of chlorinated water - Experimental studies of chlorine, chloramine, and trihalomethanes. Journal of the National Cancer Institute 85: 817–822.
16. Shannon M, Bohn P, Elimelech M, Georgiadis J, Marinas B, et al. (2008) Science and technology for water purification in the coming decades. Nature 452: 301–310.
17. Mintz E, Bartram J, Lochery P, Wegelin M (2001) Not just a drop in the bucket: Expanding access to point-of-use water treatment systems. American Journal of Public Health 91: 1565–1570.
18. Lantagne DS, Quick R, Mintz ED (2006) Household water treatment and safe storage options in developing countries: A review of current implementation practices. Woodrow Wilson International Center for Scholars. Environmental Change and Security Program.
19. Arnal J, Fernandez M, Verdu G, Garcia J (2001) Design of a membrane facility for water potabilization and its application to Third World countries. Desalination 137: 63–69.
20. Reed B, Reed B (2011) How much water is needed in emergencies. World Health Organization. Available: http://www.who.int/water_sanitation_health/publications/2011/tn9_how_much_water_en.pdf. Accessed 2014 Jan 14.

Assessing the Impact of Water Filters and Improved Cook Stoves on Drinking Water Quality and Household Air Pollution: A Randomised Controlled Trial in Rwanda

Ghislaine Rosa[1]*, Fiona Majorin[1], Sophie Boisson[1], Christina Barstow[2], Michael Johnson[3], Miles Kirby[1], Fidele Ngabo[4], Evan Thomas[5], Thomas Clasen[1,6]

1 Department of Disease Control, London School of Hygiene and Tropical Medicine, London, United Kingdom, **2** Department of Civil, Environmental and Architectural Engineering, University of Colorado at Boulder, Boulder, Colorado, United States of America, **3** Berkeley Air Monitoring Group, Berkeley, California, United States of America, **4** Ministry of Health, Government of Rwanda, Kigali, Rwanda, **5** Department of Mechanical and Materials Engineering, Portland State University, Portland, Oregon, United States of America, **6** Department of Environmental Health, Emory University, Atlanta, Georgia, United States of America

Abstract

Diarrhoea and respiratory infections remain the biggest killers of children under 5 years in developing countries. We conducted a 5-month household randomised controlled trial among 566 households in rural Rwanda to assess uptake, compliance and impact on environmental exposures of a combined intervention delivering high-performance water filters and improved stoves for free. Compliance was measured monthly by self-report and spot-check observations. Semi-continuous 24-h $PM_{2.5}$ monitoring of the cooking area was conducted in a random subsample of 121 households to assess household air pollution, while samples of drinking water from all households were collected monthly to assess the levels of thermotolerant coliforms. Adoption was generally high, with most householders reporting the filters as their primary source of drinking water and the intervention stoves as their primary cooking stove. However, some householders continued to drink untreated water and most continued to cook on traditional stoves. The intervention was associated with a 97.5% reduction in mean faecal indicator bacteria (Williams means 0.5 *vs.* 20.2 TTC/100 mL, $p<0.001$) and a median reduction of 48% of 24-h $PM_{2.5}$ concentrations in the cooking area ($p=0.005$). Further studies to increase compliance should be undertaken to better inform large-scale interventions.

Trial registration: Clinicaltrials.gov; NCT01882777; http://clinicaltrials.gov/ct2/results?term = NCT01882777&Search = Search

Editor: Robert K. Hills, Cardiff University, United Kingdom

Funding: This study was funded by DelAgua Health, a for-profit company that implements the intervention in Rwanda in conjunction with the Rwanda Ministry of Health. The funders had no role in study design, data collection and analysis, decision to publish, or preparation of the manuscript.

Competing Interests: We have the following interests: This study was funded by DelAgua Health, a for-profit company that implements the program in Rwanda in conjunction with the Rwanda Ministry of Health. Evan Thomas and Christina Barstow, co-authors of this article, are compensated consultants to DelAgua Health, and are in charge of overseeing the implementation of the program in Rwanda. Michael Johnson, co-authors of this article, is employed by the commercial company, Berkeley Air Monitoring Group, which was contracted to provide advice on household air pollution monitoring. This does not alter our adherence to all the PLoS ONE policies on sharing data and materials.

* E-mail: ghislaine.rosa@lshtm.ac.uk

Introduction

Environmental contamination at the household level is a major cause of death and disease, particularly among rural populations in low-income countries. Unsafe drinking water, together with poor sanitation, account for an estimated 0.9% of the global burden of disease and 0.3 million deaths [1]. Much of this disease burden is associated with diarrhoea, which alone accounts for 10.5% of deaths in children under 5 years in low-income countries [2]. Household air pollution (HAP) from biomass fuel smoke has been linked to increased risk of respiratory tract infections, low birth weight, exacerbations of inflammatory lung conditions, cardiac events, stroke, eye disease, tuberculosis, cancer and nutritional deficiencies [3]. The Global Burden of Disease (GBD) 2010 project found HAP from solid fuels to be responsible for 3.5 million premature deaths globally [1]. In this same assessment, smoke from household cooking fuels was also responsible for another half a million premature deaths due to contributions to outdoor air pollution [1]. These environmental hazards are aggravated among rural inhabitants of sub-Saharan Africa who are more likely to rely on unsafe water supplies and cook using biomass fuels on inefficient stoves [4–6].

Inefficient cookstoves also present substantial economic, developmental and environmental costs. At the household level, poverty is exacerbated and time spent at school is reduced by the burden of collecting more fuel for boiling drinking water and cooking [7]. Individuals, households and governments bear the cost of expenditures for seeking treatment of enteric and respiratory infections. Cookstove emissions also contribute to greenhouse gas and black carbon emissions, and in some cases the fuel harvesting can result in denuding of forests [8,9].

With a population of 10.5 million and a density of 412 persons per sq. km, Rwanda is the most densely populated country in East Africa [10]. Eighty per cent of the population of Rwanda lives in rural areas and is engaged in agriculture [11]. Despite significant progress over the last decade, 57% of the population is living below the poverty line, 37% of them living in extreme poverty [11]. While a large proportion of the rural population has access to improved water sources (71.2%), mainly through protected springs, only 2.2% of rural areas have water on their premises [12], resulting in an increased risk for drinking water contamination during transport and storage [13]. Almost all of rural Rwanda (99.0%) relies on biomass for their cooking needs [12]. Morbidity and mortality are largely dominated by communicable diseases, including HIV/AIDS, acute respiratory infections, diarrhoeal diseases, intestinal parisitoses, and malaria [14]. Among deaths of children under 5, pneumonia accounts for 20% and diarrhoea for 12% [15].

In an effort to reduce the disease burden in rural Rwanda, decrease poverty associated with expenditures for fuel, and minimize the impact of greenhouse gases from inefficient combustion of biomass in low-efficiency stoves, the Rwanda Ministry of Health (MiniSante) and the Rwanda Environmental Management Authority (REMA) have partnered with DelAgua Health (implementer) to design, deploy and evaluate the impact of a project that will deliver and promote the use of advanced water filters and high efficiency cookstoves to lower-income households in Rwanda. Prior to initiating the full campaign, the implementer with the Ministry of Health undertook a pilot distribution of filters and cookstoves to approximately 2200 households in 15 villages in 11 of the country's 30 districts. We conducted this study in three of those villages in order to assess the uptake of the intervention and its impact on drinking water quality and household air pollution.

Methods

Study setting

The study was conducted from September 2012 to April 2013 in three rural villages, Nyarutovu and Kabuga located in Muhanga district, Southern province; and Rubona, located in Gakenke district, Northern province. These villages were purposely selected from the 15 villages comprising the pilot distribution phase. The sites were changed from the original protocol, Karongi and Ngororero districts in Western province, to accommodate access to better microbiology laboratory facilities in Kigali.

Study design and sample size

The study employed a parallel, household-randomised, control trial design with a 1:1 ratio. This trial followed a non-blinded design because previous attempts to blind an earlier version of the LifeStraw Family filter in the Democratic Republic of Congo were unsuccessful [16]. The objectives of the study were to assess (i) uptake and use of the intervention by the target population when delivered programmatically, and (ii) the impact of the intervention on the microbiological quality of household drinking water and air quality near the self-reported cooking area over the 5-month follow-up period. Our primary outcomes were (i) to assess levels of faecal contamination (measured by thermotolerant coliforms, TTC) in stored water in the home that householders used for drinking, and (ii) to determine average 24-h concentrations of $PM_{2.5}$ in the main cooking area as identified by participants. Our secondary outcome was to assess use of the intervention filters and stoves based on self-report and spot-check observations.

The sample size calculation was based on $PM_{2.5}$ emissions reductions rather than TTC reductions in drinking water as the former was determined to require a larger sample. Assuming a 50% reduction in $PM_{2.5}$ emissions, 80% power, $\alpha = 0.05$ and a coefficient of variation (COV) of 1, we estimated a sample size of 63 households per arm.

The protocol of this trial and CONSORT checklist are available as supporting information; see Text S1 and S2.

Intervention

Each intervention household received one LifeStraw Family 2.0 filter and one EcoZoom Dura improved wood burning stove. The filter is the second-generation of a gravity-based water purifier that uses ultrafiltration in the form of a hollow-fibre cartridge to remove pathogens from drinking water. The first generation device has been shown in field studies to be highly effective in improving water quality and to achieve consistent (though not exclusive) use [16]. The second-generation version used in this study employs a table-top design and an integrated safe storage vessel. Untreated water is poured through a 20-μm pre-filter plastic mesh into a 6.0 L container; over time, gravity forces the water through the cartridge comprised of hollow-fibres with a 20-nm pore size. The water then passes into a 5.5 L storage vessel where it can be dispensed via a plastic tap. The device is cleaned daily by backwashing the cartridge using a squeeze-pump mounted on the back of the storage container. The device is designed to treat 18,000 L of water [17] with a flow rate of approximately 3 L per hour. In the laboratory, the filter cartridge was found to meet the USEPA standards for microbiological water purifiers by reducing bacteria by 6 logs, viruses by 5 logs and protozoa by 4 logs [18]. The filter meets the "highly protective" World Health Organization (WHO) rating for household water treatment technologies [19].

The intervention stove is based on the 'rocket' concept that uses an internal 'chimney' in the stove that directs air through the burning fuel (usually biomass), and encourages the mixing of gases and flame above it. Precise internal stove dimensions are used to achieve high combustion efficiency and transfer heat to the cooking pot. Two additional components are included with the stove, a "stick support" onto which fuel wood is placed to promote airflow and a "pot skirt" which increases fuel efficiency. A study comparing cookstoves in Uganda, Kenya and Tanzania reported that the EcoZoom (aka StoveTec) stove saved 39% to 54% of fuel compared to open fires, cooked meals faster, and was participants' most preferred stove during controlled cooking of local dishes [20–22]. In the intervention group, householders were encouraged to cook outdoors on the EcoZoom stove and to use dry wood only to increase the efficiency of the stove. Further details on the messaging used in the pilot distribution can be found elsewhere [23].

Houses that were allocated to the intervention group also received a poster with illustrations and instructions in *Kinyarwanda*, the local language, on filter and stove use, maintenance, and contact names and phone numbers for the implementer. Most households had easy access to a cellular phone for contacting the implementer. Intervention households received one-to-one training on use and maintenance in their homes by community health workers (CHWs) who were previously trained by trainers who themselves had been trained by the filter and stove manufacturers and implementer. Intervention households were then visited periodically at approximately one-month intervals by CHWs to refresh health messaging and encourage use. Households allocated to the control group were instructed to continue usual practices throughout the study. At the end of the study in April 2013, these control households received their own filters, stoves, posters and training.

Enrolment, baseline survey, randomisation and deployment of devices

Households were eligible to participate in the study if (i) they were registered as being members of the village, (ii) the head of the household was over 18 years, and (iii) no members of the household worked as a CHW. The last criterion was included after the original protocol was drafted as at the time of the design the researchers were not aware that the CHW that would deliver the intervention resided in the villages selected for the study. It was explained that while all participating households would receive filters and stoves, half would receive them at the outset of the study and the balance at the conclusion of the 5-month follow-up period. After obtaining consent from the heads of participating households, a baseline survey was undertaken in September-November 2012 to collect information on demographics, socio-economic characteristics, water, hygiene and sanitation practices as well as fuel and cooking practices. Data collection tools were translated into *Kinyarwanda* and piloted before use.

Following the baseline survey, a public lottery was organised by the implementer and research teams during a village meeting to randomly allocate an approximately equal number of households from each village to intervention or control groups. Local authorities and village chiefs were extensively engaged to assess the suitability of this randomisation approach. After the lottery, members of control households were invited to leave the venue while those of intervention households attended a demonstration on the use and maintenance of the filter and stove, collected their devices, and carried them to their homes.

Outcome assessment

Compliance. Monthly cross-sectional surveys were conducted by trained field investigators (the evaluation team) working independently of the implementation team at unannounced visits among each household. At each visit participants were asked to identify the main drinking water container in the household, whether it was the intervention filter or another container; the surveyor also recorded whether the filter contained water at the time of the visit, a possible objective indicator of filter use. The field investigators also observed the cookstove and if cooking was taking place at the time of the unannounced visit, recorded where and whether such cooking was on the intervention stove or the traditional stove. If no cooking was taking place, field investigators noted the presence of smoke marks on the intervention stove, a possible objective indicator of use. Reported measures of stove use were also collected by asking participants what stove had been used the last time cooking took place in their home.

Independent to our study, the implementers undertook a separate survey, conducted by Environmental Health Officers (EHOs), to assess use and acceptability of the intervention for their own monitoring and evaluation purposes. The details of this assessment have been presented elsewhere [23].

Additionally, to assess use of the intervention in a more objective manner, remotely reporting electronic sensors were mounted onto 23 intervention filters and 27 intervention stoves and deployed in a randomly selected sub-sample of intervention households for a two-week period. The details of the implementation of this nested study, data handling and analysis, and results are presented elsewhere [24].

Water quality. During each of the five monthly visits, field investigators took a sample from the water container identified by the householder as being used mainly for drinking by children under 5 years of age, or adults if no under 5 s resided in the household. If this was other than directly from the intervention filter, a second sample was taken directly from the filter if it contained water. All water samples were collected in sterile Whirl-Pak bags (Nasco, Fort Atkinson, WI) containing a tablet of sodium thiosulphate to neutralize any halogen disinfectant. Samples were placed on ice and processed within 6 h of collection to assess levels of TTC. Microbiological assessment was performed using the membrane filtration technique [25] on membrane lauryl sulphate medium (Oxoid Limited, Basingstoke, Hampshire, UK) using a DelAgua field incubator (Robens Institute, University of Surrey, Guilford, Surrey, UK).

Household air pollution. Monitoring of particulate matter with an aerodynamic diameter <2.5 µm ($PM_{2.5}$) in the main cooking area took place between November 2012 and March 2013. 126 households (63 control and 63 intervention households) were randomly selected for semi-continuous 24-h $PM_{2.5}$ monitoring. Households were numbered and selected by using a computerised random number generator. Upon arrival at the participant's home, the family member mainly responsible for cooking was identified and a short survey was employed to identify the area in the household where cooking primarily took place. "Stacking" of stoves (using different stoves, often in different locations) [26] was a common scenario, both in control and intervention households, though more common in the latter. In cases where the participant reported cooking equally in two locations or with two or more stoves, we sampled from indoor rather than outdoor locations and from traditional rather than intervention stove. UCB-PATS $PM_{2.5}$ monitors (described below) were placed 1.5 m above the ground and 1 m away from the stove and, whenever possible, at least 1.5 m from windows and doors by suspending the monitors from the roof beams. When cooking was reported to take place outdoors, the $PM_{2.5}$ monitor was mounted onto a vertical wooden stand and placed at the same distance and height from the stove. The location of the stand was marked on the floor and participants were advised not to touch or move the equipment.

$PM_{2.5}$ was measured using the University of California, Berkeley Particle and Temperature Sensor (UCB-PATS^{TM}), (Berkeley Air Monitoring Group, USA), a semicontinuous (1-min averages), light-scattering nephelometer [27,28]. Laboratory and field validations of the UCB-PATS have been described previously [27–30]. To take into account that nephelometer sensitivity is a function of an aerosol's specific optical properties such as size, colour, and shape [31], calibration of the UCB response with the target aerosol was undertaken by conducting 24-h $PM_{2.5}$ gravimetric co-location measurements in a sub-sample of homes ($n = 30$). Five field blanks were obtained, resulting in an adjustment of subtracting 5 µg to the final filter masses (<1% of the mean mass deposition). The UCB-PATS response was then linearly regressed against the gravimetric samples ($n = 27$, $R^2 = 0.86$), with the resulting equation then used to adjust the UCB-PATS response to the gravimetric measures (Figure S1 of supporting information). Three gravimetric samples were omitted due to incomplete sampling durations.

Gravimetric $PM_{2.5}$ samples were collected using standard air sampling pumps (PXR8, SKC Inc., USA) with $PM_{2.5}$ cyclones (SCC 1.062, BGI, USA) using a flow rate of 1.5 L/min. Flow rates were measured before and after installation of the sampling equipment in the home with a rotameter (Matheson Trigas, Montgomeryville, PA, USA) that had been calibrated using a TSI Flow Calibrator 4146 (TSI, Inc., USA). $PM_{2.5}$ was collected on 37-mm Teflon filters (Pall, USA). Filters were stored at 4°C until shipment to Berkeley Air Monitoring Group in California, USA for weighing. Filters were equilibrated for 24 h at 22±3°C and 40±5% relative humidity before being weighed on a 0.1

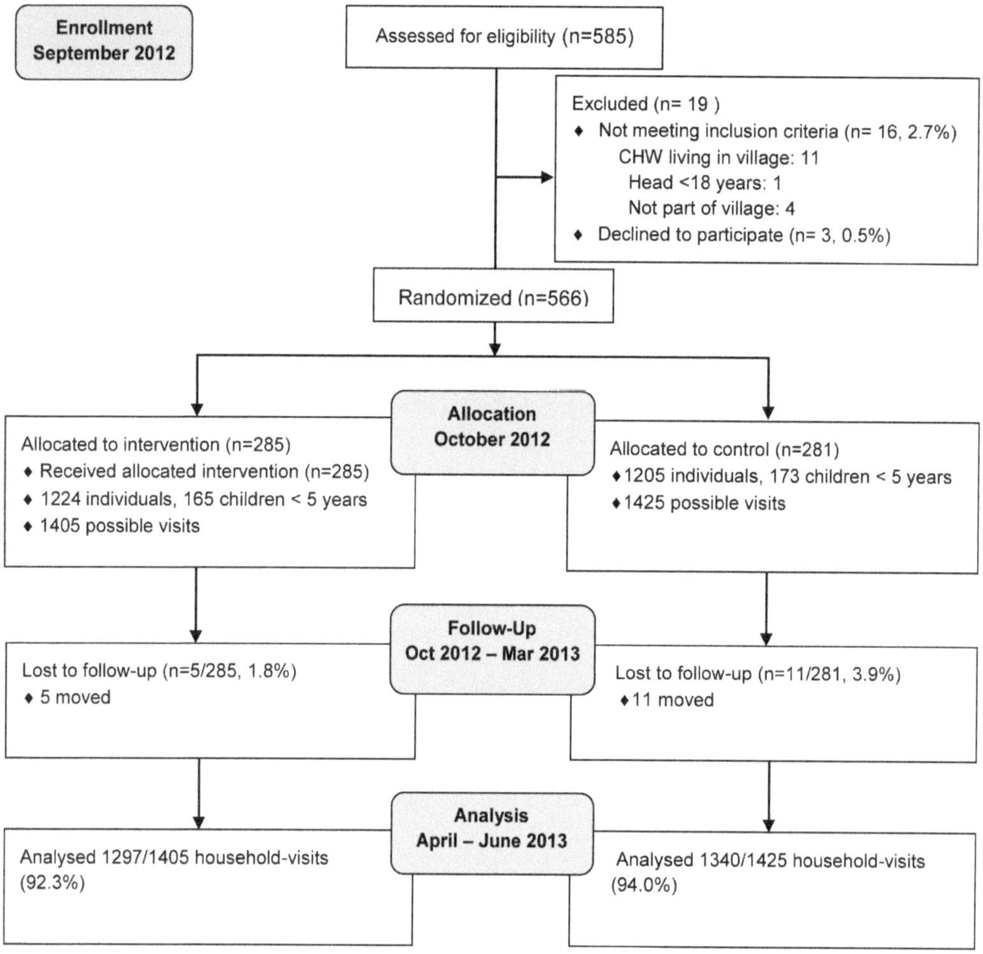

Figure 1. CONSORT diagram showing the flow of participants through the trial.

microgram resolution electro microbalance (XP2U, Mettler Toledo, USA).

Data analysis

All data were analysed using Stata 12 (Stata Corporation, College Station, TX, USA). Because both $PM_{2.5}$ concentrations and TTC counts in drinking water followed non-normal distributions, medians, geometric means and Williams means are presented together with arithmetic means. The Williams mean is calculated by adding 1 to all the data values, then taking the geometric mean, then subtracting 1 again [32]. Categorical data were compared using a Chi square or a Fisher's exact test where appropriate. The non-parametric Wilcoxon rank sum test was used to compare $PM_{2.5}$ concentrations in the main cooking area between intervention and control groups. To assess the effect of the intervention on water quality, TTC counts during follow-up were compared using random effects negative binomial regression as describe elsewhere [33] to account for (i) repeated observations within households and, (ii) the skewed distribution of the TTC counts. Model comparison was assessed by using the Bayesian information criterion (BIC), which is a well-established measure of goodness of fit that also applies to non-nested models [33,34]. For the purpose of analysis, plates that yielded coliform forming units (CFUs) that were too numerous to count (TNTC) were assigned a value of 300 TTC/100 mL. Data were analysed in an intention-to-treat basis in order to estimate the effect of the intervention regardless of compliance. Only those households with complete follow-up data were analysed.

Ethics

The study was reviewed and approved by the ethics committee at the London School of Hygiene and Tropical Medicine (No. 6239, as amended) and the Rwanda National Ethics Committee (No. 328 RNEC/2012). Written informed consent to participate in the research was obtained from the male or female head or the wife of each participating household.

Results

Study population

The three villages participating in the study comprised 585 households, all of which were screened to participate in the study, 16 (2.7%) were ineligible and 3 (0.5%) refused to participate (Figure 1). A total of 566 households with 2429 individuals were enrolled in the study. Of those 281 (49.7%) were assigned to the control group and 285 (50.4%) were assigned to receive the intervention filter and cookstove. Household loss-to-follow-up was 2.8%, primarily due to participants moving out of the study area. A total of 2737 household-visits were completed during the follow-

Table 1. Filter and stove use among intervention households: Evaluator's survey.

	All visits	
	N	%
Filter use[1]		
Reported drinking container		
Intervention filter	1210	89.2
Other container	146	10.8
Water stored in other container treated	57	39.0
Method of treatment: Intervention filter	56	98.2
No water in intervention filter among households identifying filter as drinking container	12	1.0
No water in intervention filter among households not identifying filter as drinking container	83	56.8
Stove use[2]		
Observation data on use		
Intervention household cooking at time of visit	280	26.9
Stove in current use		
Intervention stove only	152	54.3
Both stoves simultaneously	12	4.3
Traditional stove only	116	41.4
Currently cooking outdoors	59	21.1
Intervention stove users cooking outdoors[3]	49	32.2
Reported data on use		
Reported last stove used[4]		
Intervention stove only	593	78.0
Both stoves simultaneously	15	19.3
Traditional stove only	147	2.0
Reported using intervention stove in last three follow-up visits	130	47.5

[1]Based on households that completed the visits and allowed enumerators to observe the container, 1356/1393 = 97.3%.
[2]Data only available from mid follow-up 2 onwards (1040/1393 = 74.7%).
[3]Among those households cooking only on the intervention stove.
[4]Excludes those households cooking at time of home visit.

up period (96.7%) and data on one of the primary outcomes (water quality) was collected for 2637 households-visits (93.2%).

Baseline characteristics

Baseline characteristics were distributed evenly between the trial arms, with the exception of availability of soap among households with a designated hand washing area and boiling or chlorination of drinking water (see Table S1 of supporting information). At baseline, drinking water samples were obtained from 551 (97.3%) households. The median and Williams mean of drinking water was 14 and 20.2 TTC/100 mL (95% CI: 15.0–27.0 TTC/100 mL) and 22 and 30.3 TTC/100 mL (95% CI: 22.8–40.2 TTC/100 mL) for control and intervention groups, respectively.

Filter and improved stove use and compliance

Most households used the filter throughout the study period (Table 1). Intervention households identified the filter as the main drinking source in 89.2% of all household visits where drinking water was available. Visual inspection at the time of the unannounced visit was consistent with reported use, with 99% of the filters containing water. Of the 10.8% of intervention households that stored their drinking water elsewhere, overall only 39.0% of them reported that the water had been treated with the intervention filter. Over the course of the study, however, only 62.9% of intervention households identified the filter as the main drinking water storage container in all five follow-up visits with available water (n = 240, 84.2%). Of the remainder, 11.2% reported treating it and storing it elsewhere at least once during the 5-month follow-up, 25.0% reported drinking untreated water at least once during follow-up and 0.8% did not know the status of their water in at least one of the visits. During the last follow-up visit, the major reasons for not having filtered water at the time of the visit were (i) forgetting to fill the filter (48.1%), (ii) drinking mainly locally produced beer instead of water (22.2%), or (iii) having a broken or not properly functioning filter (18.5%).

The intervention stoves were also used throughout the study, though most householders also continued to use their traditional stoves. Field investigators observed actual cooking on about a quarter (26.9%) of their unannounced visits. Of these, 54.3% were cooking only with the intervention stove and 4.3% were using both the intervention and traditional stoves (Table 1). Reported use was higher, with householders claiming they last cooked solely on the intervention stove on 78.0% of visits. Use of the intervention stove was not consistent, with 47.5% of intervention households reporting to have used the intervention stove during the last cooking event at all three home visits (data not collected during initial phases of follow-up). Likewise, of the households that were cooking at all three unannounced visits (n = 8), or at two of the

Table 2. Filter and stove use among intervention households: Implementer's survey.

	N	%
Filter use		
Filter presence confirmed in households[1]	283	99.7
Tap accessible to <5 s	267	94.4
Water present in filter	269	95.1
Stove use		
Observation data on use		
Intervention household cooking at time of visit[2]	78	27.7
Stove in current use		
Intervention stove only	50	64.1
Both stoves simultaneously	4	5.1
Traditional stove only	24	30.8
Currently cooking outdoors	30	38.5
Intervention stove users cooking outdoors[3]	28	17.2
Reported data on use		
Reported last stove used[4]		
Intervention stove only	163	79.9
Both stoves simultaneously	2	1.0
Traditional stove only	39	19.1
Primary stove in current use is intervention stove[4]	253	89.1
Use intervention stove $\geq$7/week	236	93.3
Use intervention stove $\geq$14/week	137	54.2
Continue using traditional stove	217	76.4
Use traditional stove $\geq$7/week	58	26.7
Reported cooking less indoors	175	61.6
Reported main cooking is outdoors	163	57.4
Tend more the fire with the intervention stove[5]	212	83.8

[1]Observation not allowed in one household.
[2]Of those households that allowed the observation (n = 282, 99.3%).
[3]Among those households cooking only on the intervention stove.
[4]Excludes those households cooking at time of home visit.
[5]Among those households identifying intervention stove as main cooking stove.

three unannounced visits (n = 52), only 50.0% and 34.6% were using the intervention stove at all three or two visits, respectively. During the last follow-up visit, the major reasons for not using the intervention stove during the last cooking event were (i) having no time to tend the fire (34.1%), (ii) not having dry (30.7%) or the right-size wood (10.2%) or, (iii) cooking beans for which a traditional stoves was regarded most appropriate (10.2%).

Data on use of the intervention from the implementer's survey was very similar to our assessment (Table 2). A similar percentage of intervention households (27.7%) were cooking at the time of the visit. Of these, just over two thirds (64.1%) were exclusively cooking on the intervention stove, but only 17.2% of these were cooking outdoors, a figure just slightly lower than the one observed on our independent follow-up. Data from the implementer's more extensive survey confirmed the stacking of stoves in intervention households, with 76.4% of intervention households reporting to continue using their traditional stoves. Of these, 26.7% reported using it $\geq$7 times per week. Of interest was the fact that 83.8% of intervention households identifying the intervention stove as their primary cookstove reported that the intervention stove required more active tending of the fire as compared to the traditional stove.

In the last round of our follow-up, 5.4% and 5.1% of intervention households reported having problems with their filter or stove at the time of the visit, respectively. Data collected from the implementer's repair team indicates that 24.9% of filters and 6.7% of stoves had to be repaired during the study. No devices had to be fully replaced, though some repairs involved the replacement of individual components. The main reasons for filters being repaired were (i) filters being clogged (48.6%) and, (ii) tubes being damaged by rodents (27.0%). The main reasons for the intervention stoves being repaired included (i) pot skirts melting (65%), and (ii) stick supports breaking (10%).

Overall, only 1.0% of water samples collected from control households were reported to have been treated with a neighbour's intervention filter, showing low levels of cross-contamination between groups.

Water quality

The microbiological quality of the stored drinking water was significantly higher in intervention households than control households (Williams means 0.5 vs. 20.2 TTC/100 mL, respectively, $p<0.001$). Overall, 86.8% (95% CI: 84.9%–88.6%) of drinking water samples from intervention households were free of TTC compared to 22.4% (95% CI: 20.1%–24.6%) of control household samples ($p<0.001$) (Figure 2). The proportion of samples that had >100 TTC/100 mL was 3.6% (95% CI: 2.6%–4.6%) for intervention households and 31.9% (95% CI: 29.4%–34.5%) for control households. Overall, 96.6% of drinking water samples collected directly from filters were free of TTC. In intervention households, water quality was significantly higher in water samples collected directly from the filter (Williams mean 0.14 TTC/100 mL; 95% CI: 0.10–0.18) than water stored in another container (Williams mean 13.8 TTC/100 mL; 95% CI: 9.0–20.7) (see Table S2 of supporting information). The quality of the drinking water stored in other containers did not differ significantly between control and intervention households ($p = 0.07$). However, among intervention households, water that was stored in another container and was reportedly treated with the intervention filter was significantly of higher quality than reportedly non-treated stored water (Williams means 5.4 vs. 23.2 TTC/100 mL, respectively, $p<0.001$). Throughout the duration of the study, only 2.5% of control households had drinking water free of TTC on all follow-up visits as opposed to 56.5% of intervention households. Overall 15.2% of samples from control households and 5.1% of samples from intervention households yielded plates that were TNTC.

Air quality

A total of 121 households (60 intervention and 61 control) completed the 24-h $PM_{2.5}$ monitoring of the main cooking area. 66.7% of intervention households identified the intervention stove as their main cooking stove. However, only 23.3% of intervention households reported that their main cooking area was outdoors as promoted by the intervention. Of these, all households reported cooking with the intervention stove. Among the control households, the three stone fire was identified as the main cooking stove in 65.6% of cases, followed by the locally made *rondereza* stove (24.6%). Only one control household reported cooking outdoors.

Table 3 shows the $PM_{2.5}$ concentrations of the main cooking area for control and intervention households on an aggregate level and stratified by reported main area of cooking. Overall, mean and median 24-h $PM_{2.5}$ concentrations in intervention households were 0.485 mg/m^3 and 0.267 mg/m^3, respectively, compared to

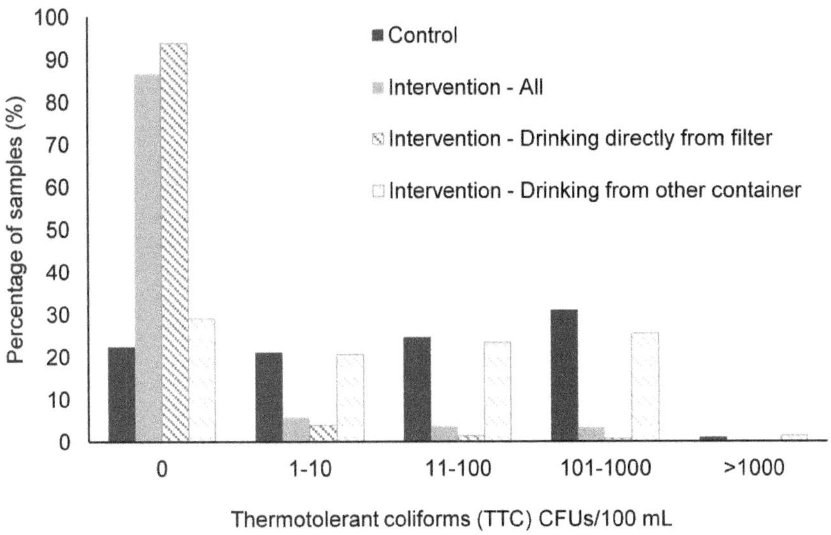

Figure 2. Percentage of water samples by level of contamination (TTC/100 mL).

0.905 mg/m^3 and 0.509 mg/m^3 for control households. This represents a 48% reduction in median 24-h concentrations ($p = 0.005$). Compared to control households that predominantly cooked indoors, intervention homes that reported indoor cooking showed a reduction in median concentrations of 37%, which was only borderline significant, possibly due to the smaller sample size ($p = 0.08$). Outdoor cooking in the intervention was associated with a median reduction of 73% when compared to control households ($p<0.001$) and 57% reduction when compared to indoor-cooking intervention homes ($p = 0.02$).

Discussion

We report on a randomised controlled trial to independently evaluate a pilot implementation program distributing free water filters and improved cooking stoves to rural homes in Rwanda. We found high reported use of the intervention filter, which was associated with significantly higher microbiological quality of drinking water when consumed directly from the filter. Nevertheless, such use was not exclusive; a sizable proportion of householders continued to drink untreated water. We also found improved household air quality among intervention households despite continued use of the traditional stove.

Filter uptake among the intervention population was high, with filters being reportedly used in 89.2% of all household visits. Similar levels of uptake of filter-based interventions have been reported elsewhere [16,35,36]. Nevertheless, we found that 25% of intervention householders were reporting untreated water in at least one of the five follow-up visits. The nested study within this RCT using remotely reporting electronic sensors that collected objective data on use of the intervention devices (mainly times and volumes of water filtered for the intervention filter and times and duration of use for the intervention stove) corroborated our findings, showing that the filters and stoves were not used in a consistent and exclusive manner [24]. Epidemiological modelling based on quantitative microbial risk assessment suggests that even occasional consumption of untreated water can vitiate the health benefits associated with improved water quality interventions [37–39]. However, the intervention did significantly improve the microbiological quality of the drinking water when the filter was used as the main storage container. Since 96.6% of drinking water samples collected directly from filters were free of TTC, the conditions for achieving health gains may be achieved with better messaging.

Exclusive use was more problematic for the intervention stove. Only half of the intervention households reported that the last

Table 3. Summary statistics for 24-h $PM_{2.5}$ concentrations in the reported main cooking area.

$PM_{2.5}$ (mg/m^3)	N	Mean	SD	Min	Median	Max	Geometric mean	% Mean reduction	% Median reduction	Wilcoxon RST[1] p-value
Control	61	0.905	1.05	0.06	0.509	4.69	0.51	-	-	
Intervention	60	0.485	0.53	0.04	0.267	2.28	0.28	46%	48%	0.005
Reported cooking location										
Control- Indoor cooking	60	0.910	1.06	0.06	0.506	4.69	0.51	-	-	
Intervention- Indoor cooking	46	0.558	0.56	0.04	0.321	2.28	0.33	39%	37%	0.08
Intervention- Outdoor cooking	14	0.243	0.34	0.05	0.139	1.40	0.16	73%	73%	<0.001

[1]Wilcoxon rank-sum (Mann-Whitney) test

cooking event was performed with the intervention stove in the last three monthly follow-up visits. Likewise, only a third of those households that were visited twice at times that cooking was taking place were using the intervention stove at both instances, showing that among the intervention arm, households continued to rely on their traditional stove. Results from the implementers' survey showed similar results, with 76.4% of households reporting the continued use of their traditional stove, 26.7% of them using it more than 7 times per week. This is consistent with other studies that have shown that the introduction of a new stove often results in "stacking" rather than an immediate complete substitution [40–43].

Households reported continuing the use of their traditional stove because the intervention stove required more tending, unavailability of the adequate fuel or personal preferences for cooking traditional dishes. Context-specific issues regarding a community's cooking needs and preferences have been commonly cited in the literature as reasons for not achieving higher uptake and/or exclusive sustained use of improved cookstoves [42,44]. Thus re-considerations of the promoted stove or more active messaging addressing each of the main barriers may be required if a switching of the stove as opposed to an addition of the intervention stove to the current cooking system is to be achieved. This is not only going to affect the potential health impact of the intervention but also its environmental impact.

The assessment of HAP among control and intervention households showed an overall reduction of 48% of 24-h $PM_{2.5}$ among intervention households, which was comparable to reductions in household air pollution for rocket stove interventions in Ghana (52%) and Kenya (33%) [22,45]. Indoor cooking with the intervention stoves as opposed to the traditional stove was associated with a 37% reduction in 24-h $PM_{2.5}$, which was of borderline significance. However, we cannot rule out that this association may be due to residual bias by comparing sub-groups. Likewise, cooking outdoors, as recommended by the implementer, doubled the reduction in 24-h $PM_{2.5}$ from 37% to 73% as compared to indoor cooking on traditional stoves. Future studies, randomising participants not only to stove technology but also to cooking location (indoors *vs.* outdoors) would be advisable. More effective messaging may increase the levels of outdoor cooking expected by the intervention, as only 57.4% of households reported that their main cooking area was outdoors. Nevertheless, both the indoor and outdoor concentrations in the cooking area were well over even the initial interim 24-h WHO target for $PM_{2.5}$ (75 $\mu g/m^3$) [46]. At the same time, it will be important to monitor personal exposure directly, as most householders that identified the intervention stove as their primary cooking stove (83.8%) reported that the intervention stove required more tending than their traditional one, which could mitigate some of the impact from the household level reductions in $PM_{2.5}$. Indeed, many studies have found that reductions in personal exposure tend to be lower than reductions of emissions in the cooking area [47,48]. Given that a recent RCT study suggested that personal exposure reductions exceeding 50% may be required to achieve meaningful health impacts [47], further assessments of the intervention stove maybe be needed to determine whether the use of the intervention stove translates into meaningful health benefits.

This study has certain limitations. First, the villages included in the RCT were not selected randomly and should not be viewed as representative of any larger population. Second, we cannot rule out the potential for reactivity due to repeated monthly follow-up visits [49]. Third, while we attempted to collect objective indicators of use, by both undertaking visual observations of the filter and stoves and cooking events, the study relied heavily on reported data, which is susceptible to reporting bias. Furthermore, in this study we failed to collect data on reported supplementation of treated water with untreated water, which would have further implications for the health impact of the study. Previous studies with the earlier version of the LifeStraw Family filter have found quite varied results. A study in the Democratic Republic of Congo showed substantial supplementation despite high levels of filter use [16]. On the other hand, a study among HIV-positive mothers, who may be more aware of their health and their children's health, reported almost no supplementation [36]. However, in the latter storage containers were provided. Fourth, budget constraints allowed only the main cooking area, as identified by the participant, to be monitored for HAP. Given the potential for reporting bias and that stacking was commonly reported among the study population, it is very likely that cooking events may have taken place during the monitoring period in areas other than the one being monitored, thus giving a misleading and probable underestimate of the actual total HAP. Likewise, budget constraints did not permit personal $PM_{2.5}$ assessment, a more reliable metric for exposures associated with health outcomes [50]. Fifth, we did not collect any self-reported or other measures of diarrhoea or respiratory infections in our study communities. Finally, the follow-up period of this evaluation was limited to 5 months. This represents under a fraction of the lifespan of both the filter and stove and provided little opportunity to assess the impact of seasonal variations that are common in water quality and HAP. It also provided no opportunity to assess long-term patterns of use, which have been shown to diminish or vary over time for both water filters and improved cookstoves [35,43]. We are endeavouring to address some of these shortcomings in a longer-term follow-up study, currently underway, that will focus on health outcomes and sustained use.

Notwithstanding these limitations, this study suggests that a combined filter/stove intervention accompanied by consistent follow-up to promote use has the potential to significantly improve drinking water quality and household air pollution among a vulnerable population in Rwanda. If the longer-term follow-up study demonstrates sustained use with more exclusive reliance on the intervention hardware and lower personal exposure to HAP, then a large-scale roll out in Rwanda could significantly reduce exposures linked to much of the country's disease burden.

Supporting Information

Figure S1 Mass calibration of UCB-PATS against co-located $PM_{2.5}$ gravimetric samples.

Table S1 Baseline characteristics of participating.

Table S2 Summary statistic of TTC/100 mL in household drinking water and filter effluent by study group.

Text S1 Protocol.

Text S2 CONSORT Checklist.

Acknowledgments

We thank all of the participants who contributed to this study. We also thank the staff for their diligent efforts in collecting the data. We thank too Wolf-Peter Schmidt and Alexandra Huttinger for their contribution to the design of the study. Michael Johnson, senior scientist at Berkeley Air Monitoring Group was responsible for providing advice on HAP monitoring and acts as lead author for the group, mjohnson@berkeleyair.com.

Author Contributions

Conceived and designed the experiments: GR SB FN TC. Performed the experiments: GR FM MK. Analyzed the data: GR MJ. Contributed reagents/materials/analysis tools: CB ET. Wrote the paper: GR TC.

References

1. Lim SS, Vos T, Flaxman AD, Danaei G, Shibuya K, et al. (2012) A comparative risk assessment of burden of disease and injury attributable to 67 risk factors and risk factor clusters in 21 regions, 1990–2010: a systematic analysis for the Global Burden of Disease Study 2010. Lancet 380: 2224–2260.
2. Liu L, Johnson HL, Cousens S, Perin J, Scott S, et al. (2012) Global, regional, and national causes of child mortality: an updated systematic analysis for 2010 with time trends since 2000. Lancet 379: 2151–2161.
3. Ezzati M, Kammen DM (2001) Quantifying the effects of exposure to indoor air pollution from biomass combustion on acute respiratory infections in developing countries. Environ Health Perspect 109: 481.
4. Rehfuess E, Mehta S, Pruss-Ustun A (2006) Assessing Household Solid Fuel Use: Multiple Implications for the Milleniem Development Goals. Environ Health Perspect 114.
5. WHO/UNICEF (2012) Progress on drinking water and sanitation: 2012 update. Geneva: WHO.
6. Bonjour S, Adair-Rohani H, Wolf J, Bruce NG, Mehta S, et al. (2013) Solid Fuel Use for Household Cooking: Country and Regional Estimates for 1980–2010. Environ Health Perspect 121: 784–790.
7. Barnes DF, Openshaw K, Smith KR, van der Plas R (1994) What Makes People Cook with Improved Biomass Stoves? A Comparative International Review of Stove Programs. World Bank.
8. Johnson M, Edwards R, Ghilardi An, Berrueta V, Gillen D, et al. (2009) Quantification of Carbon Savings from Improved Biomass Cookstove Projects. Environ Sci Technol 43: 2456–2462.
9. Bond TC, Doherty SJ, Fahey DW, Forster PM, Berntsen T, et al. (2013) Bounding the role of black carbon in the climate system: A scientific assessment. J. Geophys Res: Atmos 118: 5380–5552.
10. NISR (2012) Rwanda Ministry of Health Annual Health Statistics Booklet 2011. National Institute of Statistics of Rwanda.
11. NISR (2008) Rwanda in Statistics and Figures 2008. National Institute of Statistics of Rwand.
12. National Institute of Statistics of Rwanda (NISR) [Rwanda], Ministry of Health (MOH) [Rwanda], International I (2012) Rwanda Demographic and Health Survey 2010. Calverton, Maryland, USA: NISR, MOH, and ICF International.
13. Wright J, Gundry S, Conroy R (2004) Household drinking water in developing countries: a systematic review of microbiological contamination between source and point-of-use. Trop Med Int Health 9: 106–117.
14. RMOH (2011) Rwanda Health Statistics Booklet 2011. Rwanda Ministry of Health.
15. UNICEF (2012) Committing to Child Survival: A Promise Renewed. Progress Report 2012. New York: United Nations Children's Fund.
16. Boisson S, Kiyombo M, Sthreshley L, Tumba S, Makambo J, et al. (2010) Field Assessment of a Novel Household-Based Water Filtration Device: A Randomised, Placebo-Controlled Trial in the Democratic Republic of Congo. PLoS One 5: e12613.
17. Clasen T, Naranjo J, Frauchiger D, Gerba C (2009) Laboratory Assessment of a Gravity-Fed Ultrafiltration Water Treatment Device Designed for Household Use in Low-Income Settings. Am J Trop Med Hyg 80: 819–823.
18. Naranjo J, Gerba C (2011) Assessment of the LifeStraw Family Unit using the World Health Organization Guidelines for "Evaluating Household Water Treatment Options: Health-based Targets and Performance Specifications". University of Arizona, Department of Soil, Water and Environmental Science.
19. WHO (2011) Evaluating household water treatment options: health-based targets and performance specifications. Geneva: WHO.
20. Adkins E, Tyler E, Wang J, Siriri D, Modi V (2010) Field testing and survey evaluation of household biomass cookstoves in rural sub-Saharan Africa. Energy Sustain Dev 14: 172–185.
21. USAID (2011) In-Home Emissions of Greenhouse Pollutants from Rocket and Traditional Biomass Cooking Stoves in Uganda.
22. Pennise D, Brant S, Agbeve SM, Quaye W, Mengesha F, et al. (2009) Indoor air quality impacts of an improved wood stove in Ghana and an ethanol stove in Ethiopia. Energy Sustain Dev 13: 71–76.
23. Barstow CK, Thomas EA, Ngabo Fidele, Ghislaine R, Majorin F, et al. (2013) Environmental health product delivery and education in rural Rwanda through a public-private partnership. Unpublished.
24. Thomas EA, Barstow CK, Rosa G, Majorin F, Clasen T (2013) Use of Remotely Reporting Electronic Sensors for Assessing Use of Water Filters and Cookstoves in Rwanda. Environ Sci Technol 47: 13602–13610.
25. APHA (2001) Standard Methods for the Examination of Water and Wastewater. 21st Edition. Washington, (DC): APHA.
26. Masera OR, Saatkamp BD, Kammen DM (2000) From Linear Fuel Switching to Multiple Cooking Strategies: A Critique and Alternative to the Energy Ladder Model. World Development 28: 2083–2103.
27. Edwards R, Smith K, Kirby B, Allen T, Litton C, et al. (2006) An inexpensive dual-chamber particle monitor: laboratory characterization. J Air Waste Manag Assoc 56: 789–799.
28. Edwards RD, Liu Y, He G, Yin Z, Sinton J, et al. (2007) Household CO and PM measured as part of a review of China's National Improved Stove Program. Indoor Air 17: 189–203.
29. Chowdhury Z, Edwards RD, Johnson M, Naumoff Shields K, Allen T, et al. (2007) An inexpensive light-scattering particle monitor: field validation. J Environ Monit 9: 1099–1106.
30. Dutta K, Shields KN, Edwards R, Smith KR (2007) Impact of improved biomass cookstoves on indoor air quality near Pune, India. Energy Sustain Dev 11: 19–32.
31. Cynthia AA, Edwards RD, Johnson M, Zuk M, Rojas L, et al. (2008) Reduction in personal exposures to particulate matter and carbon monoxide as a result of the installation of a Patsari improved cook stove in Michoacan Mexico. Indoor Air 18: 93–105.
32. Alexander N (2012) Review: analysis of parasite and other skewed counts. Trop Med Int Health 17: 684–693.
33. McElduff F, Cortina-Borja M, Chan S-K, Wade A (2010) When t-tests or Wilcoxon-Mann-Whitney tests won't do. Adv Physiol Educ 34: 128–133.
34. Kuha J (2004) AIC and BIC: Comparisons of Assumptions and Performance. Sociol Methods Res 33: 188–229.
35. Hunter PR (2009) Household Water Treatment in Developing Countries: Comparing Different Intervention Types Using Meta-Regression. Environ Sci Technol 43: 8991–8997.
36. Peletz R, Simunyama M, Sarenje K, Baisley K, Filteau S, et al. (2012) Assessing Water Filtration and Safe Storage in Households with Young Children of HIV-Positive Mothers: A Randomized, Controlled Trial in Zambia. PLoS One 7: e46548.
37. Hunter PR, Zmirou-Navier D, Hartemann P (2009) Estimating the impact on health of poor reliability of drinking water interventions in developing countries. Sci Total Environ 407: 2621–2624.
38. Brown J, Clasen T (2012) High Adherence Is Necessary to Realize Health Gains from Water Quality Interventions. PLoS One 7: e36735.
39. Enger KS, Nelson KL, Rose JB, Eisenberg JNS (2013) The joint effects of efficacy and compliance: A study of household water treatment effectiveness against childhood diarrhea. Water Res 47: 1181–1190.
40. Ruiz-Mercado I, Masera O, Zamora H, Smith KR (2011) Adoption and sustained use of improved cookstoves. Energy Policy 39: 7557–7566.
41. Burwen J, Levine DI (2012) A rapid assessment randomized-controlled trial of improved cookstoves in rural Ghana. Energy Sustain Dev 16: 328–338.
42. Hanna R, Duflo E, Greenstone M (2012) Up in smoke: The influence of household behaviour on the long-run impact of improved cooking stoves. NBER Working Paper No. 18033. Cambridge, MA: National Bureau of Economic Research.
43. Pine K, Edwards R, Masera O, Schilmann A, Marrón-Mares A, et al. (2011) Adoption and use of improved biomass stoves in Rural Mexico. Energy Sustain Dev 15: 176–183.
44. Manibog FR (1984) Improved Cooking Stoves in Developing Countries: Problems and Opportunities. Annu Rev Energy 9: 199–227.
45. Ochieng CA, Vardoulakis S, Tonne C (2013) Are rocket mud stoves associated with lower indoor carbon monoxide and personal exposure in rural Kenya? Indoor Air 23: 14–24.
46. WHO (2006) Global update 2005: Particulate matter, ozone, nitrogen dioxide and sulfur dioxide, air quality guidelines. Copenhagen: WHO.
47. Smith KR, McCracken JP, Weber MW, Hubbard A, Jenny A, et al. (2011) Effect of reduction in household air pollution on childhood pneumonia in Guatemala (RESPIRE): a randomised controlled trial. Lancet 378: 1717–1726.
48. Fitzgerald C, Aguilar-Villalobos M, Eppler AR, Dorner SC, Rathbun SL, et al. (2012) Testing the effectiveness of two improved cookstove interventions in the Santiago de Chuco Province of Peru. Sci Total Environ 420: 54–64.
49. Zwane AP, Zinman J, Van Dusen E, Pariente W, Null C, et al. (2011) Being surveyed can change later behavior and related parameter estimates. Proc Natl Acad Sci U S A.
50. Naeher LP, Brauer M, Lipsett M, Zelikoff JT, Simpson CD, et al. (2007) Woodsmoke health effects: A review. Inhal Toxicol 19: 67–106.

20

A Phenological Timetable of Oak Growth under Experimental Drought and Air Warming

Thomas M. Kuster[1,2,3]*, Matthias Dobbertin[1], Madeleine S. Günthardt-Goerg[1], Marcus Schaub[1], Matthias Arend[1]

1 Forest Dynamics, Swiss Federal Research Institute WSL, Birmensdorf, Switzerland, **2** Institute of Terrestrial Ecosystems ITES, ETH Zürich, Zürich, Switzerland, **3** Soil and Ecosystem Ecology Group, University of Manchester, Manchester, United Kingdom

Abstract

Climate change is expected to increase temperature and decrease summer precipitation in Central Europe. Little is known about how warming and drought will affect phenological patterns of oaks, which are considered to possess excellent adaptability to these climatic changes. Here, we investigated bud burst and intra-annual shoot growth of *Quercus robur*, *Q. petraea* and *Q. pubescens* grown on two different forest soils and exposed to air warming and drought. Phenological development was assessed over the course of three growing seasons. Warming advanced bud burst by 1–3 days °C^{-1} and led to an earlier start of intra-annual shoot growth. Despite this phenological shift, total time span of annual growth and shoot biomass were not affected. Drought changed the frequency and intensity of intra-annual shoot growth and advanced bud burst in the subsequent spring of a severe summer drought by 1–2 days. After re-wetting, shoot growth recovered within a few days, demonstrating the superior drought tolerance of this tree genus. Our findings show that phenological patterns of oaks are modified by warming and drought but also suggest that ontogenetic factors and/or limitations of water and nutrients counteract warming effects on the biomass and the entire span of annual shoot growth.

Editor: Han Y. H. Chen, Lakehead University, Canada

Funding: For funding, the authors are gratefully indebted to the Velux foundation (project number: 489, www.veluxstiftung.ch). The funders had no role in study design, data collection and analysis, decision to publish, or preparation of the manuscript.

Competing Interests: The authors have declared that no competing interests exist.

* E-mail: thomaskuster@gmx.ch

Introduction

The growing season of deciduous trees is commonly defined as the time between bud burst and autumnal leaf senescence. More specifically, it refers to the period of annual shoot, stem and root growth. In spring, bud burst is driven by the degree of preceding winter chilling, the increasing length of the photoperiod and by rising temperatures [1,2]. Winter chilling and short photoperiod are known to delay early bud burst and prevent frost damage, whereas rising temperatures promote bud burst [3,4]. Similarly, the ontogenetically fixed cessation of tree growth and autumnal down-regulation of physiological functions are mainly controlled by photoperiod and temperature [5,6,7,8,9,10,11]. Based on the IPCC scenario A2, spring (+2.2 to 4.2°C) and autumn temperatures (+2.4 to 5.0°C) are both predicted to increase in Central Europe throughout the 21st century [12,13]. Therefore, through climate change, extended annual growing seasons may be expected due to both earlier bud burst and later cessation of growth [7,14,15]. Climate change is also predicted to reduce summer precipitation in Central Europe between 21 and 28% [12,13]. Consequently, interactions between rising temperature and reduced water availability may modify growing season's extensions.

Oaks are regarded as popular tree species in future forestry. They are considered as to be tolerant against drought and heat due to their xeromorphic adaptations in leaf and wood structure and root growth [16,17,18,19]. Among several deciduous tree species, phenological parameters of oaks have been described to be most flexible to changes in temperature [20]. Whereas positive effects of warming on bud burst in trees including oaks are known [3], the consequences of increased temperatures on intra-annual shoot growth and carbon sequestration in oaks still require further investigation. In oaks, intra-annual shoot growth is of particular importance as this tree genus is able to flush several times during a growing season. In general, an overall positive response of warming on plant growth might be expected due to accelerated plant metabolism and increased soil nutrient turnover rates [21,22]. Thus, warming and its effects on oak trees might be advantageous for future sequestration of atmospheric carbon [2,23].

Not only warming but also decreased precipitation is predicted to influence oaks as a consequence of climate change. Reduced water availability has not been found to affect leaf unfolding and senescence in *Quercus robur* and *Q. petraea* seedlings [24]. However, drought has been reported to affect intra-annual shoot growth of oaks by reducing the number of flushes and delaying the onset of a 2nd flush later in the growing season when water availability becomes more limited than in spring [25]. Only little is known about the resilience of oaks after several drought periods or how increasing temperature affects this relationship.

Based on the literature presented above, we hypothesised that (i) an increase in air temperature promotes bud burst and increases shoot growth duration leading to enhanced biomass production of oaks, but (ii) drought will counteract these effects by reducing shoot growth duration and therefore biomass. To test these hypotheses, including the resilience after re-watering and the interaction between temperature and drought, we conducted a fully-factorial model ecosystem experiment in which various provenances of *Q. robur*, *Q. petraea* and *Q. pubescens* were grown on two different forest soils and subjected to drought and air warming. Bud burst and intra-annual shoot growth were assessed over the course of three growing seasons to consider potential ontogenetic effects on these phenological traits.

Materials and Methods

Study Site & Experimental Design

The study was part of the multidisciplinary experimental setup "Querco" [26]. The experimental design of the model ecosystem facility (MODOEK) located at the Swiss Federal Research Institute WSL, Birmensdorf, Switzerland (47° 21′48″ N, 8° 27′ 23″ E, 545 m a.s.l.) is described in detail in [27]. Briefly, the facility consists of 16 hexagonal chambers of 3 m height and a surface area of 6 m^2 each. Each chamber is split below ground into two 1.5 m deep concrete-walled lysimeters. In spring 2005, the lysimeters were filled with a 0.5 m drainage packing composed of 3 layers of pure quartz gravel of decreasing grain size (from bottom to top) and a 1 m soil layer on top of this drainage layer. In one lysimeter of each chamber the soil consisted of two layers of acidic loamy sand taken from a Haplic Alisol (pH 4.0, subsoil 0.15–1.00 m, topsoil 0–0.15 m), while in the other lysimeter it consisted of a single layer of calcareous sandy loam taken from a Calcaric Fluvisol (pH 6.9). Further soil properties, including nutrient concentrations of the two soils, are presented in Table 1, [27] and [28].

In spring 2006, two-year-old saplings of *Q. robur*, *Q. petraea* and *Q. pubescens* were randomly planted and grown with sufficient water supply at ambient air temperature [29]. Each species was represented by four different provenances from Switzerland and in one case (*Q. pubescens*) from Northern Italy [17]. Climatic conditions at the 12 oak stands, where the acorns were harvested in 2003, were different for all provenances [19]. Each provenance was represented by two saplings on each lysimeter (statistical unit = oak tree, $n = 8$), leading to a total number of 768 trees (4 climatic treatments, 2 soils, 12 provenances nested in 3 species, 8 replications).

From February 2007 to October 2009, each chamber was subjected to one of the following four treatments: air warming (AW), drought (D), the combination of air warming and drought (AWD) and control (CO). Each treatment was replicated four times and statistically arranged in a Latin square design. In the air-warming treatment, day-time temperatures were increased by 1–2°C in relation to the control by reducing the opening of the side-walls (Table 2, [27]). As temperatures in forest stands are generally lower than in an open field [30], the applied warming treatment came close to a moderate temperature increase in Central Europe throughout the 21st century based on the IPCC scenario A2 [12,13].

In the drought treatment, irrigation was discontinued for several consecutive weeks during the growing seasons while it was continued without interruption in the control. Irrigation was reduced by 60% in 2007 and by 43% in 2008 and 2009 in relation to the long-term mean precipitation during the growing season at the experimental site (April to October, 1961–2009). The drought stress increased from 2007 to 2009 due to longer intervals without irrigation and the developing canopies from year-to-year leading to increased water consumption of the trees [27]. After the first drought treatment in 2007, which had only little effects on intra-annual shoot growth, longer intervals without irrigation were applied in 2008 (5 to 10 weeks) and in 2009 (6 to 16 weeks) to increase the drought intensity and stimulate a treatment effect. In 2008 and 2009, the drought treatment was not stopped before soil moisture reached a threshold value of 0.05 $m^3\ m^{-3}$ for several days [27]. At the end of each drought period, intensive irrigation was applied for a few days, simulating heavy rainfall, according to the increasing probability of extreme precipitation due to global warming. Rainfall was excluded from the chambers in all treatments during the growing seasons, while the roof tops were left open during dormant seasons to allow for natural irrigation.

Table 1. Physical and chemical soil properties.

	Acidic Soil		Calcareous Soil	
Depth	**0–0.15 m**	**0.15–1 m**	**0–0.15 m**	**0.15–1 m**
Texture (% sand, silt, clay)	85, 10, 5	87, 8, 5	71, 18, 12	71, 18, 12
pH (0.01 M $CaCl_2$)	3.93	4.00	6.85	6.89
C_{tot} (%)	2.06	0.48	2.20	1.85
N_{tot} (%)	0.11	0.03	0.07	0.05
N_{av} (mg kg^{-1})	4.10	3.30	5.91	4.93
P_{tot} (mg kg^{-1})	524.46	469.43	422.81	325.54
P_{av} (mg kg^{-1})	3.91	4.78	3.08	2.23
$Ca_{exch.}$ (mg kg^{-1})	364.07	142.21	1798.63	1544.88
$Mg_{exch.}$ (mg kg^{-1})	25.40	9.48	29.38	18.12
$K_{exch.}$ (mg kg^{-1})	30.69	18.98	31.26	21.50
$Mn_{exch.}$ (mg kg^{-1})	45.79	18.63	1.46	1.43

Soil properties of the acidic and calcareous soils were measured at different depths (topsoil: 0–0.15 m, subsoil: 0.15–1 m) at the end of the experiment in autumn 2009. Table adapted from [27] and [28].

Phenological Observations

During spring in 2007, 2008 and 2009, all oak saplings were assessed bi-weekly for phenological development (greening of the buds, bud burst and leaf unfolding, see Fig. 1). In 2007, each bud was separately evaluated for phenological development. In 2008 and 2009, there were too many buds to be assessed individually so, the phenological development was estimated in 10% classes. For statistical analyses of the phenological development, we used the day of the year (DOY) when >50% of the buds of a tree attained the respective phenological stage.

Shoot Growth Measurements

In the spring of each growing season, the shoot closest to the tip was determined as the potential main shoot. Elongation of the main shoot was measured every fourth to seventh day until the end of August and every second to third week from September until mid October. Shoot length was measured with a ruler from the last terminal bud to the newly formed bud. As oaks flush several times per growing season, we measured the shoot elongation of each flush separately.

Stem diameter increment was monitored in 2009 with one selected *Q. robur* provenance (Tägerwilen) grown on well-watered, acidic soil using automated single point dendrometers (Zweifel Consulting, Hombrechtikon, Switzerland), as described in detail in [31]. In September 2009, all aboveground wood and foliage biomass was harvested. Dry weights of shoot (aboveground wood and foliage) were determined after drying at 65°C for several days.

Table 2. Air temperature and drought stress during the experimental period from 2006 to 2009.

		ambient air temperature[1)]	effect of warming on air temperature[2)]	drought[3)]
2006	April	+0.0°C	none	none
	May to September	+0.9°C	none	none
2007	April	+5.1°C	not available	none
	May to September	−0.8°C	+1.4°C	weak
2008	April	−0.9°C	+1.2°C	none
	May to September	+0.1°C	+1.3°C	strong
2009	April	+3.2°C	+2.0°C	weak
	May to September	+1.1°C	+1.1°C	very strong

Ambient air temperature in relation to the long term mean (1981–2010), effect of warming on the daytime air temperature (8:00–18:00 h, UTC+1) and development of the drought stress from year-to-year.
[1)]difference to the norm ambient air temperatures (1981–2010) at the nearby WMO weather station Zürich/Fluntern, data provided by SwissMeteo.
[2)]in relation to control, see [27] for further details.
[3)]based on drought effects on soil water concentration [27], predawn leaf water potential [34] and foliage injury [17].

Temperature and Soil Data

Air temperature in the chambers was measured every hour at a height of 120 cm with shaded EL-USB-2 data sensors (Lascar Electronics Ltd, Whiteparish, UK), with initial measurements in June 2007 [27]. As there was no temperature data available for spring 2007, we used temperature data from the nearby WMO weather station Zurich Fluntern as a substitute in Fig. 2 (data provided by MeteoSwiss).

At the end of the experiment in 2009 (26 October – 5 November) soil samples (each pooled from five soil cores of 3 cm diameter) were taken at different depths from each lysimeter (topsoil: 0–0.15 m; subsoil: 0.15–1 m). Available soil nitrogen (N_{av}: NO_3^-+NH_4^+) was extracted with 1 M KCl and measured in fresh samples as described in [28]. Remaining soil samples were dried at 105°C, ground and sieved (>1 mm). For measuring available phosphate (P_{av}), soil samples were extracted with 0.5 M $NaHCO_3$ (Olsen-P), following [32]. The concentrations of P_{av} in the extracts were detected by spectrophotometry at 880 nm (Cary 50 UV-VIS, Varian, Palo Alto, California, USA). Exchangeable cations ($Ca_{exch.}$, $Mg_{exch.}$, $K_{exch.}$ and $Mn_{exch.}$) were extracted with 0.1 M $BaCl_2$ according to [32] and detected using an ICP-OES (Vista MPX, Varian, Palo Alto, California, USA). Total soil carbon (C_{tot}), nitrogen (N_{tot}) and phosphorous (P_{tot}) were measured by means of a dry combustion analyser (CN-2000, LECO Corp., St. Joseph, Michigan, USA) and X-ray fluorescence (X-Lab 2000, Spectro, Kleve, Germany), respectively. Soil pH was measured in 0.01 M $CaCl_2$.

Statistical Analysis

For statistical analyses, we used R 2.11.1 (R Development Core Team, Vienna, AT). The data were analysed after log-transformation by ANOVA using a linear mixed-effects model accounting for the split-plot design of the experiment with two soils in each chamber (significant at $P<0.05$). The oak provenances were nested within species and compared using the mean of each provenance versus those of all other provenances (significant at $P<0.05$). Selected differences between treatments and soils were tested pair-wise using contrasts based on *t*-tests (significant at $P<0.01$). Correlation factors (r) were calculated based on the Pearson method (significant at $P<0.05$).

Results

Timing of Bud Burst

The timing of bud burst was correlated with the preceding greening of the buds ($r=0.807$) and the subsequent unfolding of the leaves ($r=0.939$). As the effects of the climatic treatments, soils, growing seasons and species/provenances were very similar in all three phenological stages being assessed, only the results from bud burst are presented and discussed in detail.

Bud burst was mainly triggered by air temperature, which differed significantly between the three growing seasons, but also between the climatic treatments (Table 2, [27]). In April, when the growing season for oaks starts, mean ambient air temperature was lower in 2008 (8.0°C) compared to 2007 (13.9°C) and 2009

Figure 1. Examples of the phenological observations assessed during spring (2007 to 2009). A: greening of buds; B: bud burst; and C: leaf unfolding. The day of the year when 50% of a tree's buds/leaves were green, burst or unfolded was used for statistical analysis and data presentation (Table 3, Figs. 2, 3, S1 and S2).

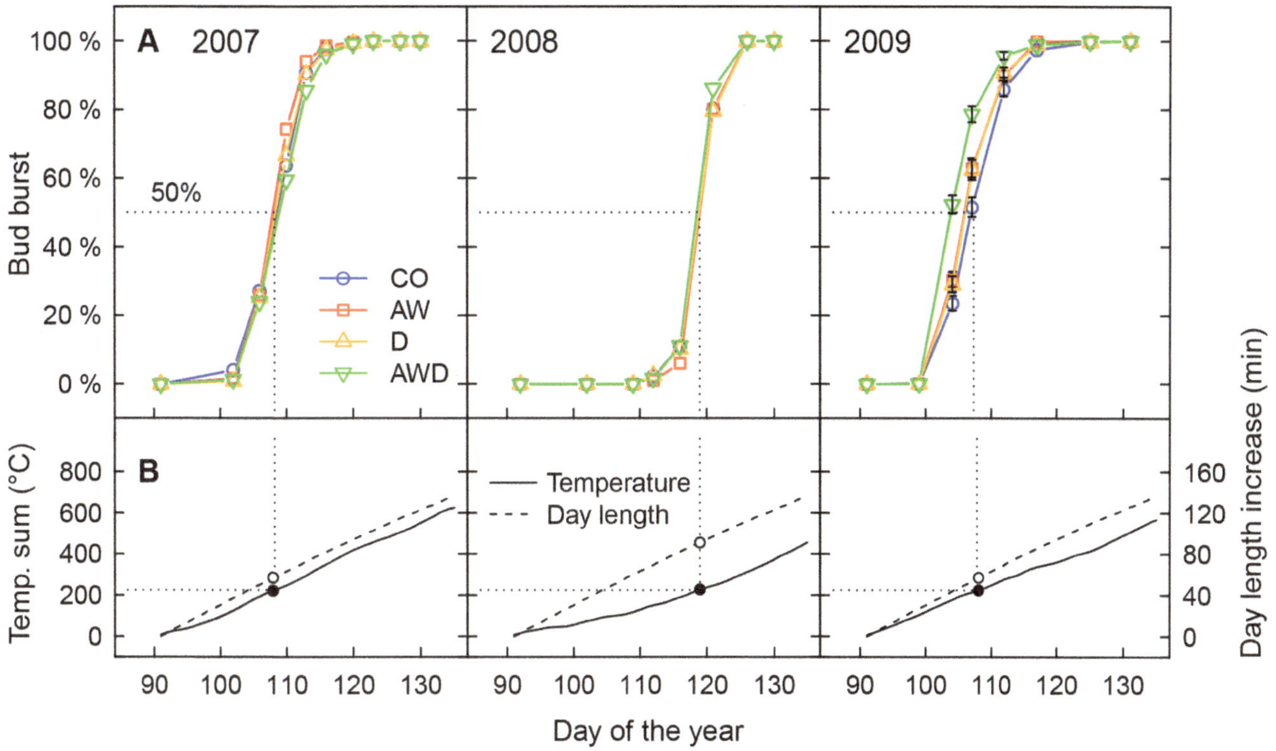

Figure 2. Average annual (2007 to 2009) development of bud burst. A: Relative bud burst development, separately shown for the control treatment (CO) and the air warming (AW) and drought (D) treatments and the combined treatment for air warming & drought (AWD). Data of the two soils and the 12 provenances were pooled ($n = 192$ for each treatment and year combination) as the influence of the soil on bud burst was only minor (<1 day, see text for more details) and the effect of the provenances on bud burst is shown separately in Fig. S2. For graphical reasons, error bars (± SE) are only partially shown for the 2009 development. Results of statistical analyses are given in Table 3. B: Sum of daily mean air temperature (°C) and day length increase (min) starting on day 91 (2007/2009 1st April; 2008 31st March). The dotted vertical line refers to the average air temperature sum (224 degree days) required for bud burst (>50% bud burst per tree). The temperature data was taken from the nearby WMO weather station Zurich Fluntern.

(12.0°C). Consequently, in 2008, bud burst (>50% of the total number of a tree's buds) of control oaks was 11 days later (DOY 119) than in 2007 and 2009 (DOY 108; Fig. 2A). Across all three growing seasons, 224 degree days (sum of mean daily temperature in °C) in April triggered >50% bud bursts per tree (Fig. 2B). We found that April temperatures were sufficient to explain patterns in bud burst and applying more complex phenological models including early spring temperature, chilling threshold [4] or day length [33] did not help explain the uncertainty left in the present study.

The air-warming treatment influenced bud burst only in 2009 (Table 3), in that increased air temperature (2.0°C) enhanced bud burst by 1 to 2 days (Fig. 3). During the warmer 2007 spring (Table 2), bud burst was not affected by the air-warming treatment. However, we observed a temperature effect on the unfolding of the leaves (AW x provenance: $P = 0.046$), indicating that there is nevertheless a warming effect on spring phenology. In contrast to 2007 and 2009, temperatures in April 2008 were relatively low (Table 2); upon warming, bud burst presumably occurred to rapidly to detect any differences between treatments by using a bi-weekly assessment (steep increase of bud burst development curve in Fig. 2A, Fig. S1). In summary, each 1°C temperature increase led to an earlier bud burst of 1 to 3 days regardless of the source of warming (different ambient temperature between years or climatic treatments, Figs. 2 & 3).

Earlier bud burst was also observed in drought-exposed oaks in 2009, although this effect was less obvious than in oaks subjected to air warming (Table 3, Fig. 3). The combination of warming and drought triggered an even earlier bud burst. No drought effect on bud burst was found in 2007 and 2008 (Table 3, Fig. S1), possibly as the oaks were not or less affected by previous years' drought (Table 2).

Differences in bud burst between the two soils and the tested provenances were only significant in 2009 (Table 3). Bud burst of oaks grown on acidic soil was marginally earlier compared to those grown on calcareous soil. The influence of soil on bud burst was minor (Δ <1 day in all treatment and provenance combinations) with no significant interactions between soil and other factors. Bud burst of the provenances *Q. robur* Magadino (Mi), *Q. petraea* Magden (M), *Q. petraea* Wädenswil (W) and *Q. pubescens* Leuk (Lk) was earlier, whereas bud burst of *Q. robur* Bonfol (B) and *Q. pubescens* Promontogno (P) was later than the mean of all provenances (S2). The variation in bud burst within species was therefore stronger than between species. Furthermore, there was no interaction between the provenances and the climatic treatments on bud burst, indicating that the different provenances responded similarly to air warming and drought.

Number of Flushes Formed within the Season

The mean number of flushes per tree decreased from 2.4 in 2007 (mean across all treatments, provenances and soils) to 2.1 in 2008 and to 1.5 in 2009 (Tables 3 & 4, Figs. 4 & 5), indicating an age effect on the capacity of oaks to form several flushes within a growing season. Drought significantly reduced the number of

Table 3. ANOVA *F*-values of bud burst, number of flushes and duration of growth.

	df1	df2	bud burst			number of flushes			duration of growth		
			2007	2008	2009	2007	2008	2009	2007	2008	2009
AW	1	12	0.0	0.0	26.9***	1.5	3.2	0.6	0.9	0	0.8
D	1	12	1.1	0.2	22.0***	9.5**	49.9***	32.4***	12.4**	59.7***	35.5***
Soil	1	12	0.5	0.2	53.8***	53.2***	79.8***	4.9*	22.0***	97.9***	23.4***
Spec	2	9	0.2	0.1	0.1	1.5	6.4*	12.5**	2.2	8.6**	7.1*
Prov	9	264	0.8	1.0	67.0***	11.2***	11.1***	10.3***	13.1***	10.7***	14.7***
AW:D	1	12	0.6	0.3	1.1	0.1	0.0	0.3	1.4	2.2	0.8
AW:Spec	2	264	2.0	0.4	0.9	0.0	3.4*	0.0	0.6	0.6	0.3
AW:Prov	9	264	1.8(*)	1.2	1.7(*)	1.4	1.9(*)	1.1	1.6	2.4*	1.6
D:Soil	1	12	0.0	0.1	0.1	2.1	0.3	16.1**	0.1	0.1	22.7***
D:Spec	2	264	0.1	0.1	3.4*	1.3	8.9***	4.1*	0.6	0.7	2.5(*)
D:Prov	9	264	0.6	0.5	1.3	0.4	3.2***	2.9**	0.3	2.3*	1.4
Soil:Spec	2	264	0.9	0.4	3.7*	8.5***	15.3***	3.8*	7.8***	13.8***	3.5*
Soil:Prov	9	264	0.2	1.2	1.5	2.4*	1.5	1.5	1.4	3.9***	1.4
AW:D:Spec	2	264	1.5	1.0	3.3*	0.7	2.1	1.2	1.2	2.4(*)	0.6
AW:Soil:Spec	2	264	0.9	0.0	1.0	2.1	0.1	0.0	3.9*	0.5	0.0
D:Soil:Spec	2	264	0.2	0.4	0.3	0.1	1.0	5.9**	0.4	0.1	2.6(*)
AW:D:Soil:Prov	9	264	0.5	0.4	1.2	1.6	0.6	2.4*	0.6	0.6	1.0

Main effects and selected two-way interactions of drought (D, continuous *vs.* discontinuous irrigation), air warming (AW, ambient *vs.* elevated air temperature), soil (acidic *vs.* calcareous), species (Spec) and provenance (Prov, nested within species) on bud burst (day of the year when 50% of buds were burst), the number of flushes per tree and duration of shoot growth, all separately shown for each year, $n = 8$. df1: degrees of freedom in the numerator; df2: degrees of freedom in the denominator; *F*-values: 0.0 refer to values <0.05; levels of significance: (*) $P<0.1$, * $P<0.05$, ** $P<0.01$, *** $P<0.001$. Non-significant interactions are not shown.

flushes by approximately 20% in 2008 and 2009, but not in 2007 when water limitation was much less pronounced due to shorter drought periods and less competition for soil water (Tables 2, 3 & 4, [27]). In contrast to drought, air warming had no influence on the number of flushes. In 2007, trees grown on the calcareous soil flushed more often than those grown on the acidic soil (*e.g.* control +10%), whereas the opposite was observed in 2008 (−15%) and 2009 (−20%).

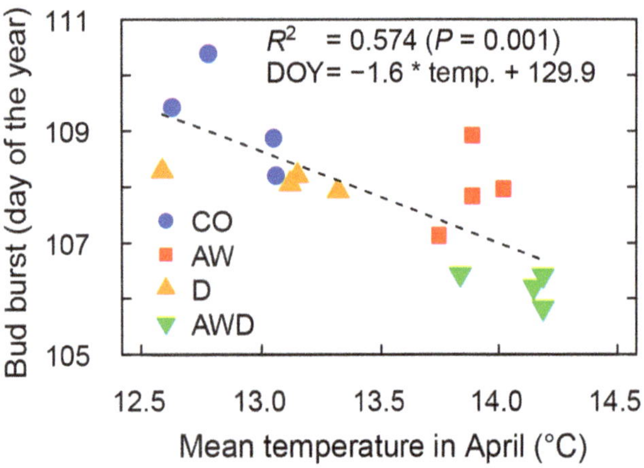

Figure 3. Relationship between mean April air temperature in 2009 and bud burst phenology. Air temperature (°C, 0:00 to 23:00, UTC+1) and the day of the year when 50% of the buds were open are separately shown for each model ecosystem chamber. Data of the two soils and the 12 provenances were pooled because air temperature was equal within a chamber. Over all treatments, the negative slope of the correlation line was significant ($R^2 = 0.574$, $P = 0.001$). See Fig. S1 for the non-significant relationship in 2008.

Shoot Elongation and Radial Stem Growth

With increasing drought stress from 2007 to 2009 (Table 2, [27]), the effect from reduced soil water availability on shoot elongation became more apparent, with the strongest effect in 2009 (Fig. 4). Mainly, the length of the 2nd flush was reduced by drought. Thus, in 2009, drought-exposed oaks completed most of their total shoot elongation (>90% shoot length) within the first few weeks after bud burst when soil water availability was still sufficient. In contrast to this pattern, shoot growth from well-watered oaks, was spread over several flushes. Additionally, we observed a carry-over effect of drought on the 1st flush in the following spring. Although soil water availability in spring was not different from the control at this point, shoot elongation of the 1st flush of drought-exposed oaks was reduced in 2008 (−12%, $P = 0.005$) and 2009 (−13%, $P = 0.002$). Unlike longitudinal shoot growth, radial stem growth was rather continuous as demonstrated for well-watered *Q. robur* Tägerwilen trees in 2009 (Fig. 4). The start and duration of shoot growth, however, were similar and there was also no air-warming effect.

Timing and Duration of Shoot Flushing within a Season

As expected, the start of the 1st flush within a given year was strongly correlated with the preceding bud burst ($r = 0.827$). There was, however, no correlation between the 2nd ($r = 0.091$) or 3rd flush ($r = 0.380$) and bud burst (more correlations are shown in Table S1). In general, air warming led to an earlier onset of intra-annual flush growth, specifically in 2007, 2008 (2nd flush) and 2009 (1st flush and 3rd flush), both under drought and well-watered

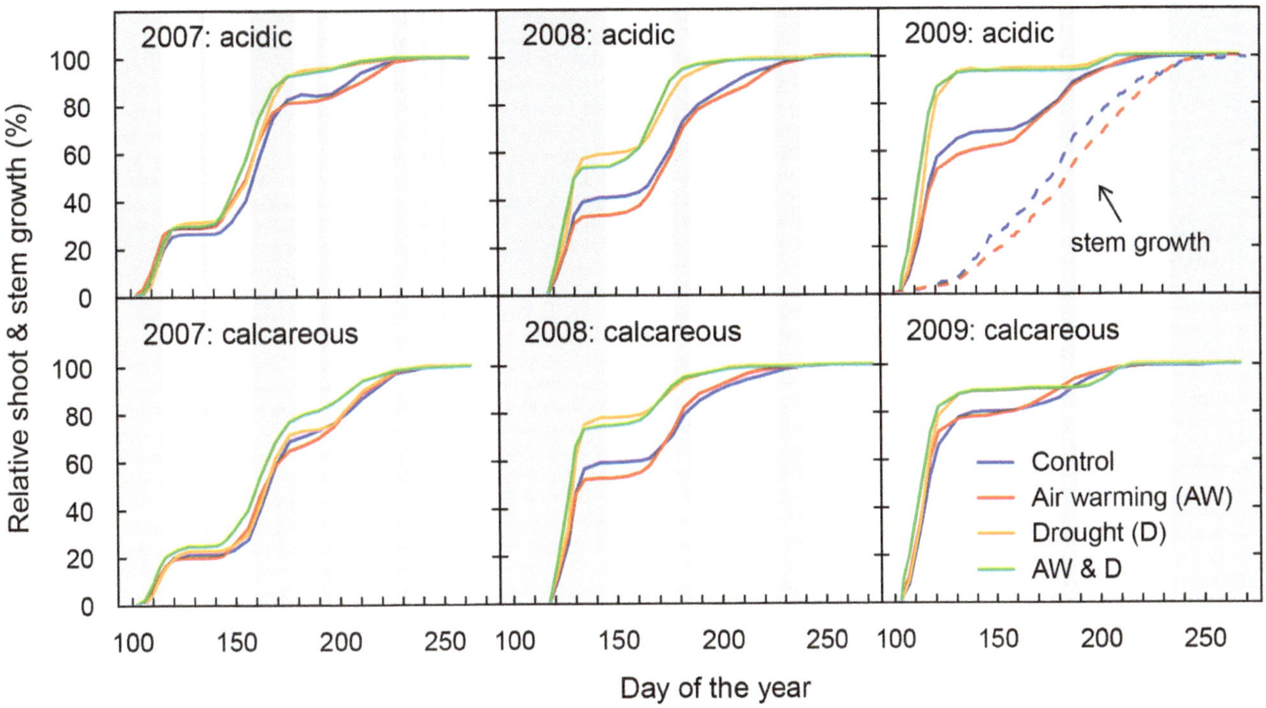

Figure 4. Averages of annual relative shoot growth. Growth development is separately shown for the three years (2007 to 2009), all treatments and the two soils (acidic *vs.* calcareous). Data of the 12 provenances were pooled ($n = 96$) as the effects of the species and the provenances are separately shown in Fig. S2. Additionally in 2009: example of relative stem diameter growth (*Q. robur* Tägerwilen (T)) on the acidic soil, separately shown for the control (CO) and air warming (AW) treatments, $n = 4$. The grey bars indicate periods when all chambers (including the drought treatment) were irrigated.

conditions (Table 5, Fig. 5). For oaks exposed to air warming however, we observed no universal prolongation of the time period for flush growth as it stopped earlier compared to the flush growth for non-warmed oaks (Table 5).

Drought only affected the start of the 2nd flush (Table 5). In 2008, the onset of the 2nd flush was advanced by the moderate drought (Fig. 5). However, in 2009 the drought was much more severe with mean predawn leaf water potentials of −1.5 and −2.9 MPa on acidic and −2.8 and −3.9 MPa on calcareous soil in D and AWD respectively (data presented in detail in [34]). Consequently, the start of the 2nd flush growth was delayed by 15 days and 22 days on the calcareous and acidic soil, respectively, until the drought was interrupted by intermediate irrigation.

Table 4. Mean number of flushes per tree 2007–2009.

	2007		2008		2009	
	acidic	**calcareous**	**acidic**	**calcareous**	**acidic**	**calcareous**
Control	*2.4^{a}±0.1	*2.7^{a}±0.1	*2.4^{a}±0.0	*2.1^{a}±0.1	*1.9^{a}±0.1	*1.5^{ab}±0.1
AW	*2.3^{a}±0.1	*2.7^{a}±0.1	*2.5^{a}±0.1	*2.2^{a}±0.1	1.9^{a}±0.1	1.6^{a}±0.1
D	*2.1^{a}±0.1	*2.7^{a}±0.0	*2.0^{b}±0.1	*1.7^{b}±0.1	1.2^{b}±0.1	1.3^{b}±0.1
AW & D	*2.1^{a}±0.0	*2.5^{a}±0.1	*2.1^{b}±0.0	*1.8^{b}±0.1	1.3^{b}±0.0	1.3^{b}±0.0

Air warming (AW), drought (D) and soil effects on the number of flushes per tree. Data are separately shown for each growing season (means ± SE). Data of the different provenances were pooled before statistical analyses ($n = 8$). Different letters indicate significant differences between respective treatment means on the same soil, an asterisk * indicates significant differences between acidic and calcareous soil for the same treatment ($P < 0.01$).

Thereafter, growth resumed within a few days after re-watering the soil. In 2008 and 2009, drought reduced average growth duration of all flushes by 2 days. As an exception, the growth of the 3rd flush in 2008 of drought-exposed oaks lasted 9 days longer compared to the well-watered trees - most likely due to re-watering during flush growth. Flush growth on acidic soil started earlier than on the calcareous soil (2nd flush 2007: 4 days earlier (mean of all treatments); 2nd flush 2008: 7 days; 1st flush 2009: 2 days). Also, the duration of growth of some flushes was longer on the acidic than on the calcareous soil (2nd flush 2008: +4 days; 1st flush 2009: +3 days).

Duration of Total Shoot Growth per Growing Season

Air warming neither increased neither the number of flushes nor the duration of single flush growth. As such, the duration of total shoot growth was not affected (Table3, Fig. S2). The duration of total shoot growth of drought-exposed oaks was shorter than that of well-watered trees (on average over all years and both soils: −18 days; Table 3, Fig. S2). This reduction in growth duration was particularly distinct on the acidic soil in 2009 when only a quarter of the trees flushed a 2nd time (−35 days, Fig. 5). Differences in the duration of total shoot growth between the two soils were determined by the development of a 2nd or 3rd flush; the period of shoot growth was longer for oaks grown on calcareous soil than for those grown on the acidic soil in 2007, whereas it was the other way around in 2008 and 2009 (Figs. 5 & S2).

The duration of the total shoot growth differed between the provenances as well as between the three oak species (Table 3, Fig. S2). In general, *Q. robur* provenances formed more flushes per season with a longer growth period than the provenances of the other two species. Shoot growth of the *Q. pubescens* provenance

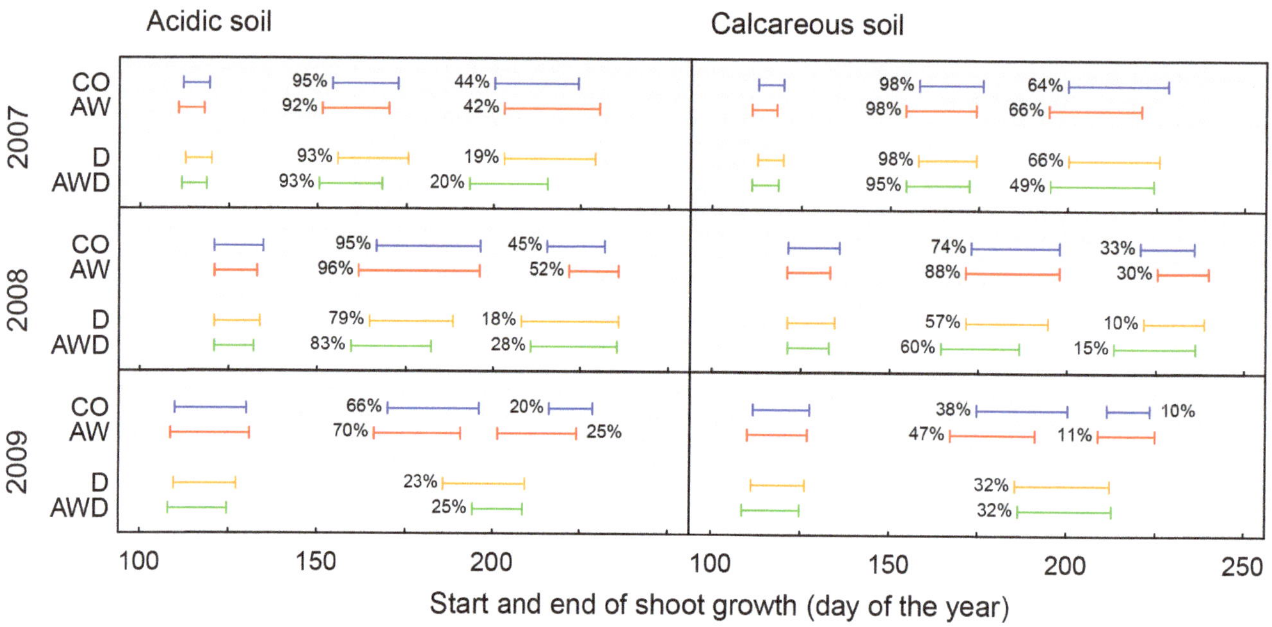

Figure 5. Average start, end (>95% of growth) and duration of shoot growth. Timetables are separately shown for each flush, the control treatment (CO), air warming (AW) and drought (D) treatments and the combined treatment for air warming & drought (AWD), the two soils (acidic *vs.* calcareous) and the three growing seasons (2007, 2008 and 2009). The numbers next to the bars refer to the ratio of trees with a 2nd and/or 3rd flush (%, 1st flush always 100%). Data of the 12 provenances were pooled ($n = 96$) as the effects of the species and the provenances are separately shown in Fig. S2. Results of statistical analyses are given in Tables 3 & 5. The grey bars indicate periods when chambers were irrigated.

Leuk (Lk) was longer, whereas the duration of growth of *Q. petraea* Wädenswil (W) and of the two *Q. pubescens* provenances Promontogno (P) and Arezzo (A) was shorter compared to the mean of all provenances.

Shoot Biomass

After three growing seasons, total shoot biomass (aboveground wood and foliage) was significantly reduced by drought (Table 6). Shoot biomass was significantly higher on the acidic than on the calcareous soil under well-watered conditions, whereas no difference was observed between the soils under drought conditions. Air warming had no effect on shoot biomass. Further details regarding shoot biomass, including differences between species and provenances, are presented in [19].

Discussion

Warming Effects on Bud Burst and Intra-annual Shoot Growth

Advanced bud burst has shown to be a specific response of temperate tree species to increasing spring temperature [1,2]. It has been hypothesised that climate warming might extend the growing season in temperate areas and thus increase seasonal biomass production in forest ecosystems [7,14,15]. In the present study, we observed advanced bud burst in oaks exposed to warming as well as an earlier start of intra-annual shoot growth. The advanced phenological development, however, was counteracted by an earlier cessation of intra-annual shoot growth. Thus, we neither observed a prolonged time period of seasonal growth nor an increased biomass production. It is therefore reasonable to assume that bud burst is triggered by environmental factors, *e.g.* temperature and photoperiod, while the duration of subsequent shoot growth is under strong ontogenetic control. Similarly, Soolanayakanahally et al. reported that bud burst of *Populus balsamifera* provenances was driven by local environmental factors whereas shoot growth duration was ontogenetically fixed as it depended on the latitude of the provenance's origin [35]. Ontogenetic control of shoot growth duration of oaks is also in line with the sudden down-regulation of photosynthetic carbon allocation in late summer in our model ecosystem experiment, which occured independent of any obvious changes in photoperiod and temperature [34]. The treatments also did not change leaf longevity [17].

Besides ontogenetic fixed shoot development, other factors, such as nutrient availability, might also have limited growth. For example, average growth duration and shoot biomass were both lower on the calcareous than on the acidic soil, most likely due to limited availability of phosphorous (P) and manganese (Mn) [36]. Averaged over all growing seasons, treatments and provenances, leaf Mn and P concentrations of trees grown on the calcareous soil (Mn: 43 mg kg^{-1}, P: 2.77 g kg^{-1}) were lower than of those on the acidic soil (Mn: 2766 mg kg^{-1}; P: 3.32 g kg^{-1}). Leaf Mn and P concentrations of trees grown on the calcareous soil were also below the deficiency level given for oak leaves (Mn: 35–100 mg kg^{-1}, P: 1.5–3 g kg^{-1}, [37]). Similarly, the number of flushes, and therefore duration of shoot growth, was reduced under nutrient limiting conditions in several other studies on *Q. petraea* [38,39,40,41]. We therefore conclude that warming can induce a phenological shift towards earlier bud burst and shoot growth, but, in contrast to our first hypothesis, that factors other than temperature, *e.g.* ontogenetic fixed shoot development and/or nutrient availability, limit the total time period of intra-annual shoot growth.

Advanced bud burst due to increased temperatures is in agreement with several other studies investigating temperature effects on spring phenology of European oaks [20,24,42,43]. As warming of each 1°C led to a 1 to 3 day earlier onset of bud burst in our experiment, increased spring temperature up to 4°C (upper boundary of the IPCC A2 scenario [12,13]) might advance bud

Table 5. ANOVA *F*-values of start and duration of growth per flush.

	AW	D	Soil	AW x D	D x Soil
Start growth					
2007					
1st Flush	28.577***	1.951	0.429	0.201	3.180
2nd Flush	17.433**	0.002	25.490***	0.511	0.151
3rd Flush	7.129*	1.870	5.359	5.359*	1.740
2008					
1st Flush	0.403	0.011	2.974	0.872	0.014
2nd Flush	12.032**	5.068*	22.013***	0.903	0.753
3rd Flush	0.187	3.413(*)	1.685	1.257	0.351
2009					
1st Flush	29.284***	4.460(*)	18.893***	1.333	0.453
2nd Flush	0.011	65.617***	0.267	4.718(*)	2.380
3rd Flush	5.234(*)	–	1.561	–	–
Duration growth					
2007					
1st Flush	0.106	0.013	0.347	0.012	0.525
2nd Flush	0.549	1.358	0.622	0.515	2.424
3rd Flush	0.682	0.004	2.214	0.070	0.192
2008					
1st Flush	23.968***	5.139*	2.231	0.017	0.124
2nd Flush	0.484	19.034***	11.032**	1.717	3.574(*)
3rd Flush	0.053	8.042*	1.737	1.841	0.018
2009					
1st Flush	1.430	19.673***	35.499***	1.784	6.738*
2nd Flush	1.705	3.972(*)	3.536(*)	0.614	5.050*
3rd Flush	6.778*	–	4.103	–	–

Main effects and selected two-way interactions of drought (D, continuous *vs.* discontinuous irrigation), air warming (AW, ambient *vs.* elevated air temperature) and soil (acidic *vs.* calcareous) on start and duration of growth. Data is separately shown for each growing season and each flush. Data of the provenances growing on the same half of each lysimeter were pooled before analysis ($n = 8$). *F*-values and level of significances: (*) $P<0.1$, * $P<0.05$, ** $P<0.01$, *** $P<0.001$. For all main and interaction effects: degrees of freedom in the numerator (df1) = 1 and denominator (df2) = 12. Non-significant interactions are not shown.

burst by up to 12 days, without consideration of inter-annual variability. However, premature bud burst might be hampered by insufficient chilling, daylight or frost [44]. As demonstrated for maple trees, atmospheric CO_2 enrichment might have no or only minimal effects on bud bust [44]. In contrast to this, we did not find any evidence for delayed growth cessation due to the applied warming in the present study; this is in line with a former study in which we failed to show delayed autumnal leaf senescence due to warming, as indicated by the loss of photosynthetic activity or chlorophyll degradation [34]. Indeed, the temperature effect on autumnal phenology has been reported to be highly variable in oak species [7,20,24,43].

Drought Effects on Bud Burst and Intra-annual Shoot Growth

Surprisingly, in 2009 the onset of bud burst for drought-exposed oaks was considerably earlier than for well-watered oaks. A direct effect of reduced water availability on bud burst can, however, be excluded as the drought in the 2009 season was not then effective during the time when buds burst and air temperatures were not increased [27]. We therefore conclude that the observed advancement of bud burst is most likely explained by a carry-over effect from the previous year's drought. Indeed, the first flush is preformed within the bud formed during the previous season [45]. As such, a drought event during bud formation might impact the date of its outgrowth in the following season, as has been shown for evergreen oaks [46]. This conclusion is in agreement with the missing carry-over effects in the 2007 and 2008 growing seasons during which bud burst was not or less influenced by the previous year's drought. Carry-over effects due to the previous year's climatic conditions have also been demonstrated for the length of newly formed shoots in *Fagus sylvatica* and oaks ([47], this study), the width of tree rings in *Fagus grandifolia* and *Quercus alba* [48] and the size of early wood vessels in *Castanea sativa* [49]. While such carry-over effects on shoot length and wood growth can be simply explained by insufficient resource storage for bud formation or xylem production [50], the nature of the carry-over effect on the timing of bud burst remains so far unknown.

In support of the second hypothesis, the growth rate of drought-exposed oak shoots was slowed down or even stopped with increasing drought stress during the course of the growing season, as indicated by decreasing leaf water potentials [34]. This effect mainly occurred in 2009 when drought stress was strongest. Hence, not all oaks formed a 2nd flush under drought conditions and, if they did, the onset was delayed until soil-water conditions became non-limiting. As discussed by [51] and [52], such a reduction of shoot phenological development may be a successful strategy for limiting leaf area and therefore transpirational water loss. Remarkably, the trees quickly recovered after re-watering and shoot growth was resumed. This recovery was in close synchrony with the fast up-regulation of photosynthetic activity [34] and increased evapotranspiration [27]. A similar recovery was also shown in a study with *Q. robur* seedlings, in that shoot growth of second and third flushes after re-watering was even stronger under drought than under control conditions [25]. These findings demonstrate the superior drought resistance and resilience of oaks, even during their most vulnerable stage of early growth, and confirm their high phenotypic plasticity [17]. It is likely that canopy trees will be able to cope even better with severe drought as they are able to tap into deeper water reservoirs with their long taproots [53,54]. Also, transpiration might be reduced with increasing atmospheric CO_2 concentrations, allowing soil moisture to sustain oak growth for an even longer period of time [55,56].

Table 6. Aboveground wood and foliage biomass after three growing seasons.

	acidic	calcareous
Control	*324^a ±17	*258^a ±10
AW	*338^a ±26	*233^a ±5
D	147^b ±3	141^b ±3
AW & D	141^b ±7	137^b ±5

Air warming (AW), drought (D) and soil effects on shoot biomass (g tree^{-1}) at the end of the experiment in autumn 2009. Data of the different provenances were pooled before calculating means and SE ($n = 8$). Different letters indicate significant differences between respective treatment means on the same soil ($P<0.01$). An asterisk (*) indicates significant differences between acidic and calcareous soil for the same treatment. More biomass and shoot length data are presented in [29] and [27].

Conclusions

In conclusion, our study demonstrates advanced bud burst and earlier onset of intra-annual shoot growth as specific responses to air warming. However, these acclimatisation mechanisms do not necessarily lead to prolonged seasonal growth, higher biomass or increased carbon sequestration due to ontogenetically fixed development and/or limited availability of nutrient and soil water. With respect to drought, we found that oaks are able to cope well with reduced soil water availability, both in terms of resistance and resilience, by delaying or suspending individual shoot growth during the season. Furthermore, we found that bud burst is advanced by a carry-over effect from the previous year's drought. We therefore strongly suggest further investigation on the nature of potential carry-over effects that might bias warming effects on trees under future climate scenarios.

Supporting Information

Figure S1 Relationship of mean April air temperature in 2008 to bud burst phenology. Air temperature (°C, 0:00 to 23:00, UTC+1) and the day of the year, when 50% of the buds were open, are separately shown for each model ecosystem chamber. Data of the two soils and the 12 provenances were pooled because air temperature was equal within a chamber.

Figure S2 Bud burst development and total duration of shoot growth separately shown for all provenances. A: Average day of the year (2007 to 2009) when 50% of a tree's buds were open ($n = 16$). B: average total duration of shoot growth in days (sum over all flushes, $n = 8$), separately shown for all treatments, the two soils (acidic and calcareous) and the 12 provenances, nested in 3 species (Qro: *Quercus robur*, T = Tägerwilen, B = Bonfol, H = Hühnenberg, Mi = Magadino; Qpe: *Q. petraea*, C = Corcelles, M = Magden, W = Wädenswil, G = Gordevio; Qpu: *Q. pubescens*, Lk = Leuk, LL = Le Landeron, P = Promontogno, A = Arezzo/Italy; a map of the provenance's sites is presented by [17], climate conditions at the provenance's sites are shown in detail by [19]). Asterisks indicate that a provenance (mean of CO and AW) is different from the mean of all other provenances (* = $P<0.05$, ** = $P<0.01$, *** = $P<0.001$). Further results of statistical analyses are given in Table 3.

Table S1 Correlation matrix between phenology and foliage biomass and stem diameter. Correlation (Pearson correlation coefficients) between bud burst, duration of growth per flush ($period_1$: 1st flush, $period_2$: 2nd flush, $period_3$: 3rd flush), total duration of growth ($period_{tot}$), shoot length per flush ($length_1$: 1st flush, $length_2$: 2nd flush, $length_3$: 3rd flush), total shoot length per growing season growth ($length_{tot}$), number of flushes (# flushes) and foliage biomass and stem diameter (Ø stem). Each parameter $n = 2304$).

Acknowledgments

This study was designed and carried out under the supervision of Matthias Dobbertin (†) and we particularly wish to acknowledge his valuable expertise and advice. We thank Peter Bleuler, Martin Keller, Björn Studer, Terry Menard, Rainer Schulin, the internship students and the garden team of WSL for their help in performing the experiment. We acknowledge Corina Del Fabbro and the Seminar for Statistics of ETHZ for their support in data analysis. We thank Terry Menard, Hayley Craig and Thomas Walker for editing this manuscript.

Author Contributions

Conceived and designed the experiments: MD MSG. Performed the experiments: TMK MD MSG MA. Analyzed the data: TMK MD MSG MS MA. Wrote the paper: TMK MSG MS MA.

References

1. Körner C, Basler D (2010) Phenology under global warming. Science 327: 1461–1462.
2. Polgar CA, Primack RB (2011) Leaf-out phenology of temperate woody plants: from trees to ecosystems. New Phytologist 191: 926–941.
3. Reyer C, Leuzinger S, Rammig A, Wolf A, Bartholomeus RP, et al. (2013) A plant's perspective of extremes: Terrestrial plant responses to changing climatic variability. Global Change Biology 19: 75–89.
4. Fu YH, Campioli M, Van Oijen M, Deckmyn G, Janssens IA (2012) Bayesian comparison of six different temperature-based budburst models for four temperate tree species. Ecological Modelling 230: 92–100.
5. Rohde A, Bastien C, Boerjan W (2011) Temperature signals contribute to the timing of photoperiodic growth cessation and bud set in poplar. Tree Physiology 31: 472–482.
6. Bohlenius H, Huang T, Charbonnel-Campaa L, Brunner AM, Jansson S, et al. (2006) *CO/FT* regulatory module controls timing of flowering and seasonal growth cessation in trees. Science 312: 1040–1043.
7. Menzel A, Fabian P (1999) Growing season extended in Europe. Nature 397: 659–659.
8. Hanninen H, Tanino K (2011) Tree seasonality in a warming climate. Trends in Plant Science 16: 412–416.
9. Jayawickrama KJS, McKeand SE, Jett JB (1998) Phenological variation in height and diameter growth in provenances and families of loblolly pine. New Forests 16: 11–25.
10. Myking T, Skroppa T (2007) Variation in phenology and height increment of northern *Ulmus glabra* populations: Implications for conservation. Scandinavian Journal of Forest Research 22: 369–374.
11. Johnsen KH, Seiler JR, Major JE (1996) Growth, shoot phenology and physiology of diverse seed sources of black spruce. 2. 23-year-old field trees. Tree Physiology 16: 375–380.
12. CH2011 (2011) Swiss climate change scenarios CH2011. Zurich, Switzerland: C2SM, MeteoSwiss, ETH, NCCR Climate, and OcCC. 88 p.
13. IPCC (2007) Climate change 2007: Synthesis report. Cambridge: Cambridge University Press. 73 p.
14. Penuelas J, Filella I (2001) Phenology - Responses to a warming world. Science 294: 793–795.
15. Penuelas J, Filella I, Comas P (2002) Changed plant and animal life cycles from 1952 to 2000 in the Mediterranean region. Global Change Biology 8: 531–544.
16. Fonti P, Heller O, Cherubini P, Rigling A, Arend M (2013) Wood anatomical responses of oak saplings exposed to air warming and soil drought. Plant Biology 15: 210–219.
17. Günthardt-Goerg MS, Kuster TM, Arend M, Vollenweider P (2013) Foliage response of young central European oaks to air warming, drought and soil type. Plant Biology 15: 185–197.
18. Abrams MD (1990) Adaptions and responses to drought in *Quercus* species of North America. Tree Physiology 7: 227–238.
19. Kuster TM, Arend M, Günthardt-Goerg MS, Schulin R (2013) Root growth of different oak provenances in two soils under drought stress and air warming conditions. Plant and Soil 369: 61–71.
20. Vitasse Y, Porte AJ, Kremer A, Michalet R, Delzon S (2009) Responses of canopy duration to temperature changes in four temperate tree species: relative contributions of spring and autumn leaf phenology. Oecologia 161: 187–198.
21. Saxe H, Cannell MGR, Johnsen B, Ryan MG, Vourlitis G (2001) Tree and forest functioning in response to global warming. New Phytologist 149: 369–399.
22. Way DA, Oren R (2010) Differential responses to changes in growth temperature between trees from different functional groups and biomes: a review and synthesis of data. Tree Physiology 30: 669–688.
23. Hlasny T, Barcza Z, Fabrika M, Balazs B, Churkina G, et al. (2011) Climate change impacts on growth and carbon balance of forests in Central Europe. Climate Research 47: 219–236.
24. Morin X, Roy J, Sonie L, Chuine I (2010) Changes in leaf phenology of three European oak species in response to experimental climate change. New Phytologist 186: 900–910.
25. Spiess N, Oufir M, Matusikova I, Stierschneider M, Kopecky D, et al. (2012) Ecophysiological and transcriptomic responses of oak (*Quercus robur*) to long-term drought exposure and rewatering. Environmental and Experimental Botany 77: 117–126.
26. Günthardt-Goerg MS, Arend M (2013) Woody plant performance in a changing climate. Plant Biology 15: 1–4.
27. Kuster TM, Arend M, Bleuler P, Günthardt-Goerg MS, Schulin R (2013) Water regime and growth of young oak stands subjected to air-warming and drought

on two different forest soils in a model ecosystem experiment. Plant Biology 15: 138–147.
28. Kuster TM, Schleppi P, Hu B, Schulin R, Günthardt-Goerg MS (2013) Nitrogen dynamics in oak model ecosystems subjected to air warming and drought on two different soils. Plant Biology 15: 220–229.
29. Arend M, Kuster T, Günthardt-Goerg MS, Dobbertin M (2011) Provenance-specific growth responses to drought and air warming in three European oak species (*Quercus robur*, *Q. petraea* and *Q. pubescens*). Tree Physiology 31: 287–297.
30. Renaud V, Rebetez M (2009) Comparison between open-site and below-canopy climatic conditions in Switzerland during the exceptionally hot summer of 2003. Agricultural and Forest Meteorology 149: 873–880.
31. Ehrenberger W, Rüger S, Fitzke R, Vollenweider P, Günthardt-Goerg M, et al. (2012) Concomitant dendrometer and leaf patch pressure probe measurements reveal the effect of microclimate and soil moisture on diurnal stem water and leaf turgor variations in young oak trees. Functional Plant Biology 39: 297–305.
32. Carter MR (2008) Soil sampling and methods of analysis. Boca Raton, FL, USA: CRC Press. 1224 p.
33. Nizinski JJ, Saugier B (1988) A model of leaf budding and development for a mature *Quercus* forest. Journal of Applied Ecology 25: 643–652.
34. Arend M, Brem A, Kuster TM, Günthardt-Goerg MS (2013) Seasonal photosynthetic responses of European oaks to drought and elevated daytime temperature. Plant Biology 15: 169–176.
35. Soolanayakanahally RY, Guy RD, Silim SN, Song MH (2013) Timing of photoperiodic competency causes phenological mismatch in balsam poplar (*Populus balsamifera* L.). Plant Cell and Environment 36: 116–127.
36. Thomas FM, Brandt T, Hartmann G (1998) Leaf chlorosis in pedunculate oaks (*Quercus robur* L.) on calcareous soils resulting from lime-induced manganese/iron-deficiency: Soil conditions and physiological reactions. Journal of Applied Botany-Angewandte Botanik 72: 28–36.
37. Bergmann W (1993) Ernährungsstörungen bei Kulturpflanzen. Jena: Fischer. 835 p.
38. Collet C, Colin F, Bernier F (1997) Height growth, shoot elongation and branch development of young *Quercus petraea* grown under different levels of resource availability. Annales Des Sciences Forestières 54: 65–81.
39. Collet C, Frochot H (1996) Effects of interspecific competition periodic shoot elongation in oak seedlings. Canadian Journal of Forest Research-Revue Canadienne De Recherche Forestière 26: 1934–1942.
40. Chaar H, Colin F, Collet C (1997) Effects of environmental factors on the shoot development of *Quercus petraea* seedlings - A methodological approach. Forest Ecology and Management 97: 119–131.
41. Harmer R (1989) The effect of mineral nutrients on growth, flushing, apical dominance and branching in *Quercus petraea* (Matt) Liebl. Forestry 62: 383–395.
42. Fu YH, Campioli M, Demaree G, Deckmyn A, Hamdi R, et al. (2012) Bayesian calibration of the Unified budburst model in six temperate tree species. International Journal of Biometeorology 56: 153–164.
43. Vitasse Y, Bresson CC, Kremer A, Michalet R, Delzon S (2010) Quantifying phenological plasticity to temperature in two temperate tree species. Functional Ecology 24: 1211–1218.
44. Norby RJ, Hartz-Rubin JS, Verbrugge MJ (2003) Phenological responses in maple to experimental atmospheric warming and CO_2 enrichment. Global Change Biology 9: 1792–1801.
45. Fontaine F, Chaar H, Colin F, Clément C, Burrus M, et al. (1999) Preformation and neoformation of growth units on 3-year-old seedlings of *Quercus petraea*. Canadian Journal of Botany-Revue Canadienne De Botanique 77: 1623–1631.
46. Sanz-Pérez V, Castro-Díez P (2010) Summer water stress and shade alter bud size and budburst date in three mediterranean *Quercus* species. Trees-Structure and Function 24: 89–97.
47. Löf M, Welander NT (2000) Carry-over effects on growth and transpiration in *Fagus sylvatica* seedlings after drought at various stages of development. Canadian Journal of Forest Research-Revue Canadienne De Recherche Forestière 30: 468–475.
48. Fritts HC (1962) The relation of growth ring widths in American beech and white oak to variations in climate. Tree-Ring Bulletin 1: 2–10.
49. Fonti P, Solomonoff N, García-González I (2007) Earlywood vessels of *Castanea sativa* record temperature before their formation. New Phytologist 173: 562–570.
50. Dobbertin M (2005) Tree growth as indicator of tree vitality and of tree reaction to environmental stress: a review. European Journal of Forest Research 124: 319–333.
51. Misson L, Degueldre D, Collin C, Rodriguez R, Rocheteau A, et al. (2011) Phenological responses to extreme droughts in a Mediterranean forest. Global Change Biology 17: 1036–1048.
52. Limousin JM, Rambal S, Ourcival JM, Rodriguez-Calcerrada J, Perez-Ramos IM, et al. (2012) Morphological and phenological shoot plasticity in a Mediterranean evergreen oak facing long-term increased drought. Oecologia 169: 565–577.
53. Hanson PJ, Todd DE, Amthor JS (2001) A six-year study of sapling and large-tree growth and mortality responses to natural and induced variability in precipitation and throughfall. Tree Physiology 21: 345–358.
54. Leuzinger S, Zotz G, Asshoff R, Körner C (2005) Responses of deciduous forest trees to severe drought in Central Europe. Tree Physiology 25: 641–650.
55. Bucher-Wallin I, Sonnleitner MA, Egli P, Günthardt-Goerg MS, Tarjan D, et al. (2000) Effects of elevated CO_2, increased nitrogen deposition and soil on evapotranspiration and water use efficiency of spruce-beech model ecosystems. Phyton (Austria) 40: 49–60.
56. Leuzinger S, Körner C (2007) Water savings in mature deciduous forest trees under elevated CO_2. Global Change Biology 13: 2498–2508.

21

Virtuous and Vicious Virtual Water Trade with Application to Italy

Julia Anna Winter*, Paola Allamano, Pierluigi Claps

Department of Environment, Land and Infrastructure Engineering, Politecnico di Torino, Torino, Italy

Abstract

The current trade of agricultural goods, with connections involving all continents, entails for global exchanges of "virtual" water, i.e. water used in the production process of alimentary products, but not contained within. Each trade link translates into a corresponding virtual water trade, allowing quantification of import and export fluxes of virtual water. The assessment of the virtual water import for a given nation, compared to the national consumption, could give an approximate idea of the country's reliance on external resources from the food and the water resources point of view. A descriptive approach to the understanding of a nation's degree of dependency from overseas food and water resources is first proposed, and indices of water trade virtuosity, as opposed to inefficiency, are devised. Such indices are based on the concepts of self-sufficiency and relative export, computed systematically on all products from the FAOSTAT database, taking Italy as the first case study. Analysis of time series of the self-sufficiency and relative export can demonstrate effects of market tendencies and influence water-related policies at the international level. The goal of this approach is highlighting incongruent terms in the virtual water balances by the viewpoint of single products. Specific products, which are here referred to as "swap products", are in fact identified as those that lead to inefficiencies in the virtual water balance due to their contemporaneously high import and export. The inefficiencies due to the exchanges of the same products between two nations are calculated in terms of virtual water volumes. Furthermore, the cases of swap products are investigated by computing two further indexes denoting the ratio of virtual water exchanged in the swap and the ratio of the economic values of the swapped products. The analysis of these figures can help examine the reasons behind the swap phenomenon in trade.

Editor: Alejandro Raul Hernandez Montoya, Universidad Veracruzana, Mexico

Funding: The first author received support from the EU Project WARECALC (2008-230845). The funders had no role in study design, data collection and analysis, decision to publish, or preparation of the manuscript.

Competing Interests: The authors have declared that no competing interests exist.

* E-mail: julia.winter@polito.it

Introduction

The virtual water content of a product is the volume of water used to produce it, measured at the place where it is actually grown or manufactured. The adjective "virtual" refers to the fact that most of the water used in the production is ultimately not contained within the product. The real water content of products is generally negligible compared with their virtual water content [1].

The virtual water content of a given product varies greatly from place to place, depending mainly on the climate and on the technology adopted for irrigation and farming [2]. The variability of virtual water contents among different products and different locations has led to a diversification of case studies and research topics. Studies have been conducted at a global scale (e.g. [3], [4], [5]) but also in depth on single countries or areas (e.g. [6], [7], [8]) or on single products (e.g. [9], [10], [11]).

The global trade of all goods can also be translated into a corresponding network of virtual water trade, allowing quantification of the import and export fluxes of virtual water (e.g. [12], [13], [14]). Nations have been classified as net importers or exporters of virtual water according to the balance of trade fluxes entering and leaving the country (e.g. [15]).

It is therefore necessary to address the problems concerning water resources, such as water scarcity, water excess and deterioration of quality, not only at the agricultural district level and at the river basin level, but also at the nation and, ultimately, at a global level [15].

A nation can be considered "wise" when it exports products that are obtained from abundantly available resources within the country and imports products that are produced with resources that are scarcely available within the country. One of these resources could clearly be water; a water-scarce country might thus aim at behaving "wisely" by importing products that require a lot of water in their production and exporting products or services that require less water. However, this "virtuous" kind of trade does not (always) correspond to what trade figures show when examined on a long term and at the nation level, most likely because other factors, such as profit for instance, have a stronger influence on trade patterns.

To better examine this concept, it is important to determine the significance of the virtual water import for a given nation compared to what is nationally consumed; this could give an approximate idea of the country's reliance and dependency on external resources. The resources taken into consideration for this study are water resources destined for food production. Using the virtual water concept, a descriptive approach of the degree of dependency of a nation from overseas water resources is expanded in the next section in which indices are created, that could suggest a degree of trade "virtuosity" or "inefficiency". This requires establishing virtual water balances between a country and the rest of the world, in terms of exchanges of crops and livestock.

The immediate goal of this approach is to highlight incongruent terms in the virtual water balances, at a nation's scale and by the viewpoint of single products. In the subsequent sections, specific incongruences emerging in the trade of single products are highlighted to allow a clearer connection between economic driving forces and water resources. Italy is the nation considered in all examples shown in the paper, because it is among the leading OECD (Organisation for Economic Co-operation and Development) countries in the list of the greatest virtual water importers [8].

Materials and Methods

Descriptive Analysis of Production and Overall Trade

Data necessary for the investigations on production and trade presented in this paper are obtained from the database FAOSTAT (www.faostat.fao.org), that provides time-series of data relating to food and agriculture for some 200 countries. For this analysis primary and processed crops and primary and processed livestock are taken into consideration: 397 items in total for Italy - from year 1961 to year 2009. Exchanges are considered in terms of quantities imported (I), exported (E) and produced (P), in tonnes. Production data are paired with data on import and export and the balance term 'Consumption' (C) is defined as $C = P - E + I$ (considering constant storage stocks).

In order to have an overview of the overall trade of the country of interest, two ratios are computed for each product throughout the available time span:

i) the ratio between 'Import' and 'Consumption' (I/C), that shows the countries' dependency on overseas products; this term, named DP after *dependency*, indicates how much of the national request for crops or livestock is satisfied by import and not by internal production (the closer DP is to 0 the more is the commodity supply provided by the country itself);
ii) the ratio between 'Export' and 'Production' (E/P), named ER for *export ratio*, that shows the relative export attitude of the country; the index ER explains how much of the internal production is destined to be exported.

Items which have a high DP are items whose national demand is not satisfied by internal production; those which have a high ER are items whose production is destined mostly to be exported instead of being nationally consumed. When combining these two metrics, products can be found which show both a high DP and a high ER. These are products which behave very particularly: whilst there is a significant import, seemingly because of an impossibility of self sufficiency of the nation, there is for the same commodity a strong export of its production. Acknowledging that further internal processing of products is impossible to track, and assuming that the country does not function as a mere transit, i.e. imports are not consumed nationally but further exported, this feature configures a "vicious" kind of trade. The interest in computing DP and ER systematically on all products and on all the available years is in demonstrating possible "vicious" and incongruent trade figures for a given nation, at least in terms of virtual water budget.

Considering several products one may want to visualize the above indices on a bubble graph where the horizontal axis shows the dependency DP and the vertical axis the export ratio ER (Fig. 1). The production amount is reflected in the size of the bubbles, that also varies among the years. The radius of the bubble is assumed to scale with production amounts according to a logarithmic growth, i.e. when the radius doubles, a factor of 10 is applied. Logarithmic scaling allows to appreciate figures for the smallest productions. Everything is then scaled by the numeric amount of the highest production (that is reported in the caption of the figure).

The main information content of fig. 1 is the position of a commodity according to the quadrant where the two indices fall. The items located in the lower left quadrant are those which have very simple behaviour: they are produced locally mainly for the local consumption, keeping export and import to the minimum. This otherwise means that the production of such items seems to be calibrated on the request of the population and does not exceed it. The commodities located towards the upper right quadrant of the (DP, ER) diagram, instead, have at the same time high import and high export amounts compared respectively to consumption and production. These positions hence exemplify the incongruent type of trade discussed above. Figure 1 shows the ER-DP representation for the totality of the products analysed for Italy in three different years (1961, 1985 and 2009) which were taken as indicative of the complete time series. The colour green indicates primary and processed crops whereas light blue stands for primary and processed livestock. The products located in the upper right quadrant have evidently increased in number from 1961 to the year 2009.

A different way of examining information on production and trade is to visualize the time trend of a single product on the ER-DP diagram. The examples chosen for this time-series representation are Beans and Tobacco, whose location in the ER-DP diagram is highlighted by an orange star and that -among the products with high DP, high ER or both- presented the most significant trends. DP and ER data have been plotted versus time in figure 2, that shows for these commodities an increase in time in both DP and ER. The radius of the bubble is again proportional to the production but is not scaled by logarithm; the numeric amount of the highest production is reported in the caption of the figure. Economic considerations are likely to justify this peculiar evolution, as will be discussed hereinafter.

Results and Discussion

Virtual Water Inefficiency Assessment: The Swap Phenomenon

The previously shown behaviour for which a commodity is at the same time highly imported and highly exported will be referred to as the "swap" attitude. Between two trading countries a product can therefore be considered a swap product when being "unreasonably" exchanged both ways.

In order for a country to quantify the actual swap, information is necessary regarding import and export considering all fluxes entering (leaving) Country #1 from (to) the individual country of origin (destination). Such information can be obtained from a further section of the FAO database (i.e. the FAO Statistics TradeSTAT). A specific evaluation of the entity and evolution of this attitude is conducted on Italy as an initial case study.

The quantification of swap for the various products is done by comparing, year by year, the products' amounts traded by Italy with each different country: the minimum between import and export of each trade connection is selected as the quantity of product which is unreasonably exchanged, as expressed in the following equation:

$$swap_{i,j} = \min(import_{i,j}, export_{i,j})$$

where i indicates the traded commodity and j the country with which Italy is trading it. This is because a product should (ideally)

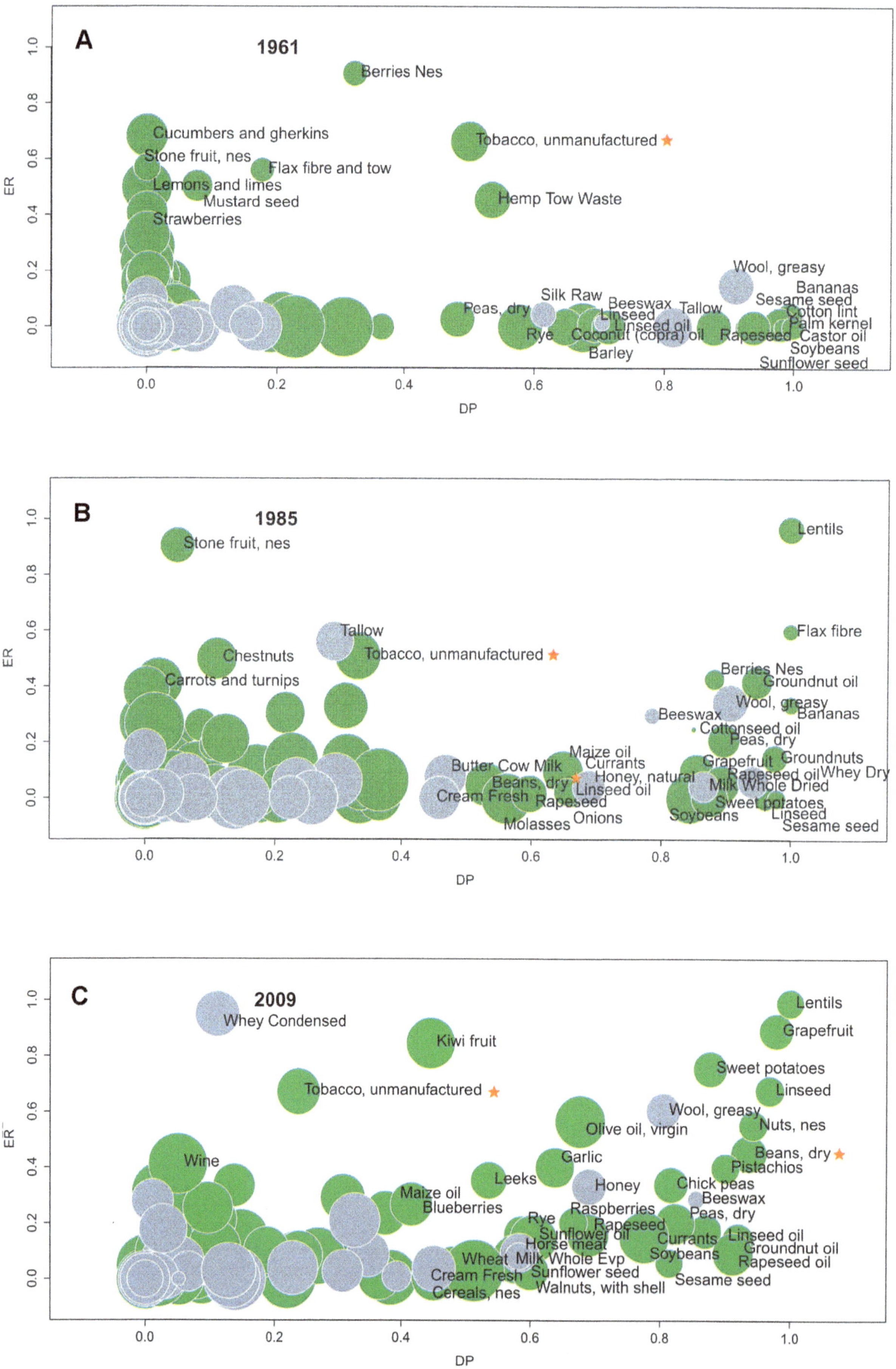
A
1961
Berries Nes
Cucumbers and gherkins
Tobacco, unmanufactured
Stone fruit, nes
Flax fibre and tow
Lemons and limes
Mustard seed
Hemp Tow Waste
Strawberries
Wool, greasy
Bananas
Silk Raw
Sesame seed
Beeswax
Tallow
Cotton lint
Peas, dry
Linseed
Linseed oil
Palm kernel
Rye
Coconut (copra) oil
Rapeseed
Castor oil
Barley
Soybeans
Sunflower seed
ER
DP
0.0
0.2
0.4
0.6
0.8
1.0
B
1985
Lentils
Stone fruit, nes
Tallow
Flax fibre
Chestnuts
Tobacco, unmanufactured
Berries Nes
Carrots and turnips
Groundnut oil
Wool, greasy
Bananas
Beeswax
Cottonseed oil
Peas, dry
Maize oil
Groundnuts
Butter Cow Milk
Currants
Grapefruit
Whey Dry
Rapeseed oil
Honey, natural
Beans, dry
Linseed oil
Milk Whole Dried
Cream Fresh
Rapeseed
Sweet potatoes
Soybeans
Linseed
Onions
Molasses
Sesame seed
ER
DP
0.0
0.2
0.4
0.6
0.8
1.0
C
2009
Lentils
Whey Condensed
Grapefruit
Kiwi fruit
Sweet potatoes
Tobacco, unmanufactured
Linseed
Wool, greasy
Olive oil, virgin
Nuts, nes
Beans, dry
Wine
Garlic
Pistachios
Leeks
Honey
Chick peas
Maize oil
Beeswax
Blueberries
Raspberries
Peas, dry
Rye
Rapeseed
Sunflower oil
Currants
Linseed oil
Horse meat
Soybeans
Groundnut oil
Wheat
Milk Whole Evp
Rapeseed oil
Cream Fresh
Sunflower seed
Sesame seed
Cereals, nes
Walnuts, with shell
ER
DP
0.0
0.2
0.4
0.6
0.8
1.0

Figure 1. ER-DP diagram for Italy for years 1961, 1985 and 2009. The indices of dependency DP and export ratio ER are computed over all products available and are visualised on a bubble graph where the horizontal axis shows DP and the vertical axis ER. The production amount is reflected in the size of the bubbles, that also varies among the years. The radius of the bubble is assumed to scale with production amounts according to a logarithmic law, i.e. when the radius doubles, a factor of 10 is applied. Logarithmic scaling allows to appreciate figures for the smallest productions. Everything is then scaled by the numeric amount of the highest production (for years 1961, 1985, 2009 respectively: Grapes 8,5E+06 tonnes, Grapes 9,6E+06 tonnes, Grapes 8,2E+06 tonnes). The colour green indicates primary and processed crops whereas light blue stands for primary and processed livestock. The orange stars indicate the position of beans and tobacco in reference to Figure 2.

only be imported or exported. When both terms are different from zero the smallest of the two can be considered "unnecessary". In other words, if the minimum is zero, there is an understandable dependency of one of the two countries (Italy in the specific case) on the other country. If this minimum is different from zero there is overlapping between I and E and this overlap is the swap amount.

The total (in tonnes) of the swap exchanges for each commodity is calculated by adding up all contributions from each trading country. The translation of these data into quantities of virtual water is subsequently done by multiplying the total swap amounts by the virtual water content per ton of each commodity, assessed for Italy [3], [8], [16]. This turns the product amounts into volumes of virtual water, in m^3, inefficiently exchanged for each commodity; for the assessment of the annual totals it is simply sufficient to add up all products contributions. This calculation is conducted for each year from 1986 to 2009 over all products reported by FAO. Results are presented as a dotted time series in Figure 3. A significant increasing trend is observed for the total volumes of virtual water traded through swap amounts at the nation's level on all trade connection. In order to understand the entity of the swap phenomenon a significant comparison with the total net flux can be done for the Italian case study: in 2009 the average net import of virtual water is 50E+09 m^3 [8] while the swap entails for 3E+09 m^3.

In the supporting information, Figure S1 and Figure S2, shown are synthetic representations of the virtual water carried from unnecessary imports, per commodity, for year 2009. The products with a null result of swap amount were omitted from this already "dense" representation. Of the totality of 397 items in year 2009, 365 undergo the swap phenomenon, with differences in quantity and in the number of countries involved. Of 365 items only for 162 was the virtual water content (m^3/ton) available and therefore the computation of the inefficiency in terms of traded virtual water possible. A logarithmic x axis is used to cover the whole range of trading amounts. For the sake of a clear representation the totality of the 162 products is divided into two categories, called 'crop' and 'livestock'. From this moment onwards the pool of products which will be taken into consideration for further analysis consists in the 162 commodities for which information on water content was obtainable.

While studying the possible explanations behind these vicious exchanges, two main considerations emerge regarding the phenomenon.

Firstly, the existence of swap entails that a country may work as a transit, i.e. products are passing through the country to reach their final destination. A method to pinpoint the manifestations of this phenomenon in the trade of a single country is to detect the products with a higher import than consumption (DP>1). This behaviour, corresponding also to higher export than production (ER>1), indicates a transportation of the product in question through the country, which may just act as a sorter. The values of the ratios ER and DP previously computed for all products and all years were scanned, allowing to identify a few products, such as Cotton lint, Mustard seed, Bananas and Flax fibre & tow, which show such a behaviour in Italy.

Secondly, many of the trade links between countries from different hemispheres are justified by the seasonality of some cultivations and allow to obtain a year-round availability of the commodities in question. The values obtained from FAO refer to the whole year which makes these trade links not immediately detectable. A more sensible analysis is needed in order to avoid considering swap for what is possibly an import or export of seasonal relevance; this could be further attempted by considering only the European countries, with which Italy shares climate and seasonality in crop production. The relevance of this subset finds support in Tamea et al. (8) from which it emerges that 60% of the total import of virtual water for Italy in 2010 comes from Europe and 72% of the export of virtual water from Italy in the same year is destined to Europe. European countries as a whole therefore represent the main trading partners for Italy. In particular, France is the major supplier of virtual water and one of the greatest importer of Italian goods, second only to Germany.

Swap Analysis

The analyses reported so far provide figures that raise questions about the sustainability of swap trade, as reasons behind the anomalies of swap products could be numerous. The attempt made is to investigate each of the previous trade connections which present a swap from an economic point of view and in terms of water resources efficiency. To this purpose, two ratios are introduced which are firstly computed for all trading links and, at a later stage, examined in depth for those products presenting the highest quotes of swap, as they result from the previous section.

The first indicator, referred to as Virtual Water Volume Ratio, is a ratio between the virtual water content (m^3/ton) of an item leaving Italy as export for another country, and the virtual water content (m^3/ton) of that item entering Italy from that same country. The ratio is computed for all items and all countries, generating a matrix of Virtual Water Volume Ratios, with a value for each trade connection. A "spatial" weighted mean is calculated for each product by using the import quantities from each country as weights, resulting into a single average value for each commodity. A Virtual Water Volume Ratio higher than 1 indicates a commodity which averagely requires more water for its production in Italy than in all other countries from which Italy is importing it. Table S1 reports the complete list of items considered in the computation. Of the 162 commodities for which the calculation was conducted, 93 present a Virtual Water Volume Ratio higher than 1.

The second indicator is the ratio between the economic unit value (1000USD/ton) of a certain commodity leaving Italy towards another country and the economic unit value (1000USD/ton) of the same commodity entering Italy from that same country. Information on such figures is obtained from the previously cited FAO Statistics TradeSTAT database. This ratio can justify the swap of products, since it is singularly calculated on each trading link between Italy and every single other nation on products which are being traded both ways. It will be referred to as Value Ratio and allows to pinpoint trade exchanges which are economically convenient in terms of unit value. As was done for

Beans

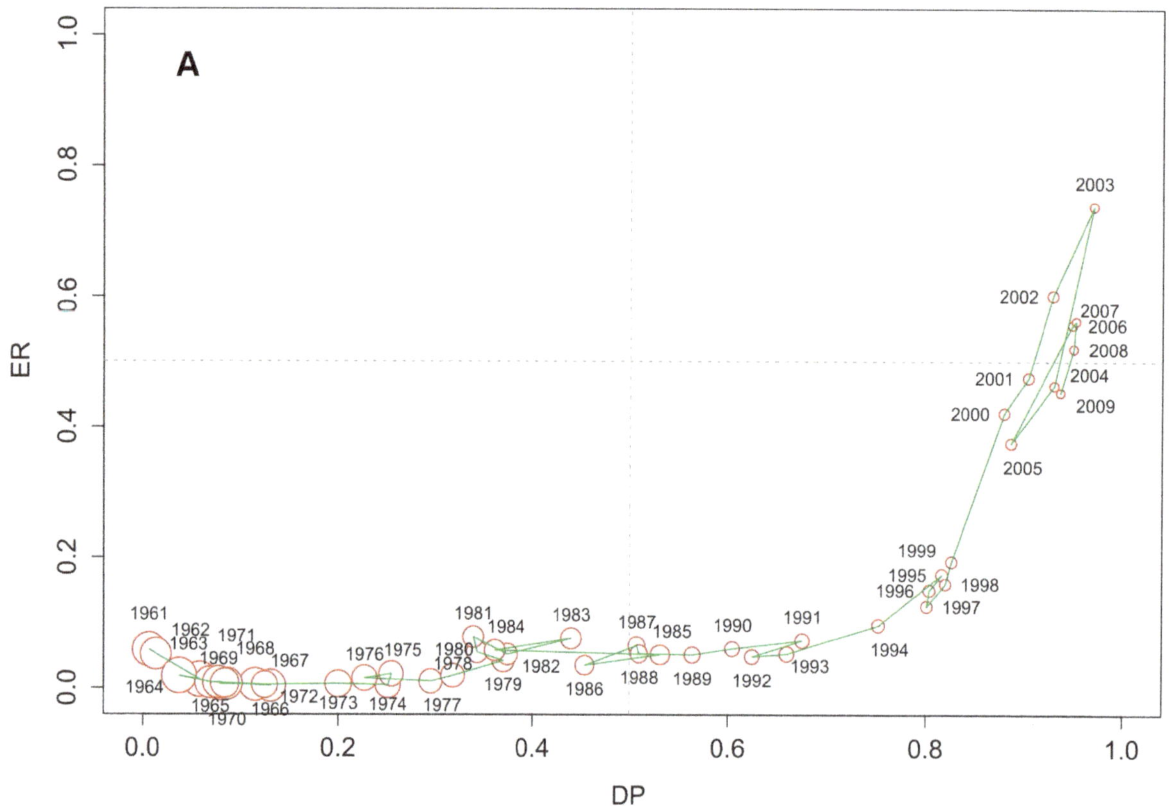

Tobacco

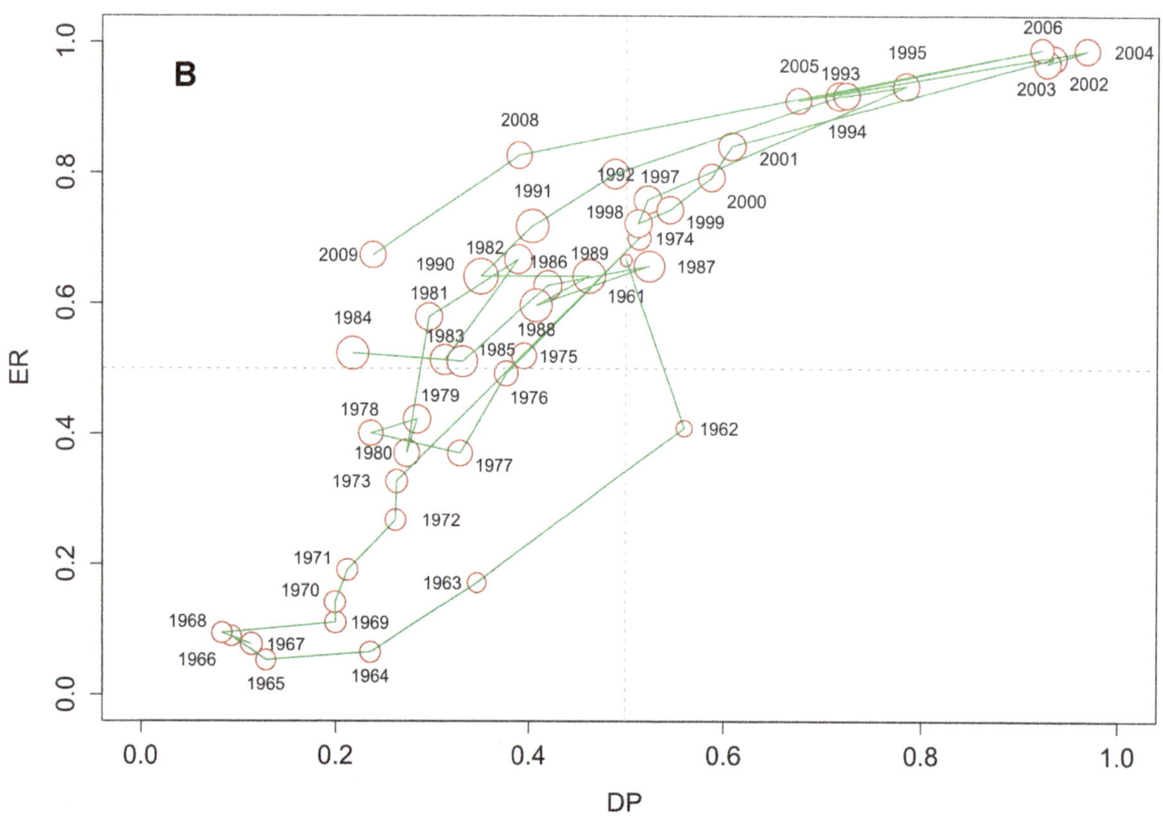

Figure 2. ER-DP diagram for Italy of Beans and Tobacco 1961–2009. Visualisation of the time trend of a single product on the ER-DP diagram. The examples chosen for this time-series representation are Beans and Tobacco, whose location in the ER-DP diagram is highlighted by an orange star in Figure 1 (beans in 1961 were below the threshold for labelling). For both of these crops, DP and ER data have been plotted versus time. It is evident that for these commodities there has been an increase in time in both DP and ER. The radius of the bubble is again proportional to the production, but is not scaled in a logarithmic way; the numeric amount of the highest production is 2,1E+05 tonnes in case of Beans in 1964 and 2,1E+05 tonnes for Tobacco in 1990.

the Virtual Water Volume Ratio, a matrix of Value Ratios is obtained when investigating each trading link and the weighted mean for each product is computed by using the swap (in tonnes) of each trade connection as weight. Table S2 presents the complete list of Value Ratio for the pool of commodities considered. The calculated Value Ratio is higher than 1 in 123 cases, meaning that Italian products are on average more valuable than the foreign product with which they are swapped. Goods exported from Italy thus present an added-value that could justify the swap.

Figure 4 presents the totality of commodities taken into consideration (162 items) plotted on a bubble graph where the horizontal axis shows the Value Ratio and the vertical axis the Virtual Water Volume Ratio (both axis are in logarithmic scale). Among those commodities produced in Italy in year 2009, those with a swap higher than 10E+07 m^3 are represented as orange bubbles, while the swap amount is reflected in the size of the bubbles (the radius is scaled according to the swap amount). It can be seen that, with the notable exception of cattle meat, the most important commodities from the point of view of swap fall in the right part of the graph, indicating a profit. The opposite behaviour could be ascribed to other drivers which determine the existence of some trade links, as for instance the ones related to the quality of imported products or to the customs and habits among neighbouring countries. These categories of drivers that influence the global food trade logics are extremely difficult to map and to quantify and are not investigated in this paper.

Data presented in tables 1 and 2 show the 20 most impacting items from the point of view respectively of m^3 of virtual water inefficiently swapped (Table 1) and of what is referred to as net flux (8) (difference between the m^3 of virtual water imported and m^3 of virtual water exported, Table 2). The reported figures are coupled with the value and ranking (in brackets) of the two previous ratios, the Virtual Water Ratio and the Value Ratio. It can be noticed that many of the products which present a high swap, have at the same time a high value of net flux, such as wheat, meat-cattle boneless(beef and veal), maize, horse meat, soybeans. For a product presenting a high value of net flux, the country is strongly relying on import and consequently on the import of virtual water in order to satisfy the national demand of such commodity. A swap in such case would be most unwelcome and unreasonable, even if small in percentage. This is the case of wheat. Wheat, in fact, represents a highly traded crop (net flux of 8.3E+09 m^3 of virtual water) and at the same time a highly swapped crop (swap of 1.3E+08 m^3). By analysing the Virtual Water Volume Ratio one realises that, on average, the virtual water content of imported wheat is higher than the virtual water content of Italian wheat (the Italian contains 0.63 times the water averagely contained by foreign wheat). However, according to the Value Ratio, the wheat imported is economically less valuable than the Italian one (the Italian wheat has an economic unit value which is 1.54 times the average incoming wheat value). One could conclude, also in this case, that Italy swaps wheat in order to profit from it.

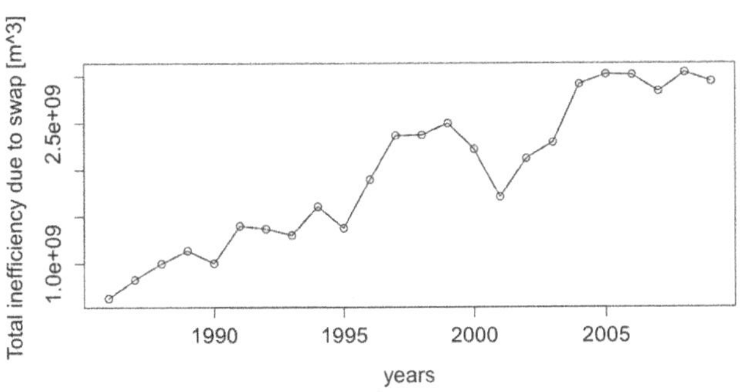

Figure 3. Total inefficiency assessment in terms of volume of virtual water traded through swap amounts. Total inefficiency assessment in terms of volume of virtual water traded through swap amounts, computed for Italy. The total (in tonnes) of the swap exchanges for each commodity is calculated by adding up all contributions from each trading country. The translation of these data into quantities of virtual water is subsequently done by multiplying the total swap amounts by the virtual water content per ton of each commodity. This turns the product amounts into volumes of virtual water inefficiently exchanged for each commodity; for the assessment of the annual totals it is simply sufficient to add up all products contributions. This calculation is conducted for each year from 1986 to 2009 over all products reported by FAO. Results are presented as the dotted time series.

Conclusions

Relations between food production and the related water consumption are investigated by examining the nature and the evolution of bilateral trade between countries, with particular reference to the virtual water efficiency or inefficiency connected to production and trade. The research presented here relates not so much to the marginal production efficiency, which will depend on the characteristics of irrigation techniques or to the variations of productivity according to climate conditions; it rather considers the influence of commercial interests on the global trade of virtual water, by means of the isolation of two-way exchanges of the same products between importers and exporters ("swap").

The Italian case, in this sense, is very representative since Italy is a strong importer of virtual water. By using the concept of swap it is highlighted that, even for products which are abundant in Italy, a substantially inefficient trade from the point of view of the budget of virtual water is practiced with increasing intensity. The swap occurs for the economic benefit of the country in question, but raises doubts about its sustainability from the point of view of global management of water resources, in particular with reference to the growing trend in these exchanges.

The considerations expressed in this paper do not contemplate the shipping costs related to the traded goods nor any aspect related to the organoleptic quality of the imported and exported food.

The extension of this work on a larger scale could allow interesting developments in the assessment of the impact of swap and its economic drivers on a global scale, for example with reference to sudden food shortages, deriving both from climate anomalies or from other human interventions. Future activities in this direction will allow us to evaluate the possible relations

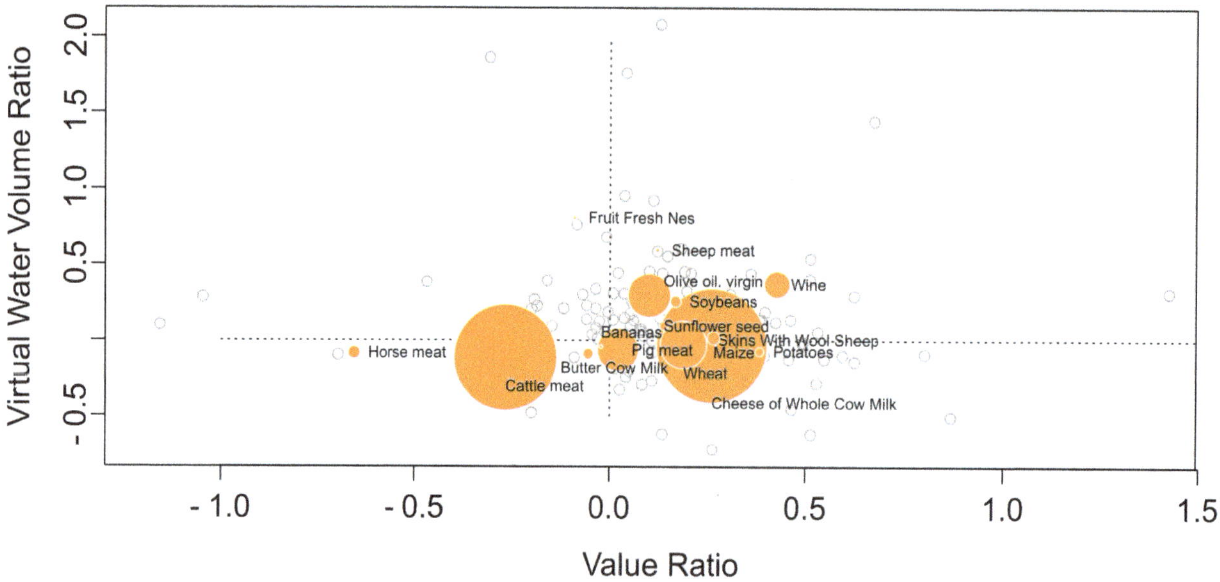

Figure 4. Value Ratio – Virtual Water Volume Ratio diagram for Italy in 2009. The indices of Value Ratio and Virtual Water Volume Ratio are computed over all products available and are presented as a scatter plot. Among the commodities produced in Italy in 2009 those with a swap higher than 10^7 m^3 are represented by bubbles whose dimension is given by the swap amount. The radius of the bubbles is scaled (non logarithmically) from the highest figure which is 3.2E+08 m^3 of Cheese of Whole Cow Milk.

Table 1. The 20 most swapped products.

FAO code	Item	Swap (m3)	Virtual Water Volume Ratio	Value Ratio
870	Meat-CattleBoneless(Beef&Veal)	4.9E+08	0.79 (129)	0.88 (136)
901	Cheese of Whole Cow Milk	3.2E+08	0.90 (106)	1.82 (44)
867	Cattle meat	3.0E+08	0.75 (134)	0.54 (154)
15	Wheat	1.3E+08	0.63 (105)	1.54 (66)
261	Olive oil, virgin	1.2E+08	1.94 (34)	1.26 (93)
1035	Pig meat	1.2E+08	0.85 (112)	1.05 (117)
1039	Bacon and Ham	9.6E+07	0.81 (117)	2.21 (30)
564	Wine	7.4E+07	2.29 (24)	2.67 (19)
920	Hides Wet Salted Cattle	7.4E+07	0.81 (120)	0.95 (128)
921	Hidesdry S.Cattle	7.0E+07	0.77 (132)	0.81 (142)
268	Sunflower oil	5.7E+07	1.35 (64)	1.15 (102)
1232	Food Prep Nes	5.6E+07	2.51 (20)	1.44 (74)
898	Milk Skimmed Dry	4.7E+07	0.84 (113)	0.42 (157)
56	Maize	3.7E+07	1.02 (90)	1.84 (42)
16	Flour of Wheat	3.6E+07	0.63 (142)	1.12 (108)
1097	Horse meat	3.5E+07	0.81 (121)	0.22 (159)
1042	Prep of Pig Meat	3.3E+07	0.82 (116)	2.31 (27)
236	Soybeans	3.2E+07	1.77 (40)	1.47 (72)
886	Butter Cow Milk	3.1E+07	0.80 (124)	0.88 (135)
18	Macaroni	3.0E+07	0.93 (102)	0.97 (126)

The 20 most swapped products in terms of virtual water inefficiently traded, with relative Virtual Water Volume Ratio and Value Ratio.

Table 2. The 20 products with the highest net flux.

FAO code	Item	Net flux (m3)	Virtual Water Volume Ratio	Value Ratio
15	Wheat	8.3E+09	0.63 (105)	1.54 (66)
236	Soybeans	3.1E+09	1.77 (40)	1.47 (72)
238	Cake of Soybeans	2.2E+09	2.02 (29)	1.09 (112)
882	Cow milk, whole, fresh	1.9E+09	4.01 (10)	1.52 (68)
867	Cattle meat	1.7E+09	0.75 (134)	0.54 (154)
56	Maize	1.4E+09	1.02 (90)	1.84 (42)
268	Sunflower oil	1.4E+09	1.35 (64)	1.15 (102)
870	Meat-CattleBoneless(Beef&Veal)	1.3E+09	0.79 (129)	0.88 (136)
271	Rapeseed oil	1.3E+09	0.55 (148)	1.78 (48)
261	Olive oil, virgin	1.1E+09	1.94 (34)	1.26 (93)
1035	Pig meat	1.1E+09	0.85 (112)	1.05 (117)
267	Sunflower seed	9.4E+08	1.22 (75)	1.37 (81)
1097	Horse meat	8.9E+08	0.81 (121)	0.22 (159)
920	Hides Wet Salted Cattle	8.7E+08	0.81 (120)	0.95 (128)
901	Cheese of Whole Cow Milk	8.2E+08	0.90 (106)	1.82 (44)
269	Sunflower Cake	6.0E+08	1.47 (55)	1.59 (61)
44	Barley	5.0E+08	0.88 (109)	1.67 (56)
231	Almonds Shelled	4.2E+08	1.46 (56)	1.45 (73)
237	Soybean oil	3.7E+08	1.86 (36)	1.35 (84)
888	Milk Skm of Cows	3.4E+08	0.81 (122)	6.32 (3)

The highest 20 products from the point of view of the net flux (difference between the virtual water totally imported and exported), with relative Virtual Water Volume Ratio and Value Ratio.

between virtuous (or vicious) trade policies and country (or global) resilience to droughts, famines etc.

Supporting Information

Figure S1 Inefficiency per CROP product in terms of volume of virtual water traded through swap amounts (in 10× units). Synthetic representations of the virtual water carried from unnecessary imports, per commodity per year, for year 2009. A logarithmic x axis is used to cover the whole range of trading amounts. The category represented is 'crop'.

Figure S2 Inefficiency per LIVESTOCK product in terms of volume of virtual water traded through swap amounts (in 10× units). Synthetic representations of the virtual water carried from unnecessary imports, per commodity per year, for year 2009. A logarithmic x axis is used to cover the whole range of trading amounts. The category represented is 'livestock'.

Table S1 Complete list of items considered in the computation of the Virtual Water Volume Ratio. Of the 162 commodities for which the calculation was possible, 93 present a Virtual Water Volume Ratio higher than 1.

Table S2 Complete list of commodities for which Italy is conducting a swap, with the referring Value Ratio. Of the 162 products analysed the calculated Value Ratio is higher than 1 in 123 cases.

Acknowledgments

The authors would like to thank Francesco Laio, Paolo D'Odorico and Hector Malano for the precious contributions and valuable comments. Two anonymous reviewers are gratefully acknowledged. Support from the EU Project WARECALC (2008-230845) to the first author is also acknowledged.

Author Contributions

Conceived and designed the experiments: JW PA PC. Performed the experiments: JW. Analyzed the data: JW PA PC. Wrote the paper: JW.

References

1. Allan JA (1998) Virtual water: A strategic resource global solutions to regional deficits. Ground Water 36: 545–546.
2. Hoekstra AY, Mekonnen MM (2012) The water footprint of humanity. P Natl Acad Sci USA 109: 3232–3237.
3. Chapagain AK, Hoekstra AY (2004) Water Footprints of Nations vol 1 Main Report. (Delft: UNESCO-IHE).
4. Mekonnen MM, Hoeckstra AY (2010) A global and high-resolution assessment of the green, blue and grey water footprint of wheat. Hydrol Earth Syst Sci. : 1259–1276.
5. D'Odorico P, Carr J, Laio F, Ridolfi L (2012) Spatial organization and drivers of the virtual water trade: a community-structure analysis. Environ Res Lett 7: 034007.
6. Van Oel PR, Mekonnen MM, Hoekstra AY (2009) The external water footprint of the Netherlands: Geographically-explicit quantification and impact assessment. Ecological Economics 69: 82–92.
7. Chapagain AK, Hoekstra AY (2007) The water footprints of Morocco and the Netherlands: global water use as a result of domestic consumption of agricultural commodities. Ecol Econ 64: 143–151.

8. Tamea S, Allamano P, Carr JA, Claps P, Laio F et al. (2013) Local and global perspectives on the virtual water trade. HESS, 17: 1205–1215.
9. Mekonnen MM, Hoekstra AY (2010) The green, blue and grey water footprint of crops and derived crop products, Value of Water Research Report Series No. 47. UNESCO-IHE, Delft, the Netherlands.
10. Salmoral G, Aldaya MM, Chico D, Garrido A, Llamas MR (2011) The water footprint of olives and olive oil in Spain. Spanish Journal of Agricultural Research 9: 1089–1104.
11. Bulsink F, Hoekstra AY, Booij MJ (2010) The water footprint of Indonesian provinces related to the consumption of crop products. Hydrol Earth Syst Sci 14: 119–128.
12. Dalin C, Konar M, Hanasaki N, Rinaldo A, Rodiguez-Iturbe I (2012) Evolution of the global virtual water trade network P. Natl Acad Sci USA 109: 5989–5994.
13. Carr J, D'Odorico P, Laio F, Ridolfi L (2012) On the temporal variability of the virtual water network. Geoph Res Letts 39 : L06404.
14. Konar M, Dalin C, Suweis S, Hanasaki N, Rinaldo A et al. (2011) Water for food: the global virtual water trade network. Water Resources Research 47 Issue 5.
15. Hoekstra AY, Chapagain AK (2008) Globalization of water: sharing the planet's freshwater resources. Blackwell, Oxford (UK).
16. Food and Agriculture Organisation, FAOSTAT commodity list. Available: http://www.fao.org/economic/ess/ess-standards/commodity/en/ (verified on February 18, 2014).

22

Overexpression of *OsMYB48-1*, a Novel MYB-Related Transcription Factor, Enhances Drought and Salinity Tolerance in Rice

Haiyan Xiong[1], Jinjie Li[1], Pengli Liu[1], Junzhi Duan[1], Yan Zhao[1], Xiao Guo[1], Yang Li[1], Hongliang Zhang[1], Jauhar Ali[2], Zichao Li[1]*

1 Key Lab of Crop Heterosis and Utilization of Ministry of Education and Beijing Key Lab of Crop Genetic Improvement, China Agricultural University, Beijing, People's Republic of China, **2** International Rice Research Institute, Metro Manila, Philippines

Abstract

MYB-type transcription factors (TFs) play essential roles in plant growth, development and respond to environmental stresses. Role of MYB-related TFs of rice in drought stress tolerance is not well documented. Here, we report the isolation and characterization of a novel MYB-related TF, *OsMYB48-1*, of rice. Expression of *OsMYB48-1* was strongly induced by polyethylene glycol (PEG), abscisic acid (ABA), H_2O_2, and dehydration, while being slightly induced by high salinity and cold treatment. The OsMYB48-1 protein was localized in the nucleus with transactivation activity at the C terminus. Overexpression of *OsMYB48-1* in rice significantly improved tolerance to simulated drought and salinity stresses caused by mannitol, PEG, and NaCl, respectively, and drought stress was caused by drying the soil. In contrast to wild type plants, the overexpression lines exhibited reduced rate of water loss, lower malondialdehyde (MDA) content and higher proline content under stress conditions. Moreover, overexpression plants were hypersensitive to ABA at both germination and post-germination stages and accumulated more endogenous ABA under drought stress conditions. Further studies demonstrated that overexpression of *OsMYB48-1* could regulate the expression of some ABA biosynthesis genes (*OsNCED4, OsNCED5*), early signaling genes (*OsPP2C68, OSRK1*) and late responsive genes (*RAB21, OsLEA3, RAB16C* and *RAB16D*) under drought stress conditions. Collectively, these results suggested that *OsMYB48-1* functions as a novel MYB-related TF which plays a positive role in drought and salinity tolerance by regulating stress-induced ABA synthesis.

Editor: Haibing Yang, Purdue University, United States of America

Funding: This work was supported by grants from National Natural Science Foundation of China (Grant No. 31061140458. http://www.nsfc.gov.cn/Portal0/default166.htm); Ministry of Agriculture of the People's Republic of China (Grant No. 2011ZX08009-003-002. http://english.agri.gov.cn/np/); National Basic Research Program of China (Grant No. 2010CB125904. http://www.973.gov.cn/English/Index.aspx); the Program of the Co-Construction with Beijing Municipal Commission of Education of China (http://english.bjedu.gov.cn/) The funders had no role in study design, data collection and analysis, decision to publish, or preparation of the manuscript.

Competing Interests: I have read the journal's policy and have the following conflicts: Co-author Jauhar Ali is a PLOS ONE Editorial Board member, but this does not alter our adherence to PLOS ONE Editorial policies and criteria.

* E-mail: lizichao@cau.edu.cn

These authors contributed equally to this paper.

Introduction

Crop plants are frequently exposed to variable abiotic stresses. Among them, drought is a major stress factor for crop growth and production. During the long evolutionary history, plants have evolved various mechanisms to survive in the stress environments, such as changes in internal molecules, cellular and physiological processes [1]. Abscisic acid (ABA) is such an essential phytohormone that regulates various aspects of plant growth and development in response to the abiotic stresses [2]. ABA content increases dramatically when plants are exposed to abiotic stresses such as drought and salinity that triggers the expression of many genes encoding various proteins important for biochemical and physiological processes [3–5]. These proteins help plants be tolerant to the stresses that includes late embryogenesis abundant (LEA) proteins, enzymes involved in osmo-protectant synthesis, protection proteins for plants from oxidative damage, and signaling proteins, such as the transcription factors (TFs) [2].

TFs like MYB, AP2/ERF, NAC, bZIP, and WRKY families act as the early responders to environmental signals and trigger the expression of stress-induced genes that are important for plants to be tolerant to abiotic stresses [6–8]. In recent years, many researchers have demonstrated the utilization of TFs in the engineering of crop plants as a powerful approach for enhanced tolerance against abiotic stresses. For instance, overexpression of *AP37*, *OsNAC10*, *OsbZIP23* and *OsbZIP46* in rice can significantly improve the tolerance to drought and high-salinity stresses [9–12].

MYB is a large TF family in plants. There are over 198 MYB genes in *Arabidopsis* and over 183 members in rice [13]. According to the number of imperfect repeats of the SANT (for SWI3, ADA2, N-CoR, and TFIIIB) (50–53 amino acids) in DNA-binding domains, plant MYB proteins can be classified into three major subfamilies: the MYB-related group (one single SANT domain), the R2R3-type group (two SANT domains) and R1R2R3-MYB group (three SANT domains) [13,14]. Over the past decade, it has been reported that the most of MYB genes, involve in response to diverse abiotic stress, belong to the R2R3-type group. For instance, *AtMYB2* functions in the ABA-mediated drought stress response in *Arabidopsis* [15]. *AtMYB102* is a key regulatory component in responses of *Arabidopsis* to wounding, osmotic stress, salinity stress and ABA application [16]. In addition, *AtMYB96* modulates ABA signaling in response to abiotic stresses in *Arabidopsis* [17]. In rice, *OsMYB4* was earlier reported to play a positive role in transgenic *Arabidopsis*, tomato, and apple to cold and drought tolerance [18–20]. *OsMYB55* was shown to be involved in tolerance to high temperature through enhanced amino acid metabolism [21]. In a recent study, molecular characteristic features of *OsMYB2* have clearly been indicating its regulatory role in salt, cold, dehydration tolerance in rice [22].

Compared with R2R3-type MYB genes, few studies of the MYB-related group genes in abiotic stress response have been reported in plants. In potato, the single MYB domain TF *StMYB1R-1* has been shown to involve in drought tolerance through activation of drought-related genes [23]. Overexpression of a single-repeat MYB TF *AmMYB1* from grey mangrove confers salt tolerance in transgenic tobacco [24]. So far, MYBS3 is the only MYB-related protein in rice, which was reported to be involved in abiotic stresses, and it is essential for cold stress tolerance of rice [25]. However, little is known about the role of MYB-related TFs of rice in drought and salt stress tolerance.

Based on our cDNA microarray data published previously [26], we isolated a novel rice MYB-related type gene, *OsMYB48-1*. In this study, we found that the expression of *OsMYB48-1* was induced by various abiotic stresses. In order to carry out functional studies on *OsMYB48-1* gene and later on to apply it for genetic engineering of crops, firstly we have developed the transgenic rice overexpressing *OsMYB48-1*. Our study clearly indicated that many ABA-related genes were up-regulated in the *OsMYB48-1*-overexpression plants especially under drought stress and the transgenic plants had shown improved tolerance to drought and high salinity stress. All these findings will be utilized for the engineering crops with drought and salinity tolerance.

Materials and Methods

Plant Growth Conditions and Stress Treatments

To detect the transcription level of *OsMYB48-1* under various abiotic stresses and phytohormone treatment, seeds of Nipponbare (*O. sativa* L. ssp. japonica) were sterilized in 75% ethanol for 2 min and in 20% NaClO for 30 min, and washed with sterile water thoroughly. The sterilized seeds were germinated at 37°C for 2 d and grown hydroponically with Hoagland solution in growth chamber (28°C/25°C (day/night) with a 12 h photoperiod). Three-week-old seedlings were treated with stress including dehydration stress (the intact seedlings were exposed in the air without water supply), high osmotic pressure using 20% (w/v) polyethylene glycol (PEG) 6000 solution, salinity stress using 150 mM NaCl solution, cold exposure at 4°C, oxidative stress using 1 mM H_2O_2 solution, and hormone treatment using 100 mM ABA solution followed by sampling in a pre-determined time manner. Total RNA was extracted from leaves and used for further analysis.

RNA Isolation and qRT-PCR Analysis

Total RNA was isolated from the rice leaves using RNAiso Plus (TaKaRa, Dalian, China) according to the manufacturer's instructions. The DNase-treated RNA was reverse-transcribed into first-strand cDNA with M-MLV reverse transcriptase (TaKaRa, Dalian, China) according to the manufacturer's instructions. A fivefold dilution of the resultant 1st strand cDNA was used as template for PCR. Quantitative real-time PCR (qRT-PCR) was performed according to described previously [27]. The rice *Actin* gene was used as the endogenous control for data normalization. Each analysis was repeated at three times. All the primers were listed in Table S1.

Sequence Analysis

The homolog genes of *OsMYB48-1* were searched through BLAST database (http://www.ncbi.nlm.nih.gov/BLAST/). The phylogenetic tree was constructed in MEGA5 software based on neighbor-joining method. Numbers indicate percentage values after 1000 replications. The multiple sequence alignment was performed with ClustalW. We searched the promoter sequence (2,000 bp upstream from the transcription start site) against the PLACE database (http://www.dna.affrc.go.jp/PLACE/signalscan.html) to detect the *cis*-acting elements for reveal the regulatory roles of *OsMYB48-1*.

Plasmid Construction and Plant Transformation

To generate *OsMYB48-1* overexpression plants, the full-length coding region of *OsMYB48-1* was amplified from the first-strand cDNA of Nipponbare using gene-specific primer pairs and cloned into the *Asc* I and *Pac* I sites of binary expression vector PMDC32 under the control of double *Cauliflower Mosaic virus* (*CaMV*) *35S* promoter. The resultant construct was transformed into *Agrobacterium tumefaciens* EH105 and then transformed into Nipponbare by *Agrobacterium*-mediated transformation method [28]. The transgenic plants were selected in 1/2 MS medium containing 50 mg/L hygromycin (Roche, Germany).

To generate the RNA interference (RNAi) plants, the 462 bp cDNA sequence was amplified and cloned into the *Kpn* I and *Bam*H I and then *Spe* I and *Sac* I sites of the pTCK303 vector as described previously [29]. Plant transformation protocols were taken as described above.

Subcellular Localization and Transactivation Assay

To determine its subcellular localization, the whole *OsMYB48-1* coding region without stop codon was amplified and sub-cloned into the binary vector PMDC83, to produce OsMYB48-1-GFP fusion construct driven by *CaMV 35S* promoter. The recombinant constructs and free GFP were introduced into onion (*Allium cepa*) epidermal cells, performed with a model PDS-1000/He Biolistic particle delivery system (Bio-Rad, CO, USA). After incubation at 25°C for 24 h, the green fluorescence signal was observed using a confocal laser scanning microscope LSM510 Meta (Carl Zeiss; http://www.zeiss.com/) with an argon laser excitation wavelength (488 nm).

The transactivation experiment was carried out according to the manual of Yeast Protocols Handbook (Clontech). The full length coding region and truncated fragments of *OsMYB48-1* generated by PCR amplification were fused in frame to the GAL4 DNA binding domain in the vector of pGBKT7 (Invitrogen). pGBKT7 was used as a negative control, while the pGBKT7-53 was used as a positive control. These constructs were transformed into Yeast strain AH109 by the Lithium acetate-method. The transformed yeast strains were screened on the selective medium plates without Tryptophan (SD/−Trp), and cultured at 30°C for 2 d. The PCR-verified positive strains were transferred to SD medium lacking Tryptophan/Histidine/Adenine (SD/−Trp/−His/−Ade) and cultured for 2 d, and the *in vivo* agar plate assay (x-α-gal in medium) was performed to analyze the transactivation activities.

Evaluation of Transgenic Plants in Response to Various Stresses

The seeds of *OsMYB48-1* overexpression lines and the wild type (WT) were germinated in 1/2 MS medium for various stress evaluations. For osmotic and high salinity treatment, the 3 d seedlings of both transgenic and WT plants (10 plants each repeat, 3 repeats) were transplanted to normal 1/2 MS medium, and 1/2 MS medium containing 200 mM mannitol and 150 mM NaCl, respectively. After 10 d at 28°C/25°C (day/night) with a 12 h photoperiod, shoot height and fresh weight of each line and WT were measured.

For dehydration treatment, three-week-old seedlings of both the *OsMYB48-1* transgenic and WT plants were grown hydroponically in bottom removed 96-well plates using Hoagland solution containing 20% (w/v) PEG6000 for 3 d. Then, the stressed plants were recovered in normal Hoagland solution for 10 d. Survival rates of each line and WT were investigated.

For drought treatment, 20 seedlings of the transgenic and WT in each flowerpot at 140 mm diameter x 160 mm deep filled with well forest soil and vermiculite (1:1) were grown for one month. After creating drought-stress by not watering for 7 days under outdoor conditions, the seedlings were followed by recovery with water supply for the next 10 days. Plants were regarded as survivals if there were green and healthy young leaves after re-watering treatment. Survival rate was calculated as the ratio of number of survived plants over the total number of treated plants in a flowerpot. The representative plants of the transgenic and WT were photographed before and after the drought treatments. Each stress test was repeated at three times.

Water Loss Assay and Quantification of Malondialdehyde (MDA) and Proline

To detect the rate of water loss under dehydration conditions, the leaves of the transgenic and WT plants grown under normal growth conditions for three weeks were detached and weighed immediately as the initial weight. The samples were placed on a laboratory bench at room temperature and weighed at the pre-determined time intervals. The rate of water loss was calculated on the basis of the initial weight of the leaves [27]. Ten plants of each transgenic and WT line were used in this assay, and three replicates were made for each line.

The malondialdehyde (MDA) content was measured following the methods described by Duan *et al* [27]. Free proline content was determined according to the methods used by Bates *et al* [30] and *Song et al* [31].

ABA Sensitivity Test

The sensitivity of seed germination to ABA was assayed in 1/2 MS medium with ABA. Seeds of the transgenic and WT were germinated in 1/2 MS medium (30 seeds each, 3 repeats) containing four concentrations of ABA (0, 2, 3 and 5 μM) and the germination rates were recorded after 7 d.

To test the ABA sensitivity at post-germination stage, the transgenic and WT were germinated in normal 1/2 MS medium plates. The seedlings (about 3 days, 10 plants each, and 3 repeats) were transferred to 1/2 MS medium containing 0 μM and 3 μM ABA. Plant height and fresh weight of each seedling were measured after 10 d of the ABA treatment in a greenhouse at 28/25°C (day/night) with a 12 h photoperiod.

Endogenous ABA Content Assay

Fresh leaves (0.2 g) from both the normal and drought stressed (without water for 5 d) plants were ground in an ice-cooled mortar, and homogenized in 4 mL 80% methanol containing 1 mM butylated hydroxytoluence as an antioxidant. The extract was incubated at 4°C for 4 h and centrifuged at 4,000 g for 20 min. The supernatant was passed through a Sep-Pak C18 cartridge (Waters Corp. Milford, MA), and prepared for endogenous ABA determination using enzyme-linked immunosorbent assay (ELISA) as described previously [32]. Each data point was the average value of three replicates.

Accession Numbers

Sequence data from this article can be found with the following accession numbers in the GenBank databases: *Actin* (AK100267), *NCED4* (AK11978), *NCED5* (AY838901), *OsPP2C68* (AK063334), *OSRK1* (AB125307), *RAB21* (AK121952), *LEA3* (AK119713), *RAB16C* (AK071366), *RAB16D* (AK109096), *P5CS1* (AK102633) and *P5CS2* (AK069425).

Results

Isolation and Characterization of *OsMYB48-1*

Based on our previous work of rice expression profiling under water stress using a cDNA microarray, an EST showing increased expression level after water stress was identified [26]. The EST showed homology to known MYB genes. Two transcripts that resulted from alternative splicing were identified from the Gramene database, designated as *OsMYB48-1* and *OsMYB48-2* (with locus name LOC_Os01g74410.2 and LOC_Os01g74410.1). OsMYB48-1 misses the first R region, and encodes a single MYB-like domain TF, while OsMYB48-2 encodes a R2R3 type MYB TF. Here, we first introduce the isolation and characterization of *OsMYB48-1*. The gene comprises 1583 bp nucleotides with a 633 bp open reading frame, 494 bp 5′ UTR, 456 bp 3′ UTR, and without intron. It encodes a putative protein of 211 amino acids with a calculated molecular mass of 24.0 kD and a pI of 6.7855. Amino acid analysis revealed that the OsMYB48-1 is a MYB-related type protein with one single conserved SANT domain at the N terminus between 45 and 90 amino acids.

The phylogenetic tree based on the amino acid sequences of plant MYB proteins indicated that the OsMYB48-1 amino acid sequence showed high similarity with other MYB proteins, including AtMYB48 and AtMYB59 in Arabidopsis, OsMYB146 and OsMYB154 in rice (Fig. S1), indicating the possibility of functional conservation between them.

Sequences of various putative stress response-related *cis*-acting elements were identified in the promoter region of *OsMYB48-1*, such as MYB, MYC recognition site, W-box element, RAV and GCC-box element, and so on. The predicted functions and frequencies of these *cis*-acting elements were summarized in Table S2. Such an enriched presence of stress response-related *cis*-acting elements may suggest a critical role of *OsMYB48-1* in stress tolerance.

OsMYB48-1 Expresses in Various Tissues and is Induced by Different Stresses

Tissue specificity may be associated with specific biological functions. We characterized the expression patterns of *OsMYB48-1* with qRT-PCR in various organs at seedling stage and productive stages. As shown in Figure 1A, *OsMYB48-1* was expressed in various tissues, including root, stem, sheath, leaf and panicle, but mainly expressed in roots at both seedling stage and reproductive stage, whereas was lowly expressed in sheath at seedling stage.

As described above, the promoter region of *OsMYB48-1* comprises of several stress response-related *cis*-elements. The results of inducible expression analysis showed that *OsMYB48-1*

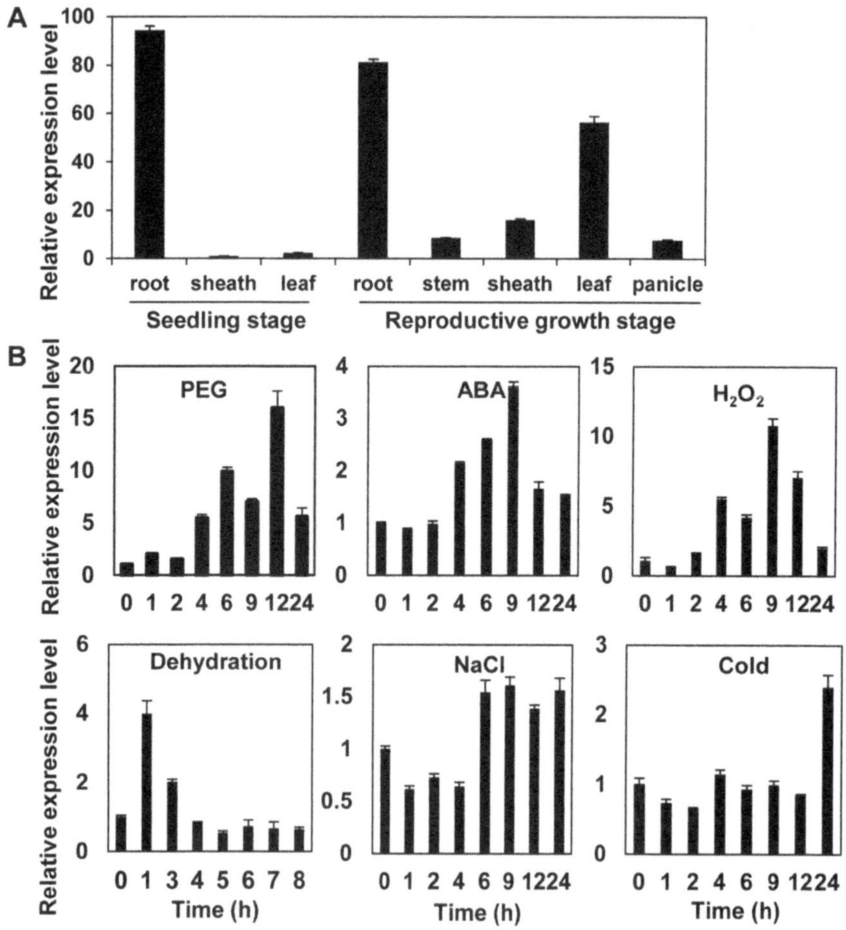

Figure 1. Expression analysis of the *OsMYB48-1* gene. (A) qRT-PCR analysis of the expression level of *OsMYB48-1* in different tissues of Nipponbare. (B) Expression patterns of *OsMYB48-1* under various stress treatments including PEG, ABA, H_2O_2, dehydration, NaCl, and cold. Error bars indicate standard error (SE) based on 3 replicates.

was strongly induced by PEG, H_2O_2, ABA, and dehydration, and slightly induced by NaCl and cold treatment (Fig. 1B). It suggested that *OsMYB48-1* may be involved in response to multiple abiotic stresses.

Nucleus Localization and Transactivation Activity of *OsMYB48-1*

As predicted that *OsMYB48-1* is a TF, a nuclear localization signal was detected in OsMYB48-1-GFP transformed cell, while the free GFP showed ubiquitous distribution of signal in the whole cell (Fig. 2A), implying the role of OsMYB48-1 as a nuclear localized protein. To check if OsMYB48-1 has transactivation activity, the full length coding region and a series of truncated fragments of *OsMYB48-1* were constructed to the pGBKT7 vector and transformed into yeast. *In vivo* plate assay showed that OsMYB48-1 has transcriptional activity, and serial deletion in the N-terminal region from the start to the position of 174 aa did not affect the activation (Fig. 2B). These results confirmed that OsMYB48-1 is a transcription activator, and the C-terminal has transactivation activity.

Overexpressing *OsMYB48-1* Enhances Drought and Salinity Tolerance

As described above, the expression of *OsMYB48-1* was induced by multiple abiotic stresses suggesting that *OsMYB48-1* may play a positive role in the regulation of stress tolerance. To confirm this, ten transgenic lines with overexpression of *OsMYB48-1* were obtained. We also obtained six RNAi transgenic lines, but the expression level of *OsMYB48-1* in all these lines showed no obvious decrease (Fig. S2). Therefore, just overexpression lines (T3, T6, and T10) with normal morphology (Fig. S3) were selected to further study the physiological function of *OsMYB48-1*.

Under high osmotic stress conditions caused by mannitol treatment, the transgenic plants showed less growth inhibition than WT (Fig. 3A). Overexpression lines showed significantly lesser suppression of relative shoot growth (63.83% to 76.45% of normal growth) than WT (only about 58.72% of the normal growth), and also lesser suppression of relative fresh weight (88.8% to 96.57% of normal growth) than WT (only about 76.93% of the normal growth) (Fig. 3B).

Considering the cross-talk between the drought and high salinity response pathway [33], the performance of *OsMYB48-1* overexpression lines on high salinity stress was also determined. The results revealed that the transgenic plants exhibited lesser growth inhibition than WT (Fig. 3C), resulting from the

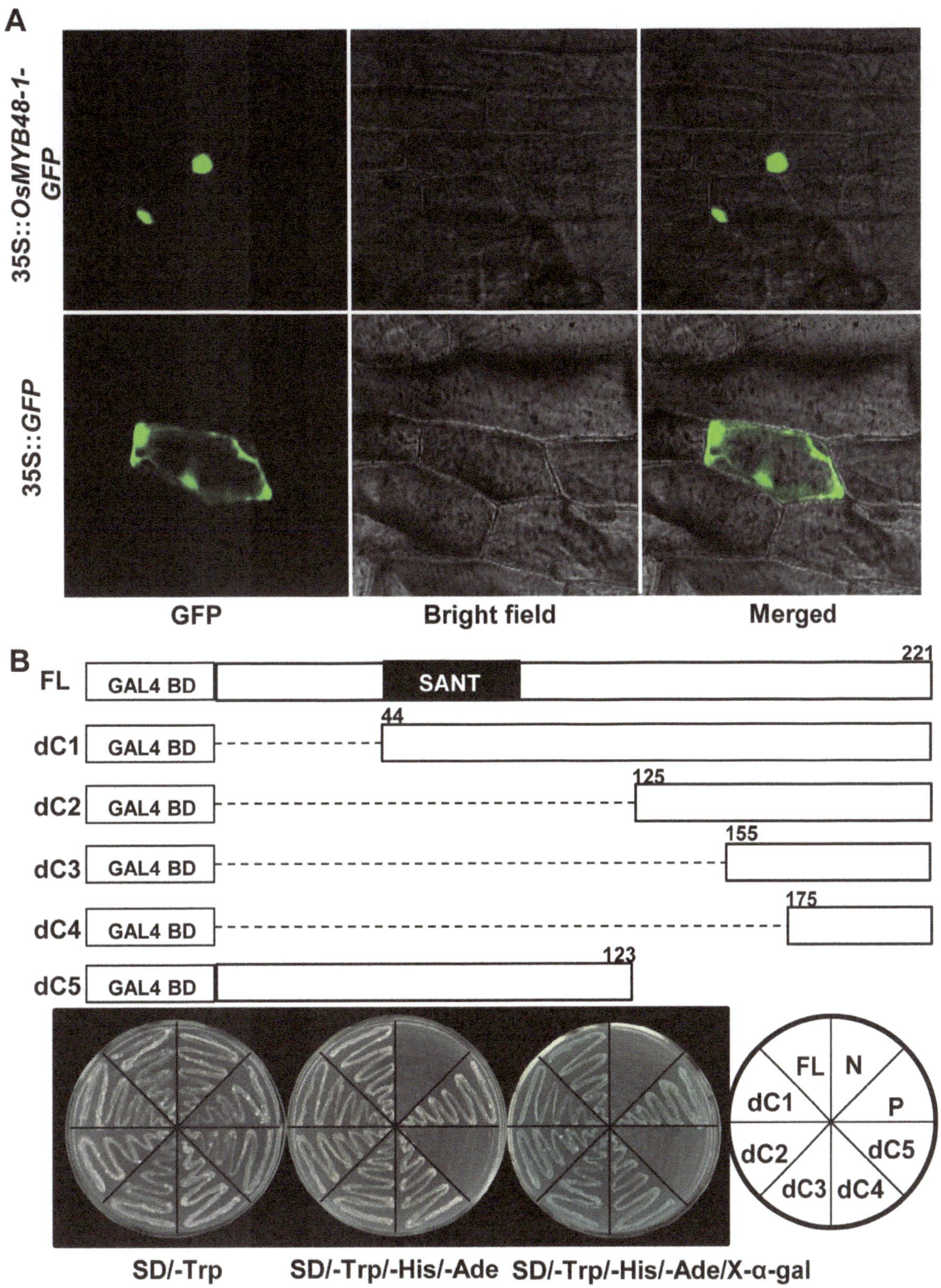

Figure 2. Subcellular localization and transactivation activity of OsMYB48-1. (A) The OsMYB48-1-GFP fusion protein was transiently expressed in onion epidermal cells and visualized by fluorescence microscopy. The upper panels show the localization of OsMYB48-1-GFP in onion cells in a transient assay, while bottom panels show the localization of GFP as a control. (B) Transactivation assay of truncated OsMYB48-1. Fusion proteins of the GAL4 DNA-binding domain and different portions of OsMYB48-1 were expressed in yeast strain AH109. FL indicates the full-length of OsMYB48-1; dC1 to dC5 indicate the mutated forms of OsMYB48-1 (nucleotide positions were labeled in the diagrams), respectively. P and N indicate the positive and negative control, respectively.

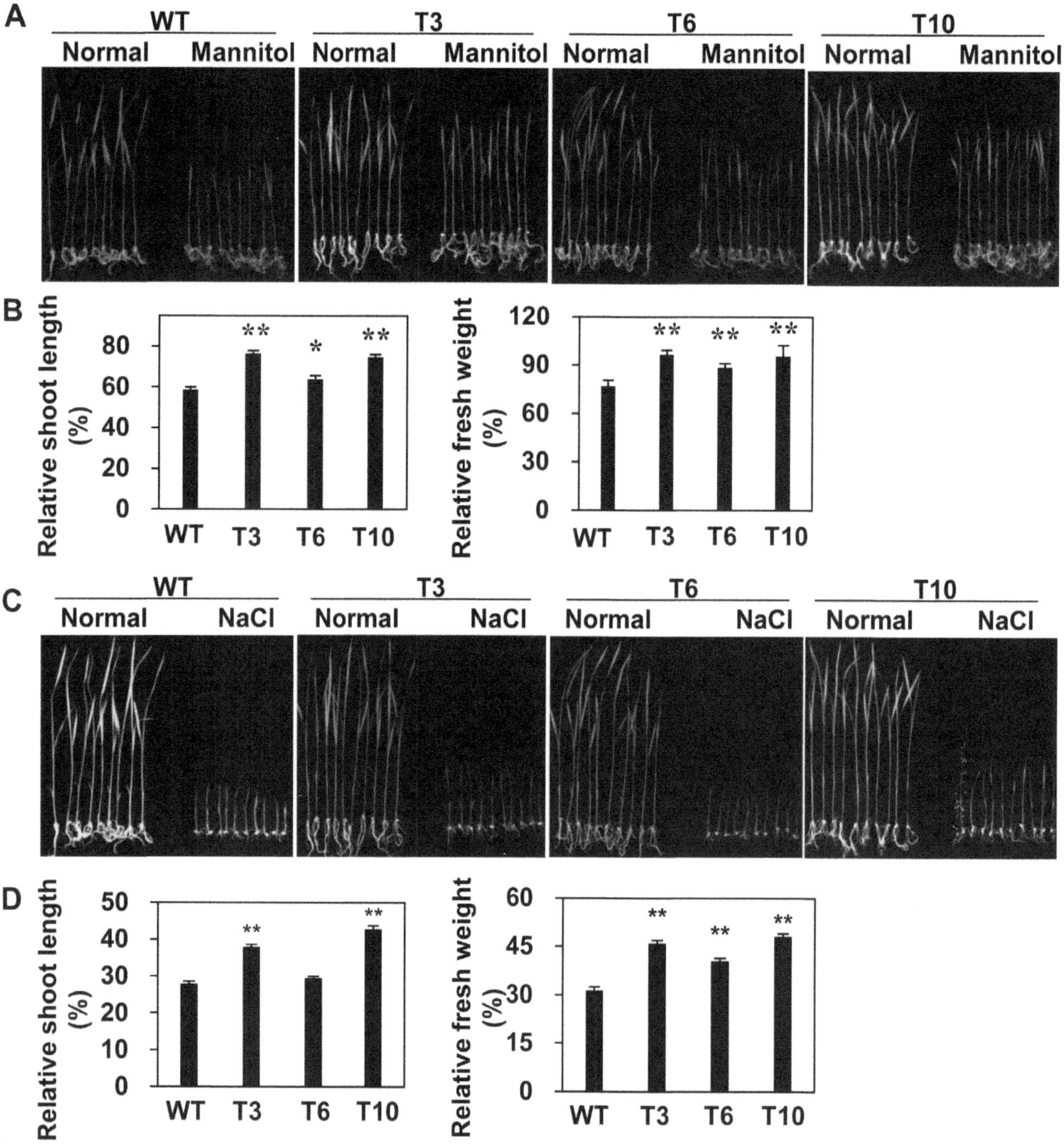

Figure 3. Osmotic and high salinity stress tolerance of *OsMYB48-1* overexpression plants. (A) Growing performance of *OsMYB48-1*-overexpression plants and WT on osmotic stress medium caused by 200 mM mannitol. (B) The relative shoot length and relative fresh weight of the transgenic and WT plants testing in (A) were compared. (C) Growing performance of transgenic and WT plants on high salinity stress medium caused by 150 mM NaCl. (D) The relative shoot length and relative fresh weight of the transgenic and WT plants testing in (C) were compared. Values are means ± SE (n = 30). * and ** indicate significant difference at P<0.05 and P<0.01 probability, respectively.

significantly higher relative shoot growth (29.25%–42.57%) and relative plant fresh weight (40.25%–47.87%) of the overexpression lines than that of WT (27.67% and 31.13%, respectively) (Fig. 3D).

Under PEG treatment that mimics the physiological dehydration stress condition, the transgenic plants had shown higher survival rate in the range of 83.33 to 100% as compared to WT with only 53.33% to 63.33% (Fig. 4A, C).

Likewise, under soil drought stress conditions caused by not watering, the transgenic plants had showed significantly higher survival rate in the range of 21.67 to 100% as compared to WT plants with 6.67 to 28.33%. (Fig. 4B, D).

Overexpression of *OsMYB48-1* showed significant increase for drought and salinity tolerance in rice based on the above findings.

Figure 4. Drought stress tolerance of *OsMYB48-1* overexpression plants. (A) Mimic physiological dehydration stress tolerance assay of the *OsMYB48-1* transgenic and WT plants under 20% PEG treatment. (B) Performance of *OsMYB48-1* transgenic and WT plants subjected to soil drought stress without water for 7 d and then recovered for 10 d. (C) Survival rates of transgenic and WT plants testing in (A). Values are means ± SE (n = 3). (D) Survival rates of the transgenic and WT plants testing in (B). Values are means ± SE (n = 3). * and ** indicate significant difference at $P<0.05$ and $P<0.01$ probability, respectively.

Physiological and Biochemical Changes in *OsMYB48-1* Overexpression Plants

Rate of water loss in detached leaves is an important trait to detect drought tolerance and the transgenic plants showed a lower rate of water loss as compared to WT (Fig. 5A), indicating *OsMYB48-1* gene's possible role in reducing water loss especially under drought stress conditions.

Under abiotic stressed conditions, ROS gets accumulated in plant tissues leading to oxidative damage, as indicated by the membrane lipid peroxidation. Malondialdehyde (MDA) is a lipid peroxidation product and serves as a biomarker for membrane lipid peroxidation. As shown in Figure 5B, drought stress conditions enhanced MDA production in the leaves of both transgenic and WT plants, however it was significantly lower in the transgenic overexpression plants (6.8–8.1 μg/g) as compared to WT (10.11 μg/g). These results suggested that transgenic lines overexpressing *OsMYB48-1* exhibited less oxidative damage under drought stress conditions.

Accumulation of proline to counteract the effects of osmotic stress is a common adaptive mechanism for plant responses to stress conditions [22]. Proline content increased under stressed conditions in both transgenic plants and WT, however, it was significantly higher in transgenic plants (701-884 μg/g) as compared to WT (581 μg/g) (Fig. 5C). Consequently, we detected the transcription levels of two proline biosynthesis genes, *OsP5CS1* and *OsP5CS2* in transgenic plants and WT. As shown in Figure 5D, the expression levels showed non-significant differences between the transgenic plants and WT under normal growth conditions. However, it was significantly higher in the transgenic plants than WT under drought stress conditions. These results clearly indicated that the overexpression of *OsMYB48-1* could increase the production of proline by regulating proline biosynthesis genes in rice especially under drought stress conditions.

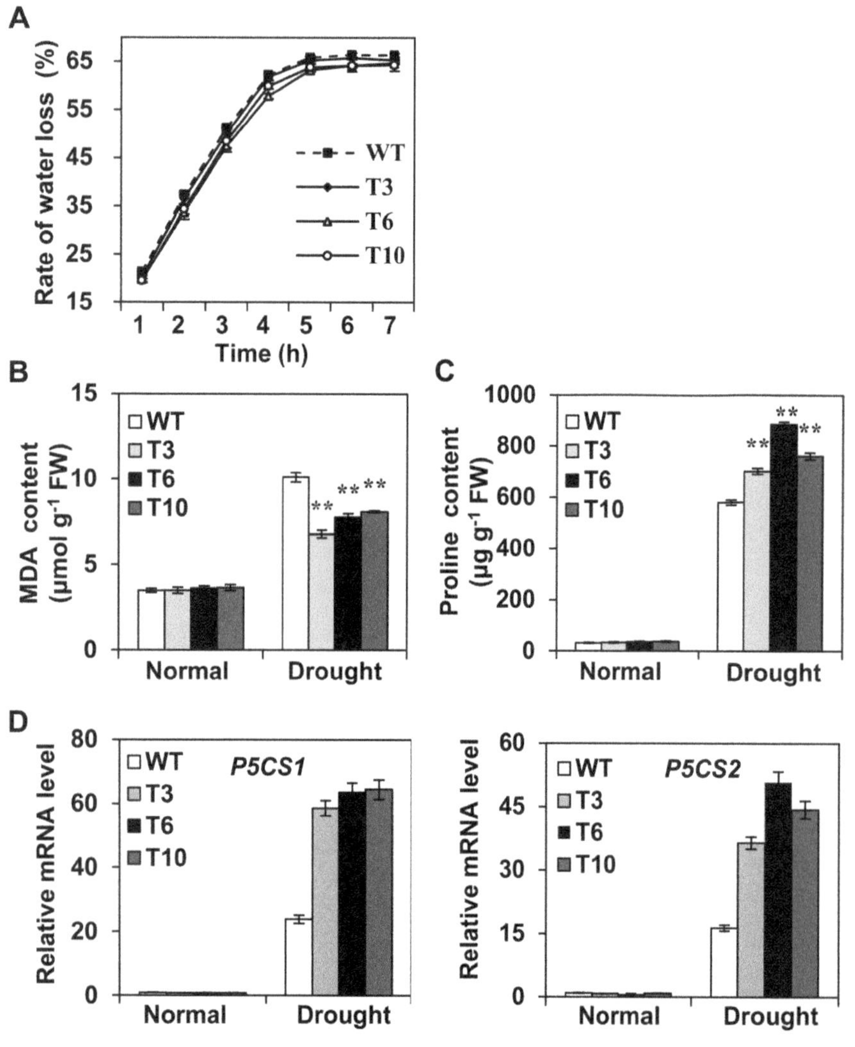

Figure 5. Physiological and biochemical changes in *OsMYB48-1* overexpression plants. (A) Rate of water loss in detached leaves cut from the *OsMYB48-1* transgenic and WT plants under normal conditions. Values are means ± SE (n = 10). (B) MDA contents. (C) Free proline contents. The MDA and free proline contents were measured in the leaves of two month old plants under natural condition and separately after creating drought treatment conditions by withholding water to plants for 5 d. Values are means ± SE (n = 4). FW, fresh weight. * and ** indicate significant difference at P<0.05 and P<0.01. (D) Expression of drought stress responsive genes in *OsMYB48-1* transgenic plants and WT. The qRT-PCR analysis was performed with total RNA from *OsMYB48-1* transgenic plants and WT under normal or drought stress treatment (withholding water for 5 d). Values are means ± SE (n = 3). * and ** indicate significant difference at P<0.05 and P<0.01 probability, respectively.

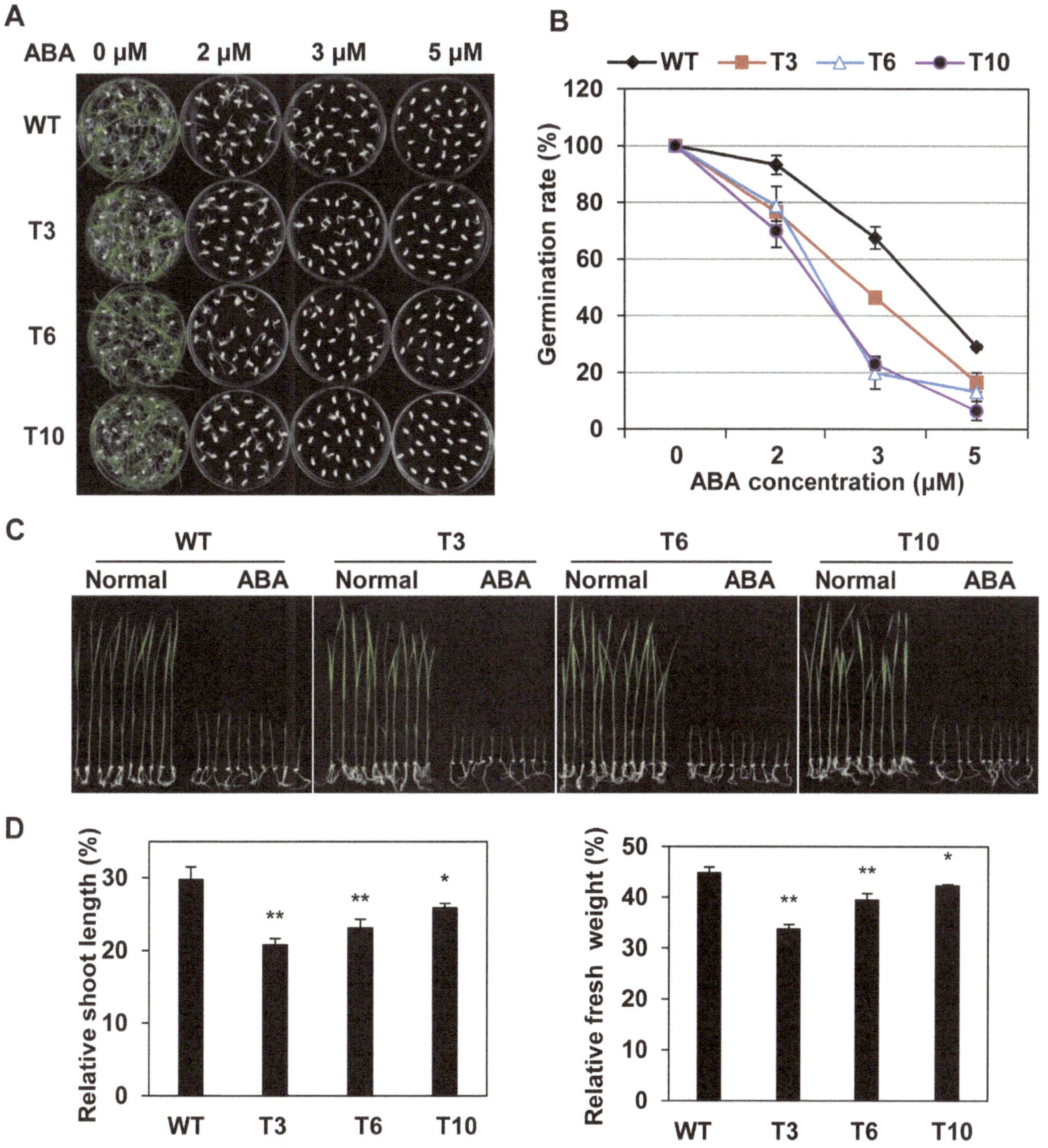

Figure 6. ABA hypersensitivity of *OsMYB48-1* overexpression plants. (A) Germination performance of *OsMYB48-1*-overexpression and WT seeds on 1/2 MS medium containing 0, 2, 3, or 5 μM/L ABA measured at 10 d after initiation. (B) Calculation of the germination rates of transgenic and WT seeds. Values are means ± SE (n = 3). (C) Performance of the *OsMYB48-1* overexpression plants and WT in 1/2 MS medium containing 3 μM/L ABA. (D) The relative shoot length and fresh weight of the transgenic and WT plants grown on normal and ABA-containing medium were compared. SE based on 30 seedlings. Values are means ± SE. * and ** indicate significant difference at $P<0.05$ and $P<0.01$ probability, respectively.

Increased ABA Sensitivity of *OsMYB48-1* Overexpression Plants

Since the expression of *OsMYB48-1* was induced by ABA, we had investigated the exogenous ABA sensitivity of transgenic plants at germination stage. Results showed non-significant differences in germination rates between the overexpression lines and WT at 0 μM ABA, however, it was significantly lower in the transgenics as compared to WT at 2, 3 and 5 μM ABA (Fig. 6A, B). The results suggested that *OsMYB48-1* overexpression plants were hypersensitive to ABA than WT at the germination stage.

Hypersensitivity response of *OsMYB48-1* overexpression plants to exogenous ABA was also investigated at the post-germination

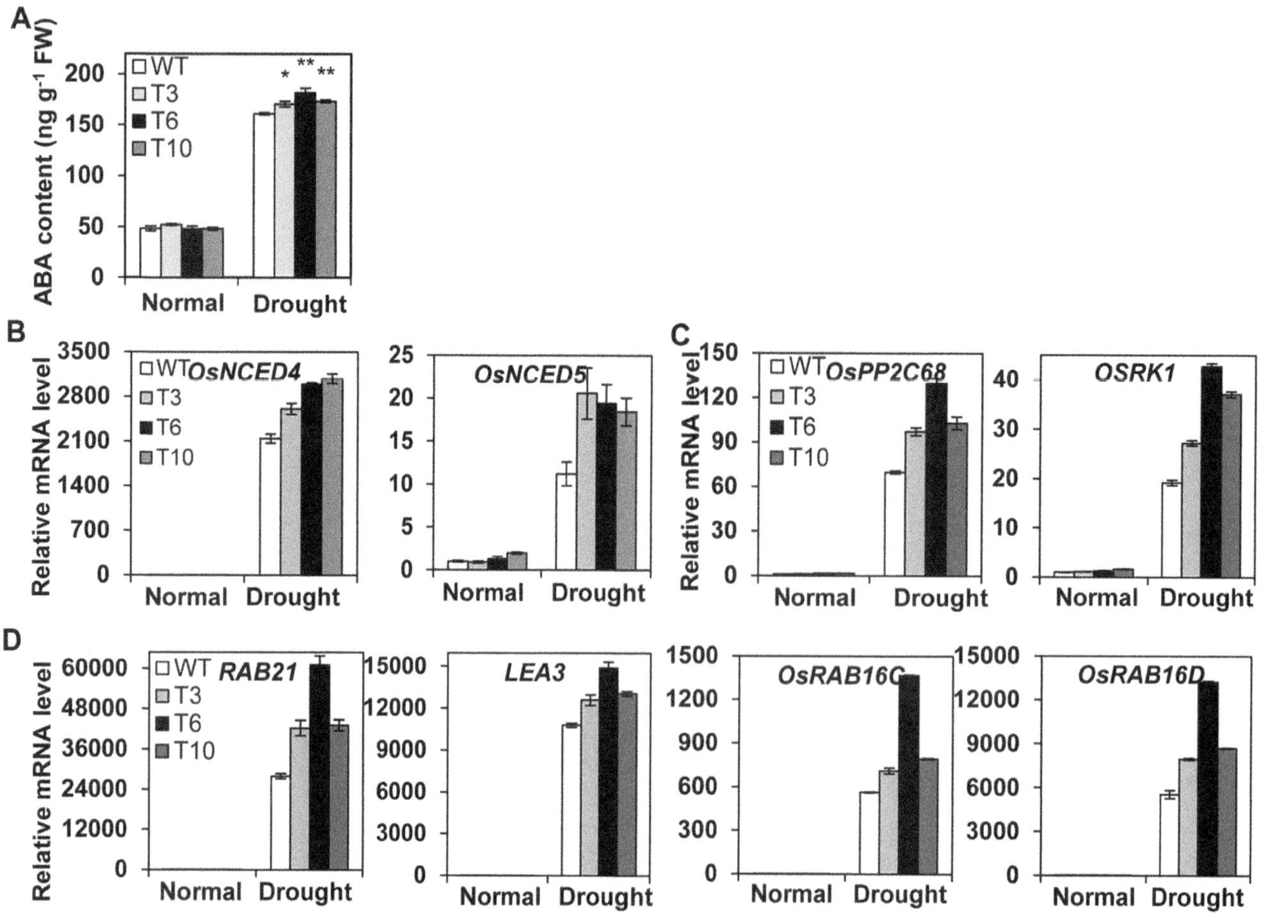

Figure 7. ABA accumulation and the expression of ABA -related genes under drought stress. (A) ABA contents in normal or drought treated seedlings. Data are means SE (n = 4). (B) qRT-PCR analysis of the expression of ABA biosynthesis genes. (C) Transcript levels of ABA early signaling genes. (D) Transcript levels of ABA later responsive genes under normal or drought stress conditions (withholding water for 5 d). Values are means ± SE (n = 3).

stage. *OsMYB48-1*-overexpression lines had significantly ($P<0.05$) more suppression of fresh shoot growth (20.78% to 25.87% of normal growth) and fresh weight growth (33.75% to 42.21% of normal growth) than that of WT (29.78% and 44.84% of the normal growth, respectively) on the ABA treatments (Fig. 6C, D). Overexpression of *OsMYB48-1* could increase exogenous ABA sensitivity at post-germination stages as observed by the results.

Endogenous ABA Accumulation and the Expression of ABA-related Genes in *OsMYB48-1*overexpression Plants

Enhanced drought tolerance and hypersensitive response of *OsMYB48-1* overexpression plants to exogenous ABA application necessitates the determination of endogenous ABA content both in the transgenic and WT plants. Transgenic and WT plants showed non-significant differences for endogenous ABA content under normal growth conditions, however, it was significantly higher in transgenic seedlings (170–182 ng/g) than the WT (161 ng/g) (Fig. 7A), indicating the role of *OsMYB48-1* in regulating ABA synthesis under drought stress conditions.

Since the transgenic plants were sensitive to exogenous ABA and accumulated more endogenous ABA under drought stress conditions, the transcription levels of some ABA biosynthesis and signaling genes were analyzed. As shown in Figure 7B, the transcript levels of two ABA biosynthesis genes, Os*NCED4* and Os*NCED5* in both the transgenic plants and WT were clearly up-regulated by drought stress, while the expression levels of these genes were clearly higher in *OsMYB48-1* overexpression lines as compared to WT. As shown in Fig. 7C, two early ABA signaling genes, *OsPP2C68* and *OSRK1*, showed much higher expression levels in transgenic plants than that in WT under drought stress conditions. The expression levels of four well-documented ABA late responsive genes, *RAB21*, *OsLEA3*, *RAB16C* and *RAB16D*, showed non-significant differences between the transgenic plants and WT under normal growth conditions. However, under drought stress conditions significant differences were observed for higher expression levels of these genes in transgenic plants as compared to WT were detected (Fig. 7D). Together, these data suggested that *OsMYB48-1* play an important role in drought stress-induced ABA-biosynthesis through regulating the expression of ABA-biosynthesis genes, such as *OsNCED4* and *OsNCED5*, consequently, regulating early ABA signaling genes, such as *OsPP2C68* and *OSRK1*, and late responsive genes, such as *RAB21*, *OsLEA3*, *RAB16C* and *RAB16D*.

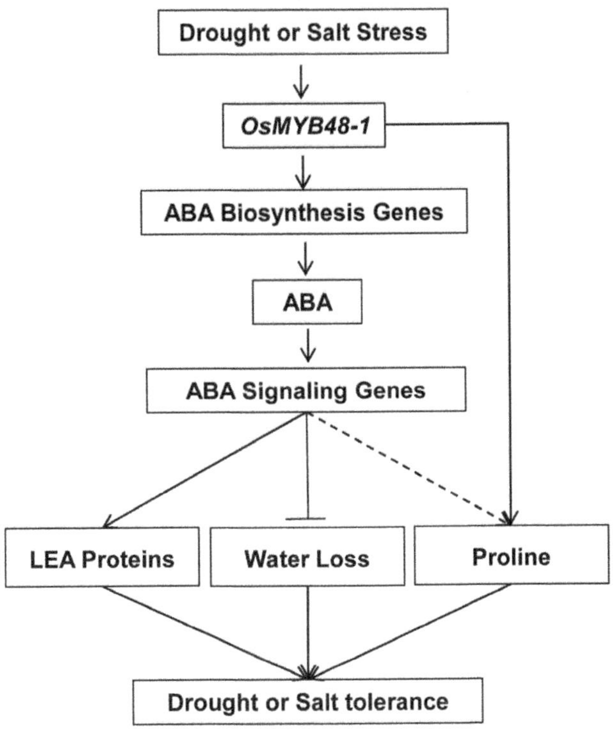

Figure 8. Proposed model depicting the function of *OsMYB48-1* in the regulation of drought and salinity stress tolerance.

Discussion

OsMYB48-1, a Novel MYB-related TF of Rice in Abiotic Stress Tolerance

The identification of some MYB TFs in rice and their roles in abiotic stress response has been reported recently [20,22,34], however, our results indicated that *OsMYB48-1* to have unique features as compared to other MYB members in rice.

The expression of *OsMYB48-1* was found to be tissue specific that showed highest expression levels in roots at both seedling and reproductive stages, followed by leaves and sheath at reproductive stages. However, *OsMYB2* was detected to have a greatest expression levels in leaves, followed by roots and shoots [22]. The highest level of transcripts of *OsMYB3R-2* was found in young stems while the lowest was in spikes [34]. Differential tissue expression patterns of *OsMYB48-1* in comparison to other MYB genes involved in abiotic stress tolerance indicates a possibility of different role for it.

The expression profiles of *OsMYB48-1* under abiotic stress treatments were much different from those of *Osmyb4*, *OsMYB2*, and *OsMYB3R-2*, which were reported to be involved in drought tolerance in rice. In contrast, the expression of *OsMYB2* was significantly induced by salt, cold, and PEG treatments [22]. *OsMYB3R-2* was induced by cold, drought, and salt stress [35]. The expression of *Osmyb4* in rice was only induced by cold but not by other stresses [20]. Our results showed that the expression of *OsMYB48-1* was strongly induced by PEG, H_2O_2, ABA, and dehydration treatments, while slightly induced by salt and cold treatments. Therefore, based on the specific stress inductive patterns of *OsMYB48-1* as compared to other MYB proteins suggest it to play a special role in response to abiotic stresses in rice.

Moreover, four single MYB-like domain proteins in rice have been functionally studied. *OsMYBS1*, *OsMYBS2* and *OsMYBS3* mediate α-amylase gene expression [36], and *OsMYBS3* plays a role in cold tolerance in rice [25]. *OsAID1* involves in spikelet sterility [37]. To our knowledge, it is the first report that MYB-related type TF, *OsMYB48-1*, involved in drought and salt tolerance in rice. Placing all the results together we could deduce OsMYB48-1 to be a novel MYB-related TF of rice for abiotic stress tolerance.

Positive Effect of *OsMYB48-1* in Drought and Salt Stress Tolerance

Growth retardation has been frequently observed in rice plants overexpressing stress-responsive TFs, such as *OsDREB1A*, *OsDREB1B*, *OsWRKY89* and *OsNAC6* in rice [38–40]. However, the overexpression plants of *OsMYB48-1* showed no obvious phenotypic changes from that of the WT under normal growth conditions. Overexpression of *OsMYB48-1* in rice resulted in increased sensitivity to exogenous ABA at both germination and post-germination stages and accumulated more endogenous ABA content under drought stress conditions, suggesting *OsMYB48-1* to possibly play an important role in ABA-signaling transduction pathway. It has been widely accepted that a positive regulator of ABA signaling could play a positive role in the stress tolerance of plants. Consequently, transgenic plants overexpressing *OsMYB48-1* showed improved growth performance under simulated drought stress conditions created by mannitol and PEG treatments, and also under dry soil conditions with enhanced survival rate. Interestingly, *OsMYB48-1* was only slightly up-regulated by high salinity, overexpression plants showed tolerance to salt stress, which suggested that *OsMYB48-1* may be involved in salt stress tolerance through a cross-talk between drought and high salinity stress response pathway. Overall, *OsMYB48-1* overexpression plants demonstrated improved drought and salt tolerance besides possessing an unaltered phenotype under normal conditions in comparison to WT implies the usefulness of *OsMYB48-1* in genetic improvement of abiotic stress tolerance in rice.

Possible Mechanisms for the Drought Tolerance of *OsMYB48-1* Overexpression Plants

ABA plays a key role in abiotic stress response. In response to abiotic stresses, transgenic plants overexpressing some ABA biosynthesis-related genes increased ABA content dramatically to cope with the stress, such as transgenic cottons overexpressing *AtLOS5* [41], transgenic *Arabidopsis* overexpressing *OsNCED3* [42]. Here, we identified that the expression of *OsMYB48-1* could be induced by exogenous ABA, and *OsMYB48-1* overexpression plants showed hypersensitivity to exogenous ABA. Accumulation of the endogenous ABA content in transgenic plants was higher than in WT to confer drought tolerance, implicating *OsMYB48-1* possible role in stress-induced ABA accumulation. The expression of 9-*cis* epoxycarotenoid dioxygenase (*NCED*) is generally considered to be the rate-limiting step among all the steps of drought stress-induced ABA biosynthesis pathway [43]. Expression levels of rice *NCED* genes in transgenic plants and WT under normal and drought stress conditions were comprehensively studied by us. Results showed similar low expression levels of *OsNCED4* and *OsNCED5* in transgenic plants and WT under normal conditions which suggested *OsMYB48-1* alone may not be sufficient to regulate the ABA biosynthesis. However, the high expression levels of *OsNCED4* and *OsNCED5* in transgenic plants under drought stress conditions clearly suggested that *OsMYB48-1* possible role in stress-induced ABA-biosynthesis along with additional factors.

The increased endogenous ABA levels in response to drought stress results in further modifying the expression of many stress-responsive genes [41], we also detected the expression levels of some early ABA signaling genes and late ABA-responsive genes in the transgenic plants and WT under normal and drought stress conditions. OsPP2C68 encodes a protein phosphatase 2C, which plays a prime role in ABA-mediated signaling network related to stress responses [44]. OSRK1 encodes a protein kinase of SnRK2 family, which is associated with ABA signaling [45]. Under drought stress conditions, *OsPP2C68* and *OSRK1* had higher expression levels in the transgenic plants than that in WT. Four LEA genes, *RAB21*, *OsLEA3*, *RAB16C* and *RAB16D* have been reported to be marker genes in ABA-dependent stress response way [12,46,47]. We identified upon drought stress, the expression levels of these genes were significantly higher in the transgenic plants than WT. Overall the results suggested that the overexpression of *OsMYB48-1* had further induced ABA synthesis in transgenic plants under drought stress conditions, thus, leading to increased expression levels of ABA signaling and -responsive genes.

Proline acts as one of the most common osmolytes in plants. The step catalyzed by enzyme pyrroline-5-carboxylate synthase (P5CS) is the key step in proline synthesis [48]. Overexpression of a *P5CS* gene in tobacco (*Nicotiana tabacum*) demonstrated the causal relationship between proline synthesis and tolerance of drought and salt stress [49]. A greater accumulation of free proline was found in the *OsMYB48-1* overexpression plants than that in WT under drought stress conditions. Meanwhile, higher expression levels of *OsP5CS1* and *OsP5CS2* were detected in the transgenic plants as compared to WT under drought stress conditions. These results suggested that the accumulated proline might be seen as a response of *OsMYB48-1* transgenic plants towards abiotic stresses.

Furthermore, the transgenic plants overexpressing drought-tolerance related genes showed slower rate of water loss under dehydration conditions, as observed in transgenic Arabidopsis overexpressing *TaMYB30-B* [50] and *OsMYB3R-2* [34] and transgenic rice overexpressing *OsbZIP23* [12] and *OsMIOX* [27]. Based on the earlier reports, overexpression of *OsMYB48-1* transgenic rice had showed lower rate of water loss and improved drought tolerance in comparison to WT indicating the vital role of *OsMYB48-1* in protecting water loss especially under drought stress conditions.

We tried to summarize all the results obtained into a model to explain the potential role of *OsMYB48-1* in regulating drought stress tolerance in rice (Fig. 8). In conclusion, this study isolated and characterized a novel MYB-related type TF *OsMYB48-1*, which regulates the expression of ABA synthesis genes thus leading to increased endogenous ABA accumulation under drought stress conditions. Moreover, overexpression of *OsMYB48-1* increased LEA protein and proline content while reduced rate of water loss in transgenic plants to confer drought tolerance. In addition, the overexpression of *OsMYB48-1* did not alter the phenotypes of the transgenic plants carrying them. Therefore, we like to conclude that *OsMYB48-1* may be an effective gene for improving drought and salinity tolerance in rice.

Supporting Information

Figure S1 Protein sequence analysis of OsMYB48-1. (A) Phylogenetic tree of the MYB members in Arabidopsis and rice. The phylogenetic tree was constructed in MEGA5 software based on neighbor-joining method. Numbers indicate percentage values after 1000 replications. **(B)** Multiple sequence alignment of OsMYB48-1, OsMYB146, OsMYB154, AtMYB48, AtMYB59 with DNAMAN software. Conserved SANT DNA-binding domain is indicated by red boxes.

Figure S2 RNA interference of *OsMYB48-1*. (A) RNAi construct of *OsMYB48-1* for rice transformation. **(B)** Expression level of *OsMYB48-1* in RNAi transgenic lines analyzed by qRT-PCR.

Figure S3 Overexpression of *OsMYB48-1*. (A) Overexpression construct of *OsMYB48-1* for rice transformation. **(B)** The expression level of *OsMYB48-1* in WT and overexpression transgenic lines analyzed by qRT-PCR. Transgenic lines T3, T6, T10 showed the highest expression level, were selected for further study.

Table S1 Primer sequences used in this study.

Table S2 Potential stress-related *cis*-elements in the promoter of *OsMYB48-1*.

Author Contributions

Conceived and designed the experiments: ZCL JJL HLZ JA. Performed the experiments: HYX PLL JZD YZ XG YL. Analyzed the data: HYX JJL. Contributed reagents/materials/analysis tools: ZCL JA. Wrote the paper: HYX JJL ZCL.

References

1. Hou X, Xie K, Yao J, Qi Z, Xiong L (2009) A homolog of human ski-interacting protein in rice positively regulates cell viability and stress tolerance. Proc Natl Acad Sci U S A 106: 6410–6415.
2. Fujita Y, Fujita M, Shinozaki K, Yamaguchi-Shinozaki K (2011) ABA-mediated transcriptional regulation in response to osmotic stress in plants. J Plant Res 124: 509–525.
3. Finkelstein RR, Gampala SSL, Rock CD (2002) Abscisic acid signaling in seeds and seedlings. Plant Cell 14: S15–S45.
4. Wang ZY, Xiong L, Li W, Zhu JK, Zhu J (2011) The plant cuticle is required for osmotic stress regulation of abscisic acid biosynthesis and osmotic stress tolerance in Arabidopsis. Plant Cell 23: 1971–1984.
5. Yamaguchi-Shinozaki K, Shinozaki K (2006) Transcriptional regulatory networks in cellular responses and tolerance to dehydration and cold stresses. Annu Rev Plant Biol 57: 781–803.
6. Hu H, You J, Fang Y, Zhu X, Qi Z, et al. (2008) Characterization of transcription factor gene *SNAC2* conferring cold and salt tolerance in rice. Plant Mol Biol 67: 169–181.
7. Huang GT, Ma SL, Bai LP, Zhang L, Ma H, et al. (2012) Signal transduction during cold, salt, and drought stresses in plants. Mol Biol Rep 39: 969–987.
8. Singh K, Foley R, Oñate-Sánchez L (2002) Transcription factors in plant defense and stress responses. Curr Opin Plant Biol 5: 430–436.
9. Jeong JS, Kim YS, Baek KH, Jung H, Ha S-H, et al. (2010) Root-specific expression of *OsNAC10* improves drought tolerance and grain yield in rice under field drought conditions. Plant Physiol 153: 185–197.
10. Oh S-J, Kim YS, Kwon C-W, Park HK, Jeong JS, et al. (2009) Overexpression of the transcription factor *AP37* in rice improves grain yield under drought conditions. Plant Physiol 150: 1368–1379.
11. Tang N, Zhang H, Li X, Xiao J, Xiong L (2012) Constitutive activation of transcription factor OsbZIP46 improves drought tolerance in rice. Plant Physiol 158: 1755–1768.
12. Xiang Y, Tang N, Du H, Ye H, Xiong L (2008) Characterization of OsbZIP23 as a key player of the basic leucine zipper transcription factor family for conferring abscisic acid sensitivity and salinity and drought tolerance in rice. Plant Physiol 148: 1938–1952.
13. Chen Y, Yang X, He K, Liu M, Li J, et al. (2006) The MYB transcription factor superfamily of Arabidopsis: expression analysis and phylogenetic comparison with the rice MYB family. Plant Mol Biol 60: 107–124.

14. Du H, Zhang L, Liu L, Tang X-F, Yang W-J, et al. (2009) Biochemical and molecular characterization of plant MYB transcription factor family. Biochemistry (Moscow) 74: 1–11.
15. Abe H, Urao T, Ito T, Seki M, Shinozaki K, et al. (2003) Arabidopsis AtMYC2 (bHLH) and AtMYB2 (MYB) function as transcriptional activators in abscisic acid signaling. Plant Cell 15: 63–78.
16. Denekamp M (2003) Integration of wounding and osmotic stress signals determines the expression of the *AtMYB102* transcription factor gene. Plant Physiol 132: 1415–1423.
17. Seo PJ, Xiang F, Qiao M, Park JY, Lee YN, et al. (2009) The MYB96 transcription factor mediates abscisic acid signaling during drought stress response in Arabidopsis. Plant Physiol 151: 275–289.
18. Pasquali G, Biricolti S, Locatelli F, Baldoni E, Mattana M (2008) *Osmyb4* expression improves adaptive responses to drought and cold stress in transgenic apples. Plant Cell Rep 27: 1677–1686.
19. Vannini C, Campa M, Iriti M, Genga A, Faoro F, et al. (2007) Evaluation of transgenic tomato plants ectopically expressing the rice *Osmyb4* gene. Plant Sci 173: 231–239.
20. Vannini C, Locatelli F, Bracale M, Magnani E, Marsoni M, et al. (2004) Overexpression of the rice *Osmyb4* gene increases chilling and freezing tolerance of *Arabidopsis thaliana* plants. Plant J 37: 115–127.
21. El-Kereamy A, Bi YM, Ranathunge K, Beatty PH, Good AG, et al. (2012) The rice R2R3-MYB transcription factor OsMYB55 is involved in the tolerance to high temperature and modulates amino acid metabolism. PLoS One 7: e52030.
22. Yang A, Dai X, Zhang W-H (2012) A R2R3-type MYB gene, *OsMYB2*, is involved in salt, cold, and dehydration tolerance in rice. J Exp Bot 63: 2541–2556.
23. Shin D, Moon SJ, Han S, Kim BG, Park SR, et al. (2011) Expression of *StMYB1R-1*, a novel potato single MYB-like domain transcription factor, increases drought tolerance. Plant Physiol 155: 421–432.
24. Ganesan G, Sankararamasubramanian HM, Harikrishnan M, Ashwin G, Parida A (2012) A MYB transcription factor from the grey mangrove is induced by stress and confers NaCl tolerance in tobacco. J Exp Bot 63: 4549–4561.
25. Su CF, Wang YC, Hsieh TH, Lu CA, Tseng TH, et al. (2010) A novel MYBS3-dependent pathway confers cold tolerance in rice. Plant Physiol 153: 145–158.
26. Wang H, Zhang H, Gao F, Li J, Li Z (2007) Comparison of gene expression between upland and lowland rice cultivars under water stress using cDNA microarray. Theor Appl Genet 115: 1109–1126.
27. Duan J, Zhang M, Zhang H, Xiong H, Liu P, et al. (2012) *OsMIOX*, a *myo*-inositol oxygenase gene, improves drought tolerance through scavenging of reactive oxygen species in rice (*Oryza sativa* L.). Plant Sci 196: 143–151.
28. Hiei Y, Ohta S, Komari T, Kumashiro T (1994) Efficient transformation of rice (*Oryza sativa* L.) mediated by Agrobacterium and sequence analysis of the boundaries of the T-DNA. Plant J 6: 271–282.
29. Wang Z, Chen C, Xu Y, Jiang R, Han Y, et al. (2004) A practical vector for efficient knockdown of gene expression in rice (*Oryza sativa* L.). Plant Mol Biol Rep 22: 409–417.
30. Bates LS, Waldren RP, Teare ID (1973) Rapid determination of free proline for water-stress studies. Plant Soil 39: 205–207.
31. Song SY, Chen Y, Chen J, Dai XY, Zhang WH (2011) Physiological mechanisms underlying OsNAC5-dependent tolerance of rice plants to abiotic stress. Planta 234: 331–345.
32. Yang J, Zhang J, Wang Z, Zhu Q, Wang W (2001) Hormonal changes in the grains of rice subjected to water stress during grain filling. Plant Physiol 127: 315–323.
33. Shinozaki K, Yamaguchi-Shinozaki K (2007) Gene networks involved in drought stress response and tolerance. J Exp Bot 58: 221–227.
34. Dai X, Xu Y, Ma Q, Xu W, Wang T, et al. (2007) Overexpression of an R1R2R3 MYB Gene, *OsMYB3R-2*, Increases Tolerance to Freezing, Drought, and Salt Stress in Transgenic Arabidopsis. Plant Physiol 143: 1739–1751.
35. Ma Q, Dai X, Xu Y, Guo J, Liu Y, et al. (2009) Enhanced tolerance to chilling stress in *OsMYB3R-2* transgenic rice is mediated by alteration in cell cycle and ectopic expression of stress genes. Plant Physiol 150: 244–256.
36. Lu C-A, Ho T-hD, Ho S-L, Yu S-M (2002) Three novel MYB proteins with one DNA binding repeat mediate sugar and hormone regulation of α-amylase gene expression. Plant Cell 14: 1963–1980.
37. Zhu QH, Ramm K, Shivakkumar R, Dennis ES, Upadhyaya NM (2004) The *ANTHER INDEHISCENCE1* gene encoding a single MYB domain protein is involved in anther development in rice. Plant Physiol 135: 1514–1525.
38. Ito Y, Katsura K, Maruyama K, Taji T, Kobayashi M, et al. (2006) Functional analysis of rice DREB1/CBF-type transcription factors involved in cold-responsive gene expression in transgenic rice. Plant Cell Physiol 47: 141–153.
39. Nakashima K, Tran LS, Van Nguyen D, Fujita M, Maruyama K, et al. (2007) Functional analysis of a NAC-type transcription factor OsNAC6 involved in abiotic and biotic stress-responsive gene expression in rice. Plant J 51: 617–630.
40. Wang H, Hao J, Chen X, Hao Z, Wang X, et al. (2007) Overexpression of rice WRKY89 enhances ultraviolet B tolerance and disease resistance in rice plants. Plant Mol Biol 65: 799–815.
41. Yue Y, Zhang M, Zhang J, Tian X, Duan L, et al. (2012) Overexpression of the *AtLOS5* gene increased abscisic acid level and drought tolerance in transgenic cotton. J Exp Bot 63: 3741–3748.
42. Hwang S-G, Chen H-C, Huang W-Y, Chu Y-C, Shii C-T, et al. (2010) Ectopic expression of rice *OsNCED3* in Arabidopsis increases ABA level and alters leaf morphology. Plant Sci 178: 12–22.
43. Iuchi S, Kobayashi M, Taji T, Naramoto M, Seki M, et al. (2001) Regulation of drought tolerance by gene manipulation of 9-*cis*-epoxycarotenoid dioxygenase, a key enzyme in abscisic acid biosynthesis in *Arabidopsis*. Plant J 27: 325–333.
44. Xue T, Wang D, Zhang S, Ehlting J, Ni F, et al. (2008) Genome-wide and expression analysis of protein phosphatase 2C in rice and Arabidopsis. BMC Genomics 9: 1–21.
45. Chae M-J, Lee J-S, Nam M-H, Cho K, Hong J-Y, et al. (2007) A rice dehydration-inducible SNF1-related protein kinase 2 phosphorylates an abscisic acid responsive element-binding factor and associates with ABA signaling. Plant Mol Biol 63: 151–169.
46. RoyChoudhury A, Roy C, Sengupta DN (2007) Transgenic tobacco plants overexpressing the heterologous *lea* gene *Rab16A* from rice during high salt and water deficit display enhanced tolerance to salinity stress. Plant Cell Rep 26: 1839–1859.
47. Xiao B, Huang Y, Tang N, Xiong L (2007) Over-expression of a *LEA* gene in rice improves drought resistance under the field conditions. Theor Appl Genet 115: 35–46.
48. Chen Z, Hong X, Zhang H, Wang Y, Li X, et al. (2005) Disruption of the cellulose synthase gene, *AtCesA8/IRX1*, enhances drought and osmotic stress tolerance in Arabidopsis. Plant J 43: 273–283.
49. Kishor P, Hong Z, Miao GH, Hu C, Verma D (1995) Overexpression of [delta]-Pyrroline-5-Carboxylate Synthetase Increases Proline Production and Confers Osmotolerance in Transgenic Plants. Plant Physiol 108: 1387–1394.
50. Zhang L, Zhao G, Xia C, Jia J, Liu X, et al. (2012) A wheat R2R3-MYB gene, *TaMYB30-B*, improves drought stress tolerance in transgenic *Arabidopsis*. J Exp Bot 63: 5873–5885.

Impact of the Provision of Safe Drinking Water on School Absence Rates in Cambodia: A Quasi-Experimental Study

Paul R. Hunter[1]*, Helen Risebro[1], Marie Yen[2], Hélène Lefebvre[2], Chay Lo[3], Philippe Hartemann[4], Christophe Longuet[5], François Jaquenoud[2]

1 Norwich School of Medicine, University of East Anglia, Norwich, United Kingdom, **2** 1001 fontaines pour demain, Caluire et Cuire, France, **3** Teuk Saat 1001, Phnom Penh, Cambodia, **4** Département Environnement et Santé Publique, Faculté de médecine de Nancy - Université de Lorraine, Nancy, France, **5** Fondation Mérieux, Lyon, France

Abstract

Background: Education is one of the most important drivers behind helping people in developing countries lift themselves out of poverty. However, even when schooling is available absenteeism rates can be high. Recently interest has focussed on whether or not WASH interventions can help reduce absenteeism in developing countries. However, none has focused exclusively on the role of drinking water provision. We report a study of the association between absenteeism and provision of treated water in containers into schools.

Methods and Findings: We undertook a quasi-experimental longitudinal study of absenteeism rates in 8 schools, 4 of which received one 20 L container of treated drinking water per day. The water had been treated by filtration and ultraviolet disinfection. Weekly absenteeism rates were compared across all schools using negative binomial model in generalized estimating equations. There was a strong association with provision of free water and reduced absenteeism (Incidence rate ratio = 0.39 (95% Confidence Intervals 0.27–0.56)). However there was also a strong association with season (wet versus dry) and a significant interaction between receiving free water and season. In one of the intervention schools it was discovered that the water supplier was not fulfilling his contract and was not delivering sufficient water each week. In this school we showed a significant association between the number of water containers delivered each week and absenteeism (IRR = 0.98 95%CI 0.96–1.00).

Conclusion: There appears to be a strong association between providing free safe drinking water and reduced absenteeism, though only in the dry season. The mechanism for this association is not clear but may in part be due to improved hydration leading to improved school experience for the children.

Editor: C. Mary Schooling, CUNY, United States of America

Funding: The study received financial support as charitable donations from Fondation Ensemble, Danone Group, Agence Française de Développement, Fondation Mérieux, and Fondation Avenir Finance. The funders had no role in study design, data collection and analysis, decision to publish, or preparation of the manuscript.

Competing Interests: In 2010 PRH received travelling expenses and a small honorarium from Danone to give a talk at a science meeting organised by the company. This does not alter the authors' adherence to all the PLOS ONE policies on sharing data and materials.

* E-mail: Paul.Hunter@uea.ac.uk

Introduction

The receipt of a good quality education is one of the most important factors in enabling children to fulfil potential in later life and reduce poverty [1]. Increased educational attainment is also associated with substantial health gains especially on child health in future generations including reduction in child mortality [2,3]. Important gains in child health may be associated even with future mothers improved access to primary education alone [3]. The importance of access to education is reflected within the Millennium Development Goals of the commitment to ensure that all children can complete a course of primary education [4]. In an earlier review of studies from developing countries, the author pointed out that time spent learning being linked to educational achievement is one of the most consistent findings [5]. However, as pointed out by Abadzi [1], instructional time available to children in many developing countries is often markedly reduced. Indeed Abadzi concluded that "assumptions about the pro-poor poverty alleviation effect of education may be unrealistic", and that additional public investment may fail to mitigate poverty, unless it improves instructional delivery [1]. There are many reasons for this reduced educational contact time in low income countries, some of which are institutional such as teacher absenteeism, frequent school closures, etc [1]. However, even when schools are open, pupil absenteeism can be high [1]. Clearly reducing student absenteeism is important to improving educational attainment and consequent poverty alleviation.

Recent interest has turned towards the potential role of improving water and sanitation provision in schools as a tool towards improving children's health and educational achievements. In a recent systematic review, the authors identified 41 studies that reported on the impact of water, hygiene and sanitation interventions on health and educational outcomes of which 8 were concerned with absenteeism [6]. Most of these studies were from developed nations. Although there was some indication of links between water and sanitation prevision and absenteeism, this was strongest around sanitation provision and absenteeism in menstruating girls. No strong conclusions could be made around the importance of drinking water. In probably the largest study of its type Freeman et al. reported on a large cluster randomised trial in Kenya on the impact of water, sanitation and hygiene interventions on absenteeism [7]. This study included over 6000 pupils from 135 schools in three study arms: a control with no intervention, a second study arm with hygiene promotion and chlorine for drinking water treatment and a third study arm with sanitation improvement in addition to the hygiene promotion and water treatment. The authors found no significant difference between any of the study arms, unless they did further sub-group analysis. Furthermore, by restricting their drinking water intervention to provision of near-to-use chlorination, Freeman et al. only tested the impact of drinking water disinfection and did not undertake an adequate assessment of the value of provision of drinking water per se in schools on educational outcomes [7]. Furthermore, serious doubts have been raised about the health value of household chlorination of drinking water as blinded studies have repeatedly found no benefit [8,9]. The available evidence of the benefits or otherwise of drinking water interventions targeted at the level of the school in developing countries is, therefore, weak. We report a quasi-experimental study of the impact of provision of treated water in containers to schools on recorded absenteeism.

Methods

This study was approved by the University of East Anglia Faculty of Health ethical committee and the Cambodian National Ethics Committee for Health Research. Given that we did not introduce any intervention, only summaries of routinely collected data were obtained and no person specific data was collected, there was no requirement for informed consent to be obtained.

The intervention being investigated was delivery of treated water in containers to schools. These water containers were provided free of charge to schools and funded by "1001 fontaines pour demain" (1001F), a non-governmental and not-for-profit organization based in Caluire, France. 1001F has been working in Cambodia since 2005. The basic model is to identify local entrepreneurs and financially support them to build a local plant to bottle filtered and ultraviolet disinfected water in cleaned and disinfected 20 L containers. Most of these containers are then sold to local customers. During and after start-up 1001F technical staff provide training and an ongoing quality assurance scheme. Funding for 1001F is mainly from private donors, though it has also received financial support from French Embassies in the countries where it works. A video highlighting 1001F's work can be seen at the following link: http://fr.youtube.com/watch?v=8bykbVECVrE. In some of the villages, 1001F paid the entrepreneur to provide free water to the village school. Each participating school was provided with 1001F water in containers to be placed in the classroom so that each child could take water whenever they wished. For those schools participating in the scheme one 20 L bottle of water was delivered to each class each day. Given that the average class size was 38 children, this equates to approximately 0.53 L per child per day. The overall cost of the scheme was US$1.4 per child per year.

In this study, we obtained absenteeism data from the four schools where 1001F were providing free water. In addition we obtained this same data from four schools not in receipt of free water. In a related community study of childhood diarrhoea we were conducting a longitudinal study of childhood diarrhoea and water use in 25 villages. These villages had been chosen at random from all villages with an established 1001F presence or through a process of propensity score matching, the details of which is described elsewhere [10]. Four schools from these 25 villages were in receipt of the free school water scheme and willing and able to provide absenteeism data. Four control schools were chosen from the other 25 villages based on number of registered students present and the proportion of students under 14 years closest to those values of the intervention schools. The head teacher was then approached and invited to participate.

Data collection was based on routinely collected absenteeism data provided to the study team by the head teacher. Data was provided from the week beginning 4th December 2011 to 31st May 2012. This period spread over two school terms one of which was in the dry season and the other the wet season.

Data analysis was done using STATA version 11. Absenteeism rates per week were calculated as the number of days absent/(5×children registered). Random effects negative binomial regression analyses were done using a generalized linear model with a random intercept for school. The outcome variable was the number of days lost in each week from absenteeism and the number of children enrolled in the school was the exposure variable. The predictor variables were whether or not the school received water and season. Interaction terms were included for intervention and season.

In one school it was discovered that the number of water containers delivered fell short of the contracted amount. A further regression analysis was done for this school with days missed in the week being the dependent variable. The actual number of water containers delivered in the week and days missed at all the other schools combined were predictor variables. The analysis was restricted to the dry season and excluded holiday weeks.

Results

Table 1 shows certain key characteristics of to the villages where each of the eight schools were based. It can be seen that across most characteristics the intervention and control schools were generally very similar. The main exception is that very few people in intervention villages have access to improved water or sanitation compared to the control villages. This is not too surprising as the 1001F had primarily targeted its intervention at schools in areas where it was known that the local community had poor access to improved drinking water. Also of note was that rather more of the populations of the intervention villages were reported as being migrants. The predominant source of drinking water in the control schools was whatever the children brought in from their home. In one control school (C2) children also had access to a hand pump and jar in a pagoda about 100 m from the school and in another (C3) there was a rainwater harvesting tank for which children were reported to have some use.

Data was collected for 26 consecutive weeks. Three schools were closed during week 18, all schools were closed during week 19, and all but one in week 20. The dry season was taken to include all the weeks before the break in week 19 and the wet season in weeks subsequent to this holiday. Across all eight schools this represented 60,194 child weeks of follow-up. The overall absenteeism rate was 5.57%. Figure 1 shows the absenteeism rate for each school by week. The most obvious finding was the dramatic increase in absenteeism during the wet season, towards the end of the study period. This was not surprising given the fact that in many villages, children would be kept off school at this time to help in the fields.

Table 2 shows the results of the negative binomial regression analysis comparing absenteeism rates using whether or not the

Table 1. Characteristics of the villages in which the schools were based.

School code	C1	C2	C3	C4	I1	I2	I3	I4
Number of children in school	379	438	267	450	587	271	954	174
Number of households in village	318	320	405	191	331	256	227	192
Population	1634	1706	1951	942	1378	1126	1075	873
% population male	49	50	48	51	51	50	49	49
% population <5 yrs old in village	8	10	11	13	8	12	12	10
% population 5 to 14 yrs old	20	27	25	26	19	28	25	24
% population with lower secondary education or greater	96	96	95	98	92	98	99	98
% adult female literacy rate	84	58	72	34	66	50	73	64
% working in primary sector (mainly agricultural)	68	96	81	87	85	94	66	94
% population migrants	23	4	32	20	57	48	36	65
% with access to improved water	48	72	1	63	5	0	2	2
% with access to sanitation	84	65	28	19	14	6	16	38

School codes prefixed with I received the free water and those prefixed by C did not. Data from The National Institute of Statistics, Ministry of Planning, Royal Government of Cambodia (http://www.nis.gov.kh/index.php/online-statistics/resultonline).

school received free 1001F water and season as predictor variables. In addition we investigated the interaction between season and receipt of 1001F water. It can be seen that absenteeism was less than half in the intervention schools compared to those who did not receive 1001F water. Given the significant interaction term the association between having 1001F water and reduced absenteeism was restricted to the dry season with no such association in the wet season (as was also clear in figure 1).

At the end of the study period it became clear that one of the suppliers was not fully fulfilling their contract as they did not have sufficient capacity to provide water to the school and to their paying customers. Although container water was provided this fell short of the contracted amount. The remaining three schools received all their assigned supplies. Table 3 shows the results of the regression analysis of absenteeism in the school with incomplete water delivery adjusted for within week absenteeism in other participating schools. There was a significant association between the number of containers of water delivered in the week and reported absenteeism. For every extra container delivered there was a 2.9% reduction in absenteeism (95% confidence intervals (CI) 0.5 to 5.1%). The association was also tested between absenteeism and delivery in the previous week. Absenteeism was not associated with the number of containers delivered in the previous week.

Discussion

In this study we have shown lower absenteeism in schools receiving free containers of 1001F water. However, this association was only seen in the dry season and not in the wet season. There were also strong seasonal effects as absenteeism in several of the schools increased dramatically during the wet season, irrespective of water delivery. We were informed that this increase in absenteeism during the early wet season was partly because children were frequently kept off school to help in the fields. We have, furthermore, shown that in one school where delivery of water containers fell short of the contracted amount, absenteeism rates were associated with the number of containers delivered in the week. As far as we are aware this is the first study to show that provision of adequate safe drinking water in school can affect attendance in a developing country.

Clearly one has to be cautious when interpreting the results of an observational study like this. Nevertheless, taking both analyses together, this gives a fairly strong indication that provision of safe palatable drinking water is indeed associated with reduced absenteeism. Firstly although this study was not blinded and so potentially open to some form of reporting bias, school absenteeism rates are not subjective and so our results should not be as at risk of reporting bias that has affected many other studies of water and health in low income countries [8]. We cannot, of course, exclude bias in the way the classroom teacher records the daily attendance register or in how the school compiles absenteeism data from the class registers. However, any such bias is far less likely when based on register records than may be expected by asking children to recall their absence history during interview as was done in the only other study of school absence and WASH [7]. Secondly, although it is plausible that selection of schools for the intervention may have led to a degree of bias, it is difficult to see how this would have affected the association found between number of containers delivered and absenteeism in the school with incomplete contract fulfilment. Of particular note here was that the intervention schools were generally in areas with poor domestic access to improved drinking water supplies and sanitation. If inadequate drinking water and sanitation does

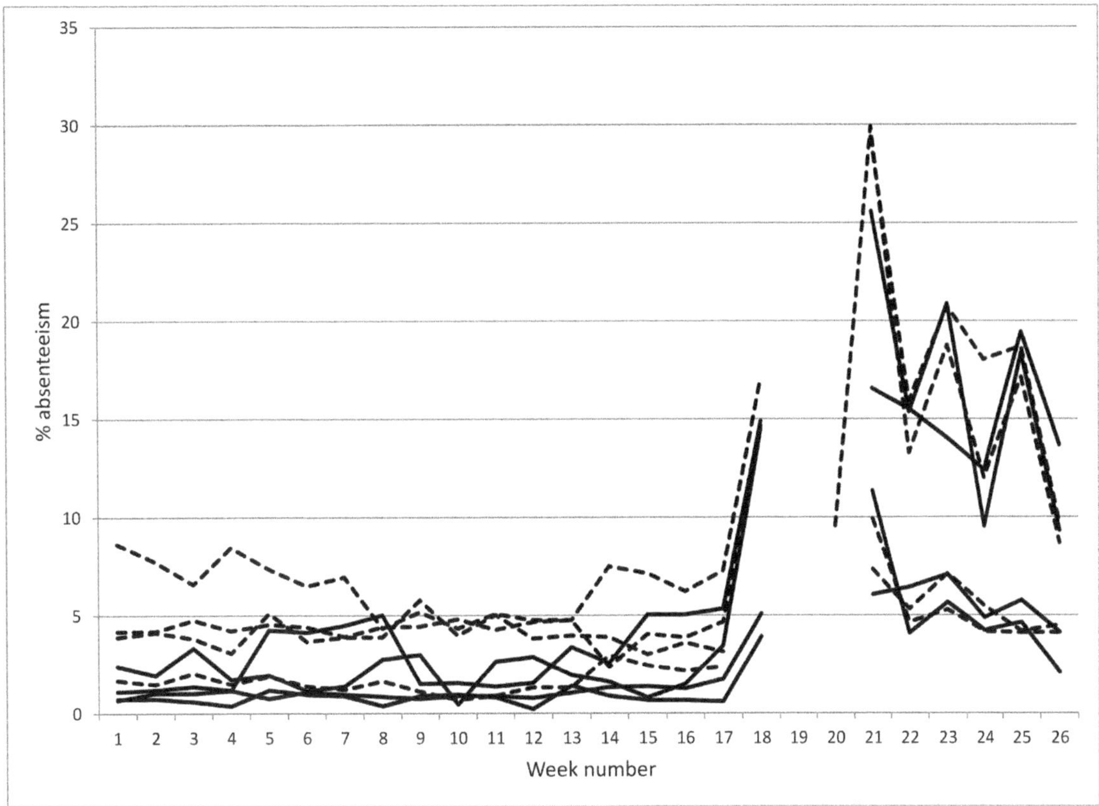

Figure 1. Absenteeism rate by school and week. Solid line shows rates for intervention schools and broken line for control schools.

impact of school absenteeism, then if anything this source of bias would be expected to increase absenteeism in the intervention schools rather than reduce it. It is not clear what effect if any the greater number of migrants in some villages would have on absenteeism in school. We would however suggest that further randomised studies are required before a more definitive conclusion can be made.

This leaves the question of what was the mechanism between water supply provision and absenteeism. In this study we were not able to collect any data on the reasons for the absenteeism. Given the fact that the association was between absenteeism rates and water delivery in the same week and not the previous week, we are not suggesting that this association was primarily due to a reduction in waterborne infectious disease. A possible explanation in our view may be that by providing readily available palatable and safe water in the classroom, children are more likely to drink during the school day and so not become dehydrated. Even mild dehydration in vulnerable groups such as young children has been suggested as being associated with various adverse health effects [11]. Furthermore, in a recent study from a hot dry region of Italy, the authors showed that supplementary drinking water was associated with improved cognition and an improved subjective sense of vigour [12]. This Italian study is in line with similar findings from several previous researchers [13,14]. What this suggests therefore is that provision of supplementary water sufficiently improves the child's general wellbeing as well as the learning and experience of the school day as he/she is better hydrated. Consequently they are more likely to attend school the following day if they had felt good at school the previous day.

Even if, as we suspect, the main reason for the reduced absenteeism in the intervention group is due to improved hydration rather than a reduction in waterborne disease, this should not be taken as an indication that the provision of water of any quality would be acceptable. The link between contaminated

Table 2. Risk factors for absenteeism rates in schools.

Predictor		Incidence Rate Ratio	Lower 95% CI	Upper 95% CI	P
Receives 1001F Water	No	1			<0.001
	Yes	0.386	0.266	0.560	
Season	Dry	1			<0.001
	Wet	2.618	2.279	3.008	
Season-1001F water interaction	Yes-Wet	1.991	1.603	2.472	<0.001

Table 3. Risk factors for absenteeism in school with incomplete delivery of water containers during the dry season.

Predictor		Incidence Rate Ratio	Lower 95% CI	Upper 95% CI	P
Water delivered in week	/container	0.971	0.949	0.995	0.016
Absenteeism days in other schools	/days missed in week	1.000	0.992	1.007	0.976

drinking water and disease risk is well accepted and it is clear that the main risk falls on young children [15]. Any scheme to increase drinking water provision in the classroom that does not ensure that that water is safe to drink is likely to put the children at risk of waterborne disease. However, providing safe water in the school environment does not necessarily mean children will drink it. Indeed taste appears to be a major determinant affecting whether or not people continue to use safe drinking water sources [16,17]. Chlorination of drinking water is associated with poorer taste for many people [18,19]. On the other hand filtration can be associated with improved taste [20]. The fact that 1001F water uses filtration and Ultraviolet disinfection but not chlorination would mean that it would have better taste qualities than other safe water sources and so may be more likely to be used by children.

In conclusion, we have shown a significant association between provision of supplementary water in the classroom and reduced absenteeism rates. With the delivery mechanism in this study the cost per child is modest, but the potential benefit to children's education and subsequent life potential could be extremely large. There is a great need for further research in this area, especially randomised control trials and studies aimed at determining the biological mechanisms behind this reduction in absenteeism.

Acknowledgments

We would like to thank Indochina Research of Phnom Penh, Cambodia who undertook all the field work.

Author Contributions

Conceived and designed the experiments: PRH HR FJ HL MY PH C. Longuet. Performed the experiments: MY HL C. Lo. Analyzed the data: PRH FJ. Wrote the paper: PRH HR MY HL C. Longuet PH C. Lo FJ.

References

1. Abadzi H (2009). Instructional Time Loss in Developing Countries: Concepts, Measurement, and Implications. World Bank Res Obser 24: 267–290.
2. Gakidou E, Cowling K, Lozano R, Murray CJL (2010) Increased educational attainment and its effect on child mortality in 175 countries between 1970 and 2009: a systematic analysis. Lancet 376: 959–974.
3. Basu AM, Stephenson R (2005) Low levels of maternal education and the proximate determinants of childhood mortality: a little learning is not a dangerous thing. Soc Sci Med 60: 2011–2023.
4. United Nations (UN) (2011). The Millennium Development Goals Report 2011; United Nations: New York, NY, USA.
5. Fuller B (1987) What School Factors Raise Achievement in the Third World? Review of Educat Res 57: 255–292.
6. Jasper C, Le T-T, Bartram J (2012) Water and Sanitation in Schools: A Systematic Review of the Health and Educational Outcomes. Int. J. Environ. Res. Public Health 9: 2772–2787.
7. Freeman MC, Greene LE, Dreibelbis R, Saboori S, Muga R, et al. (2012) Assessing the impact of a school-based water treatment, hygiene and sanitation programme on pupil absence in Nyanza Province, Kenya: a cluster-randomized trial. Trop Med Int Health 17: 380–391.
8. Schmidt W-P, Cairncross S (2009) Household water treatment in poor populations: Is there enough evidence for scaling up now? Environm Sci Technol 43: 986–992.
9. Hunter PR (2009) House-hold water treatment in developing countries comparing different intervention types using meta-regression. Environm Sci Technol 43: 8991–8997.
10. Hunter PR, Risebro H, Yen M, Lefebvre H, Lo C, et al. (2013) Water source and diarrhoeal disease risk in children under 5 years old in Cambodia: a prospective diary based study. BMC Public Health 13: 1145.
11. Maughan RJ (2012) Hydration, morbidity, and mortality in vulnerable populations. Nutr Rev 70 (Suppl. 2): S152–S155.
12. Fadda R, Rapinett G, Grathwohl D, Parisi M, Fanari R, et al. (2012)Effects of drinking supplementary water at school on cognitive performance in children. Appetite 59: 730–737.
13. Bar-David Y, Urkin J, Kozminsky E (2005) The effect of voluntary dehydration on cognitive functions of elementary school children. Acta Paediatrica 94: 1667–1673.
14. Benton D, Burgess N (2009) The effect of the consumption of water on the memory and attention of children. Appetite 53: 143–146.
15. Risebro H, Breton L, Aird H, Hooper A, Hunter PR (2012) Risk of infectious intestinal disease in consumers drinking from private water supplies: A prospective cohort study. PLOS One 7(8): e42762.
16. Huber AC, Mosler HJ (2013) Determining behavioral factors for interventions to increase safe water consumption: a cross-sectional field study in rural Ethiopia. Int J Environ Health Res 23: 96–107.
17. Tamas A, Mosler HJ (2011) Why Do People Stop Treating Contaminated Drinking Water With Solar Water Disinfection (SODIS)? Health Educ Behav 38: 357–366.
18. Heiner JD, Simmons EA, Hile DC, Wedmore IS (2011) A Blinded, Randomized, Palatability Study Comparing Variations of 2 Popular Field Water Disinfection Tablets. Wild Environ Med 22: 329–332.
19. Doria MD, Pidgeon N, Hunter PR (2009) Perceptions of drinking water quality and risk and its effect on behaviour: A cross-national study. Sci Total Environ 407: 5455–5464.
20. Ngai TKK, Shrestha RR, Dangol B, Maharjan M, Murcott SE (2007) Design for sustainable development - Household drinking water filter for arsenic and pathogen treatment in Nepal. J Environ Sci Health A 42: 1879–1888.

24

Leaf Area Index Drives Soil Water Availability and Extreme Drought-Related Mortality under Elevated CO_2 in a Temperate Grassland Model System

Anthony Manea*, Michelle R. Leishman

Department of Biological Sciences, Macquarie University, North Ryde, NSW, Australia

Abstract

The magnitude and frequency of climatic extremes, such as drought, are predicted to increase under future climate change conditions. However, little is known about how other factors such as CO_2 concentration will modify plant community responses to these extreme climatic events, even though such modifications are highly likely. We asked whether the response of grasslands to repeat extreme drought events is modified by elevated CO_2, and if so, what are the underlying mechanisms? We grew grassland mesocosms consisting of 10 co-occurring grass species common to the Cumberland Plain Woodland of western Sydney under ambient and elevated CO_2 and subjected them to repeated extreme drought treatments. The 10 species included a mix of C_3, C_4, native and exotic species. We hypothesized that a reduction in the stomatal conductance of the grasses under elevated CO_2 would be offset by increases in the leaf area index thus the retention of soil water and the consequent vulnerability of the grasses to extreme drought would not differ between the CO_2 treatments. Our results did not support this hypothesis: soil water content was significantly lower in the mesocosms grown under elevated CO_2 and extreme drought-related mortality of the grasses was greater. The C_4 and native grasses had significantly higher leaf area index under elevated CO_2 levels. This offset the reduction in the stomatal conductance of the exotic grasses as well as increased rainfall interception, resulting in reduced soil water content in the elevated CO_2 mesocosms. Our results suggest that projected increases in net primary productivity globally of grasslands in a high CO_2 world may be limited by reduced soil water availability in the future.

Editor: Gerrit T. S. Beemster, University of Antwerp, Belgium

Funding: This research was funded by a Macquarie University Research Excellence and National Climate Change Adaptation Research Flagship scholarship to AM. The funders had no role in study design, data collection and analysis, decision to publish, or preparation of the manuscript.

Competing Interests: The authors have declared that no competing interests exist.

* E-mail: anthony.manea@students.mq.edu.au

Introduction

A major driver that shapes the physiology, ecology and evolution of terrestrial plants is climatic extremes [1]. It is widely acknowledged that the magnitude and frequency of climatic extremes, such as drought, are likely to increase under future climate change conditions [2]. The IPCC [2] defines a climatic extreme, such as extreme drought, as an event that occurs once every 20 years, on average. The potential for climatic extremes to alter the structural and functional dynamics of ecological communities [2], [3], coupled with their increasing magnitude and frequency in the future, suggests research on climatic extremes should be a high priority.

Grass-dominated systems (grasslands, savannas and open grassy woodlands) occupy more than 30% of the global terrestrial landscape [4] and play an important role in the global carbon cycle [5]. The productivity of grass-dominated systems (referred to as grasslands from now on) is strongly mediated by soil water availability [4], [6], [7]. For example, Fay et al [8] increased the mean annual rainfall to an experimental grassland community in Kansas by 250% and found that both soil water content (SWC) and aboveground net primary productivity significantly increased (see exception [9]). Consequently it is likely that soil water availability prior to an extreme drought event will be a major driver in the response of the grassland to the event.

It has rarely been considered how CO_2 concentration will alter soil water availability in grasslands and thus modify grassland responses to extreme drought. One of the major drivers of soil water availability in grasslands is canopy transpiration [6]. The amount of water that is lost through canopy transpiration depends on the stomatal conductance and leaf area index (LAI) of the grasses [10]. Reduction in stomatal conductance of grasses under elevated CO_2 has been well-documented [6], [11], [12]. For example, Morgan et al [7] found that in the Wyoming mixed-grass prairies in the United States, the annual SWC increased on average by 17.3% over a three year period due partly to reductions in stomatal conductance under elevated CO_2 levels. The physiology (C_3 or C_4) and origin (native or exotic) of the grasses play an important role in this reduction of stomatal conductance under elevated CO_2 levels. C_4 plants evolved in a low CO_2 environment, allowing high rates of photosynthesis at low stomatal conductance [13]. This suggests that C_4 plants should have greater reductions in stomatal conductance compared to C_3 plants under elevated CO_2 levels. In addition, meta-analysis studies have shown that at the leaf-level natives tend to have lower stomatal conductance than exotics [14].

In contrast to reductions in stomatal conductance, grasses often have higher leaf area index (LAI) under elevated compared to ambient CO_2 levels, resulting in increased canopy transpiration. For example, the LAI of the C_4 bunchgrass *Pleuraphis rigida* significantly increased under elevated CO_2 in the 10 year Mojave Desert FACE experiment [15]. In a meta-analysis of semi-wild and wild C_3 and C_4 grasses, leaf area (related to LAI) increased under elevated CO_2 by 15% and 25% respectively [16]. Therefore reductions in the stomatal conductance of grasses under elevated CO_2 may be offset by increases in their LAI [17].

In this study we asked whether the response of grasslands to repeat extreme drought events is modified by elevated CO_2, and if so, what are the underlying mechanisms? Experimental mesocosms containing common co-occurring native and exotic grass species of the Cumberland Plain Woodland of western Sydney, Australia were grown under ambient and elevated CO_2 levels and exposed to repeated one in 20 year extreme drought events. We hypothesized that a reduction in the stomatal conductance of the grasses, particularly the C_4 and native grasses, grown under elevated CO_2 levels would be offset by increases in the LAI of the grasses. That is, the total canopy transpiration of the grasses would not differ between the CO_2 treatments. Therefore the retention of soil water in the mesocosms and consequently the vulnerability of the grasses to extreme drought would not differ between the CO_2 treatments.

Methods

Species selection

We selected five native and five exotic perennial grass species which commonly co-occur in a grassy open woodland community known as Cumberland Plain Woodland that occurs in western Sydney, Australia. All the exotic species are considered to be invasive rather than simply exotics that have become naturalized in Cumberland Plain Woodland [18]. Within both the native and exotic groups we included two C_3 and three C_4 species. Seeds for the 10 grass species were obtained from a commercial supplier (Nindethana Seed Service, Albany, WA, Australia). Information on the biology of each grass species is provided in table 1. Once collected, the seeds for each of the 10 grass species were germinated on moist paper towels within covered aluminium trays. To spread the risk of germination failure, each grass species was germinated in a number of different aluminium trays.

Experimental design and treatments

The native and exotic grass species were grown together in mesocosms using a fully factorial experimental design with two factors: CO_2 concentration and drought treatment. The mesocosms consisted of 65 L tubs (60 cm long ×40 cm wide ×28 cm deep), each tub containing 55 L of soil mixture consisting of field-collected Cumberland Plain Woodland soil and coarse river sand in a ratio of 3:1. The Cumberland Plain Woodland soil was obtained from Mt Annan (34.07°S, 150.76°E) and Luddenham (33.88°S, 150.69°E) in western Sydney and was homogenized into a single batch. No specific approvals or permits were required for soil collection. The soil collection locations were not privately-owned or protected in any way and did not endanger any protected species. The river sand was obtained from a commercial supplier (Australian Native Landscapes, North Ryde, NSW, Australia). Seedlings were transplanted from the germination trays into the CO_2 treatment mesocosms at the stage of second true leaf emergence. All seedlings were planted within 24 hours of each other. The seedlings were planted in two rows of five with each species allocated a position randomly within each mesocosm. For each grass species multiple seedlings were transplanted into each mesocosm as insurance against seedling mortality. After six days, the excess seedlings were removed from the mesocosms, leaving one individual per species per mesocosm.

CO_2 treatments were set to two levels: ambient (380–420 ppm) and elevated (530–570 ppm) CO_2. These CO_2 concentration ranges were maintained and monitored daily by a CO_2 dosing and monitoring system (Canary Company Pty Ltd, Lane Cove, NSW, Australia). The lower concentrations of these ranges tended to occur at night-time while the higher concentrations occurred during the daytime. The elevated CO_2 treatment represents the predicted atmospheric CO_2 concentration by 2050 [19].

We defined a drought as the number of consecutive days with < 1 mm of rainfall. This definition is a part of the ETCCDI/CRD climate change indices [20]. We used the IPCC [2] definition of climatic extreme which is an extreme that occurs once every 20 years, on average. Gumbel I distributions were fitted to the annual drought extremes of the Cumberland Plain for each year from 1867–2010. Rainfall on the Cumberland Plain is not seasonal so

Table 1. Grass species used in the study.

Species	Origin	Seed mass (mg)	Photosynthetic pathway	Longevity
Chloris gayana Kunth	Exotic	0.4	C_4	Perennial
Eragostis curvula (Schrad.) Nees	Exotic	0.2	C_4	Perennial
Pennisetum clandestinum Hochst. Ex Chiov	Exotic	7.0	C_4	Perennial
Bromus catharticus Vahl	Exotic	10.7	C_3	Short lived perennial
Ehrharta erecta Lam.	Exotic	2.0	C_3	Perennial
Bothriochloa macra (Steud.) S.T.Blake	Native	1.2	C_4	Perennial
Chloris truncata R.Br.	Native	0.3	C_4	Perennial
Themeda australis (R.Br.) Stapf	Native	2.6	C_4	Perennial
Austrodanthonia racemosa (R.Br.) H.P.Linder	Native	0.7	C_3	Perennial
Microlaena stipoides (Labill.) R.Br.	Native	4.3	C_3	Perennial

Grass species (family: *Poaceae*) used in this study, with information on the origin, seed mass, photosynthetic pathway and longevity of each species. Average seed mass was obtained by oven drying 50 seeds from each species at 60°C for 48 hours and then weighing them. Data on taxonomy, physiology and longevity were obtained from PlantNET (www.plantnet.rbgsyd.nsw.gov.au).

the time of year that the annual drought extremes occurred differed from year to year. The data used were obtained from the Australian Bureau of Meteorology historical records of Brownlow Hill (34.03°S, 150.65°E, 1867–1969), Kentlyn (34.05°S, 155.88°E; 1970–1971), Camden Airport (34.04°S, 150.69°E; 1972–1992, 1998–2001), Ruse (34.06°S, 150.85°E; 1993–1997) and Mt Annan Botanical Gardens (34.07°S, 150.76°S; 2002–2010). A one in 20 year extreme drought event for the Cumberland Plain was calculated to last for a period of 53 days. The extreme drought was simulated by turning off the watering system for the treatment period.

The extreme drought treatment mesocosms were replicated five times at each CO_2 level. These were called the 'drought treatment' mesocosms. In addition, five extra mesocosms were grown under each CO_2 level. These were called the 'before treatment' mesocosms. All mesocosms were mist watered for one minute twice daily which is representative of the average daily amount of rainfall (828 mm annually) on the Cumberland Plain. This daily rainfall average was based on the same Australian Bureau of Meteorology historical records from Camden airport (1943–2004) as described in the above paragraph. This design resulted in a total of 20 mesocosms each containing 10 grass species (i.e. [2 CO_2 treatments ×5 replicates]+10 extra mesocosms). The mesocosms within each CO_2 treatment were split between two glasshouses. The mesocosms were switched between glasshouses within each CO_2 treatment once during the growth period and once during the treatment period to reduce any glasshouse effect. The temperature of the glasshouses was set for a maximum of 24°C and a minimum of 16°C. During the entire duration of the experiment the relative humidity within each glasshouse was measured every day at 9am and 3pm using a HOBO temperature/RH/2 external channel data logger (OneTemp, Parramatta, NSW, Australia).

The grasses were grown for 12 weeks at which stage they were mature and seeding. After the growth period the 'before treatment' mesocosms were harvested, washed free of soil and separated into their following components: leaf biomass, stem biomass and belowground biomass. The total leaf area of the leaf biomass was measured using a LI-3100C Area Meter (Li-Cor, Lincoln, NE, United States). The plant components were then oven-dried at 60°C for 72 hours and weighed using a Mettler Toledo B-S electronic balance.

The extreme drought treatment was then applied to the five 'drought treatment' mesocosms at each CO_2 level. After the 53 day treatment period the mesocosms were mist-watered twice daily as previously for four weeks to allow a recovery period. In field conditions on the Cumberland Plain four weeks would be ample time for the grasses to recover (i.e. resprout) from a dry period. For 30 days prior to and during the drought/recovery cycles the SWC of each mesocosm was measured at a depth of 15 cm every 10 days using a Hydrosense II Portable Soil Moisture System (Campbell Scientific Australia Pty Ltd, Garbutt, QLD, Australia). In addition, for the 30 days prior to the extreme drought treatment the stomatal conductance of each grass was measured every five days using a Porometer AP4 (Delta-T Devices, Burwell, CB, Uniting Kingdom). Measurements would begin at 8.30 am and would take approximately three and a half hours to complete. After the four week recovery period, the mortality of the grasses was recorded. Each grass was classified as dead or alive depending on if it showed signs of regeneration by the end of the recovery period. The grasses were then exposed to two more cycles of drought and recovery with mortality of the grasses recorded after each cycle. At the end of the recovery period of the final cycle, every mesocosm was dug up to ensure that the grasses classified as dead had no living root material.

Data analysis

Leaf area index and biomass allocation analysis. We calculated LAI and root to shoot ratio (R:S) based on the harvested biomass data of the 'before treatment' mesocosms. LAI for each individual grass was calculated as total leaf area divided by ground area in the mesocosm (0.24 m^2). R:S for each individual grass was calculated as total root mass divided by total shoot mass. We used two-way ANOVAs to test for a CO_2 and species effect on LAI, R:S and total biomass across all grass species in the 'before treatment' mesocosms.

We used two-way ANOVAs to test for a CO_2 and plant type (physiology and origin) effect on LAI across all grass species in the 'before treatment' mesocosms.

CO2 level and plant type survival analysis. Kaplan-Meier survival curves were generated to determine the survival function across all the grass species in relation to (1) CO_2 level, (2) physiology, (3) origin and (4) CO_2 level combined with each plant type (origin and physiology). If there was a significant difference between the Kaplan-Meier survival curves within each CO_2 level/plant type combination then the survival distributions within the combination were tested for a significant difference. All survival distributions were compared using log rank tests, with significance set at $P<0.05$.

Stomatal conductance, total canopy transpiration and soil water content analysis. We used General Estimating Equation (GEE) models with a Bonferroni adjustment to determine if there was a difference in the stomatal conductance between CO_2 level and each plant type (origin and physiology) combination during the 30 days prior to the extreme drought treatment. The same analyses were performed for SWC with the addition of testing for a difference between the CO_2 levels for the three drought/recovery cycles. For all GEE models we specified a Gamma with log-link model using an exchangeable correlation matrix for the continuous variables (stomatal conductance and SWC).

Total canopy transpiration for each species was calculated as the average stomatal conductance multiplied by the average LAI. We carried out species-pair comparisons using paired t-tests with a Bonferroni adjustment to test for a CO_2 effect on total canopy transpiration.

Relative humidity analysis. We used a paired t-test with a Bonferroni adjustment to determine if there was difference in the mean daily relative humidity between the ambient and elevated CO_2 glasshouses. The relative humidity data was then averaged within each CO_2 level and compared to the historical average relative humidity (9am-74%, 3pm-49%) of the Cumberland Plain using one-sample t-tests. This involved separate analyses for the relative humidity data obtained at 9am and 3pm. The historical average used was obtained from the Australian Bureau of Meteorology historical records of Camden Airport (34.04°S, 150.69°E; 1943–2010).

All data analyses were performed using IBM SPSS statistical software, Version 21.0.0 (SPSS Inc., 2012, IBM, Illinois, United States, http://www.spss.com) with the significance level set at 0.05. Data were $\log_{10}$ transformed when necessary to fulfil the assumptions of ANOVA.

Results

Humidity analysis

There was no significant difference in the relative humidity between and ambient and elevated CO_2 glasshouses ($t_{1,38} = 1.00$, $p = 0.326$). However the relative humidity in the glasshouses at 9am ($t = 10.01$, $p < 0.001$) and 3pm ($t = 17.78$, $p < 0.001$) was significantly higher than the historical average for the same times of day.

Leaf area index and biomass allocation analysis

There was no significant interaction between CO_2 and species for LAI, R:S or total biomass but there was a significant difference between species for all traits. LAI ($F_{1,99} = 14.20$, $p < 0.001$) and total biomass ($F_{1,99} = 19.97$, $p < 0.001$) were significantly higher under elevated CO_2 compared to ambient CO_2 across all the grass species. R:S ($F_{1,99} = 0.26$, $p = 0.615$) did not significantly differ between ambient and elevated CO_2.

There was a significant interaction between CO_2 and physiology for LAI ($F_{1,99} = 5.15$, $p = 0.026$) with the C_4 grasses grown under ambient CO_2 having significantly lower LAI than all of the other CO_2×physiology combinations (figure 1). There was also a significant interaction between CO_2 and origin for LAI ($F_{1,99} = 4.21$, $p = 0.043$) with the native grasses grown under ambient CO_2 having significantly lower LAI than all of the other CO_2×physiology combinations (figure 1).

CO_2 level and plant type survival analysis

The average survival rate across all grass species was significantly higher under ambient compared to elevated CO_2 levels ($\chi^2 = 8.58$, df = 1, $p = 0.003$). The physiology ($\chi^2 = 3.24$, df = 1, $p = 0.072$) or origin ($\chi^2 = 0.03$, df = 1, $p = 0.855$) of the grasses did not significantly influence their survival rates.

There was a significant difference in survival between the different CO_2×physiology combinations ($\chi^2 = 13.29$, df = 3, $p = 0.004$; figure 2a). Both the C_3 ($\chi^2 = 10.01$, df = 1, $p = 0.002$) and C_4 ($\chi^2 = 7.91$, df = 1, $p = 0.005$) grasses grown under ambient CO_2 had significantly higher survival rates than the C_4 grasses grown under elevated CO_2.

There was a significant difference in survival between the different CO_2×origin combinations ($\chi^2 = 16.64$, df = 3, $p = 0.001$; figure 2b). The native grasses grown under ambient CO_2 had a significantly higher survival rate than the native ($\chi^2 = 17.45$, df = 1, $p < 0.001$) and exotic ($\chi^2 = 6.52$, df = 1, $p = 0.011$) grasses grown under elevated CO_2. The exotic grasses grown under ambient CO_2 had a significantly higher survival rate than the native grasses grown under elevated CO_2 ($\chi^2 = 4.13$, df = 1, $p = 0.042$).

Stomatal conductance, total canopy transpiration and soil water content analysis

Stomatal conductance across all grass species was significantly lower under elevated CO_2 compared to ambient CO_2 (Wald $\chi^2 = 7.254$, df = 1, $p < 0.007$).

There was no significant interaction between CO_2 and physiology for stomatal conductance (Wald $\chi^2 = 1.786$, df = 1, $p = 0.181$). The C_3 grasses had significantly higher stomatal conductance than the C_4 grasses (Wald $\chi^2 = 46.00$, df = 1, $p < 0.001$; figure 3a). As described above, stomatal conductance was significantly lower under elevated CO_2 compared to ambient CO_2.

There was a significant interaction between CO_2 and origin for stomatal conductance (Wald $\chi^2 = 7.369$, df = 1, $p = 0.007$; figure 3b). The exotics grown under ambient CO_2 had significantly higher stomatal conductance than all of the other CO_2×origin combinations.

There was no significant difference in the total canopy transpiration between ambient and elevated CO_2 treatments ($t_8 = -0.93$, $p = 0.375$).

The SWC during the 30 days prior to the drought/recovery cycles was significantly higher under ambient CO_2 compared to elevated CO_2 (Wald $\chi^2 = 5.91$, df = 1, $p = 0.015$). The overall SWC of the mesocosms during the experiment drought/recovery cycles was significantly higher under ambient CO_2 compared to elevated CO_2 (Wald $\chi^2 = 11.48$, df = 1, $p = 0.001$; figure 4). However SWC differences varied between drought/recovery cycles. During the first drought/recovery cycle SWC was significantly higher in the ambient compared with elevated CO_2 treatment (Wald $\chi^2 = 8.98$, df = 1, $p = 0.003$). During the second and third drought/recovery cycles SWC did not significantly differ

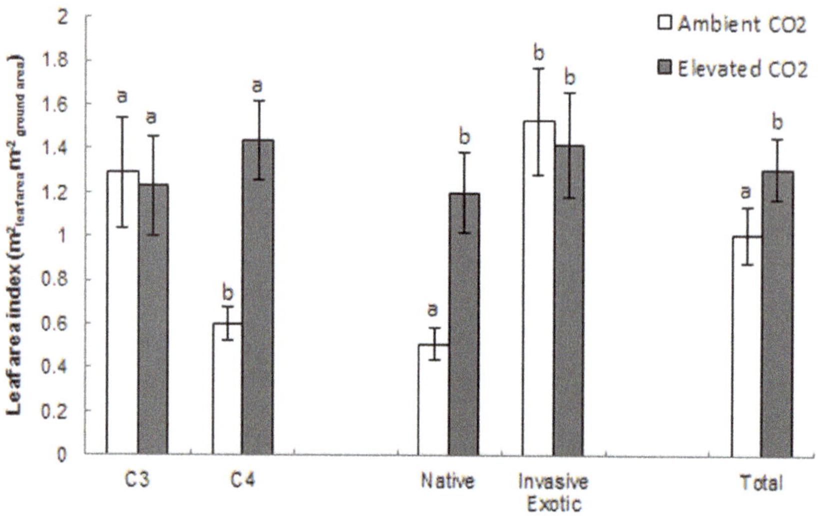

Figure 1. Leaf area index across all grass species for each CO_2×plant type combination. Mean leaf area index across all grass species for each CO_2×physiology and CO_2×origin combination. Error bars represent one standard error. Letters indicate significant differences at $p < 0.05$.

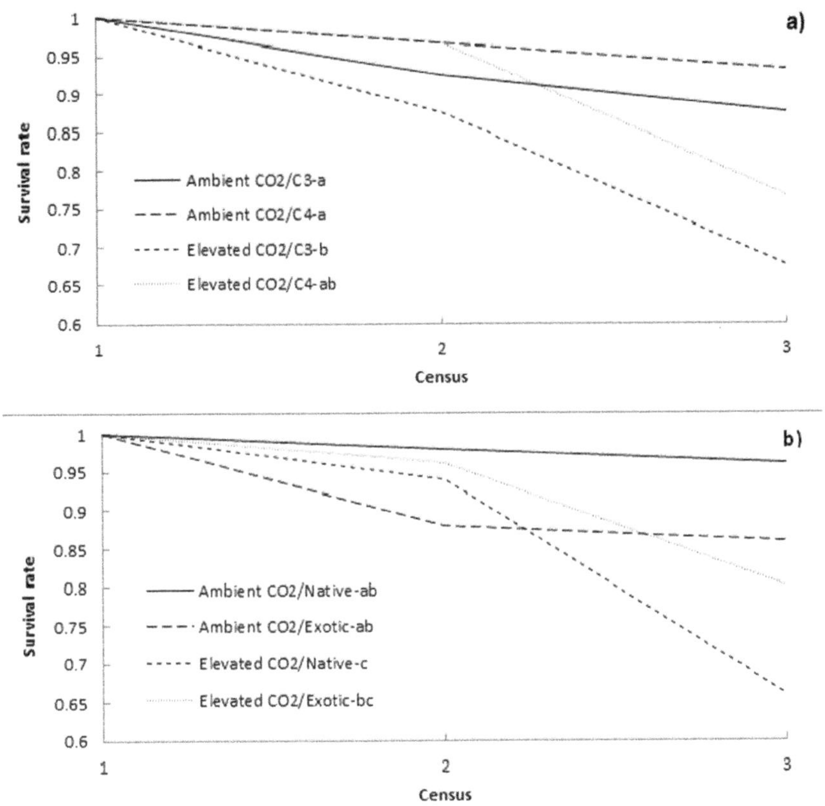

Figure 2. Survival rates across all grass species for each CO_2×plant type combination. The average survival rate across all grass species of each (a) CO_2×physiology and (b) CO_2×origin combination, at each survival census. A survival census was carried out after each extreme drought cycle which consisted of a 53 extreme drought period and 28 day recovery period. Letters indicate significant differences at $p<0.05$.

between CO_2 treatments (second cycle Wald $\chi^2=0.71$, df = 1, p = 0.400; third cycle Wald $\chi^2=2.504$, df = 1, p = 0.114).

Discussion

In this study we tested the vulnerability of an experimental grassland community to repeat extreme drought events under ambient and elevated CO_2 levels. Our results show that grasses grown under elevated CO_2 had significantly higher mortality in response to extreme drought than those grown under ambient CO_2. Our original hypothesis was that SWC of the mesocosms and thus vulnerability of the grasses to extreme drought would not differ between the CO_2 treatments because the reductions in stomatal conductance would be offset by increases in LAI. Our results did not support this hypothesis: soil water content was significantly lower in the elevated CO_2 treatment thus increasing extreme drought-related mortality in the grasses. The 32% reduction in stomatal conductance under elevated CO_2 was offset by a 30% increase in LAI. This offset is shown by the non-significant difference in total canopy transpiration of the grasses between the CO_2 treatments. These results contrast the results of previous studies that have found that the SWC in semi-arid grasslands increased under elevated CO_2 because the increases in total leaf area (related to LAI) were insufficient to offset the decreases in stomatal conductance [7], [21].

As the amount of water lost through total canopy transpiration did not differ between the CO_2 treatments, differences in SWC must have been due to the amount of water reaching the soil surface. Rain throughfall decreases with LAI because of increased interception and subsequent evaporation of rainfall from the surfaces of leaves [10]. In this experiment the grasses had significantly higher LAI under elevated CO_2. This would have increased rainfall interception consequently causing differences in the SWC between the CO_2 treatments prior to the extreme drought treatments. As the experiment progressed through each drought/recovery cycle, the initial difference in SWC between the CO_2 treatments converged. This convergence of SWC between the CO_2 treatments coincided with the mortality of grasses in the elevated CO_2 mesocosms. We suggest that this is because there were fewer individuals (due to the extreme drought-related mortality) in the elevated CO_2 mesocosms which would have reduced the total canopy transpiration and rainfall interception in those mesocosms.

We hypothesised that decreases in stomatal conductance under elevated CO_2 would be greater for the C_4 and native grasses. Surprisingly we found that the exotics had the greatest reduction in stomatal conductance among the grasses under elevated CO_2. In contrast the C_4 and native grasses had greater increases in LAI under elevated CO_2 compared to the other grass species. It is often assumed that native plants are less influenced by water limited conditions (e.g. extreme drought) in comparison to exotic plants [22], [23]. Our findings suggest that the exotic grasses would be less influenced by extreme drought than C_4 and native grasses in a high CO_2 world. However it is difficult to suggest if this could significantly alter the species composition of grasslands. This is because changes in SWC can alter the structure and function of grasslands in aspects other than productivity that may influence species composition. SWC can contribute to species shifts within grasslands by changing seed production and seedling recruitment

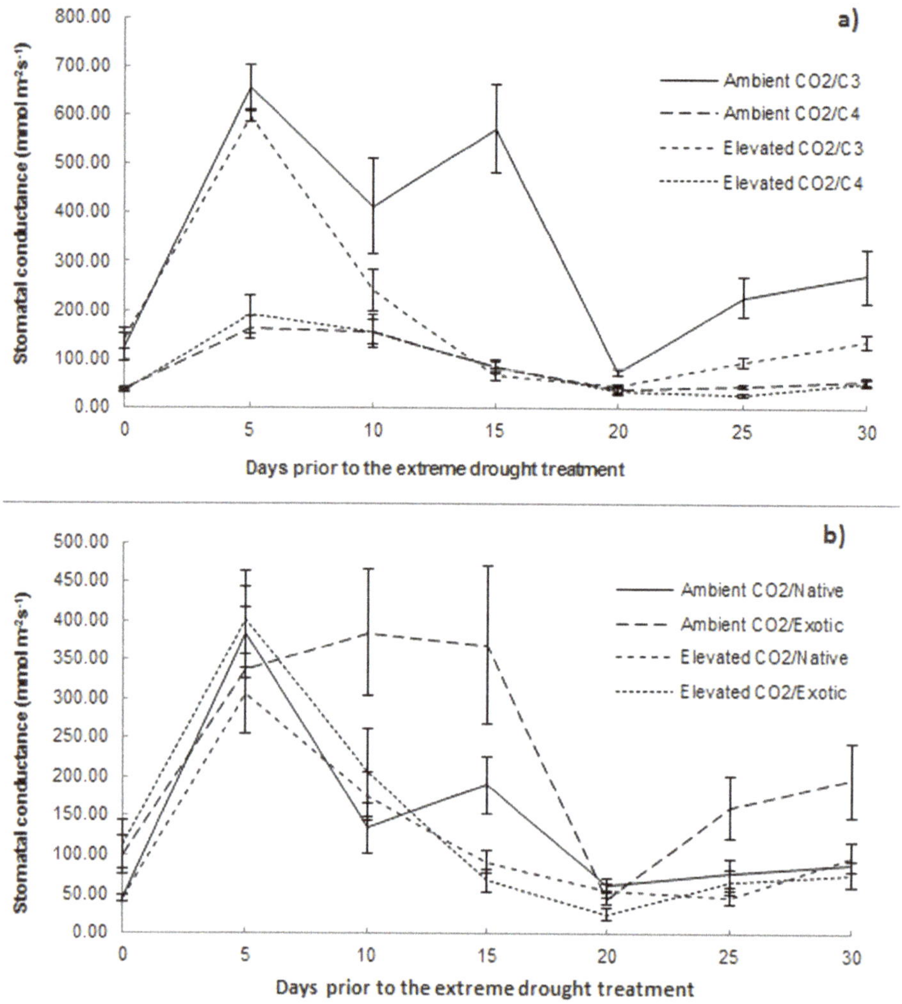

Figure 3. Stomatal conductance across all grass species for each CO_2×plant type combination. Mean stomatal conductance across all grass species for each (a) CO_2×physiology and (b) CO_2×origin combination, during the 30 days prior to the extreme drought treatment. Error bars represent one standard error.

among species [6], [24] and altering competitive interactions among established plants [25], [26], [27].

The findings from our study suggest that CO_2 concentration and soil water availability are important in mediating grassland productivity through grass dieback and mortality. It has been projected that a drastic shift in the annual global precipitation patterns may result in up to a 20% loss in soil water [28]. This is reinforced by global climate models which predict that large areas on every inhabitable continent will experience intensified droughts and widespread decreases in soil water [2]. However, net primary productivity in grasslands is projected to increase under elevated CO_2 conditions. Parton et al [29] modelled the effects of increased CO_2 for 31 temperate and tropical grassland sites, using the CENTURY model. They found that with the exception of cold desert steppe regions the net primary productivity of grasslands increased under elevated CO_2 conditions [29]. From our results we suggest that this projected global increase in the productivity of grasslands under elevated CO_2 levels may be negated (by grass dieback and mortality) due to the projected soil water constraints in the future, both as a direct consequence of changed precipitation and changes in plant-level traits such as stomatal conductance and LAI.

This study only examined the response of a single growth stage (i.e. mature) of the grasses to repeated extreme drought events under different CO_2 treatments. However, the vulnerability of plants to climatic extremes may be different in the early stages of their growth and development in comparison to mature plants. Therefore it is important to consider also the effect of climatic extremes on seed germination and seedling establishment as these stages are likely to strongly influence vegetation dynamics [30].

Predicting responses at local and regional scales to global change is often dependent on scaling up from plant-level mechanisms [10]. Although we attempted to make our glasshouse based mesocosm experiment as realistic a representation of the field conditions as possible, it is almost an impossible exercise to exactly recreate these conditions. For example, the soil depth in our mesocosms was 28 cm which is shallow compared to soil depth in the field. This may have caused higher mortality of the grasses in the mesocosms as they did not have access to deeper soil water as grasses in field conditions may have. Mesocosm studies such as this are important for understanding mechanisms and stimulating further research, rather than simply assessing outcomes [31]. It is important to note that we did not try to quantify exact mortality of grasses in response to extreme drought, but rather we

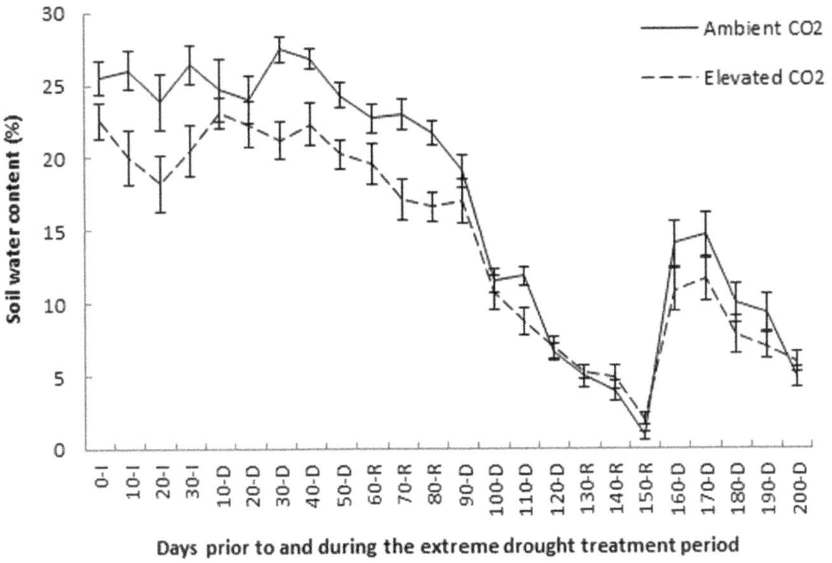

Figure 4. Soil water content across all mesocosms under ambient and elevated CO_2 levels. Mean soil water content across all mesocosms under ambient and elevated CO_2 levels over the 30 days prior to the extreme drought treatment and the drought/recovery cycles. The letters on the x-axis represent different phases of the experiment with I = the 30 day initial period prior to the extreme drought treatment, D = drought period and R = recovery period. Error bars represent one standard error.

aimed to test a potential mechanism that may be important in determining mortality which can then be scaled up to a local and regional scale.

This study has shown that vegetation response to climatic extremes is likely to be affected by CO_2 concentration, with extreme drought-related mortality of grasses increasing under elevated CO_2 levels. The suggested mechanism for this increased mortality is an increase in the LAI of the grasses which offset reductions in stomatal conductance and increased rainfall interception prior to extreme drought events. This highlights the importance of considering the interactions between climatic extremes and other aspects of climate change such as CO_2 concentration on plant communities. Our results also highlight the importance of better quantification of soil water in global climate and vegetation models as this may be a key driver affecting global vegetation patterns and responses to climate change.

Acknowledgments

We gratefully acknowledge the Plant Invasion and Restoration Ecology Laboratory (PIREL) of Macquarie University for their input throughout the experiment and Muhammad Masood for assistance in the glasshouses. The experiments conducted complied with all current laws and regulations of Australia, where they were conducted.

Author Contributions

Conceived and designed the experiments: AM ML. Performed the experiments: AM. Analyzed the data: AM. Wrote the paper: AM ML.

References

1. Gutschick VP, BassiriRad H (2003) Extreme events as shaping physiology, ecology, and evolution of plants: Toward a unified definition and evaluation of their consequences. New Phytologist 160: 21–42.
2. IPCC (2011) Summary for policymakers. In Field CB, Barros V, Stocker TF, Qin D, Dokken D, Ebi KL, Mastrandrea MD, Mach KJ, Plattner G-K, Allen SK, Tignor M, Midgley PM, editors. Managing the risks of extreme events and disasters to advance climate change adaptation.
3. Easterling DR, Meehl GA, Parmesan C, Changnon SA, Karl TR, et al. (2000) Climate extremes: Observations, modelling, and impacts. Science 289: 2068–2074.
4. Asner GP, Elmore AJ, Olander LP, Martin RE, Harris AT (2004) Grazing systems, ecosystem responses and global change. Annual Review of Environment and Resources 29: 261–299.
5. Briggs JM, Knapp AK, Blair JM, Heisler JL, Hoch GA, et al. (2005) An ecosystem in transition: Causes and consequences of the conversion of mesic grassland to shrubland. BioScience 55: 243–254.
6. Morgan JA, Pataki DE, Korner C, Clark H, Del Grosso SJ, et al. (2004) Water relations in grassland and desert ecosystems exposed to elevated atmospheric CO_2. Oecologia 140: 11–25.
7. Morgan JA, LeCain DR, Pendall E, Blumenthal DM, Kimball BA, et al. (2011) C_4 grasses prosper as carbon dioxide eliminates desiccation in warmed semi-arid grassland. Nature 476: 202–205.
8. Fay PA, Kaufman DM, Nippert JB, Carlisle JD, Harper CW (2008) Changes in grassland ecosystem function due to extreme rainfall events: Implications for responses to climate change. Global Change Biology 14: 1600–1608.
9. Jentsch A, Kreyling J, Elmer M, Gellesch E, Glaser B, et al. (2011) Climate extremes initiate ecosystem-regulating functions while maintaining productivity. Journal of Ecology 99: 689–702.
10. Woodward FI (1990) Global change: Translating plant ecophysiological responses to ecosystems. Trends in Ecology and Evolution 5: 308–311.
11. Anderson LJ, Maherali H, Johnson HB, Polley HW, Jackson RB (2001) Gas exchange and photosynthetic acclimation over subambient to elevated CO_2 in a C_3–C_4 grassland. Global Change Biology 7: 693–707.
12. Morgan JA, LeCain DR, Mosier AR, Milchunas DG (2001) Elevated CO_2 enhances water relations and productivity and affects gas exchange in C_3 and C_4 grasses of the Colorado shortgrass steppe. Global Change Biology 7: 451–466.
13. Osborne CP, Sack L (2012) Evolution of C_4 plants: A new hypothesis for an interaction of CO_2 and water relations mediated by plant hydraulics. Philosophical Transactions of the Royal Society of London B: Biological Sciences 367: 583–600.
14. Cavaleri MA, Sack L (2010) Comparative water use of native and invasive plants at multiple scales: A global meta-analysis. Ecology 91: 2705–2715.
15. Newingham BA, Vanier CH, Charlet TN, Ogle K, Smith SD, et al. (2013) No cumulative effect of 10 years of elevated $[CO_2]$ on perennial plant biomass components in the Mojave Desert. Global Change Biology 19: 2168–2181.
16. Wand SJE, Midgley GF, Jones MH, Curtis PS (1999) Responses of wild C_4 and C_3 grass (Poaceae) species to elevated atmospheric CO_2 concentration: A meta-analytic test of current theories and perceptions. Global Change Biology 5: 723–741.
17. Piao S, Friedlingstein P, Ciais P, de Noblet-Ducoudre N, Labat D, et al. (2007) Changes in climate and land use have a larger direct impact than rising CO_2 on

global river runoff trends. Proceedings of the National Academy of Sciences of the United States of America 104: 15242–15247.
18. Department of Infrastructure, Planning and Natural Resources (2003) Major weeds of the Cumberland Plain. In: Little D, editor. Bringing the bush back to western Sydney. pp. 47–55.
19. IPCC (2013) Climate change 2013: A physical science basis. In: Joussaume S, Penner J, Tangang F, editors. Working Group I Contribution to the IPCC 5th Assessment Report - Changes to the Underlying Scientific/Technical Assessment.
20. ETCCDI/CRD (2012) Climate change indices: Definitions of the 27 core indices. In: CC1/CLIVAR/JCOMM Expert Team, editors. Climate Change Detection and Indices.
21. LeCain DR, Morgan JA, Mosier AR, Nelson JA (2003) Soil and plant water relations determine photosynthetic responses of C_3 and C_4 grasses in a semi-arid ecosystem under elevated CO_2. Annals of Botany 92: 41–52.
22. Funk J, Zachary V (2010) Physiological responses to short-term water and light stress in native and invasive plant species in southern California. Biological Invasions 12: 1685–1694.
23. Diez JM, D'Antonio CM, Dukes JS, Grosholz ED, Olden JD, et al. (2012) Will extreme climatic events facilitate biological invasions? Frontiers in Ecology and the Environment 10: 249–257.
24. Niklaus PA, Leadley PW, Schmid B, Korner C (2001) A long-term field study on biodiversity x elevated CO_2 interactions in grassland. Ecological Monographs 71: 341–356.
25. Polley HW, Johnson HB, Derner JD (2003) Increasing CO_2 from subambient to superambient concentrations alters species composition and increases above-ground biomass in a C_3/C_4 grassland. New Phytologist 160: 319–327.
26. Dijkstra FA, Blumenthal D, Morgan JA, LeCain DR, Follett RF (2010) Elevated CO_2 effects on semi-arid grassland plants in relation to water availability and competition. Functional Ecology 24: 1152–1161.
27. Polley HW, Jin VL, Fay PA (2012) CO_2-caused change in plant species composition rivals the shift in vegetation between mid-grass and tallgrass prairies. Global Change Biology 18: 700–710.
28. Schiermeier Q (2008) Water: A long dry summer. Nature 452: 270–273.
29. Parton WJ, Scurlock JMO, Ojima DS, Schimel DS, Hall DO (1995) Impact of climate change on grassland production and soil carbon worldwide. Global Change Biology 1: 13–22.
30. Allen CD, Macalady AK, Chenchouni H, Bachelet D, McDowell N, et al. (2010) A global overview of drought and heat-induced tree mortality reveals emerging climate change risks for forests. Forest Ecology and Management 259: 660–684.
31. Benton TG, Solan M, Travis JMJ, Sait SM (2007) Microcosm experiments can inform global ecological problems. Trends in Ecology and Evolution 22: 516–521.

Contamination of Groundwater Systems in the US and Canada by Enteric Pathogens, 1990–2013: A Review and Pooled-Analysis

Paul Dylan Hynds, M. Kate Thomas, Katarina Dorothy Milena Pintar*

Centre for Food-borne, Environmental and Zoonotic Infectious Diseases, Public Health Agency of Canada, Ottawa, Ontario, Canada

Abstract

Background: Up to 150 million North Americans currently use a groundwater system as their principal drinking water source. These systems are a potential source of exposure to enteric pathogens, contributing to the burden of waterborne disease. Waterborne disease outbreaks have been associated with US and Canadian groundwater systems over the past two decades. However, to date, this literature has not been reviewed in a comprehensive manner.

Methods and Principal Findings: A combined review and pooled-analysis approach was used to investigate groundwater contamination in Canada and the US from 1990 to 2013; fifty-five studies met eligibility criteria. Four study types were identified. It was found that study location affects study design, sample rate and studied pathogen category. Approximately 15% (316/2210) of samples from Canadian and US groundwater sources were positive for enteric pathogens, with no difference observed based on system type. Knowledge gaps exist, particularly in exposure assessment for attributing disease to groundwater supplies. Furthermore, there is a lack of consistency in risk factor reporting (local hydrogeology, well type, well use, etc). The widespread use of fecal indicator organisms in reported studies does not inform the assessment of human health risks associated with groundwater supplies.

Conclusions: This review illustrates how groundwater study design and location are critical for subsequent data interpretation and use. Knowledge gaps exist related to data on bacterial, viral and protozoan pathogen prevalence in Canadian and US groundwater systems, as well as a need for standardized approaches for reporting study design and results. Fecal indicators are examined as a surrogate for health risk assessments; caution is advised in their widespread use. Study findings may be useful during suspected waterborne outbreaks linked with a groundwater supply to identify the likely etiological agent and potential transport pathway.

Editor: Martyn Kirk, The Australian National University, Australia

Funding: Funding was provided by core funding within the Public Health Agency of Canada, a Canadian government department. As such, the funders had no role in study design, data collection and analysis, decision to publish, or preparation of the manuscript.

Competing Interests: The authors have declared that no competing interests exist.

* E-mail: Katarina.pintar@phac-aspc.gc.ca

Introduction

The diversity of North America's hydrology, and by extension, hydrogeology, is a reflection of its bio-physical diversity including extremely variable physiography, climate, bedrock geology and quaternary geology. Accordingly, many regional hydrological issues exist related to microbial groundwater contamination [1].

Canada is comprised of nine hydrogeological regions, ranging from extensive areas of permafrost in the north to Cordillera, the Western Canadian sedimentary basin and the St Lawrence platform in the south [2]. Approximately 8.9 million Canadians (30.3% of the population) currently rely on groundwater for domestic use; with an estimated 3.8 million Canadians served by private supplies [3–6]. Hydrogeological units within the United States (US) are similarly varied with approximately 69.5% of groundwater in the US derived from unconsolidated aquifers (alluvial sand and gravel, coastal plain, glacial outwash and fluvial/eolian) [7]. Approximately 44% of the U.S. population depends on groundwater for its drinking water supply [8]. Private wells constitute the largest proportion of water wells in the United States – more than 13.2 million households have their own well, supplying 45 million people [8]. Typically, groundwater supplies in Canada and the United States are categorized into one of three supply types; large-scale, disinfected municipal systems supplying cities and municipalities with >1000 residents, community systems of varying scale with or without disinfection and typically supplying <1000 people, and largely untreated private domestic wells, often serving single households.

Groundwater contamination occurs as a direct result of human or animal waste ingress to groundwater resources (aquifers or wells). Enteric pathogens are prevalent in both developing and developed countries, including the United States and Canada [9–10], where septic tanks [11–13], sewage and municipal wastewater treatment [14–16], wildlife [17], grazing animals [18] and other agricultural activities [19–22] can be sources of groundwater contamination. Natural attenuation occurs by overlying subsoil strata; however, age deterioration, poor wellhead hygiene or design defects provide localized pollution pathways [13,23]. Bedrock fracturing and faulting, unconsolidated bedrock materials and thin or absent overlying subsoil layers may also provide pathways for contamination transport. Where subsurface contamination does occur, adverse human health effects may result, in part due to inadequate treatment, particularly with private supplies [24,25].

Waterborne illness is a public health issue in North America and around the world. Sensitive sub-populations including the young, the elderly, pregnant women, and the immune-compromised are particularly susceptible to enteric infections [26]. An estimated 45% of all waterborne outbreaks in Canada involve non-municipal systems, largely in rural or remote areas [3,4,27]. While private supplies serve about 13% of Canada's population, >20% of Canada's reported waterborne disease outbreaks were associated with private water supplies between 1974 and 1996 [28]. Consumption from private wells has been cited as a potentially significant risk factor for contracting campylobacteriosis and other enteric diseases in Canada [29–31]. A recent study from New Zealand predicts that consumers sourcing their drinking water supply from shallow groundwater in an intensive dairy region (Waikakahi catchment, Canterbury) had an estimated 60–75% risk of *Campylobacter* infection per irrigation season [22]. Consumption of contaminated groundwater is estimated to account for approximately 6.5 million illnesses per year in the US (34% of all waterborne infections) [32]. A recent review by Wallender *et al.* [16] found that 30.3% of 818 drinking water outbreaks reported to the United States CDC Waterborne Disease and Outbreak Surveillance Systems during 1971 to 2008 were attributable to untreated groundwater sources. Exposure to shallow groundwater sources contaminated with animal faeces has been identified as a risk factor for *Escherichia coli* O157:H7 infection in the US [33].

In their summation of a 2005 workshop entitled '*Estimating Waterborne Disease Risks in the United States*' Craun et al. [34] state that significant data and method-based uncertainties remain in the improved estimation of enteric diseases attributed to drinking water, particularly with respect to risks associated with wells and groundwater. There is currently a paucity of integrated data pertaining to the presence and source of enteric pathogens in North American groundwater systems. Accordingly, the purpose of this work was to identify and review available literature relating to contamination of Canadian and US groundwater sources by enteric pathogens between 1990 and 2013. The overarching objective of this review was a comprehensive collation of the current state of knowledge relating to the presence of enteric pathogens in Canadian and US groundwater and an examination of related patterns (e.g. use of FIOs, enteric pathogen detection, study location, source types, etc.). To date, similar work has not been undertaken; it is considered that the results could be used by stakeholders to prioritize research and fill existing data gaps, thus supporting the effective management of groundwater supplies in Canada and the US.

Methods

2.1 Literature Identification and Screening

The research question that guided the review was:

> *What has been the prevalence of contamination by enteric pathogens and fecal indicator organisms in Canadian and US groundwater systems during the period 1990-present?*

A review protocol was developed in order to capture existing, published references to contamination events in Canadian and US groundwater systems (Figure 1). The review protocol was adapted from previous studies [35–37]. Literature for inclusion was identified from four bibliographic databases (PubMed, MedLine Plus, Science Direct, and Scopus) with key search terms (Table 1). Overall, 21,300 article titles that included ≥2 key search terms were located within bibliographic databases; article titles were imported to an Excel spreadsheet and de-duplicated using a replacement macro, resulting in 3400 article titles (Figure 1a). All literature scans were undertaken using Boolean positional operators ("AND", "OR", "SAME", "WITH", "ADJ") to focus literature searches. Study screening 1 (citation title) was subsequently undertaken; all articles published prior to 1990, relating to groundwater chemistry, non-enteric pathogens and/or groundwater systems outside of Canada and the US were excluded. This resulted in the inclusion of 1028 non-duplicate articles (Figure 1b).

Study screening 2 (citation abstract) was based upon explicit inclusion/exclusion criteria (Table 2). No study types were excluded, however, at this stage all non-English articles were excluded, as were articles which examined surface, recreational or coastal waterbodies. Articles reporting groundwater contamination by non-enteric pathogens were also excluded during this screening phase. This screening phase was undertaken by one researcher, with 270 article included for further consideration at this stage (Figure 1c).

Study screening 3 was completed by two review teams who independently assessed each publication (year, title, and abstract), based upon explicit eligibility criteria (Table 2). Articles that met all initial screening criteria, as defined by agreement between reviewers were included. Disagreements (approximately 15%) were resolved by discussion among reviewers and an additional round of screening. As shown (Figure 1d), 165 articles were included in the final database. Upon data extraction, 52 of 165 included articles were found to contain extractable homogenous data, with an additional three relevant articles located via article bibliographies added at this stage (Figure 1e).

2.2 Data Extraction

Relevant data were extracted to a spreadsheet (MS Office Excel 2007) and verified. Three primary data fields were extracted and exported, including:

1. Citation-specific data: Title, Author(s), Publication Year, Publication Type, Study Country.
2. Groundwater contamination data: Fecal Indicator Organism (FIO) Data (type, presence, magnitude), Pathogen Data (type, presence, magnitude), system type (private, municipal, etc.), Detection Method, Outbreak Association, Sampling Period/Season.
3. "Pathway" data: Livestock/Agriculture Associated, Wild Animal Associated, Septic Tank/Sewage Associated, Precipitation/Climate Associated, Geology Associated, Source Associated, Local (Hydro)Geological Characteristics.

Pathogen records were extracted from all included studies, with each record representing analysis for one pathogen type or strain (regardless of actual presence). For example, if groundwater samples were analysed for both *E. coli* O157 and *Cryptosporidium* spp. within one study, two records were extracted. Study design was classified into four categories, including: repeat sampling (sources sampled more than once over extended time period), "snapshot" sampling (sources sampled on a "one off" basis), outbreak investigations and laboratory-based studies (studies whose primary objective was the development or refinement of a laboratory technique relating to enteric pathogen detection in

Table 1. Review primary terms, term descriptions and search terms used for literature identification, study screening and final database development.

Primary Term	Term Description	Search Terms
Population		
Groundwater	Environment/Pathway of interest	Groundwater; Aquifer; Well; Waterborne
Jurisdiction	Geographical area of interest	USA/Canada/North America
Outcome		
Contamination	Outcome of interest	Contamination; Pollution; Runoff; Faecal Ingress, Presence
Agent		
Pathogen	Organisms of interest	Pathogen(ic); Enteric; Virus; Protozoa; Bacteria; Microbial, Indicator, *Giardia; Cryptosporidium; Shigella;* STEC; VTEC; *E. coli*; Adenovirus; Enterovirus *Campylobacter;* Norovirus; Rotavirus

environmental samples). Data pertaining to pathogen detection method were categorised into three groups, molecular diagnostics, microscopy and culture-based techniques. System type (private, community, municipal) was categorized based on explicit classification in included studies. Finally, pathogens were classified on the basis of being either potentially zoonotic or non-zoonotic, based on current CDC classification.

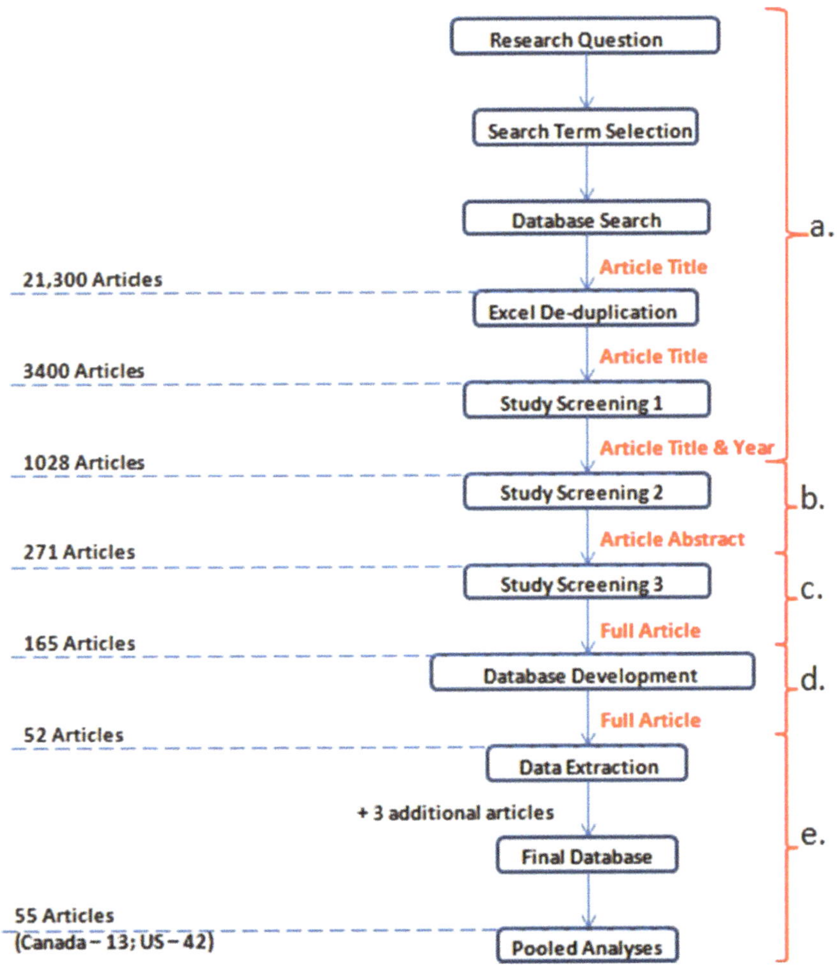

Figure 1. Review protocol employed throughout the current study, including results of literature identification, literature scans and data extraction processes.

Table 2. Inclusion and exclusion criteria for literature screening.

Inclusion Criteria	Exclusion Criteria
Study Type: All	Study Type: None
Language: English	Language: Non English
Population: Groundwater Systems (Municipal, Private, Community, Non-community, Mixed)	Population: Surface water bodies/watersheds, coastal/marine water bodies, recreational water
Events/Outcomes: Confirmed contamination of groundwater systems by enteric pathogens in North America; Confirmed presence of fecal indicator organisms in North American groundwater systems	Events/Outcomes: Groundwater contamination occurring outside North America, groundwater contamination by "non-enteric" pathogens
Period: 1990 – present	Period: pre-1990

2.3 Statistical Analysis

All data analyses were performed with IBM SPSS Statistics 20. Chi-square tests were employed to examine significance between categorical variables, using R×L contingency tables and variable categories (dichotomous, ordinal). Continuous variables were subjected to one-way ANOVA (both weighted and un-weighted) analyses for both between- and within-group differences. One-way ANOVA was undertaken based upon prior Levene tests for the assessment of both between group and within group equality of variance. Bonferroni "post-hoc" multiple comparisons analysis was utilized after all ANOVA analyses to elucidate multi-criteria variance and exhibited association "directionality" (I–J) and to ensure highlighted differences ($p<0.05$) were not systematically dispersed among categories. Two-tailed Spearman's' rho was used to investigate correlation between continuous variables. Tests for variance and categorical homogeneity were undertaken prior to all analyses via both Welch and Brown-Forsythe robust tests for mean equality. A significance level of $p<0.05$ was used throughout the study [38].

Results

3.1 Descriptive Analysis of Literature

A total of 55 published studies included extractable data, of which 23.6% (n = 13) were Canadian studies, while the remaining 76.4% (n = 42) were from the United States (Appendix S1). Included studies spanned the review time period, with a peak (n = 7) noted in 2002. Overall, 45.4% of studies (n = 25) contained both fecal indicator organism (FIO) and enteric pathogen detection data, while a further 29.1% (n = 16) contained enteric pathogen data only. The remaining 25.4% of studies (n = 14) solely reported on FIO detection. Of those studies that included extractable data pertaining to enteric pathogen detection (n = 41), 80.5% (n = 33) were from the US, while the remaining studies (n = 8) were from Canada. From these 41 studies, 102 extractable records were found (Tables 3 & 4); most were derived from US studies (90/102). All extracted enteric pathogen records were included for pooled-analyses.

3.2 Fecal Indicator Organisms

In total, 39 studies reported FIO data related to six organisms. Based on the definition of FIO as "non-pathogenic, exclusively intestinal in habit and therefore exclusively fecal in origin if found outside the intestine" [39], total coliforms are not considered a FIO, however their use as indicator organisms remains widespread and thus are included here. A median of 2 FIOs were reported per study, up to a maximum of 5. *E. coli* was the FIO analysed most frequently, while heterotrophic plate counts were employed least often (Table 4a). Two-tailed Spearman's' rho was used to examine correlations between individual FIOs and the presence of enteric pathogens, in studies where both data were reported (Table 5). The presence of total coliforms, *E. coli*, intestinal Enterococci and coliphage (Male-specific (F+) and Somatic) were all strongly correlated with each other ($p<0.01$). No significant correlation between the total percentage of sample wells with enteric pathogens present (regardless of pathogen type) and the percentage of wells with fecal indicators present (individually or combined) was found. Weak ($-0.7< r <0.7$) negative correlations were found between pathogen contamination percentage (% of sampled wells with enteric pathogens present, irrespective of pathogen category) and total coliforms (% of sampled wells with total coliforms present) ($r = -0.289$, $p = 0.034$) and intestinal enterococci ($r = -0.435$, $p = 0.006$). A weak positive correlation between bacterial pathogen presence and *E. coli* ($r = 0.636$, $p = 0.02$). Both protozoan ($r = 0.866$, $p = 0.026$) and viral pathogen presence ($r = -0.494$, $p = 0.023$) were correlated with the presence of intestinal enterococci.

3.3 Study Design & Pathogen Detection

Of 102 identified records, 74 (72.5%) reported confirmed pathogen presence, employing four main study designs. The most frequent study design focusing on enteric pathogen detection was (i) outbreak investigations (n = 18, 40%, 8.5 month mean duration), followed by (ii) repeat sampling based fieldwork (n = 14, 31.1%, 12.7 month mean duration), (iii) "snapshot" (i.e. "one-off") sampling studies (n = 9, 20%, 12 month mean duration) and (iv) laboratory method-based studies (n = 2, 4.4%, 12 month mean duration). A significant association was noted between pathogen detection percentage (% of records in which enteric pathogens were positively found) and study design ($\chi^2(3) = 7.749$, $p = 0.05$), with a higher prevalence of pathogen detection found among "snapshot" type studies (Table 4b). Source contamination percentages ranged from 7.3% (lab-based studies) to 19.5% (outbreak investigations) of analysed groundwater samples, however, this was not found to be statistically significant. No correlation was found between either source or sample number analysed per study and source or sample contamination percentage; repeat sampling studies analysed a mean well and sample

Table 3. Enteric pathogen records from included US and Canadian studies (1990–2013) (n = 102); record number, percentage of total records, record number with specified pathogen present, percentage of records with specified pathogen present.

Pathogen included in study	N	%	Presence of pathogen	
			N	%
Bacteria				
VTEC[1] (O157)	5	4.9	3	60
Salmonella spp.	5	4.9	4	80
EPEC[2] (non O157)	4	3.9	4	100
Campylobacter jejuni	4	3.9	2	50
Arcobacter spp.	1	1	1	100
Shigella spp.	1	1	0	0
Helicobacter spp.	1	1	1	100
Leptospira spp.	1	1	0	0
Yersinia spp.	1	1	1	100
Protozoa				
Cryptosporidium spp.	9	8.8	6	66.7
Giardia spp.	10	9.8	3	30
Naegleria fowleri	1	1	1	100
Virus				
Enterovirus	14	13.7	14	100
Norovirus	12	11.8	9	75
Hepatitis A (HAV)	8	7.8	7	87.5
Rotavirus	7	6.9	4	57.1
Adenovirus	6	5.9	5	83.3
Human Enteric Viruses	5	4.9	3	60
Total Culturable Viruses	4	3.9	3	75
Reovirus	2	2	2	100
SRSV[3]	1	1	1	100
Total	102	100	74	72.5

[1]Vero-toxigenic *Escherichia coli.*
[2]Enteropathogenic *Escherichia coli.*
[3]Small Round Structured Viruses.

number of 23/130, equivalent mean sample numbers were 210/219, 25/269 and 219/263 for outbreak investigations, lab-based studies and "snapshot" studies, respectively. A significant difference was noted in relation to study design and analysed well number ($F(3) = 3.468$, $p = 0.019$), however, no significant difference was found between study type and sample number. Sample season was reported within 31 studies (Table 4a). While not statistically significant at the 0.05 level, enteric pathogens were more prevalent during autumn sampling events and year round repeat sampling studies. Study design and sampling season were significantly associated ($\chi^2(12) = 34.594$, $p = 0.001$); almost 60% of outbreak investigations took place during autumn, while 17.8%, 21% and 3.5% occurred during spring, summer and winter, respectively. Moreover, 100% of lab-based studies (5/5) comprised year round repeat sampling. Study design and enteric pathogen type were also associated ($\chi^2(6) = 24.162$, $p<0.001$); all (6/6) lab-based records were associated with waterborne enteric viruses, as were the majority (26/32) of repeat sampling studies. Outbreak investigations were split with respect to pathogen type (bacteria 14/42, protozoa 9/42, viral 19/42).

3.4 Pathogen Detection Method

The majority (72.5%) of pathogen detection was undertaken using molecular techniques (Table 4a). Microscopy was not utilised in any studies for the detection of bacterial pathogens ($\chi^2(2) = 6.438$, $p = 0.04$), which were most commonly analysed using molecular diagnostics (9/14). Similarly, molecular methods were favoured for the detection of protozoan pathogens ($\chi^2(2) = 13.365$, $p = 0.001$) (7/11). Studies reporting pathogen detection were approximately equally split between mono- (51.2%) and multi-pathogen (48.8%) detection. The presence of bacterial pathogens was analysed in 35.7% of studies, while protozoan and viral detection took place in 28.6% and 54.8% of studies, respectively. Two studies reported on all three enteric pathogen categories.

3.5 Sample Sources

As shown, 60% of studies (33/55) focused on private domestic wells, with a furter 24.2% (15/55) focusing on municipal groundwater systems (Table 4a). Further, 89.2% of samples were extracted from boreholes, while 7.8%, 2.0% and 1.0% were taken from spring sources, mixed sources and hand-dug wells, respectively. No statistical association was found between groundwater

Table 4. Summary statistics: a. Extracted Studies (n = 55), b. Extracted Records (n = 102), c. Extracted records associated with confirmed enteric pathogen presence (n = 74), d. Sampled well numbers (n = 52) and analysed groundwater sample numbers (n = 39).

Variable	Canada		United States	
	Total	Pathogen Present	Total	Pathogen Present
	N (%)	N (%)	N (%)	N (%)
a.				
Supply Type (n = 54)				
Municipal	4 (7.4)	3 (75)	11 (20.3)	9 (81.8)
Private	9 (16.6)	3 (33.3)	24 (44.4)	15 (62.5)
Community	0 (0)	-	5 (9.2)	4 (80)
Mixed	0 (0)	-	1 (1.8)	1 (100)
Fecal Indicator Organisms (n = 39)				
Total Coliforms	7 (17.9)	n/a	25 (54.1)	n/a
E. coli	10 (25.6)	n/a	24 (51.5)	n/a
Intestinal *enterococci*	5 (12.8)	n/a	11 (28.2)	n/a
Coliphage	2 (7.5)	n/a	6 (15.3)	n/a
C. perfringens	0 (0)	n/a	5 (12.8)	n/a
Heterotrophic Plate Counts	0 (0)	n/a	2 (7.5)	n/a
Sample Season (n = 31)				
Spring	0 (0)	-	4 (12.9)	2 (50)
Summer	2 (6.4)	1 (50)	8 (25.9)	2 (25)
Autumn	2 (6.4)	2 (100)	9 (29)	6 (66.7)
Winter	1 (3.2)	0 (0)	0 (0)	-
Year Round	0 (0)	-	5 (16.1)	4 (80)
Pathogen Detection Method (n = 40)				
Molecular Techniques	7 (17.5)	7 (100)	22 (55)	18 (81.8)
Microscopy	1 (2.5)	0 (0)	3 (7.5)	2 (66.7)
Culture-based	0 (0)	-	7 (17.5)	5 (71.4)
b.				
Aquifer/Bedrock Type (n = 61)				
Karstified	0 (0)	-	13 (21.3)	12 (92.3)
Unconsolidated	0 (0)	-	17 (27.8)	15 (88.2)
Fractured Bedrock	1 (1.6)	1 (100)	9 (14.7)	4 (44.4)
Un-fractured Bedrock	4 (6.5)	2 (50)	9 (14.7)	7 (77.7)
Diverse (Large Area)	0 (0)	-	8 (13.1)	7 (87.5)
Study Design (n = 102)				
Repeat Sampling	7 (6.8)	3 (42.8)	25 (24.5)	21 (84)
Snapshot Sampling	4 (3.9)	2 (50)	18 (17.6)	18 (100)
Outbreak Investigation	1 (1)	1 (100)	41 (40.2)	24 (58.5)
Laboratory-based	0 (0)	-	6 (5.8)	5 (83.3)
c.				
Pathogen Type (n = 102)				
Bacterial	2 (2)	2 (100)	20 (19.6)	12 (60)
Protozoan	6 (5.8)	2 (33)	15 (14.7)	9 (60)
Viral	4 (3.9)	2 (50)	55 (53.9)	46 (88.5)
Pathogen Source (n = 74)				
Agriculture/Livestock	n/a	4 (5.4)	n/a	5 (6.7)
Wild Animals	n/a	1 (1.3)	n/a	2 (2.7)
Septic tank/Sewage system	n/a	2 (2.7)	n/a	43 (58.1)
Surface Water Body	n/a	2 (2.7)	n/a	33 (44.6)

Table 4. Cont.

Variable	Canada		United States	
	Total	Pathogen Present	Total	Pathogen Present
	N (%)	N (%)	N (%)	N (%)
Contamination Pathway (n = 74)				
Precipitation	n/a	3	n/a	14 (18.9)
Geological	n/a	1	n/a	43 (58.1)
Source-specific	n/a	2	n/a	23 (31.1)
	Canada		**United States**	
	Mean	**Range**	**Mean**	**Range**
d.				
Well Number	15	1–1,208	160	1–5,520
Sample Number	41	1–1,208	225	1–5,520
Samples/well	2.73		1.4	

system type and groundwater source type at a 95% confidence level, however, pathogen presence and system type were associated at a 90% level ($\chi^2(3) = 6.957$, $p = 0.074$); a lower mean prevalence of enteric pathogen presence was found among community systems (Table 4a). In accordance with best practise guidelines, 87.3% of all groundwater samples were extracted "pre-treatment", with a further 5.5% extracted both "pre-" and "post-treatment". Just under half of the included studies (49.1%, n = 27) also included hydrogeological data including local hydrogeological setting (bedrock type, aquifer type, subsoil type/strata thickness, porosity, lithology, faulting, etc.).

The majority of records (69.6%, n = 71) were associated with "non-outbreak" conditions, while the remaining were instigated as a direct result of a confirmed waterborne outbreak. A significant association was found between the prevalence of outbreak-related studies and groundwater system type ($\chi^2(3) = 9.644$, $p = 0.022$); community systems were significantly more likely to be associated with waterborne epidemic conditions. No association was found between the instance of waterborne outbreaks and groundwater source type, nor between outbreak-related studies and pathogen strain or type. Viral presence was significantly higher than either bacterial or protozoan presence ($\chi^2(2) = 9.505$, $p = 0.009$); overall, viral pathogens were found in 85.7% of studies in which viral detection methods were employed (Table 4c) (regardless of outbreak/non-outbreak conditions).

3.6 Hazard/Pathogen Source

Data extraction was carried out for potential pathogen sources (agriculture/livestock, wild animals, septic tanks/sewage systems, surface water bodies) within all waterborne pathogen records. Data were extracted based upon being a confirmed (e.g. use of tracer study) or likely (e.g. population cohort study) pathogen source. Septic tanks/sewage systems were the most frequently reported confirmed or likely pathogen source (Table 4c). No statistically significant association was found to exist between pathogen presence (Yes/No) and any specific potential pathogen source. Overall, 52% of records (53/102) referred to potentially zoonotic pathogens, which were over 5 times more likely to be present when adjacent agricultural activities were recorded (OR = 5.465, 95% CI 1.133, 26.361; $\chi^2(1) = 5.362$, $p = 0.021$) (Table 4c). Additionally, potentially zoonotic pathogens were over twice as likely to be present where the groundwater source was located adjacently to a surface water body (OR = 2.739, 95% CI 1.214, 6.178; $\chi^2(1) = 6.031$, $p = 0.017$). Conversely, potentially zoonotic pathogens were approximately 3 times less likely to be detected where septic tanks and/or municipal sewage systems were noted as the primary local pathogen source (OR = 0.337, 95% CI 0.145, 0.784; $\chi^2(1) = 6.574$, $p = 0.014$). Potentially zoonotic pathogens were more likely to be detected in shallow hand-dug and spring sources than deeper boreholes ($\chi^2(3) = 11.751$, $p = 0.008$). Pathogens were categorised based upon pathogen type (bacterial, protozoan, viral or multiple) and similar analysis undertaken with respect to pathogen presence and potential pathogen source. Only one significant association resulted, with bacterial presence 75% more likely in areas where livestock presence was recorded (OR = 1.75, 95% CI 1.112, 2.775; $\chi^2(1) = 4.714$, $p = 0.03$).

3.7 Contamination Pathways

Three potential contamination pathways were identified during data extraction, including precipitation patterns (leading to contamination ingress at the wellhead or rapid subsurface infiltration), geological pathways (increased contaminant transmissivity via unconsolidated/fractured aquifer materials or increased runoff coefficients) and inadequate source design/construction (e.g. uncovered wellhead, cracked jointing, absence of sanitary seal, etc.) (Table 4c). It is important to note that while these pathways are typically interconnected, with the presence of one typically amplifying the effects of others, however, the purpose of the current review is only to report on patterns within the included literature. No association was found between pathogen presence and either increased precipitation or geological pathways, however, inadequately designed or constructed wells were found to have pathogens present in 50% more records than adequately designed sources as noted by authors (OR = 1.51, 95% CI 1.283, 1.777; $\chi^2(1) = 12.531$, $p<0.001$). Overall, 59.8% of records included data pertaining to aquifer/bedrock type (n = 61) (Table 4c).

No association was noted between the presence of bacterial pathogens and precipitation patterns or geological pathways;

Table 5. Correlation matrix; fecal indictor organisms and total source contamination percentage (n = 25) (*Clostridium perfringens* and Heterotropic plate counts were excluded due to low record numbers).

			Total Coliforms	*E. coli*	Enterococci	Coliphage	% Contamination
Spearman's rho	**Total Coliforms**	Correlation Coefficient	1	0.907**	0.923**	0.608**	0.289*
		Significance (2-tailed)	.	0	0	0.001	0.034
		N	54	41	38	28	54
	E. coli	Correlation Coefficient	0.907**	1	0.843**	0.627**	−0.161
		Significance (2-tailed)	0	.	0.001	0.001	0.269
		N	41	49	32	23	49
	Enterococci	Correlation Coefficient	0.923**	0.843**	1	0.601**	−0.435**
		Significance (2-tailed)	0	0		0.001	0.006
		N	38	32	39	29	39
	Coliphage	Correlation Coefficient	0.608**	0.627**	0.601**	1	−0.032
		Significance (2-tailed)	0.001	0.001	0.001	.	0.856
		N	28	23	29	30	30
	% Contamination	Correlation Coefficient	−0.289*	−0.161	−0.435**	−0.32	1
		Significance (2-tailed)	54	0.269	0.006	0.866	.
		N		49	39	30	62

**Correlation is significant at the 0.01 level (2-tailed).
*Correlation is significant at the 0.05 level (2-tailed).

however, 100% of sources with confirmed bacterial pathogen presence were noted as being poorly designed/constructed, indicating "source-specific" contaminant pathways (OR = 2.333, 95% CI 1.274, 4.272; $\chi^2(1) = 7.184$, $p = 0.007$). Likewise, inadequately designed/constructed wells were almost 3 times more likely to have enteric protozoa present (OR = 2.75, 95% CI 1.258, 6.01; $\chi^2(1) = 9.545$, $p = 0.004$). Viral pathogen presence was less likely to be associated with precipitation patterns (OR = 0.091, 95% CI 0.016, 0.509; $\chi^2(1) = 9.722$, $p = 0.002$), with no association noted regarding well design/construction or local geological characteristics.

3.8 Study Location

Canadian and US groundwater contamination studies over the past two decades were similar based on study years, use of and FIO types tested, frequency of enteric pathogen sampling and analysis, types of wells and geological areas sampled, in addition to utilised sampling and pathogen detection methods (Table 4a–b). No significant mean differences were noted between Canadian and US studies with respect to either enteric pathogen presence (well %, sample %) or pathogen presence category (bacteria/protozoa/virus). Canadian studies included in data extraction were derived from 5 (50%) provinces and 1 territory (Yukon), while US studies included data from 21 (42%) mainland states and 5 large multi-state (>10) studies. Ontario (n = 5) and Wisconsin (n = 8) were the most frequently represented Canadian province and US state, respectively.

Some notable differences were found with respect to study country; while only 23% (n = 3) of Canadian studies noted local (hydro)geological characteristics. These parameters were reported in 57% of US studies (24/42) ($\chi^2(1) = 4.61$, $p = 0.032$). Included Canadian studies differed from US studies with respect to study design ($\chi^2(3) = 8.386$, $p = 0.039$). For example, 100% of lab-based study records (6/6) were derived from the US, while a higher proportion of repeat sampling records were from Canadian studies (Table 4b). US studies were associated with both higher mean sampled well numbers ($t = 4.369$, $p < 0.001$) and higher mean sample numbers ($t = 5.389$, $p < 0.001$), while a higher mean number of samples per well was attributed to Canadian studies (Table 4d). Studied pathogen category differed with respect to study location ($\chi^2(2) = 7.278$, $p = 0.026$); while just over 12% of extracted pathogen records were derived from Canadian studies, 29% (n = 6) of protozoan records were associated with Canada.

Tests of independence were used to ascertain whether specific contaminant sources or pathways were equally reported based on study location. Both agricultural sources (point and diffuse) and septic tanks/municipal sewage treatment reporting differed significantly (Table 4c). While agricultural livestock were cited as potential pathogen sources in 34% (n = 4) of Canadian records, only 6% of US records (n = 5) did likewise ($\chi^2(1) = 19.153$, $p < 0.001$). Conversely, while only 16% of Canadian records mentioned septic tanks/municipal sewage systems with respect to pathogen presence, 58% of US studies (n = 43) reported this link ($\chi^2(1) = 8.289$, $p = 0.008$). Geology was cited as a likely contaminant pathway in only 8.3% (n = 1) of Canadian records and in 58% (n = 43) of US records ($\chi^2(1) = 11.376$, $p = 0.001$).

Discussion

A review of groundwater pathogen contamination studies in Canada and the US was conducted, locating 102 enteric pathogen records from 55 studies. The majority of studies were from the United States. While almost 24% of studies were Canadian, only 12% of enteric pathogen records were from Canadian studies. Typically, US studies were more likely to include "multi-pathogen" data (>1 enteric pathogen per study) and therefore >1 record, while Canadian studies were more likely to include FIO data with no accompanying enteric pathogen data. There is a paucity of studies focusing on direct pathogen detection in groundwater systems in both countries. Given concerns regarding the use of FIOs as surrogate organisms [40–42], the suitability of FIOs for use in the regulatory standards for protection of water quality is a particular concern. Overall, 38.5% of Canadian studies contained FIO data in the absence of enteric pathogen data, potentially resulting in significant under-estimates of enteric pathogen presence and/or magnitude. Future work should focus on direct detection of enteric pathogens; this data may be used to improve microbial risk assessment efforts via significantly improved exposure assessment.

Just over 42% of extracted records in this review were for private domestic wells. Currently 34–45% of groundwater users in Canada and approximately 33% of groundwater users in the US employ a private water source [5,8]. There is a lack of enteric pathogen data from municipal supplies, reflecting a potentially critical knowledge gap. The inclusion of municipal groundwater supplies in future North American groundwater sampling studies would help inform this identified gap.

In Canada, the focus within scientific studies in the reviewed literature has been on protozoan detection in groundwater, while in the US, it has been on bacterial and viral detection, with viruses being the most commonly studied pathogen in this review. Pathogenic viruses are a particular concern in groundwater systems as their size enables them to evade natural attenuation processes and enter the aquifer [43,44].

No correlation was found between viral presence and either total coliforms or *E. coli*, the two most frequently employed FIOs for groundwater system surveillance. Moreover no association between coliphage detection and viral presence was observed in this review. A 1-year study of 12 municipal wells in Quebec, Canada reported similar results [45]. As previously noted by Borchardt *et al.* [44], the majority of groundwater-associated municipal systems in the US produce water without disinfection, populations served by these systems are therefore potentially exposed to waterborne viruses and consequent health risks. The limited value of FIOs in predicting the presence of enteric viruses in groundwater highlights the need for direct enteric viral detection at groundwater sources used for drinking water [46–48]. The lack of significance between viral presence and source design, in concurrence with an association between viral presence and precipitation patterns, suggests a generalised mode of contamination associated with viral pathogens [13].

Geology was rarely recorded or presented as a potential contaminant pathway in Canadian studies, while 60% of US studies stated that local geological conditions were a potential cause for concern. Awareness of local geological and hydrogeological characteristics is critical in understanding likely contamination mechanisms or pathways, resource management and remediation and, by extension, human health risk assessment [49–50]. Future research relating to microbial contamination of North American groundwater systems, that adequately describes geological and hydrogeological characteristics of the groundwater source will be more useful for informing our understanding of public health risk. Future work related to the assessment of treatment requirements in geologically vulnerable regions based on the likelihood of pathogen presence would be particularly useful.

Bacterial pathogens accounted for 22.6% of records; all studies identified poorly designed or constructed well sources as a likely pathway. In contrast to the viral pathogen findings, this suggests localized, source-specific contamination as a predominant ingress mechanism e.g. groundwater contamination occurring at source due to poor wellhead protection as opposed to generalised contamination of all or part of the aquifer [13,51]. This is further reflected by the lack of association between bacterial presence and local hydrogeological characteristics.

The increased likelihood of potential zoonotic pathogen presence in agricultural areas was anticipated; low travel times in the subsurface make it unlikely to find these pathogens in areas without animals (livestock or wildlife) [18,22,52]. Our review also found bacterial pathogen presence more likely than either viral or protozoan presence in rural areas with livestock present. Potentially zoonotic pathogens were also more likely to be present where the sampled well was located in a surface water catchment, suggesting the potential for direct (and potentially rapid) groundwater-surface water interaction as a contamination mechanism. Potentially zoonotic pathogens were significantly less likely to be present in urban areas. Studies outside North America have previously reported pathogens as more likely present in shallow hand-dug and spring sources; for example, Hynds *et al.* [13] found a higher prevalence of pathogen presence in shallow groundwater sources in the Republic of Ireland. Hand-dug wells and springs are typically associated with inferior design and construction, leading to an increased likelihood of direct pathogen ingress at the wellhead [51,53–55]. It is important to note that both study design and reporting are likely to introduce bias in these results, for example, studies tended to focus on groundwater systems in rural/agricultural areas, while few examined urban systems. Accordingly, interpretation of these results may not adequately reflect hazard sources or contamination pathways, being are based solely upon available data.

Much of North America, particularly the United States, is characterised by unconsolidated aquifers [7]. Accordingly, unprotected shallow groundwater systems are regularly used, particularly for domestic supply. Wallender *et al.* [16] have reported that 26.2% of waterborne outbreaks from 1971 to 2008 were associated with karst limestone bedrock. Consumers and water managers in these areas would benefit from a clear understanding of the potential contamination risks, and consequent human health burden posed by using untreated shallow systems, and intervention options that are most effective. These findings may be useful for informing future groundwater surveillance/monitoring study design with respect to the most appropriate pathogen (category/species) to include for analysis, and where to focus the monitoring efforts. Moreover, where a human enteric disease outbreak is suspected, these criteria may be used to elucidate the potential infection etiology.

Based on this review, we note an inconsistency between studied groundwater pathogen (category) and reported groundwater-related human health burden in North America. A knowledge gap exists with respect to information on both prevalence and concentration of pathogen contamination of groundwater systems, particularly enteric bacteria. For example, it has been reported that 11% of waterborne outbreaks were caused by *Escherichia coli* O157 (VTEC) as a confirmed etiological agent [56]; however, only 4.9% of pathogen records in this review relate to VTEC (Table 3).

Little efficacy was found with respect to the predictive ability of FIOs and protozoan or viral contamination of groundwater systems. The lack of association between protozoan presence and local hydrogeological characteristics (as for bacterial pathogens) reinforces the importance of rapid by-pass mechanisms [13], with poorly designed wells as much as three times more likely to have protozoan pathogens present. Consequently, correctly implemented source management, including appropriate well location, design, construction and maintenance, in concurrence with periodic direct pathogen analysis is of paramount importance. This is particularly significant in light of the resistance of *Cryptosporidium* spp. to conventional chlorination [57]. It is considered that an increase in direct pathogen data analysis and reporting would prove useful for future risk management and communication strategies to inform intervention decisions and source water protection initiatives. For example, Morris & Foster [58] have noted that protozoan pathogens are commonly believed to be absent from "true" groundwater due to their size; however, with better data on groundwater contamination this assumption could be evaluated to further inform risk identification and mitigation strategies.

Results and conclusions presented here are based on the extracted data from published studies and are subject to the inherent biases of the original study objectives, designs and what was reported. Few Canadian studies referenced local hydrogeological characteristics as a potential contamination pathway when reporting on contamination events or groundwater-associated outbreaks, representing a limitation in study reporting. This lack of reporting may impact subsequent analyses of contamination pathways, for example, findings may reflect a lack of reporting or examination of these factors in a rigorous manner. Furthermore, only one included study occurred in winter, representing a potential limitation.

Study design differed significantly between US and Canadian studies. The majority of lab-based studies and outbreak investigations were in the US, while a higher proportion of repeat sampling records were from Canada. Repeat sampling studies integrated low well numbers with multiple samples from each well (mean = 6 samples). Outbreak investigations and "snapshot" studies comprised significantly higher well numbers, with wells typically sampled once. Outbreak investigations are an essential epidemiological tool; however, they are reactive and therefore not suitable for assessing endemic groundwater risks. Numerous studies report that multiple samples are necessary to positively determine contamination of a groundwater source [59–60]. Accordingly, repeat sampling studies are more likely to reflect contamination rates in Canadian and US groundwater systems; however few were identified in this review.

Enteric pathogens were present in 72.5% of records (74/102) and 16.1% of wells. Upon exclusion of outbreak investigation data, 77.1% of records (54/70) and 14.7% of wells (n = 1855/12,616; Std. Dev. 20.5%) were positive for at least one enteric pathogen. These figures may provide a useful foundation for endemic groundwater-associated exposure assessment, until more direct enteric pathogen data become available.

The lack of association between the incidence of outbreak investigations and municipal and private groundwater systems was not anticipated; however, as the majority of groundwater sampling was undertaken pre-treatment, this does not accurately reflect the public health burden posed by specific system types (e.g. many private sources remain untreated or poorly treated, thus raw water samples may be more reflective of drinking water quality). Large municipal systems are more likely to be adequately maintained and appropriately treated [24–25]. Smaller community systems were represented by a significantly higher proportion of outbreak investigations. A recent review of waterborne outbreaks by Hrudey [61] found that all, with the exception of one (Milwaukee, 1993), occurred in community systems. Furthermore, Craun *et al.* [62] have reported on ten cryptosporidiosis outbreaks in the US directly attributable to community water systems. Accordingly, the characteristics of a robust groundwater supply should include an understanding of the source, delivery of adequate treatment, adequate system monitoring and effective response to monitoring results when required, particularly in the case of private and community systems where the primary responsibility for source quality lies with owners and users [63].

Our review has highlighted a number of knowledge gaps and requirements relating to contamination of Canadian and US groundwater systems over the past two decades. Improved endemic risk assessment will be possible with increased direct enteric pathogen detection and reporting. Repeat-sampling studies represent a priority for the design of future studies for the collation of these data, as they are considered more representative with respect to human endemic risk. More data are required for groundwater-associated municipal systems, both pre- and post-treatment, as currently these systems are not adequately represented in the literature. Additionally, more data are required for bacterial pathogens, as these microorganisms are associated with a large public health burden. These data would also provide a foundation for improved risk management and communication via increased awareness and understanding among well owners, well users, system operators and policy makers.

Conclusions

- Within the literature the four study designs that were identified were of a proactive (pre-waterborne illness event) or reactive (post-waterborne illness event) nature, with the proactive designs being more suitable for background characterisation. Pooled analyses of reported data found that enteric pathogens were present in approximately 15% of sampled wells; this may be useful for future microbial risk assessments of human health relating to Canadian and US groundwater systems, although it should only be employed in the absence of area-specific data, and then, only with caution due to inherent limitations associated with the current study (e.g. reporting biases associated with amalgamation of differing study designs, lack of data relating to specific hydrogeological settings, lack of data associated with winter sampling and weather events, etc.). Accordingly, the authors recommend future research be focused on addressing this current lack of temporal and spatial data.
- It is critically important to account for study design when presenting and interpreting results of a groundwater study, given that groundwater is not similar to surface water and many confounding factors will influence the presence of pathogens in one system versus another, as well as the occurrence of an associated outbreak.
- Few studies have been published on pathogen presence in municipal groundwater systems. By addressing this knowledge gap, evidence would be developed to help support local decision-making around appropriate treatment or other interventions to further promote the safety of the drinking water supply. It is recommended that future health-related studies, particularly those associated with sporadic waterborne infection and quantitative health risk assessments, focus on direct enteric pathogen detection where possible, given the limited efficacy of fecal indicator organisms for enteric pathogen presence in groundwater systems. These data will aid more focused, pathogen-specific exposure assessment and hazard characterisation, therefore providing more data for evidence-based groundwater and human-health protection measures.
- Local (hydro)geology is an important risk factor in the likelihood of pathogen presence in a groundwater supply and was not reported in a substantial proportion of study records included in this review. The availability and reporting of these data could be used in combination with land-use, source type and seasonality to more accurately inform outbreak investigators, epidemiologists and sanitary engineers.
- Bacterial and protozoan contamination appear to be associated with localised contamination mechanisms (well-specific), lending credence to the concept of "consumer control" (education of private well users may directly affect contamination rates and health burden). Viral contamination may be more generalised (aquifer scale) but intervention at the source could be targeted to address all enteric pathogens (e.g. ultraviolet disinfection).
- Our findings further reflect the importance of frequent inspection and maintenance of both private and municipal groundwater supplies in Canada and the US. Additionally, it is important that potential system vulnerabilities including climate, local hydrogeological setting and adjacent hazard sources are considered during study design.

Author Contributions

Conceived and designed the experiments: KP MKT PH. Performed the experiments: PH. Analyzed the data: PH. Contributed reagents/materials/analysis tools: KP MKT. Wrote the paper: PH MKT KP.

References

1. Environment Canada (2013) Hydrology of Canada. Available: http://www.ec.gc.ca/rhc-wsc/default.asp?lang = En&n = E94719C8-1 Accessed: 2014 April 4.
2. Sharpe D, Russell H, Grasby S (2007) Hydrogeological regions of Canada. Available: http://www.nrcan.gc.ca/sites/www.nrcan.gc.ca.earth-sciences/files/pdf/gm-ces/reports/pdf/sharpe_hg_regions1_canada_07_e.pdf Accessed: 2014 April 4.
3. Robertson W, Neil D (2001) Microbiological quality of drinking water in Canada: An overview of the Health Canada programme. In Mahmound M and van Everdingen R (eds) 2001 An Earth Odyssey, Proceedings of the 54th Canadian Geotechnical Conference, September 16–19, 2001, Calgary, AB, Canada, Bitech, Richmond, BC, Canada, 45–49.
4. Corkal DR, Schutzman WC, Hilliard CR (2004) Rural Water Safety from the Source to the On-Farm Tap. Journal of Toxicology and Environmental Health, Part A. 67: 1619–1642.
5. Charrois JWA (2010) Private drinking water supplies: challenges for public health. Canadian Medical Association Journal. 182(10)
6. Statistics Canada. (2013) Census of Canada. Available: http://www12.statcan.gc.ca/census-recensement/index-eng.cfm. Accessed: 2014 April 4.
7. Konikow LF (2013) Groundwater Depletion in the United States (1900–2008) (Report). Scientific Investigations Report 2013–5079. Reston, Virginia: U.S. Department of the Interior, U.S. Geological Survey. 63.
8. Kistemann T, Classen T, Koch C, Dangendorf F, Fischeder R, et al. (2002) Microbial load of drinking water reservoir tributaries during extreme rainfall and runoff. Appl. Environ. Microbiol. 68: 2188–2197.
9. National Groundwater Association (NGWA) (2010) Groundwater Use for America; NGWA US Factsheet 9/2010; Available: http://www.ngwa.org/Documents/Awareness/usfactsheet.pdf.

10. Ramirez E, Robles E, Gonzalez ME, Martinez MEl (2010) Microbiological and Physicochemical Quality of Well Water Used as a Source of Public Supply. Air, Soil and Water Research, 3: 105–112.
11. Spruill TB, Showers WJ, Howe SS (2002) Application of Classification-Tree Methods to Identify Nitrate Sources in Groundwater. Journal of Environmental Quality. 31: 1538–1549.
12. Gardner KK, Vogel RM (2005) Predicting ground water nitrate concentration from land use. Ground Water. 43: 343–352.
13. Hynds PD, Misstear BD, Gill LW (2012) Development of a microbial contamination susceptibility model for private domestic groundwater sources. Water Resources Research. 48(12).
14. Smith HV, Grimason AM (2003) *Giardia and Cryptosporidium* in water and wastewater. In: Mara D, Horan N (eds) The handbook of water and wastewater microbiology. Elsevier, Oxford, UK, 619–781.
15. Baum R, Luh J, Bartram J (2013) Sanitation: a global estimate of sewerage connections without treatment and the resulting impact on MDG progress. *Env.* Science Tech. 47(4)1994–2000.
16. Wallender EK, Ailes EC, Yoder JS, Roberts VA, Brunkard JM (2013) Contributing factors to disease outbreaks associated with untreated groundwater. Groundwater. doi:10.1111/gwat.12121.
17. Aramini J, McLean M, Wilson J, Allen B, Sears W, et al. (2000) Drinking Water Quality and Health Care Utilization for Gastrointestinal Illness in Greater Vancouver. Centre for Infectious Disease Prevention and Control, Foodborne, Waterborne and Zoonotic Infections Division, Health Canada, Ottawa, 79.
18. Kabore H, Levallois P, Michel P, Payment P, Dery P, et al. (2010) Association between Potential Zoonotic Enteric Infections in Children and Environmental Risk Factors in Quebec, 1999–2006. Zoonoses and Public Health. 57(7–8): 195–205.
19. Mawdsley JL, Bardgett RD, Merry RJ, Pain BF, Theodorou MK (1995) Pathogens in Livestock Waste, Their Potential for Movement through Soil and Environmental Pollution. Applied Soil Ecology. 2: 1–15.
20. Gagliardi JV, Karns JS (2000) Leaching of *Escherichia coli* O157:H7 in diverse soils under various agricultural management practices. Appl Environ Microbiol. 66: 877–883.
21. Gerba CP, Smith JE (2005) Sources of pathogenic microorganisms and their fate during land application of wastes. J Environ Qual. 34: 42–48.
22. Close M, Dann R, Ball A, Pirie R, Savill M, et al. (2008) Microbial groundwater quality and its health implications for a border-strip irrigated dairy farm catchment, South Island, New Zealand. Jour. Water & Health. 6(1): 83–98.
23. Morris BL, Tyson G (2003) Private water supplies and groundwater: an aquifer guide and pathogen risk assessment toolbox. On Chartered Institute of Environmental Health Available: http://www.cieh.org/policy/background_groundwater.html?terms = groundwater Accessed: 2014 April 4.
24. Hexemer AM, Bird TM, Garcia HP, Pintar K, Pollari F, et al. (2008) An investigation of bacteriological and chemical water quality and the barriers to private well water sampling in a Southwestern Ontario Community. *Jour.* Water Health. 6(4).
25. Kreutzwiser RD, de Loë RC, Imgrund K, Conboy MJ, Simpson H, et al. (2011) Understanding stewardship behaviour: factors facilitating and constraining private water well stewardship. *Jour. Environ.* Man. 92: 1104–1114.
26. Gerba CP, Rose J, Haas C (1996) Sensitive populations: who is at the great risk? Food Microbiology. 30: 113–123.
27. Jones AQ, Dewey CE, Dore K, Majowicz SE, McEwen SA, et al. (2006) Public perceptions of drinking water: a postal survey of residents with private water supplies. BMC Public Health. 6(94).
28. Schuster CJ, Aramini JJ, Ellis AG, Marshall BJ, Robertson WJ, et al. (2005) Infectious disease outbreaks related drinking water in Canada, 1974–2001. Canadian Journal of Public Health. 96: 254–258.
29. Michaud S, Menard S, Arbeit (2004) Campylobacteriosis, Eastern Townships, Quebec. Emerg. Infect. Dis. 10(10): 1844–1847.
30. Carrique-Mas J, Andersson Y, Hjertqvist M, Svensson A, Torner A, et al. (2005) Risk factors for domestic sporadic campylobacteriosis among young children in Sweden. Scand. J. Infect. Dis. 37: 101–110.
31. Uhlmann S, Galanis E, Takaro T, Mak S, Gustafson L, et al. (2009) Where's the pump? Associating sporadic enteric disease with drinking water using a geographic information system, in British Columbia, Canada, 1996–2005. Journal of Water and Health. 07(4): 692–698.
32. Reynolds KA, Mena KD, Gerba CP. (2008) Risk of waterborne illness via drinking water in the United States. *Rev. Environ. Contam.* Toxicol. 192: 117–58.
33. Balbus J, Embrey M (2002) Risk factors for waterborne enteric infections. Current Opinion in Gastroenterology. 18: 46–50.
34. Craun GF, Calderon RF, Craun MF (2005) Outbreaks associated with recreational water in the United States. Int J Environ Health Res. 15: 243–262.
35. Sargeant JM, Rajic A, Read S, Ohlsson A (2006) The process of systematic review and its application in agri-food public health. Prev. Vet. Med. 71: 141–151.
36. Lapworth DJ, Baran N, Stuart ME, Ward RS (2012) Emerging organic contaminants in groundwater: A review of sources, fate and occurrence. Environmental Pollution. 163: 287–303.
37. Graham JP, Polizzotto ML (2013) Pit Latrines and Their Impacts on Groundwater Quality: A Systematic Review. Environ Health Perspect. 121(5): 521–530.
38. Agresti A (2010) *Analysis of Ordinal Categorical Data*, 2nd ed., Wiley & Sons (2010).
39. Gray N (2004) Biology and Wastewater Treatment (2nd Edition), London: Imperial College Press.
40. Leclerc H, Mossel D, Edberg SC, Struijk C (2001) Advances in the bacteriology of the coliform group: their suitability as markers of microbial water safety. Annu Rev Microbiol. 55: 201–234.
41. Sinton L, Hall C, Braithwaite R (2007) Sunlight inactivation of *Campylobacter jejuni* and *Salmonella enterica*, compared with *Escherichia coli*, in seawater and river water. J. Water Health. 5: 357–365.
42. Anderson KL, Whitlock JE, Harwood VJ (2005) Persistence and differential survival of fecal indicator bacteria in subtropical waters and sediments. *Appl. Environ.* Microbiol. 71: 3041–3048.
43. Macler BA, Merkle JC (2000) Current knowledge on groundwater microbial pathogens and their control. Hydrogeology Journal. 8: 29–40.
44. Borchardt MA, Spencer SK, Kieke BA, Lambertini E, Loge F (2012) Viruses in Non-disinfected Drinking Water from Municipal Wells and Community Incidence of Acute Gastrointestinal Illness. Environ. Health. Perspect. 120(9).
45. Locas A, Barthe C, Margolin AB, Payment P (2008) Groundwater microbiological quality in Canadian drinking water municipal wells. Canadian Journal of Microbiology, 54(6): 472–478.
46. Conboy MJ, Goss MJ (2001) Identification of an assemblage of indicator organisms to assess timing and source of bacterial contamination in groundwater. *Wat.* Air Soil Pollut. 129: 101–118.
47. Sinton LW, Hall CH, Lynch PA, Davies-Colley RJ (2002) Sunlight inactivation of fecal indicator bacteria and bacteriophages from waste stabilization pond effluent in fresh and saline waters. Appl. Environ. Microbiol. 68: 1122–1131.
48. Ferguson AS, Layton AC, Mailloux BJ, Culligan PJ, Williams DE, et al. (2012) Comparison of fecal indicators with pathogenic bacteria and rotavirus in groundwater. Sci Total Environ. 431: 314–322.
49. Gaut S (2005) Factors Influencing Microbiological Quality of Groundwater from Potable Water Supply Wells in Norwegian Crystalline Bedrock Aquifers. PhD thesis, Norwegian University of Science and Technology, Faculty of Engineering Science and Technology.
50. Fong TT, Mansfield LS, Wilson DL, Schwab DJ, Molloy SL, et al. (2007) Massive microbial groundwater contamination associated with a waterborne outbreak in Lake Erie, South Bass Island, Ohio. Environ. Health Perspect. 115(6): 856–864.
51. Howard G, Pedley S, Barrett M, Nalubega M, Johal K (2003) Risk factors contributing to microbiological contamination of shallow groundwater in Kampala, Uganda. Water Research. 37, Issue 14, 3421–3429.
52. Schets FM, During M, Italiaander R, Heijnen L, Rutjes SA, et al. (2005) *Escherichia coli O157:H7* in drinking water from private water supplies in the Netherlands. Water Research. 39: 4485–4493.
53. Hibler CP (1988) Analysis of municipal water samples for cysts of *Giardia*. In: Advances in *Giardia* Research, Wallis PM., Hammond BR. eds., University of Calgary Press, Calgary, 237–245.
54. Rose JB, Gerba CP, Jakubowski W (1991) Survey of potable water supplies for *Cryptosporidium* and *Giardia*. Environ. Sci. Technol. 25(8): 1393–1400.
55. Hancock CM, Rose JB, Callahan M (1998) *Cryptosporidium* and *Giardia* in US groundwater. *Jour. Am.* Water Works Assoc. 90: 58–61.
56. Craun MF, Craun GF, Calderon RL, Beach M (2006) Waterborne outbreaks reported in the United States. Journal of Water and Health. 04(Suppl. 2).
57. Rose JB, Huffman DE, Gennaccaro A (2002) Risk and control of waterborne cryptosporidiosis. FEMS Microbiol. Rev. 26: 113–123.
58. Morris BL, Foster SD (2000) Cryptosporidium contamination hazard assessment and risk management for British groundwater sources. Water Science and Technology. 41(7): 67–77.
59. Lieberman RJ, Shadix LC, Newport BS, Frebis CP, Moyer MWN, et al. (2002) Microbial monitoring of vulnerable public groundwater supplies. American Water Works Association Research Foundation. Report 90914.
60. Hänninen ML, Haajanen H, Pummi T, Wermundsen K, Katila ML, et al. (2003) Detection and typing of *Campylobacter jejuni* and *Campylobacter coli* and analysis of indicator organisms in three waterborne outbreaks in Finland. Appl Environ Microbiol. 69: 1391–1396.
61. Hrudey SE (2011) Safe Drinking Water Policy for Canada - Turning Hindsight into Foresight. *C.D. Howe Institute Commentary, No. 323.* February 28, 2011.
62. Craun GF, Brunkard JM, Yoder JS, Roberts VA, Carpenter J, et al. (2010) Causes of outbreaks associated with drinking water in the United States from 1971 to 2006. *Clin. Microbio.* Rev. 23(3): 507.
63. Huck PM, Coffey BM (2004) The importance of robustness in drinking-water systems. Journal of Toxicology and Environmental Health, Part A. 67: 1581–1590.

PERMISSIONS

All chapters in this book were first published in PLOS ONE, by The Public Library of Science; hereby published with permission under the Creative Commons Attribution License or equivalent. Every chapter published in this book has been scrutinized by our experts. Their significance has been extensively debated. The topics covered herein carry significant findings which will fuel the growth of the discipline. They may even be implemented as practical applications or may be referred to as a beginning point for another development.

The contributors of this book come from diverse backgrounds, making this book a truly international effort. This book will bring forth new frontiers with its revolutionizing research information and detailed analysis of the nascent developments around the world.

We would like to thank all the contributing authors for lending their expertise to make the book truly unique. They have played a crucial role in the development of this book. Without their invaluable contributions this book wouldn't have been possible. They have made vital efforts to compile up to date information on the varied aspects of this subject to make this book a valuable addition to the collection of many professionals and students.

This book was conceptualized with the vision of imparting up-to-date information and advanced data in this field. To ensure the same, a matchless editorial board was set up. Every individual on the board went through rigorous rounds of assessment to prove their worth. After which they invested a large part of their time researching and compiling the most relevant data for our readers.

The editorial board has been involved in producing this book since its inception. They have spent rigorous hours researching and exploring the diverse topics which have resulted in the successful publishing of this book. They have passed on their knowledge of decades through this book. To expedite this challenging task, the publisher supported the team at every step. A small team of assistant editors was also appointed to further simplify the editing procedure and attain best results for the readers.

Apart from the editorial board, the designing team has also invested a significant amount of their time in understanding the subject and creating the most relevant covers. They scrutinized every image to scout for the most suitable representation of the subject and create an appropriate cover for the book.

The publishing team has been an ardent support to the editorial, designing and production team. Their endless efforts to recruit the best for this project, has resulted in the accomplishment of this book. They are a veteran in the field of academics and their pool of knowledge is as vast as their experience in printing. Their expertise and guidance has proved useful at every step. Their uncompromising quality standards have made this book an exceptional effort. Their encouragement from time to time has been an inspiration for everyone.

The publisher and the editorial board hope that this book will prove to be a valuable piece of knowledge for researchers, students, practitioners and scholars across the globe.

LIST OF CONTRIBUTORS

Joel A. Carr, Paolo D'Odorico
Department of Environmental Sciences, University of Virginia, Charlottesville, Virginia, United States of America

Francesco Laio, Luca Ridolfi
Department of Environmental, Land and Infrastructure Engineering, Politecnico di Torino, Turin, Italy

Jeffrey S. Evans
The Nature Conservancy, Fort Collins, Colorado, United States of America
Department of Zoology and Physiology, University of Wyoming, Laramie, Wyoming, United States of America

Joseph M. Kiesecker
The Nature Conservancy, Fort Collins, Colorado, United States of America

Sergio Saia, Gaetano Amato, Alfonso Salvatore Frenda, Dario Giambalvo and Paolo Ruisi
Dipartimento di Scienze Agrarie e Forestali, Universita` degli Studi di Palermo, Palermo, Italy

Liping Li
Key Laboratory of Bioactive Substances and Resources Utilization of Chinese Herbal Medicine (Peking Union Medical College), Ministry of Education, Institute of Medicinal Plant Development, Chinese Academy of Medical Sciences, Peking Union Medical College, Beijing, China
Faculty of Science and Technology, Free University of Bozen-Bolzano, Bolzano, Italy

Armin O. Schmitt, Giuseppe Tommaso Cirella and Stefan Zerbe
Faculty of Science and Technology, Free University of Bozen-Bolzano, Bolzano, Italy

Wenxuan Han
Key Laboratory of Plant-Soil Interactions, Ministry of Education, College of Resources and Environmental Sciences, China Agricultural University, Beijing, China

Niels Thevs
Institute of Botany and Landscape Ecology, University of Greifswald, Greifswald, Germany

Xiuhong Jia
College of Horticulture and Forestry, Huazhong Agricultural University, Wuhan, China

Chengjun Ji and Dongmei Jin
Department of Ecology, College of Urban and Environmental Sciences, Key Laboratory for Earth Surface Processes of the Ministry of Education, Peking University, Beijing, China

Fidele Ngabo
Maternal and Child Health, Ministry of Health, Kigali, Republic of Rwanda

Ping He
State Environmental Protection Key Laboratory of Regional Eco-Process and Function Assessment, Chinese Research Academy of Environmental Sciences, Beijing, China

Yubing Liu and Xinrong Li
Shapotou Desert Research & Experiment Station, Cold and Arid Regions Environmental and Engineering Research Institute, Chinese Academy of Sciences, Lanzhou, P. R. China
Key Laboratory of Stress Physiology and Ecology in Cold and Arid Regions of Gansu Province, Cold and Arid Regions Environmental and Engineering Research Institute, Chinese Academy of Sciences, Lanzhou, P. R. China

Meiling Liu and Bo Cao
Shapotou Desert Research & Experiment Station, Cold and Arid Regions Environmental and Engineering Research Institute, Chinese Academy of Sciences, Lanzhou, P. R. China
Key Laboratory of Stress Physiology and Ecology in Cold and Arid Regions of Gansu Province, Cold and Arid Regions Environmental and Engineering Research Institute, Chinese Academy of Sciences, Lanzhou, P. R. China
University of Chinese Academy of Sciences, Beijing, China

Xiaofei Ma
Key Laboratory of Stress Physiology and Ecology in Cold and Arid Regions of Gansu Province, Cold and Arid Regions Environmental and Engineering Research Institute, Chinese Academy of Sciences, Lanzhou, P. R. China

Amy J. Pickering and Jennifer Davis
Woods Institute for the Environment, Stanford University, Stanford, California, United States of America
Environment and Water Studies, Civil and Environmental Engineering, Stanford University, Stanford, California, United States of America

Annalise G. Blum
Environment and Water Studies, Civil and Environmental Engineering, Stanford University, Stanford, California, United States of America

Robert F. Breiman
Global Disease Detection Program, Kenya Office of the Centers for Disease Control and Prevention, Nairobi, Kenya
Global Disease Detection Branch, Division of Global Health Protection, Center for Global Health, Centers for Disease Control and Prevention, Atlanta, Georgia, United States of America

Pavani K. Ram
Social and Preventative Medicine, State University of New York at Buffalo, Buffalo, New York, United States of America

Ferdinando Villa
Basque Centre for Climate Change (BC3), IKERBASQUE, Basque Foundation for Science, Bilbao, Bizkaia, Spain

Kenneth J. Bagstad
Geosciences & Environmental Change Science Center, U.S. Geological Survey, Denver, Colorado, United States of America

Brian Voigt
Gund Institute for Ecological Economics, Rubenstein School of Environment and Natural Resources, University of Vermont, Burlington, Vermont, United States of America

Gary W. Johnson
Department of Computer Science, University of Vermont, Burlington, Vermont, United States of America

Rosimeiry Portela and Miroslav Honzák
Conservation International, Arlington, Virginia, United States of America

David Batker
Earth Economics, Tacoma, Washington, United States of America

Muhammad T. Amin and Mohsin Nawaz
Alamoudi Water Research Chair, King Saud University, Riyadh, Kingdom of Saudi Arabia

Muhammad N. Amin
Department of Civil and Environmental Engineering, College of Engineering, King Faisal University, Al-Hofuf Al-Ahsa, Kingdom of Saudi Arabia

Mooyoung Han
Civil and Environmental Engineering Department, Seoul National University, Shinrimdong, KwanakGu, Seoul, South Korea

Christina K. Barstow
Civil, Environmental and Architectural Engineering, University of Colorado, Boulder, Colorado, United States of America

Ghislaine Rosa, Fiona Majorin and Sophie Boisson
Department of Infectious and Tropical Diseases, London School of Hygiene and Tropical Medicine, London, United Kingdom

Thomas Clasen
Department of Infectious and Tropical Diseases, London School of Hygiene and Tropical Medicine, London, United Kingdom
Department of Environmental Health, Rollins School of Public Health, Emory University, Atlanta, Georgia, United States of America

Evan A. Thomas
Mechanical and Materials Engineering, Portland State University, Portland, Oregon, United States of America

Eduardo R. M. Barbosa
Resource Ecology Group, Wageningen University, Wageningen, The Netherlands
Departamento de Botânica, Laboratório de Termobiologia, Instituto de Ciências Biológicas, Universidade de Brasília, Brasília, DF, Brazil

Kyle W. Tomlinson
Resource Ecology Group, Wageningen University, Wageningen, The Netherlands
Community Ecology & Conservation Group, Xishuangbanna Tropical Botanical Garden, Chinese Academy of Sciences, Yunnan, China

Luísa G. Carvalheiro
School of Biology, University of Leeds, Leeds, the United Kingdom

Terestrial Zoology, Naturalis Biodiversity Center, Leiden, The Netherlands

Kevin Kirkman
School of Life Sciences, University of KwaZulu-Natal, Scottsville, South Africa

Steven de Bie and Frank van Langevelde
Resource Ecology Group, Wageningen University, Wageningen, The Netherlands

Herbert H. T. Prins
Resource Ecology Group, Wageningen University, Wageningen, The Netherlands
School of Life Sciences, University of KwaZulu-Natal, Scottsville, South Africa

Arturo Bravo-Nuevo, Minzhou Huang, Jennifer Mulgrew, Lisa Laury-Kleintop and George C. Prendergast
Lankenau Institute for Medical Research (LIMR), Wynnewood, Pennsylvania, United States of America

Alice Marcy, Frank Kappler and Annette Tobia
Dynamis Pharmaceuticals Co. Inc., Jenkintown, Pennsylvania, United States of America

Melvin Reichman
LIMR Chemical Genomics Center Inc., Wynnewood, Pennsylvania, United States of America

Meng Yang, Xuemeng Geng, Cai Lu, Yi Zhu, Yan Zhou and Guangchun Lei
School of Nature Conservation, Beijing Forestry University, Beijing, China

John Grace
School of Geosciences, The University of Edinburgh, Edinburgh, United Kingdom

Peace C. Ezeh, Fredine T. Lauer, Debra MacKenzie, Shea McClain, Ke Jian Liu, Laurie G. Hudson and Scott W. Burchiel
Department of Pharmaceutical Sciences, College of Pharmacy, University of New Mexico, Albuquerque, New Mexico, United States of America

A. Jay Gandolfi
Department of Pharmacology and Toxicology, College of Pharmacy, University of Arizona, Tucson, Arizona, United States of America

Flávia Thiebaut, Clícia Grativol, Mariana Carnavale-Bottino, Tauan Vieira, Mariana Romeiro Motta, Adriana Silva Hemerly and Paulo Cavalcanti Gomes Ferreira
Laboratório de Biologia Molecular de Plantas, Instituto de Bioquímica Médica, Universidade Federal do Rio de Janeiro, Rio de Janeiro, Rio de Janeiro, Brazil

Milos Tanurdzic
School of Biological Sciences, The University of Queensland, Brisbane St Lucia, Queensland, Australia

Cristian Rojas
Universidade Federal da INTEGRAÇÃO Latino-Americana, Foz do Iguaçu, Paraná, Brazil

Renato Vincentini
Centro de Biologia Molecular e Engenharia Genética, Universidade Estadual de Campinas, Campinas, São Paulo, Brazil

Sabrina Moutinho Chabregas
Centro de Tecnologia Canavieira, Fazenda Santo Antânio - Laboratório de Biologia Molecular, Piracicaba, São Paulo, Brazil

Robert A. Martienssen
Howard Hughes Medical Institute and Gordon and Betty Moore Foundation, Cold Spring Harbor Laboratory, Cold Spring Harbor, New York, United States of America

Dana R. Thomson
Department of Global Health and Social Medicine, Harvard Medical School, Boston, Massachusetts, United States of America

Shrutika Shitole, Tejal Shitole, Kiran Sawant and Anita Patil-Deshmukh
Partners for Urban Knowledge, Action, and Research, Mumbai, India

Ramnath Subbaraman
Partners for Urban Knowledge, Action, and Research, Mumbai, India
Division of Infectious Diseases, Brigham and Women's Hospital, Boston, Massachusetts, United States of America

David E. Bloom
Department of Global Health and Population, Harvard School of Public Health, Boston Massachusetts, United States of America

Qing Shuang, Mingyuan Zhang and Yongbo Yuan
Department of Construction Management, Dalian University of Technology, Dalian, Liaoning, China

Peng Liu, Lixiang Liu, Hongmei Shen, Dan Zhou, Shoujun Liu and Xiaohui Su
Centre for Endemic Disease Control, Chinese Centre for Disease Control and Prevention, Harbin Medical University, Harbin, Heilongjiang, China

Qingzhen Jia
Institute for Endemic Disease Control of Shanxi Province, Linfen, Shanxi, China

Jinbiao Wang
Institute for Endemic Disease Control of Shandong Province, Jinan, Shandong, China

Heming Zheng
Centre for Disease Control of Henan Province, Zhengzhou, Henan, China

Jing Ma
Centre for Disease Control of Hebei Province, Shijiazhuang, Hebei, China

Michael S. H. Boutilier, Jongho Lee, Valerie Chambers, Varsha Venkatesh and Rohit Karnik
Department of Mechanical Engineering, Massachusetts Institute of Technology, Cambridge, Massachusetts, United States of America

Ghislaine Rosa, Fiona Majorin, Sophie Boisson and Miles Kirby
Department of Disease Control, London School of Hygiene and Tropical Medicine, London, United Kingdom

Christina Barstow
Department of Civil, Environmental and Architectural Engineering, University of Colorado at Boulder, Boulder, Colorado, United States of America

Michael Johnson
Berkeley Air Monitoring Group, Berkeley, California, United States of America

Fidele Ngabo
Ministry of Health, Government of Rwanda, Kigali, Rwanda

Evan Thomas
Department of Mechanical and Materials Engineering, Portland State University, Portland, Oregon, United States of America

Thomas Clasen
Department of Disease Control, London School of Hygiene and Tropical Medicine, London, United Kingdom
Department of Environmental Health, Emory University, Atlanta, Georgia, United States of America

Thomas M. Kuster
Forest Dynamics, Swiss Federal Research Institute WSL, Birmensdorf, Switzerland
Institute of Terrestrial Ecosystems ITES, ETH Zürich, Zürich, Switzerland
Soil and Ecosystem Ecology Group, University of Manchester, Manchester, United Kingdom

Matthias Dobbertin, Madeleine S. Günthardt-Goerg, Marcus Schaub and Matthias Arend
Forest Dynamics, Swiss Federal Research Institute WSL, Birmensdorf, Switzerland
Institute of Terrestrial Ecosystems ITES, ETH Zürich, Zürich, Switzerland
Soil and Ecosystem Ecology Group, University of Manchester, Manchester, United Kingdom

Julia Anna Winter, Paola Allamano and Pierluigi Claps
Department of Environment, Land and Infrastructure Engineering, Politecnico di Torino, Torino, Italy

Haiyan Xiong, Jinjie Li, Pengli Liu, Junzhi Duan, Yan Zhao, Xiao Guo, Yang Li, Hongliang Zhang and Zichao Li
Key Lab of Crop Heterosis and Utilization of Ministry of Education and Beijing Key Lab of Crop Genetic Improvement, China Agricultural University, Beijing, People's Republic of China

Jauhar Ali
International Rice Research Institute, Metro Manila, Philippines

Paul R. Hunter and Helen Risebro
Norwich School of Medicine, University of East Anglia, Norwich, United Kingdom

Marie Yen, Hélène Lefebvre and François Jaquenoud
1001 fontaines pour demain, Caluire et Cuire, France

Chay Lo
Teuk Saat 1001, Phnom Penh, Cambodia

Philippe Hartemann
Département Environnement et Santé Publique, Faculté de médecine de Nancy - Université de Lorraine, Nancy, France

Christophe Longuet
Fondation Mérieux, Lyon, France

Anthony Manea and Michelle R. Leishman
Department of Biological Sciences, Macquarie University, North Ryde, NSW, Australia

Paul Dylan Hynds, M. Kate Thomas and Katarina Dorothy Milena Pintar
Centre for Food-borne, Environmental and Zoonotic Infectious Diseases, Public Health Agency of Canada, Ottawa, Ontario, Canada

Index